Second Edition

ELECTRONIC, MAGNETIC, AND OPTICAL MATERIALS

Advanced Materials and Technologies Series

Series Editor
Yury Gogotsi
Drexel University
Philadelphia, Pennsylvania, U.S.A.

Second Edition

ELECTRONIC, MAGNETIC, AND OPTICAL MATERIALS

Pradeep Fulay

Jung-Kun Lee

CRC Press
Taylor & Francis Group
Boca Raton London New York

CRC Press is an imprint of the
Taylor & Francis Group, an **informa** business

CRC Press
Taylor & Francis Group
6000 Broken Sound Parkway NW, Suite 300
Boca Raton, FL 33487-2742

© 2017 by Taylor & Francis Group, LLC
CRC Press is an imprint of Taylor & Francis Group, an Informa business

No claim to original U.S. Government works

Printed on acid-free paper
Version Date: 20160915

International Standard Book Number-13: 978-1-4987-0169-3 (Hardback)

Library of Congress Cataloging-in-Publication Data

Names: Fulay, Pradeep P., 1960- author. | Lee, Jung-Kun, author.
Title: Electronic, magnetic, and optical materials / Pradeep Fulay and Jung-Kun Lee.
Description: Second edition. | Boca Raton : Taylor & Francis, CRC Press, 2016. | Series: Advanced materials and technologies series | Includes bibliographical references and index.
Identifiers: LCCN 2016024617 | ISBN 9781498701693 (alk. paper)
Subjects: LCSH: Electronics--Materials. | Magnetic materials. | Optical materials.
Classification: LCC TK7871 .F85 2016 | DDC 621.381--dc23
LC record available at https://lccn.loc.gov/2016024617

Visit the Taylor & Francis Web site at
http://www.taylorandfrancis.com

and the CRC Press Web site at
http://www.crcpress.com

Printed and bound in the United States of America by Publishers Graphics, LLC on sustainably sourced paper.

Contents

Preface

The society we live in today relies on and is built upon our ability to almost instantly communicate and share information at virtually any location. In the past 10 to 15 years, we have witnessed unprecedented advances in the areas of computer and information technology, health care, biotechnology, environmental sciences and engineering, and clean energy technologies such as solar and wind power. All of these disciplines, as well as the challenges they face and the opportunities they create, are interrelated and depend on our ability to send, store, and convert information and energy.

The amount of information that we exchange using the Internet, computers, and other personal communication devices such as smart phones, laptops, high-definition televisions, fitness and activity trackers, electronic book readers, global positioning systems, autonomous cars, and other technologies continues to increase very rapidly. At the very core of these technologies lie very important devices made using a number of materials that have unique electronic, magnetic, and optical properties. Thus, in this age of information technology, any scientist or engineer must have some knowledge of the basic science and engineering concepts that enable these technologies. This is especially important as we move to technologies that cut across many different disciplinary boundaries.

A good portion of information-related technologies developed to date can be traced to micro-electronic devices based on silicon. Of course, information storage and processing has also been supported by the availability of magnetic materials for data storage, optical fibers based on ultra-high-purity silica, and advances in sensors and detectors. However, silicon has been at the core of microelectronics-related technologies and has served us extremely well for almost 70 years since the invention of the transistor on December 16, 1947. There is no doubt that silicon-based microelectronic devices that enable the fabrication of computer chips from integrated circuits will continue to serve us well for many more years to come.

The first transistor, which was based on germanium, used a bulky crystal. Now, in 2017, state-of-the-art transistors are made at a length scale of 14 nm. Both the size and the cost associated with the manufacturing of the transistors have decreased significantly. The size reduction has followed the famous Moore's law, which predicted that the number of chip-based transistors would double every two years (Moore 1965). This has stayed true for nearly 50 years.

We are now at a point where we are starting to explore technologies such as quantum computing that take us into regimes that span well beyond Moore's law. This means that simply scaling down the currently dominant Si transistors and devices would no longer meet the need. There are opportunities to create novel electronic, optical, and magnetic devices that can compete with or enhance silicon-based electronics.

Similarly, new materials other than silicon, for example, carbon nanotubes (CNT), graphene, silicon carbide, and gallium nitride, are also emerging! With the advent of nanowires, quantum dots, and so on, the distinction between a "device" and a "material" has been steadily fading away. Technologies such as those related to flexible electronics, organic electronics, photovoltaics, fuel cells, biomedical implantable devices, tissue engineering, and new sensors and actuators are evolving rapidly. Many of these technological developments have brought the fields of electrical engineering, materials science, physics, chemistry, biomedical engineering, chemical engineering, and mechanical engineering closer together—perhaps more than ever before.

The increased interdisciplinarity and interdependencies of technologies mean that engineers and scientists whose primary training or specialization is in one discipline can benefit tremendously by learning some of the fundamental aspects of other disciplines. For example, it would be useful for an electrical engineering student to not only be trained in silicon semiconductor processing but also to gain some insights into other new materials, devices, and functionalities that will likely be integrated with traditional silicon-based microelectronics. In some applications, hybrid approaches can be developed in which semiconductors such as silicon are integrated with gallium arsenide and other materials such as CNT and graphene.

The primary motivation for this book stems from the need for an introductory textbook that captures the fundamentals as well as the applications of the electronic, magnetic, and optical properties of materials. The subject matter has grown significantly more interdisciplinary and is of interest to many different academic disciplines.

There are several undergraduate engineering and science textbooks devoted to specific topics such as electromagnetics, semiconductors, optoelectronics, fiber optics, microelectronic circuit design, photovoltaics, superconductors, electronic ceramics, and magnetic materials. However, most of these books are geared toward specialists. We were not able to find a single introductory textbook that takes an interdisciplinary approach that is not only critical but also of interest to students from various disciplines. This book is an attempt to fill this gap, which is important to so many disciplines but not adequately covered in any one of them.

We have written this book for typical junior (third-year) or senior (fourth-year) students of science, such as physics and chemistry, or engineering, such as materials science and electrical, chemical, and mechanical engineering, and we have used an interdisciplinary approach. The book is also appropriate for graduate students from different disciplines and for those who do not have a significant background in solid-state physics, electrical engineering, materials science and engineering, or related disciplines. This book builds upon our almost 25 years of combined experience in teaching an introductory undergraduate course on electrical, magnetic, and optical properties of materials to both engineering and engineering physics students.

We faced three major challenges. First, we had to select a few topics that are of central importance, in my opinion, to engineers and scientists interested in the electrical, magnetic, and optical properties of materials. This means that many other important topics are not addressed in detail or are left out altogether. For example, topics such as high-temperature superconductors and ionic conductors are not discussed in detail, and some are not discussed at all. We also have not discussed some of the newest and "hottest" topics such as graphene or CNT-based devices and spintronics in detail in order to maintain this as an *introductory* textbook. Interested instructors may develop some of the topics not covered here as needed, since the book covers the fundamental framework very well.

The second challenge was maintaining a relative balance between the *fundamentals* (with respect to the underlying physics-related concepts) and the *technological aspects* (production of devices, manufacturing, materials processing, etc.). The approach we have adopted here was to provide as many real-world and interesting technological examples as possible whenever there was an opportunity to do so. For example, while discussing piezoelectric materials, we provide many examples of technologies such as smart materials and ultrasound imaging. We have found that the inclusion of examples of real-world technologies helps maintain a high level of interest as students relate to the topics easily (e.g., how is a Blu-ray disc different from a DVD?). Students also tend to retain what they have learned for a longer period when these real-world connections are made.

The third challenge was to determine the level at which different topics are to be covered. For example, electrical engineering students who have some exposure to circuits may find the device-related problems to be rather simple, such as calculating limiting resistance for a light-emitting diode (LED). However, this may not be the case for many other students from other disciplines. Similarly, students who have had a class in introductory materials science may find some of the concepts related to materials synthesis, processing, and structure–property relationships to be somewhat straightforward. However, many other students may not have such a background and may be confused by the terminology used. We have therefore maintained an introductory level for all the topics covered and have tried to make the concepts as interesting as possible while challenging the students' ability and piquing their curiosity.

In the second edition of this book, we have strengthened the optical materials section significantly. Two new chapters have been added to address the fundamentals of optical materials as well as their applications to technologies such as solar cells and LEDs.

We take full responsibility for any mistakes or errors this book may have. Please contact us as necessary so that they can be corrected as soon as possible. We welcome any other suggestions you may have as well.

Any book such as this is a team effort, and this book is no exception. In this regard, we appreciate the assistance of the Taylor & Francis staff who worked with us. In particular, we thank Allison Shatkin for helping us develop this textbook. We are thankful to Todd Perry and Viswanath Prasanna for their assistance and patience in working with us on this book. We are also grateful to a number of colleagues and corporations who have provided many of the illustrations.

Pradeep Fulay is thankful to his mother, Pratibha Fulay, and father, Prabhakar Fulay, for all they have done for him and the values they have taught him, and to his wife, Dr. Jyotsna Fulay, his daughter, Aarohee, and his son, Suyash, for the support, patience, understanding, and encouragement they have always provided him.

Jung-Kun Lee expresses his appreciation to his father and mother for their love and unconditional support. He is also very thankful to his wife, Heajin, his daughter, Grace, and his son, Noah. They have been a constant source of joy, strength, and encouragement, without whom nothing would have been accomplished.

Pradeep P. Fulay
Jung-Kun Lee

REFERENCE

Moore, G.E. 1965. Cramming more components onto integrated circuits. *Electronics* 38(8): 114–17.

Authors

Pradeep P. Fulay is a professor and associate dean for research at the Benjamin M. Statler College of Engineering, West Virginia University, Morgantown, West Virginia. Prior to joining Statler College, he was a professor in the Swanson School of Engineering at the University of Pittsburgh, Pittsburgh, Pennsylvania, from 1989 to 2012. Dr. Fulay earned his PhD degree from the University of Arizona, Tucson, Arizona. From 2008 to 2011, Dr. Fulay served as a program director in the electrical, communications, and cybersystems division in the Engineering Directorate of the National Science Foundation (NSF). At NSF, he managed the electronic, magnetic, and photonic devices program and helped initiate the BioFlex program aimed at flexible electronics-based devices for biomedical applications. Dr. Fulay has published several journal articles, holds US patents, and is an elected Fellow of the American Ceramic Society. He has authored a number of textbooks that are used around the world.

Jung-Kun Lee is an associate professor and William Kepler Whiteford Faculty Fellow in the Department of Mechanical Engineering and Materials Science at the University of Pittsburgh (Pitt), Pittsburgh, Pennsylvania. He joined Pitt in September 2007 after more than 5 years of service at Los Alamos National Laboratory as a technical staff member as well as a Director's-funded Postdoctoral Fellow. He received his PhD degree from the Department of Materials Science and Engineering at Seoul National University, Seoul, South Korea. His research interests include material processing of electronic and optical materials, ion-beam synthesis of materials, and domain engineering of ferroic materials. Currently, he is focusing on the electronic and optical properties of semiconductor materials and their energy application with emphasis on solar energy harvesting. He was a recipient of the NSF CAREER Award in 2009 and has attracted multiple external grants from NSF, DOE, and NRC. The quality of his research is validated by more than 150 journal publications and multiple patents in various areas of functional properties of materials.

1 Introduction

KEY TOPICS

- Ways to classify materials
- Atomic-level bonding in materials
- Crystal structures of materials
- Effects of imperfections on atomic arrangements
- Microstructure–property relationships

1.1 INTRODUCTION

The goal of this chapter is to recapitulate some of the basic concepts in materials science and engineering as they relate to electronic, magnetic, and optical materials and devices. We will learn the different ways in which technologically useful electronic, magnetic, or optical materials are classified. We will examine the different ways in which the atoms or ions are arranged in these materials, the imperfections they contain, and the concept of the microstructure–property relationship. You may have studied some of these—and perhaps more advanced—concepts in an introductory course in materials science and engineering, physics, or chemistry.

1.2 CLASSIFICATION OF MATERIALS

An important way to classify materials is to do so based on the arrangements of atoms or ions in the material (Figure 1.1). The manner in which atoms or ions are arranged in a material has a significant effect on its properties. For example, Si single crystals and amorphous Si film are made of same silicon atoms. Though their compositions are same, they exhibit very different electric and optical properties. While single crystal Si has higher electric carrier mobility, amorphous Si exhibits higher optical band gap. This is because of different Si atom arrangement. While atoms are perfectly aligned in single crystals, they are randomly distributed in amorphous film.

1.3 CRYSTALLINE MATERIALS

A *crystalline material* is defined as a material in which atoms or ions are arranged in a particular order that repeats itself in all three dimensions. The *unit cell* is the smallest group of atoms and ions that represents how atoms and ions are arranged in crystals. Repetition of a unit cell in three dimension leads to the formation of a three-dimensional crystal. The *crystal structure* is the specific geometrical order by which atoms and ions are arranged within the unit cell. Crystal structures can be expressed using a concept of *Bravais lattices*, or simply *lattices*. A lattice is an infinite array of points which fills space without a gap. Auguste Bravais, a French physicist, showed that there are only 14 independent ways of repeatedly placing a basic unit of points in space without leaving a hole. A concept of lattice is analogous to that of a crystal consisting of unit cells (Figure 1.2).

Note that a sphere in Figure 1.2, which is called a lattice point, does *not* refer to the location of a single atom or ion. The lattice point is a mathematical idea that refers only to points in space, which is applied to several fields of science and engineering. In crystallography, the lattice

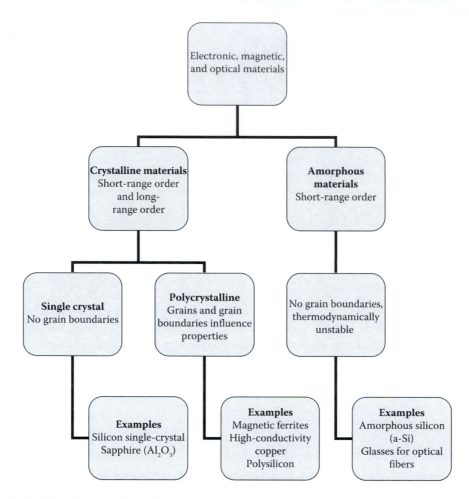

FIGURE 1.1 Classification of materials based on arrangements of atoms or ions.

represents the crystal and the lattice points are associated with the location of atoms and ions. Simply speaking, we can think that either a single atom/ion or a group of atoms/ions occupy a lattice point. When a group of atoms/ions take up the lattice point, their configuration exhibits a certain symmetry around the lattice point. The symmetry of atoms and ions occupying the same lattice point is called basis. A combination of the lattice and the basis determines the crystal structure. In other words:

$$\text{Crystal structure} = \text{lattice} + \text{basis} \tag{1.1}$$

We have only 14 Bravais lattices (Figure 1.2). However, there are many possible bases for the same Bravais lattice. The symmetry operation of bases includes angular rotation, reflection at mirror plane, center-symmetric inversion, and gliding, which in turn leads to 230 possible crystal structures out of 14 Bravais lattices. We also have many materials that exhibit the same crystal structure but have different compositions, that is, the chemical makeup of these materials. For example, silver (Ag), copper (Cu), and gold (Au) have the same crystal structure (see Section 1.6).

Similarly, materials often exhibit different crystal structures, depending upon the temperature (T) and pressure (P) to which they are subjected. In some cases, changes in crystal structures may also result from the application of other stimuli, such as mechanical stress (σ or τ), electric field (E), or magnetic field (H), or a combination of such stimuli. The different crystal structures exhibited by

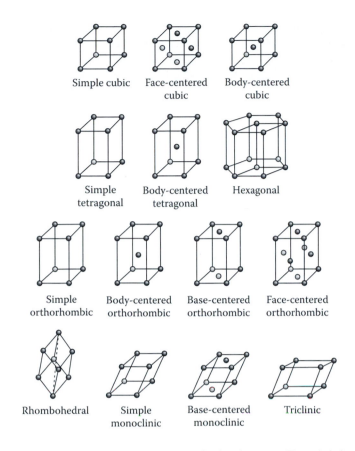

FIGURE 1.2 Bravais lattices showing the arrangements of points in space. (From Askeland, D. and Fulay, P., *The Science and Engineering of Materials*, Thomson, Washington, DC, 2006. With permission.)

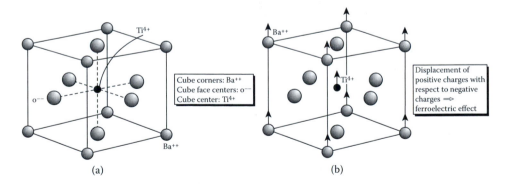

FIGURE 1.3 (a) Cubic and (b) tetragonal structures of barium titanate ($BaTiO_3$). (From Singh, J., *Optoelectronics: An Introduction to Materials and Devices*, McGraw Hill, New York, 1996. With permission.)

the same *compound* are known as *polymorphs*. For example, at room temperature and atmospheric pressure, barium titanate ($BaTiO_3$) exhibits a tetragonal structure (Figure 1.3). However, at temperatures slightly higher than 120°C, $BaTiO_3$ exhibits a cubic structure, and at even higher temperatures, this structure changes into a hexagonal structure. The different crystal structures exhibited by an *element* are known as *allotropes*.

A *phase* is defined as any portion of a system—including the whole—that is physically homogeneous and bounded by a surface so that it is mechanically separable from the other portions.

One of the simplest examples is ice, which represents a solid phase of water (H_2O). A *phase diagram* indicates the phase or phases that can be expected in a given system of materials at thermodynamic equilibrium. The phase diagram also shows a specific condition at which multiple phases coexist. For instance, a stable regime of a liquid phase and that of a solid phase meet at 0°C and 1 atm in the phase diagram of water (H_2O) at which both water and ice are found. Thus, amorphous and crystalline materials formed under nonequilibrium conditions are not shown in a phase diagram.

The phase diagram of a binary lead–tin (Pb–Sn) system is shown in Figure 1.4. Three phases, α, β, and *L*, are seen at different compositions and temperature ranges. The liquidus represents the traces of temperature above which the material is in the liquid phase. Similarly, the solidus is the trace of temperature below which the material is completely solid. In a region between liquidus and solidus lines, both solid and liquid phases are thermodynamically stable. Note that, although most alloys melt over a range of temperatures (i.e., coexistence of liquid and solid phases between the solidus and the liquidus), some specific compositions (e.g., 61.9% tin in the lead–tin system), which are known as eutectic compositions, melt and solidify at a single temperature, which is known as the eutectic temperature. The eutectic composition of lead–tin alloy has been used for soldering electronic components, because the eutectic composition alloy melts at lower temperature than any other Pb–Sn alloys. Currently, lead-free substitutes (developed because of the toxicity of lead) are increasingly being used.

The electrical, magnetic, and optical properties of a material can significantly change if the crystal structure changes. For example, the tetragonal form of $BaTiO_3$ is *ferroelectric* and *piezoelectric* (Figure 1.3). However, the cubic form of $BaTiO_3$ is neither ferroelectric nor piezoelectric. We will learn more about these materials in later chapters.

Crystalline materials can be further classified into single-crystal and polycrystalline materials (Figure 1.1). A single-crystal material, as the name suggests, is made up of one crystal in which the atomic arrangements of that particular crystal structure are followed at any location, except in the external surfaces of the crystal. A photograph of a large single crystal of silicon (Si) is shown in Figure 1.5. Single crystals are not always large; some are only a few millimeters in size.

Large single crystals of silicon (up to 18 inches in diameter and several feet in height) are sliced into thin wafers (thickness <1 mm). These silicon wafers are then used for manufacturing integrated

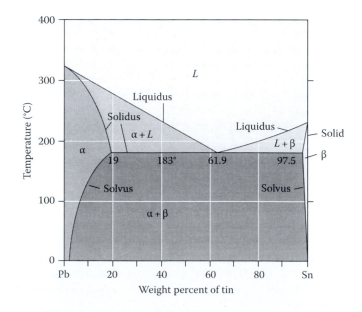

FIGURE 1.4 Lead–tin phase diagram. (From Askeland, D. and Fulay, P., *The Science and Engineering of Materials*, Thomson, Washington, DC, 2006. With permission.)

FIGURE 1.5 Single crystal of silicon. (From Askeland, D. and Fulay, P., *The Science and Engineering of Materials*, Thomson, Washington, DC, 2006. With permission.)

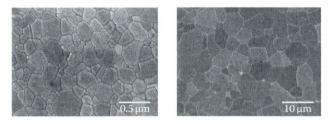

FIGURE 1.6 Microstructures of alumina ceramics. (From Kim, B.-N., et al., *Acta Mater.*, 57(5), 1319–1326, 2009. With permission.)

circuits (ICs) that are packaged into computer chips. We prefer to use the largest possible crystals for this application because it reduces the total cost of producing ICs.

In polycrystalline materials, small single crystals are put together to form bulk. A *long-range order* (LRO) of atoms or ions is present in each small single crystal; that is, atoms or ions are arranged in a particular geometric arrangement within each crystal or grain. The term *grain* refers to a relatively small, single-crystal region within a polycrystalline material. The LRO can exist across relatively larger distances, ranging from a few micrometers up to centimeters in a grain. A polycrystalline material comprises many smaller grains. The LRO in a grain ends at the boundaries of that grain. The regions or interfaces between adjacent grains are known as *grain boundaries*. The *microstructure* of a material is examined in a two-dimensional cross section in which the grain boundaries are seen as lines or curves. As we will learn in Section 1.17, grain boundaries can significantly affect some, but not all, properties.

The concept of microstructure is central to many of the ideas and technologies we will discuss throughout this book. The term microstructure is used to describe the arrangement of grains in a polycrystalline material. It also includes a description of the average grain size, grain-size distribution, grain shape, and whether the grains show a preferred orientation. In addition, other features such as imperfections in atomic arrangements also are often considered part of the microstructure.

The microstructure of a polycrystalline alumina (Al_2O_3) ceramic material showing grains and grain boundaries is shown in Figure 1.6. The relationships among the microstructure and properties of materials are explored in Section 1.17.

The atoms or ions of many materials do not show an LRO. Such materials are known as *amorphous materials*. One application of amorphous materials is window glass and lenses. We will discuss these in Section 1.18.

1.4   CERAMICS, METALS AND ALLOYS, AND POLYMERS

Another way of classifying engineered materials is based on their general behavior as metals, ceramics, or polymers/plastics. For example, stainless steels, copper, platinum (Pt), and tungsten (W) are metallic materials that exhibit high electric conductivity. In contrast, materials such as polyethylene and Teflon™ are considered synthetic polymers or plastics that are not as electrically conductive as metals. A *plastic* is a synthetic polymer-based material that is formulated with one or many polymers (i.e., large molecules or macromolecules with a carbon backbone) and other additives (e.g., carbon black, conductive or dielectric particles, and glass fibers). Since loosely packed light atoms compose polymers, plastics are lightweight. Depending on the strength of a polymer–polymer linkage, plastics can be flexible or hard. In recent years, significant research and development efforts have been made to create microelectronic devices, such as transistors, solar cells, and light-emitting diodes (LEDs), based entirely on polymers or carbon-based materials, such as poly(p-phenylene vinylene), poly-3-hexyl thiophene (P3HT), phenyl-C61-butyric acid methyl ester (PCBM), carbon nanotubes and graphene. You may have read or heard that some of the best displays for televisions and cell phones are based on organic LEDs (OLEDs). This growing field is based on the use of polymers for electronics and is known as *organic electronics.*

Efforts are underway that are aimed at integrating conventional silicon-based electronic devices with polymers to take advantage of the polymer's low density and flexibility, in addition to the electrical and optical properties. This relatively new field of research is known as *flexible electronics* or wearable electronics.

Materials such as Al_2O_3 and silica (SiO_2) are inorganic solids and are considered ceramics. Ceramics have at least partial ionic bonding characteristics that make them hard as well as brittle. This type of classification does not always distinguish among the details of atomic arrangements. For example, SiO_2 is considered a ceramic material regardless of whether it is in an amorphous or crystalline form. Due to unique structural, electric and optical properties, ceramics are widely used in machine, electronic and chemical engineering industries.

1.4.1   INTERATOMIC BONDS IN MATERIALS

The types of interatomic bonds that exist in materials are metallic (for metals and alloys), covalent, and ionic bonds (Figure 1.7). A major difference is how valence electrons of atoms are distributed in interatomic bonds. In metallic bonds, valence electrons are delocalized and do not belong to specific atoms. Two atoms share valence electrons in covalent bonds, whereas two atoms donate and accept valence electrons to become cations and anions in ionic bonds.

Most ceramics (e.g., Al_2O_3 and SiO_2) tend to exhibit a mixed ionic and covalent bond. Typically, the greater the difference between the electronegativity of the atoms that form ceramics, the higher the ionic character of the mixed bond.

Pure covalent bonds are found in technologically important semiconductors consisting of group IV elements (e.g., silicon, germanium [Ge]). In most polymers, the primary bonds (such as carbon–carbon or carbon–hydrogen) are also covalent, and atoms share valence electrons.

A secondary bond between molecules or atomic groups (i.e., a bond with lower energy), known as the *van der Waals bond,* is present in all materials and is caused by interactions among induced dipoles. The van der Waals forces and the resultant bonds are especially important in materials that have polar molecules, atoms, or groups (e.g., water [H_2O], hydroxyl group [OH^-], amine group [NH_2], chlorine atom [Cl], fluorine atom [F] and long-chain polymer). A special type of van der Waals interaction originating from the intermolecular forces among molecules with a permanent dipole moment is known as the *hydrogen bond.* It occurs in water (hence the name hydrogen bond) and in many other solids and liquids. In water (i.e., a liquid

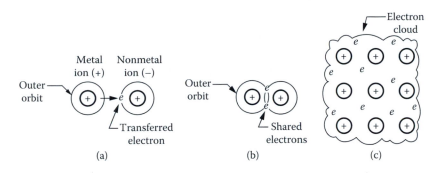

FIGURE 1.7 Different types of bonds in materials: (a) ionic, (b) covalent, and (c) metallic. (From Groover, M.P., *Fundamentals of Modern Manufacturing: Materials, Processes, and Systems*, Wiley, New York, 2007. With permission.)

phase of H_2O), local positive charge of hydrogen and local negative charge of oxygen form an intermolecular bond.

The van der Waals bonds play an important role in modifying the properties of many materials, such as water, polyvinyl chloride (PVC), graphite, and clay. For example, graphite functions as a solid lubricant. On the contrary, diamond, which is also a form of carbon, is one of the hardest naturally occurring materials. In both, the primary bonds are covalent carbon–carbon bonds. However, graphite has a layered structure and the layers of carbon atoms are bonded by relatively weak van der Waals forces. Thus, the van der Waals bonds between carbon layers can be easily broken when external force is applied.

Similarly, water has a relatively high boiling point and surface tension compared to other liquids with similar molecular weight, and it is denser than ice (a solid phase of H_2O). This is because the hydrogen bond of water is stronger than the rest of van der Waals bonds and holds H_2O molecules tightly. PVC is more brittle than other polymers, because the van der Waals forces between the chlorine and hydrogen atoms of the adjacent molecular chains makes the intermolecular linkage stronger.

1.5 FUNCTIONAL CLASSIFICATION OF MATERIALS

Materials can also be classified by their functional properties (Figure 1.8). In this classification, the *primary functionality* that a material provides is highlighted. For example, one form of iron oxide (Fe_3O_4), which is a kind of ceramics, and another metallic material, such as permendur (an alloy of iron [Fe] and cobalt [Co]) which is a kind of metals, are classified as magnetic materials, though they have different atomic bonds.

1.6 CRYSTAL STRUCTURES

We will now describe the crystal structures of some materials that have useful electrical, magnetic, or optical properties. Many materials that we will encounter later in this book exhibit a cubic structure, comprising *simple cubic* (SC), *face-centered cubic* (FCC), or *body-centered cubic* (BCC) arrangements of atoms or ions (Figure 1.9). In FCC, the same kind of atoms is additionally placed at the center of each face of an SC structure. In BCC arrangements, the same kind of atoms is additionally placed at the center of a cube of an SC structure.

In some materials, atoms are packed in a hexagonal arrangement (Figure 1.10). The hexagonal close-packed (HCP) arrangements shown in Figures 1.10 and 1.11 are the same. Similarly, the FCC structure can be schematically plotted in two different ways, which are shown in Figure 1.12. When you compare Figures 1.11 and 1.12, you find that FCC and HCP arrangements have the same packing density of atoms and that the only difference is the packing sequence of closely packed layers. Atomic locations of an upper layer are slightly shifted in a lower layer in different manners

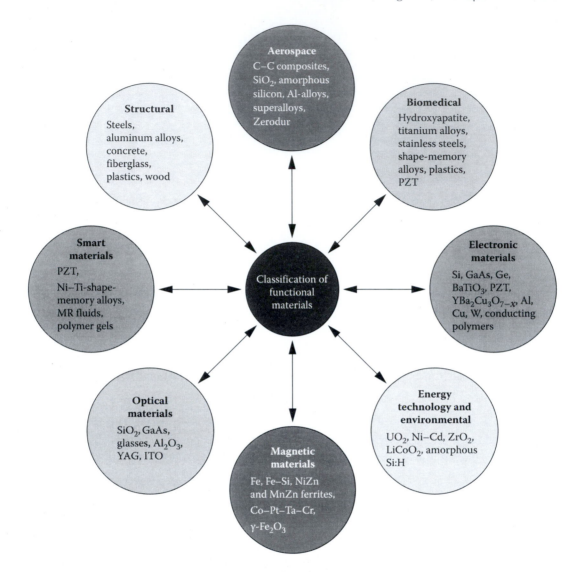

FIGURE 1.8 Functional classification of materials. (From Askeland, D. and Fulay, P., *The Science and Engineering of Materials*, Thomson, Washington, DC, 2006. With permission.)

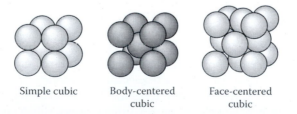

Simple cubic   Body-centered   Face-centered
                      cubic   cubic

FIGURE 1.9 Simple cubic, body-centered cubic, and face-centered cubic structures. (From Askeland, D. and Fulay, P., *The Science and Engineering of Materials*, Thomson, Washington, DC, 2006. With permission.)

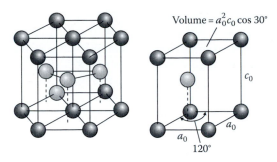

FIGURE 1.10 A unit cell of a hexagonal close-packed structure. (From Askeland, D. and Fulay, P., *The Science and Engineering of Materials*, Thomson, Washington, DC, 2006. With permission.)

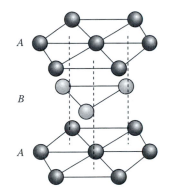

FIGURE 1.11 The packing sequence of a hexagonal close-packed structure. (From Askeland, D. and Fulay, P., *The Science and Engineering of Materials*, Thomson, Washington, DC, 2006. With permission.)

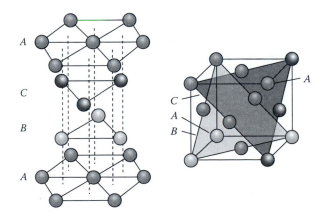

FIGURE 1.12 The packing sequence of a face-centered cubic structure. (From Askeland, D. and Fulay, P., *The Science and Engineering of Materials*, Thomson, Washington, DC, 2006. With permission.)

(ABCABC… for FCC vs. ABABAB for HCP). Both FCC and HCP arrangements lead to the maximum possible *packing fraction* of 0.74 for atoms or spheres of the same size. The term packing fraction refers to the ratio of the volume occupied by the atoms to the volume of the unit cell. It can be shown that, if we have a cube-shaped box, the best we can do is fill up 74% of the space available with spheres of a given radius. The packing fraction does *not* depend upon the radius of the sphere (i.e., whether it is a basketball or a tennis ball), as long as we have spheres of the same size.

When atoms are packed such that the structure exhibits the maximum possible packing fraction, the structure is referred to as a *close-packed* (CP) *structure*. Thus, a hexagonal structure with a maximum possible packing fraction of 0.74 is known as a close-HCP structure (Figure 1.11). Although metals with a hexagonal structure have a nearly CP structure, a hexagonal structure of ceramics (e.g., Al_2O_3) is not necessarily closely packed. As you can guess, for a ratio of Al and O atoms, a plane of Al that is not fully packed has a periodic opening.

The following example illustrates the calculation of packing fractions for cubic CP structures.

Example 1.1: Calculation of Packing Fractions

Calculate the maximum possible packing fractions for the (a) SC, (b) FCC, and (c) BCC structures. Assume that all atoms have a radius *r* and the unit-cell parameter is *a*.

Solution

1. As shown in Figure 1.13, in the SC structure, atoms touch along the cube edges, that is,

$$a = 2r \tag{1.2}$$

Thus, the packing fraction is

$$\text{Packing fraction} = \frac{\text{Volume of atoms in the unit cell}}{\text{Volume of unit cell}}$$

$$= \frac{(8 \text{ atoms} \times 1/8) \times (4/3 \pi \times r^3)}{a^3} = \frac{(4/3)\pi \times r^3}{8r^3} = \frac{\pi}{6} \approx 0.52 \tag{1.3}$$

Thus, for a CP–SC structure based on atoms of a single size, regardless of whether the atoms are big or small, the packing fraction is 0.52. This means that 48% of the space in the unit cell is empty. We can introduce other smaller atoms in the voids, or so-called interstitial sites, found within the crystal structure (Section 1.8).

2. In the CP–FCC structure, atoms touch along the face diagonals (Figure 1.13), that is,

$$\sqrt{2}a = 4r \tag{1.4}$$

or

$$a = 2\sqrt{2}r \tag{1.5}$$

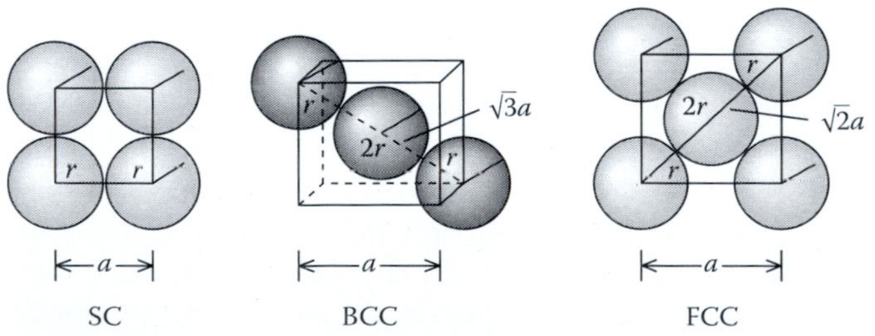

FIGURE 1.13 Packing of atoms in simple cubic (SC), body-centered cubic (BCC), and face-centered cubic (FCC) crystal structures. (From Askeland, D. and Fulay, P., *The Science and Engineering of Materials*, Thomson, Washington, DC, 2006. With permission.)

$$\text{FCC packing fraction} = \frac{\text{Volume of atoms in the unit cell}}{\text{Volume of unit cell}}$$

$$= \frac{\left[\left(6 \text{ face atoms} \times 1/2\right) + \left(8 \text{ face atoms} \times 1/8\right)\right] \times \left(4/3\pi \times r^3\right)}{a^3}$$

$$= \frac{4 \times \left(4/3\right)\pi \times r^3}{8 \times 2^{3/2} r^3} = \frac{\pi}{3\sqrt{2}} = \frac{\pi}{\sqrt{18}} \approx 0.74$$

This is the packing fraction for an FCC–CP structure. This structure has only 26% empty space. Both the CP–FCC structure and the HCP offer the same and the highest possible volume-packing fraction $\left(\pi/\sqrt{18}\right)$ for spheres of uniform size.

Note that, if the empty spaces or voids in the CP structure are filled with other smaller atoms, the packing fraction will be higher. In fact, many ceramic crystal structures are rationalized using the close packing of bigger *anions* (negatively charged ions) and then stuffing the voids with smaller, positively charged *cations* (see Section 1.11).

3. Atoms in the CP–BCC structure touch along the body diagonal (Figure 1.13). Thus, the relationship between r and a is

$$\sqrt{3}a = 4r \tag{1.6}$$

or

$$a = \frac{4r}{\sqrt{3}} \tag{1.7}$$

Thus, the packing fraction for a CP–BCC structure is

$$\text{BCC packing fraction} = \frac{\text{Volume of atoms in the unit cell}}{\text{Volume of unit cell}}$$

$$= \frac{\left[\left(1 \text{ cube center atom}\right) + \left(8 \text{ face center atoms} \times 1/8\right)\right] \times \left(4/3\pi \times r^3\right)}{a^3}$$

$$= \frac{2 \times \left(4/3\right)\pi \times r^3}{\left(4r/\sqrt{3}\right)} = \frac{\sqrt{3} \times \pi}{8} \approx 0.68$$

Therefore, the BCC structure has a packing fraction of 0.68, which is in between the values for the FCC and SC structures. This structure has 32% void space available.

1.7 DIRECTIONS AND PLANES IN CRYSTAL STRUCTURES

There is a need to specify the crystallographic directions and planes in a unit cell in many applications involving the magnetic, electronic, and optical properties of materials. We use a notation known as *Miller indices* to designate specific directions and planes in cubic unit cells. In describing these, we always use a right-handed coordinate system.

1.7.1 MILLER INDICES FOR DIRECTIONS

Let us think about how to express a direction between two lattice points A and B. To obtain the Miller indices of a direction that starts at Point A (illustrated as an arrow tail) and ends at Point B (illustrated as an arrow head), we subtract the coordinates of Point A from those of Point B. We then clear the fractions and reduce the results to the lowest integers. The results on the Miller indices of directions are included in square brackets (e.g., [*hkl*]). If a negative sign is needed to express the direction, we insert a bar on top of that index. Details are found in Example 1.2.

1.7.2 MILLER INDICES FOR PLANES

To obtain the Miller indices for planes touching multiple lattice points, we start by identifying the intercepts of a plane on the three axes (i.e., *x*, *y*, and *z*). We then take the reciprocals of the intercepts and clear the fractions by multiplying the least common multiples of denominators. If a plane is parallel to an axis, the intercept for that axis is infinity (∞) and its reciprocal is zero.

For the Miller indices of planes, we do *not* reduce these numbers to the lowest integers. Similar to the results for directions, we put a bar above any negative index. The Miller indices of planes are written in parentheses ( ). In some cases, the intercepts of planes may not be easy to identify (e.g., a plane that passes through the origin). In this case, we can move the origin of the unit cell (i.e., shift the plane in parallel to touch a set of neighboring lattice points). In the cubic system, the direction [*hkl*] is perpendicular to the plane (*hkl*).

1.7.3 MILLER–BRAVAIS INDICES FOR HEXAGONAL SYSTEMS

In a hexagonal unit cell, the *Miller–Bravais indices* for directions and planes are designated as [*hkil*] and (*hkil*), respectively; *hki* in the Miller–Bravais indices are associated with the three axes in one plane (a_1, a_2, a_3) where atoms are closely packed. The angle between these axes is 120°. In the Miller–Bravais indices, *l* is related to the *c*-axis that is perpendicular to a plane (a_1, a_2, a_3). Because the three axes a_1, a_2, and a_3 are in one plane, the Miller–Bravais indices corresponding to them cannot all be independent. The relationship among them is given by

$$h + k = -i \tag{1.8}$$

Thus, a direction [010] in the orthogonal coordinate system (i.e., [*hkl*]) will be equivalent to the direction $\left[\bar{1}2\bar{1}0\right]$ of a hexagonal coordinate system which is beneficial for expressing HCP structure. To get the Miller–Bravais indices for direction [010] of the orthogonal system, we move a point of interest by one lattice constant along the negative a_1 direction of the hexagonal coordinate system, then by two lattice constants along the positive a_2 direction, and finally by one lattice constant along the negative a_3 direction. Thus, the condition given by Equation 1.8 is satisfied (Figure 1.14).

We can also verify that the direction [100] of the orthogonal coordinate system is equivalent to $\left[2\bar{1}\bar{1}0\right]$ of a hexagonal coordinate system.

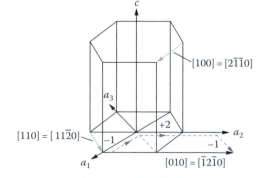

FIGURE 1.14 Typical directions in a hexagonal unit cell. The major axes are a_1, a_2, a_3, and *c*. (From Askeland, D. and Fulay, P., *The Science and Engineering of Materials*, Thomson, Washington, DC, 2006. With permission.)

Similar considerations apply to the Miller–Bravais indices for planes. We locate the intercepts on the three axes (a_1, a_2, a_3), in addition to that on the c-axis. Then, we take the inverses of these values and clear the fractions without reducing the integers. Note that the rule ($h + k$) = $-i$ holds because the a-axes lie in the plane.

In the HCP structure, atoms are packed in the closest way along $\left[11\bar{2}0\right]$-type directions. In addition, planes (0001) and (0002) are the closest-packed planes in materials with HCP structure. (i.e., they have the highest number of atoms per unit area).

1.7.4 INTERPLANAR SPACING

There are multiple planes with the same Miller indices (hkl) and the distance between neighbor planes is the same. In a cubic system with a unit-cell size a (i.e., lattice constant a), the distance d between a set of parallel planes with Miller indices (hkl) is given by the following equation:

$$d_{hkl} = \frac{a}{\sqrt{h^2 + k^2 + l^2}} \left(\text{for cubic system}\right) \tag{1.9}$$

We can measure d_{hkl} experimentally using X-ray diffraction or electron diffraction techniques and deduce a from d_{hkl}.

Examples 1.2 through 1.4 show how to obtain the Miller indices of directions and a plane and also how to calculate interplanar distances and connect them with the lattice constant.

Example 1.2: Miller Indices for Directions

What are the Miller indices for any one of the CP directions in FCC and BCC structures?

Solution
In an FCC structure, the atoms touch along the face diagonal. Therefore, the face diagonals are the CP directions.

One face diagonal (marked as OA) is shown in Figure 1.15. We follow the procedure described earlier for obtaining the Miller indices of direction:

1. The coordinates of the "head" point A are 1, 1, 0.
2. The coordinates of the "tail" point B (in this case, the origin point O) are 0, 0, 0.
3. Subtracting the coordinates of the tail from those of the head, that is, 1 – 0, 1 – 0, 0 – 0, we get 1, 1, 0.

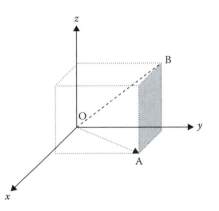

FIGURE 1.15 One of the close-packed directions in an FCC unit cell is shown as OA here. In BCC, atoms are closely packed along direction OB. Atoms are not shown.

There are no fractions to clear and no negative integers. Therefore, the Miller indices of this direction OA are [110]. Note that the opposite direction, AO, will have Miller indices of $\left[\bar{1}\,\bar{1}0\right]$.

In an FCC structure, all the face diagonals are close-packed with atoms. All directions along which atoms are arranged in the same way are said to be crystallographically equivalent. A family of such equivalent directions is known as the *directions of a form* and is designated as < >. For an FCC, the directions of a form for CP directions will be <110>. There would be 12 total directions in this family of CP directions for an FCC (six face diagonals and corresponding opposite directions). [110], [101] and [011] are included in <110> of FCC structure.

You can also see that, similarly, for a BCC structure (atoms not shown in Figure 1.15), the atoms touch along the body diagonal. The Miller indices for one such direction OB (Figure 1.15) are [111]. Note that the direction BO is also a CP direction for the BCC structure, and its Miller indices are $\left[\bar{1}\,\bar{1}\,\bar{1}\right]$. Both OB and BO directions are a part of <111> in the orthogonal coordinate system.

Example 1.3: Miller Indices for a Plane

What are the Miller indices for the plane P shown in Figure 1.16?

Solution
We see that this plane, shown in Figure 1.16, intersects the x- and y-axes at a length of "1x" lattice parameter.

The plane is parallel to the z-axis, that is, it does not intersect the z-axis at all; hence, this intercept is ∞.

We follow the directions for establishing the Miller indices of a plane as

1. The intercepts on the x-, y-, and z-axes are 1, 1, and ∞, respectively.
2. The reciprocals of these are 1, 1, and 0. There are no fractions to clear. Therefore, the Miller indices of this plane are (110).
3. Similar to directions of a form, there are also *planes of a form*. These planes are equivalent and are shown in curly brackets { }. The planes of a form for {110} will include the following: (110), (101), (011), $\left(1\,\bar{1}0\right)$, $\left(10\bar{1}\right)$, and $\left(01\,\bar{1}\right)$.

Example 1.4: Interplanar Spacing in Materials

X-rays of a single wavelength (λ) were used to analyze a glittering sample suspected to be Au. The X-ray diffraction (XRD) analysis, by which the lattice constant of an unit cell is determined, showed that the (400) planes in this sample were separated by a distance of 0.717 Å. Is the sample made of Au?

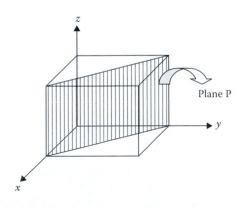

FIGURE 1.16   Plane for Example 1.3.

Assume that the radius (r) of Au atoms is 144 pm. Also assume that, in this hypothetical example, no other obvious measurements, such as density, can be made. Au has an FCC structure.

Solution

We are given d_{400} as 0.717 Å. From Equation 1.9, we get:

$$d_{400} = \frac{a}{\sqrt{4^2 + 0^2 + 0^2}} = 0.717\,\text{Å}$$

Therefore, the lattice constant for this sample is $a = (0.717)\,(4) = 2.868$ Å.

Now, it is given that the radius (r) of Au atoms is 144 pm = 144 pm × 10^{-12} m/pm × 10^{10} Å/m. This is 1.44 Å.

Because Au has an FCC crystal structure, the lattice constant (a) and the atomic radius (r) are related by Equation 1.4.

From this equation, we get $a = 2\sqrt{2} \times (1.44\,\text{Å}) = 4.07\,\text{Å}$. This is the lattice constant of Au.

The sample of this example has a lattice constant of 2.868 Å. Thus, the sample is not Au!

By comparing the lattice constant of 2.868 Å to that of other elements, we will see that it is close to the lattice constant of BCC Fe. Thus, it is likely that the sample we have is BCC Fe. Because this sample glittered like Au, it may have been plated with a thin layer of Au. Thus, the lesson to be learned here is that all that glitters is not gold! In this example, we also learned that the unknown sample has a cubic structure.

1.8 INTERSTITIAL SITES OR HOLES IN CRYSTAL STRUCTURES

As mentioned in the previous section, the packing fraction of even the CP structures is smaller than 80%, and voids, known as interstitial sites, are found in different CP structures. Smaller atoms or ions can enter these interstitial sites or holes in a crystal structure, which changes the crystal structure by varying the basis of that structure. In many ceramics, interstitial sites of the anion lattice are filled with cations smaller than the anions. Different types of interstitial sites are illustrated in Figure 1.17. (The coordination number in Figure 1.17 is explained in the next section.)

Coordination number	Location of interstitial	Radius ratio	Representation
2	Linear	0–0.155	
3	Center of triangle	0.155–0.225	
4	Center of tetrahedron	0.225–0.414	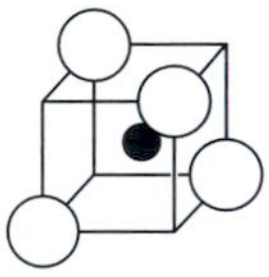
6	Center of octahedron	0.414–0.732	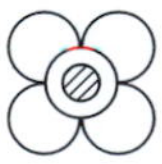
8	Center of cube	0.732–1.000	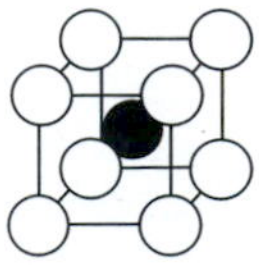

FIGURE 1.17 Different types of interstitial sites. (From Askeland, D. and Fulay, P., *The Science and Engineering of Materials*, Thomson, Washington, DC, 2006. With permission.)

1.9 COORDINATION NUMBERS

The total number of nearest-neighbor atoms that surround an atom (occupying a lattice point or an interstitial site) is often referred to as the *coordination number* (CN). For example, in the FCC and HCP structures, the CN of an atom taking up the lattice point is 12. This means that, for any atom in a specific position of the lattice, there are 12 nearest-neighbor atoms surrounding it. In the SC structure, the CN is 6. For a BCC structure, the CN is 8. This can be easily seen by looking at the atom located at the cube center. It is surrounded by eight corner atoms. We must also realize that, in a periodic structure, each atom has identical surroundings; thus, atoms in lattice points have the same CN whether we examine the atom at the cube center or at any other position in the structure.

The CNs of atoms occupying the lattice points of different structures, the corresponding relationship between the unit-cell parameter (a) and the atomic radius (r), and the volume-packing fractions are given in Table 1.1. The CNs of the interstitial sites are also shown in Figure 1.17 with illustrations.

The most common types of interstitial sites are the so-called tetrahedral and octahedral sites. A *tetrahedral site* means that an atom or ion in that site is surrounded by four other atoms or ions (CN = 4) that are located at the corners of a tetrahedron—hence the name tetrahedral site. Similarly, for an *octahedral site*, an atom or ion in this site is surrounded by six (*not* eight) atoms (CN = 6). These atoms or ions form an octahedron around the site center—hence the name of the interstitial site is octahedral site. There are several equivalent interstitial sites within a single unit cell. For example, in an FCC unit cell, the center of the cube is an octahedral site. Centers of 12 edges of an FCC unit cell are also octahedral sites. Since each edge of the FCC unit cell is shared with a total of four neighboring unit cells, there are a total of [(12 × 1/4) + 1)] = 4 octahedral sites in an FCC unit cell.

1.10 RADIUS RATIO CONCEPT

The concept of radius ratio is useful to rationalize or to guess whether a guest atom or ion is likely to enter a particular type of an interstitial site. The radius ratio is the ratio of the radius of the guest atoms/ions (occupying interstitial sites) to that of the host atoms/ions (taking up lattice points). The radius ratio ranges and corresponding sites are shown in Figure 1.17.

The general idea behind the radius ratio concept is that we can view crystal structures of many ceramic materials as being made up of a close packing of anions. These are the negatively charged ions and are typically larger because of the extra electrons. This is conceptually followed by the *stuffing* of smaller cations into the interstitial sites.

Note that not all the available interstitial sites in a unit cell need to be occupied. The sites to be occupied (e.g., centers of tetrahedron or octahedron or cube) depend upon the radius ratio, that is, in the case of most ceramics, r_{cation}/r_{anion}. Larger cations tend to fill larger interstitial sites. The fraction

TABLE 1.1

Relationships between the Unit-Cell Parameter (a) and Radius (r), the Coordination Number, and the Volume-Packing Fraction for Different Unit Cells

Structure	Relationship of a and r	Atoms per Unit Cell	Coordination Number	Packing Fraction
Simple cubic	$a = 2r$	1	6	0.52
Body-centered cubic	$\sqrt{3}a = 4r$	2	8	0.68
Face-centered cubic	$\sqrt{2}a = 4r$	4	12	0.74
Hexagonal close-packed	$a = 2r$	2	12	0.74
	$c \approx 1.633a$			

of the sites that are occupied depends upon the stoichiometry of the compound. This is illustrated in the following section.

1.11 CRYSTAL STRUCTURES OF DIFFERENT MATERIALS

We will now discuss some of simple crystal structures exhibited by electronic, magnetic, and optical materials that are of interest to us. Many materials have far more complex crystal structures, and a discussion of these is beyond the scope of this book.

1.11.1 STRUCTURE OF SODIUM CHLORIDE

The structure of a sodium chloride (NaCl) crystal (Figure 1.18), which is shared by many ceramic materials—such as magnesium oxide (MgO), can be rationalized as follows: The structure is obtained by closely packing chlorine anions (Cl^{1-}) that have a radius of 0.181 nm. The radius of sodium cations (Na^{1+}) is 0.097 nm. This results in a radius ratio of r_{Na}^+/r_{Cl}^-, or ~0.536. Therefore, we expect that sodium ions will exhibit an octahedral coordination (Figure 1.17).

Thus, we can assume that this structure is obtained by creating an FCC arrangement of chlorine ions. All the octahedral sites (i.e., cube edge centers and cube center) can then be stuffed with sodium ions.

1.11.2 STRUCTURE OF CESIUM CHLORIDE

The atomic radius of a cesium ion (Cs^+) is 0.167 nm. The radius ratio of 0.92 (= 0.167 nm/0.181 nm) for cesium chloride (CsCl) suggests an eightfold coordination of Cs ions. This is achieved by locating the Cs^+ ion at the cube center. In this structure, we have eight chlorine ions at the eight corners of the cube (Figure 1.19). Each of the anions at the corner is shared with eight other

NaCl

FIGURE 1.18 Crystal structure of sodium chloride (NaCl). (From Askeland, D. and Fulay, P., *The Science and Engineering of Materials*, Thomson, Washington, DC, 2006. With permission.)

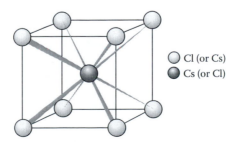

FIGURE 1.19 Cesium chloride (CsCl) structure, showing the eightfold coordination of cesium ions. (From Smart, L. and Moore, E., *Solid State Chemistry: An Introduction*, Chapman and Hall, Boca Raton, FL, 1992. With permission.)

neighboring unit cells. The number of chlorine ions per unit cell is $(8 \times 1/8) = 1$. Thus, the stoichiometry of one chlorine ion for every cesium ion is maintained, and electrical neutrality is assured.

As shown in Figure 1.19, the structure can also be represented as the chlorine ion at the center and an FCC packing of cesium ions. Example 1.5 illustrates the use of the radius ratio concept.

Example 1.5: Application of the Radius Ratio Concept

NaCl and potassium chloride (KCl) have the same stoichiometry. The radius of the potassium ion (K^+) is 0.133 nm. The radius of the chlorine ion (Cl^+) is 0.181 nm. What will be the expected CN for K^+ ions? What will be the expected structure of KCl? Is this structure consistent with the stoichiometry?

Solution

The r_{cation}/r_{anion} for KCl is $0.133/0.181 = 0.735$. This suggests that the CN for K^+ ions will be 8. This CN is possible if the K^+ ions assume the location at the cube center. Then, they are next to eight other Cl^- ions. Thus, KCl exhibits a structure similar to that of cesium chloride (CsCl) (Figure 1.19). The structure is consistent with the 1:1 stoichiometry of KCl.

1.11.3 Diamond Cubic Structure

One of the most important crystal structures of many semiconductor materials is the diamond cubic (DC) crystal structure (Figure 1.20). This is the crystal structure exhibited by semiconductors, such as silicon, germanium, and carbon in the form of diamond.

In this crystal structure, atoms are first arranged in an FCC arrangement and additional atoms are placed at half of the tetrahedral sites. Note that in Figure 1.20, for the sake of clarity, the atoms are not shown touching one another in the FCC and tetrahedral arrangements. Here is how the tetrahedral sites are filled: First, the cubic unit cell is divided into eight smaller cubes known as *octants*. Then, two smaller nonadjacent octants on the top are selected (the centers of these cubes will be at a point three-fourths the unit-cell height). Similarly, two smaller nonadjacent octants at the bottom are selected (centers of these cubes will be at one-fourth of the height of the unit cell). Additional atoms are placed into four of the octants. Thus, in this crystal structure, atoms exhibit what is called a *tetrahedral coordination*. It is easier to visualize this tetrahedral arrangement by examining the atoms located in the centers of the octants. However, every atom in this structure ultimately has the same tetrahedral coordination.

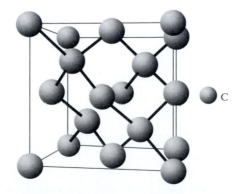

FIGURE 1.20 Diamond cubic crystal structure. (From Askeland, D. and Fulay, P., *The Science and Engineering of Materials*, Thomson, Washington, DC, 2006. With permission.)

A DC unit cell has a total of eight atoms in the unit cell. Four atoms are from the FCC arrangement [8 × (1/8) + 6 × (1/2)]. Four more atoms are derived from the inside of the unit cell (i.e., half of the tetrahedral sites). Such a description allows us to calculate the theoretical density of materials, such as silicon and germanium. This is illustrated in Example 1.6.

Example 1.6: Theoretical Density of Silicon (Si)

The radius of Si in a covalent structure is 1.176 Å. What is the lattice constant of Si? What is the theoretical density of Si if its atomic mass is 28.1?

Solution
If we examine the body diagonal in the DC structure exhibited by Si, we see that there are two atoms at the corner and an atom within an octant. There is space for two more atoms to be accommodated along the diagonal (Figure 1.21).

If the length of the unit cell is a, then the body diagonal is $\sqrt{3}a$. If r is the radius of the atoms in the DC structure, then from Figure 1.21, we get:

$$\sqrt{3} \times a = 8r \qquad (1.10)$$

For Si,

$$a = \frac{8 \times 1.176}{\sqrt{3}} = 5.4317\,\text{Å}$$

In addition, as described earlier, there is the equivalent of eight atoms of Si per unit cell. Recall that one mole of an element has an Avogadro number of atoms (6.023×10^{23} atoms). Thus, in this case, 28.1 g of Si would have 6.023×10^{23} atoms, and the theoretical density of Si will be

$$\text{Density} = \frac{\text{Mass of 8 silicon atoms}}{\text{Volume of the unit cell}} = \frac{8 \times 28.1}{6.023 \times 10^{23} \times \left(5.4317 \times 10^{-8}\,\text{cm}\right)^3} = 2.33\,\text{g/cm}^3$$

This value matches with the experimentally observed values.

1.11.4 Zinc Blende Structure

Zinc sulfide (ZnS) exhibits different polymorphic forms known as zinc blende and wurtzite. The structure of zinc blende is cubic (Figure 1.22), whereas that of wurtzite is hexagonal (Figure 1.23). Many compound semiconductors, such as gallium arsenide (GaAs), show the zinc blende structure.

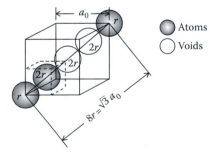

FIGURE 1.21 Schematic representation showing the arrangement of atoms and voids along a body diagonal in a diamond cubic crystal structure. (From Askeland, D. and Fulay, P., *The Science and Engineering of Materials*, Thomson, Washington, DC, 2006. With permission.)

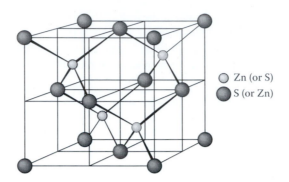

FIGURE 1.22 Schematic representation of a zinc blende structure. (From Smart, L. and Moore, E., *Solid State Chemistry: An Introduction*, Chapman and Hall, Boca Raton, FL, 1992. With permission.)

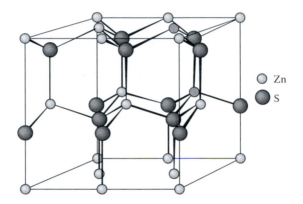

FIGURE 1.23 Structure of a wurtzite crystal. (From Smart, L. and Moore, E., *Solid State Chemistry: An Introduction*, Chapman and Hall, Boca Raton, FL, 1992. With permission.)

The zinc blende structure (Figure 1.22) is similar to the DC structure (Figure 1.20). The radius of divalent zinc ions (Zn^{2+}) is 0.074 nm. The radius of the larger sulfur anion (S^{2-}) is 0.184 nm. The radius ratio of 0.402 suggests a tetrahedral coordination for Zn^{2+} ions (Figure 1.17).

We start with an FCC arrangement of S^{2-} ions. The Zn^{2+} ions enter at the four tetrahedral sites, that is, they occupy the centers of the octants inside the main unit cell (similar to the diamond-cubic unit cell).

Note that not *all* tetrahedral sites are occupied because of the requirement of a balance of stoichiometry and electrical neutrality. In addition, each S^{2-} ion is coordinated with four Zn^{2+} ions. Note also that if all ions in the zinc blende structure (Figure 1.22) were identical, we would get a DC structure (Figure 1.20).

GaAs, an important semiconductor, exhibits the zinc blende structure. The structure of GaAs can be understood by replacing the sulfur atoms in ZnS with arsenic atoms and the zinc atoms with gallium atoms. One of the polymorphs of silicon carbide (SiC) and other materials, such as indium phosphide (InP), indium antimonide (InSb), and gallium phosphide (GaP), also exhibit this type of crystal structure. Example 1.7 explores the zinc blende structure in more detail.

Example 1.7: Density of Indium Phosphide (InP)

If the lattice constant (a) of InP is about 5.8687 Å, what is its theoretical density? The atomic masses of indium (In) and phosphorus (P) are 114.81 and 31, respectively.

Solution

Recognize that there are four In atoms and four P atoms inside the InP unit cell (Figure 1.22). Also recall that one mole of an element has an Avogadro number, that is, 6.023×10^{23} atoms.

Thus, the theoretical density of InP will be the mass of four In atoms + mass of four P atoms divided by the volume of the cubic unit cell.

$$\text{Density of indium phosphide} = \frac{\text{Mass of four In atoms} + \text{mass of four P atoms}}{\text{Volume of the unit cell}}$$

$$= \frac{\left[4 \times (114.81) + 4 \times (31)\right]}{6.023 \times 10^{23} \times \left(5.8687 \times 10^{-8}\,\text{cm}\right)^3}$$

$$= 4.79\,\text{g/cm}^3$$

What we calculated is the so-called theoretical density. The actual density is comparable but can be a little different because the arrangement of atoms in real materials is never perfect (see Section 1.12).

1.11.5 Wurtzite Structure

According to the radius ratio, the Zn^{2+} ions have a CN of 4, that is, they are surrounded by four sulfur ions. Similarly, each sulfur ion (S^{2-}) is coordinated with four zinc ions (Zn^{2+}). This tetrahedral coordination can be found in an HCP structure as well as an FCC structure. The wurtzite structure is based on an HCP array of sulfur ions. If S^{2-} ions form an HCP array, two tetrahedral sites are found per S^{2-} ion. Tetrahedral holes in alternate octants are occupied by the zinc ions so that half of the tetrahedral sites of an HCP array of sulfur ions are filled. Many semiconductors and dielectrics, such as gallium nitride (GaN), zinc oxide (ZnO), aluminum nitride (AlN), cadmium telluride (CdTe), and cadmium sulfide (CdS), show this type of polymorph.

This structure can also be visualized by considering hexagonal close packing of zinc ions, followed by the stuffing of sulfide anions in the tetrahedral sites (Figure 1.23).

1.11.6 Fluorite and Antifluorite Structure

To visualize a crystal structure, it is sometimes easier to conceptually consider that the cations (i.e., smaller ions) form a CP array and anions fill interstitial sites. For the fluorite (CaF_2) structure, we start with a CP array (FCC packing) of cations. Then, all the tetrahedral holes are filled with larger fluorine anions (F^{1-}). In this representation, the 4-fold coordination of anions is clearly seen (Figure 1.24a).

To better understand this, we can extend the structure so that the cubes have fluoride ions at the corners (Figure 1.24b). In this representation, we can easily recognize the eightfold coordination of the cations. The same structure is redrawn in Figure 1.24c by shifting the origin, and in this figure, eight octants can be seen. Every other octant is occupied by a Ca^{2+} ion. The relative distances among the ion centers are shown in one of the octants in Figure 1.24d.

Ceramic materials that show a fluorite structure include CeO_2, PbO_2, UO_2, and ThO_2.

In the antifluorite structure, the positions of cations and anions are reversed. Materials that show an antifluorite structure include Li_2O, Na_2O, Rb_2O, K_2O, and Li_2S.

1.11.7 Corundum Structure

A corundum (α-Al_2O_3) structure can be described by picturing a hexagonal close packing of oxygen anions (O^{2-}). In the hexagonal CP, the number ratio of packing atoms and octahedral sites is 1. Therefore, in the corundum structure, two-thirds of the octahedral sites are filled by the aluminum cations (Al^{3+}; Figure 1.25) to maintain the number ratio of cation/anion as 2/3. In other words, Al layer and O layer are alternately packed and holes are periodically found in 1/3 of atomic sites in Al layer.

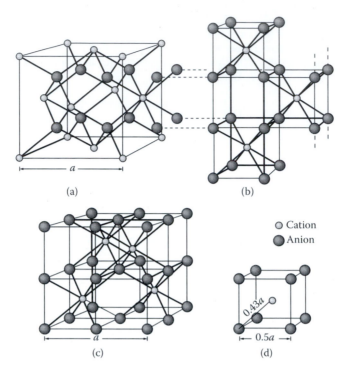

(a)   (b)

○ Cation
● Anion

(c)   (d)

FIGURE 1.24 Schematic illustration of fluorite structure. (From Smart, L. and Moore, E., *Solid State Chemistry: An Introduction*, Chapman and Hall, Boca Raton, FL, 1992. With permission.)

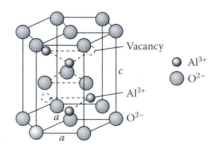

FIGURE 1.25 Crystal structure of alpha-alumina (α-Al_2O_3 or corundum). (From Askeland, D. and Fulay, P., *The Science and Engineering of Materials*, Thomson, Washington, DC, 2006. With permission.)

1.11.8 PEROVSKITE CRYSTAL STRUCTURE

A perovskite crystal structure is one of the mixed oxide structures, that is, a structure of an oxide containing different kinds of cations. The name *perovskite* comes from the calcium titanate ($CaTiO_3$) mineral. A perovskite structure is often described as ABO_3. In this notation, the oxygen ions (O^{2-}) occupy the face center positions on the cube. Then, the A-site cations are divalent (e.g., Ba^{2+}, Pb^{2+}, Sr^{2+}) and occupy the cube corners, as shown in Figure 1.3. The B-site cations (e.g., Ti^{4+}, Zr^{4+}) occupy the cube centers (octahedral site).

Many ceramics, with useful ferroelectric, piezoelectric, and other properties, exhibit the perovskite crystal structure. Examples include barium titanate ($BaTiO_3$), lead zirconium titanate ($Pb(Zr,Ti)O_3$), and strontium titanate ($SrTiO_3$).

Important functional properties of the perovskite materials are developed during their polymorphic phase transition. The cubic form of $BaTiO_3$ is similar to a tetragonal form of $BaTiO_3$ and two structures differ very slightly (~ less than 1%) in terms of their dimensions. The lattice constant along c-axis of tetragonal $BaTiO_3$ is ~4.01 Å. The lattice constant of cubic $BaTiO_3$ is only 4.00 Å (Figure 1.3). This very slight difference of ~0.01 Å in the unit-cell dimension accompanies a shift of Ti^{4+} ion from the cube center and produces very significant changes in the dielectric properties of $BaTiO_3$. For example, the cubic form of $BaTiO_3$ is not ferroelectric, whereas the tetragonal form is ferroelectric.

1.11.9 SPINEL AND INVERSE SPINEL STRUCTURES

The spinel structure is another example of a mixed oxide structure. Many magnetic materials, known as ferrites, exhibit normal or inverse spinel structures. The general formula is AB_2O_4, where A is the divalent cation (e.g., Mg^{2+}) and B is the trivalent cation (e.g., Al^{3+}). A basic frame of the spinel structure is the FCC structure of oxygen ions (O^{2-}). Each edge center and cube center of the FCC unit cell are octahedral sites. Because the edges are shared among four unit cells, there are a total of $(12 \times 1/4) + 1 = 4$ octahedral sites. Similarly, as seen in a diamond crystal structure (Figure 1.20), there are a total of eight tetrahedral sites (center of each octant) in the FCC structure.

To understand the spinel crystal structure, we will consider n formula units. In the n units, there are $4n$ O^{2-} ions, $8n$ tetrahedral holes, and $4n$ octahedral holes. In the normal spinel structure, only one-eighth of the tetrahedral sites are filled by A-type ions. This is because we have n A-type ions and $8n$ sites are available. Similarly, there are $2n$ B-type ions and $4n$ octahedral sites. Thus, one-half of the octahedral sites are occupied by B-type ions, leading to the general formula of AB_2O_4. As shown in Figure 1.26, we can break down the spinel unit cell into eight octants.

In one type of octant, called an A-type octant, the A ions occupy the tetrahedral sites. These atoms are coordinated with the corner and face center anions (oxygen ions) of the unit cell. In another type of octant, known as a B-type octant, the trivalent B cations are located at half of the corners of an octant. Note that out of a total of eight octants, four are A-type and four are B-type. Of the four A-type octants, only two actually contain A ions. Similarly, of the four B-type octants, only two contain B ions (Figure 1.26). Examples of ceramics with this crystal structure include magnesium aluminate ($MgAl_2O_4$) and zinc aluminate ($ZnAl_2O_4$).

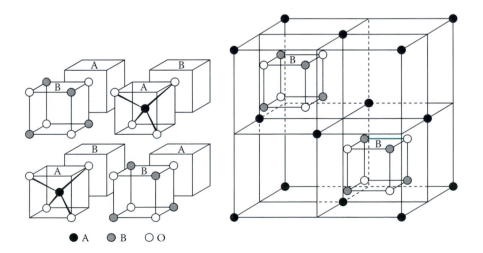

● A   ● B   ○ O

FIGURE 1.26 A spinel crystal structure. (From Smart, L. and Moore, E., *Solid State Chemistry: An Introduction*, Chapman and Hall, Boca Raton, FL, 1992. With permission.)

In some materials such as Fe_3O_4 (the formula can also be written as $FeFe_2O_4$ to emphasize the divalent and trivalent forms of iron), an inverse spinel structure is observed. In this well-known magnetic material, Fe exists in both divalent (Fe^{2+}) and trivalent (Fe^{3+}) forms. The inverse spinel structure is written as $B(AB)O_4$, suggesting that half of the trivalent B-type cations occupy *tetrahedral* sites. All the A-type atoms occupy *octahedral* sites; therefore, the structure is known as an inverse spinel structure. The other half of the B cations continue to occupy the octahedral sites. The type of trivalent ions occupying both tetrahedral and octahedral positions has a large effect on the magnetic coupling between them. This, in turn, has a significant effect on the ferrimagnetic or antiferromagnetic properties of materials known as ceramic ferrites. Nickel ferrite ($NiFe_2O_4$) is another example of a material that shows the inverse spinel structure with a strong magnetic response.

1.12   DEFECTS IN MATERIALS

Arrangements of atoms or ions in real materials are never perfect. In some cases, atoms are missing from sites at which they are supposed to be present. This creates a defect known as a *vacancy*. The presence of vacancies can be useful in many applications of electronic ceramics because vacancies enhance the bulk or volume diffusion of specific types of ions. *Diffusion* is a process by which atoms, ions, or other species move owing to a gradient in the chemical potential (equivalent of concentration). Diffusion continues until the concentration gradient disappears. The process of diffusion can occur through many pathways (e.g., within the grain, bulk, or along the grain boundaries, surfaces, etc.), which is facilitated by the presence of vacancies. Diffusion also plays a key role in enabling semiconductor device processing as well as in other materials—for example, in processing steps such as the sintering of metals and ceramics.

We sometimes deliberately add different atoms to a material. For example, we add boron (B) or antimony (Sb) atoms to silicon to change and better control the electrical properties (e.g., electric conductivity) of the silicon. The atoms that we add (or that are sometimes introduced inadvertently during processing) may take up the positions of the host atoms. This type of defect is known as a *substitutional atom* defect. Typically, this occurs if the radii of both the host and guest atoms are similar. In some other cases, the atoms we add or those introduced inadvertently may end up in the interstitial sites. This is known as an *interstitial atom* defect. These different types of point defects are shown in Figure 1.27.

When atoms of one element *dissolve* in another element as substitutional or interstitial atoms, we get what is called a *solid solution*. This is similar to how sugar dissolves in water. Formation of solid solutions strengthens metallic materials. In Figure 1.4, α and β phases represent the solid solutions of the lead–tin system.

Solid solution formation also occurs among compounds of similar crystal structures (e.g., $BaTiO_3$ and $SrTiO_3$). The formation of solid solutions is used to *tune* the electrical and magnetic properties of ceramics, metals, and semiconductors. For example, by forming solid solutions of GaAs and aluminum arsenide (AlAs), we can produce LEDs that emit light with different colors.

One of the best examples of the usefulness of point defects and formation of solid solutions is the doping of Si to make an *n-type semiconductor* or a *p-type semiconductor*. As illustrated in Figure 1.28, when a pentavalent element, such as antimony (Sb) or phosphorus, is added to Si, we create an n-type semiconductor. This is because substitutional phosphorus or antimony atoms occupy the silicon sites and thus provide an *extra* electron. At temperatures greater than ~50 K, this electron *breaks free* from the phosphorus or antimony atoms and becomes available for conduction. This leads to an n-type semiconductor (thus named because most of the charge carriers are *negatively* charged electrons).

Similarly, when atoms of trivalent elements, such as boron or aluminum, are added to silicon, there is the deficit of an electron. This is because each silicon atom contributes four electrons for covalent bonding. However, an aluminum or boron atom can contribute only three electrons from

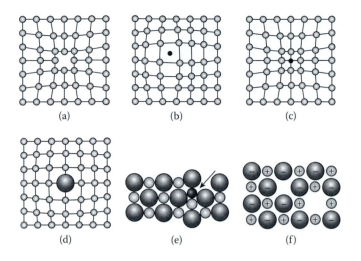

FIGURE 1.27 Illustration of point defects in materials: (a) vacancy, (b) interstitial atom, (c) small substitutional atom, (d) large substitutional atom, (e) Frenkel defect, and (f) Schottky defect. All of these defects disrupt the perfect arrangement of the surrounding atoms. (From Askeland, D. and Fulay, P., *The Science and Engineering of Materials*, Thomson, Washington, DC, 2006. With permission.)

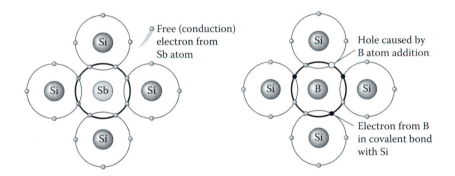

FIGURE 1.28 Creation of antimony (Sb) doped n-type and boron (B) doped p-type silicon by the introduction of substitutional atoms. (From Askeland, D. and Fulay, P., *The Science and Engineering of Materials*, Thomson, Washington, DC, 2006. With permission.)

its outermost shell. This leads to the creation of a hole, which is basically a missing electron. Such semiconductors are known as p-type (because of the *positive* effective charge on a hole).

Atoms of elements that are added purposefully and in controlled concentrations, with the assumption that they will have a useful effect, are known as *dopants*. Different types of semiconductors (i.e., n-type and p-type) form the fundamental basis of many electronic devices, such as transistors, diodes, and solar cells. We will learn in Chapter 3 that the level of conductivity of a semiconductor can be changed by controlling the concentration of dopant atoms.

Note that changes in the conductivity of a semiconductor will occur regardless of whether the different atoms are added on purpose or are introduced inadvertently. Atoms of elements that find their way (usually inadvertently) into a material of interest during the synthesis or fabrication of that material are considered *impurities*. For example, when silicon crystals are grown, oxygen atoms are introduced as impurities that come from the contact of molten silicon with the quartz (SiO_2) crucibles. Typically, we want to minimize the levels of impurities in any material. This is especially the case for semiconductors, in which we want to keep the impurity levels to a minimum (parts per million [ppm] to parts per billion [ppb] for some elements).

1.13 POINT DEFECTS IN CERAMIC MATERIALS

Point defects occur in ceramic materials as well. Many ceramic materials are based on ions. It is not possible to remove a certain number of cations or anions from these materials without causing an imbalance in the net electrical charge. Overall, the electrical neutrality and stoichiometry of a material have to be maintained. There are several ways to meet this requirement. A *Schottky defect* is a defect in which a certain number of cations and a stoichiometrically equivalent number of anions are missing together (Figure 1.27). For example, in NaCl, a Schottky defect is one in which one sodium ion (Na^+) and one chlorine ion (Cl^{1-}) are missing. For a Schottky defect in Al_2O_3, two Al^{3+} and three O^{2-} are missing. A *Frenkel defect* is a type of point defect in which an ion (often a cation) leaves its original site and enters an interstitial site (Figure 1.27). In Section 1.14, we discuss the notation used for point defects in ceramic materials.

1.14 KRÖGER–VINK NOTATION FOR POINT DEFECTS

Defects in ceramic materials are important for many applications involving the electronic, optical, and magnetic properties of ceramics. The following rules must be observed in regard to the presence of point defects in ceramic materials:

1. *Electrical neutrality*—The material, as a whole, must be electrically neutral.
2. *Mass balance*—While introducing any defects, the mass balance must be maintained. This means that no atoms or ions disappear when defects are formed.
3. *Site balance*—When introducing defects, the overall stoichiometry of sites must be maintained.

The notation used to describe point defects in ceramics and the equations that govern their relative concentrations is called the *Kröger–Vink notation*.

For example, consider an oxygen vacancy in MgO. In the Kröger–Vink notation, the vacancy in the oxygen site of MgO is shown as $V_O^{\cdot\cdot}$. The symbol V represents a vacancy. The subscript (in this case, O) indicates the location of the defect (i.e., where the defect occurs). When an oxygen ion (O^{2-}) is missing, a negative charge of two is missing. This means that the empty oxygen site has an *effective* positive charge of two in comparison with a filled oxygen site with a negative charge, -2. An effective positive charge of one unit is indicated by the dot symbol ($\cdot$), which is placed as a superscript. Because the oxygen vacancy has an effective charge of $+2$, we use two dots in the superscript. Therefore, we describe the presence of an oxygen ion vacancy as $V_O^{\cdot\cdot}$. Similarly, a Mg^{2+} ion vacancy is written as V_{Mg}''. In this case, the vacancy defect at the Mg site has an effective negative charge of -2, which is shown using two dashes in the superscript. Example 1.8 illustrates the use of the Kröger–Vink notation.

Example 1.8: Yttrium Oxide (Y_2O_3)-Zirconia (ZrO_2) for Oxygen Sensors and Solid Oxide Fuel Cells

1. Write down the equation that expresses the incorporation of Y_2O_3 in ZrO_2 to form a solid solution. Assume that the concentration of Y_2O_3 is small enough so that new compounds are formed.
2. What defects are created by adding Y_2O_3 to ZrO_2? Are these defects useful?

Solution
1. We assume that yttrium ions (Y^{3+}) occupy the sites of a zirconium ion (Zr^{4+}). The basis for this assumption is that both are cations. We also consider the relative radii of the ions. Yttrium (Y) is a trivalent ion; when it occupies a Zr^{4+} site, there will be a deficit of one positive effective charge. Thus, this defect will have an effective charge of -1.

This defect is written as

$$Y'_{Zr} \tag{1.11}$$

In ZrO_2, for each Zr^{4+} atom, there are two oxygen atoms. Therefore, to add two Y ions on two Zr^{4+} sites, we must use four oxygen sites in setting up a reaction equation. This is for site balance. However, one Y_2O_3 molecule provides only three oxygen ions and the fourth oxygen site remains empty.

Hence, the defect reaction for incorporation of Y_2O_3 in ZrO_2 is written as follows:

$$Y_2O_3 \xrightarrow{ZrO_2} 2Y'_{Zr} + 3O_O^x + V_O^{\cdot\cdot} \tag{1.12}$$

In this equation, the ZrO_2 above the arrow shows that Y_2O_3 (a solute) is being added to ZrO_2 (a solvent).

We check Equation 1.12 for mass balance, site balance, and electrical neutrality. We can see that for the site balance, we have used two zirconium sites and four oxygen sites (three have oxygen ions derived from yttria, and one site has an oxygen vacancy, $V_O^{\cdot\cdot}$). We also have charge balance—the defect caused by the presence of the Y ion in the Zr^{4+} site has an effective charge of −1, and we have two of these. This is balanced by one oxygen ion vacancy $\left(V_O^{\cdot\cdot}\right)$, which has an effective charge of +2. We also have mass balance.

2. As shown in equation 1.12, point defects are formed in Zr and O sites. Oxygen vacancies significantly increase the oxygen ion conductivity of ZrO_2. For each mole of Y_2O_3 added, we get one mole of oxygen vacancy (Equation 1.12). This provides a significant amount of *room* for the oxygen ions to move around or diffuse. This is why zirconium oxide (ZrO_2) containing small amounts of Y_2O_3 (~8–10 mol %) is an oxygen ion conductor. The rate at which oxygen ions can diffuse is considerably increased in Y_2O_3 added ZrO_2.

This material is known as yttria-stabilized zirconia (YSZ). It functions as a solid electrolyte. Thus, by adding Y_2O_3 to ZrO_2, we have converted a dielectric (nonconducting) material into an ionic conductor! Applications of this material include solid oxide fuel cells (SOFCs; Figure 1.29) and oxygen sensors that are used in cars and trucks (Figure 1.30). In applications of YSZ for SOFCs or oxygen gas sensors, what matters the most is the ionic conductivity of YSZ that is controlled by introducing oxygen vacancies.

Another effect of adding Y_2O_3 is that when the concentration of oxygen vacancies increases, a cubic polymorph of ZrO_2 which is found at high temperature becomes stabilized at room temperature. This is why this material is known as YSZ. In some

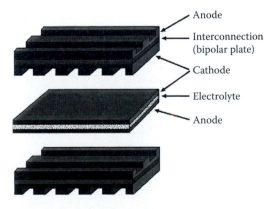

FIGURE 1.29 Schematic representation of a planar-type solid oxide fuel cell. The electrolyte typically is yttria-stabilized zirconia (YSZ). Calcium-doped lanthanum manganite ($LaMnO_3$) is used as the cathode, and YSZ-containing nickel (Ni) is used as the anode. The interconnecting material is lanthanum chromite ($LaCrO_3$). (From Singhal, S.C., *Solid State Ionics.*, 152–153, 405–10, 2002. With permission.)

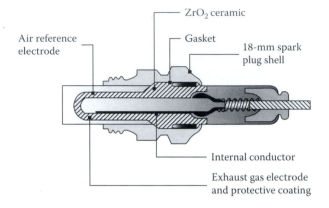

FIGURE 1.30   Schematic representation of a zirconia oxygen sensor used in cars and trucks. (Courtesy of Dynamic-Ceramic Ltd, UK.)

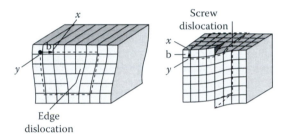

FIGURE 1.31   Illustration of an edge dislocation and a screw dislocation. (From Askeland, D. and Fulay, P., *The Science and Engineering of Materials*, Thomson, Washington, DC, 2006. With permission.)

applications, the primary interest in using YSZ is for its mechanical properties and high-temperature stability. When the cubic phase is stabilized by adding Y_2O_3, ZrO_2 does not show phase changes from cubic to tetragonal to monoclinic forms during cooling of sintered bodies. This avoids the strain induced by the transformations, which would otherwise cause zirconia ceramics to break during fabrication or thermal cycling (heating and cooling).

1.15   DISLOCATIONS

A *dislocation* is a line defect that represents half a plane of atoms missing from an otherwise perfect crystal structure. The two types of dislocations—an edge dislocation and a screw dislocation—are shown in Figure 1.31. A main difference is the shape of the defective plane. In the edge dislocation, a flat defective plane is inserted between two crystal planes. In the screw dislocation, an array of atoms slips on the defective plane. In many cases, dislocations in real materials have characteristics of both edge dislocation and screw dislocation.

Dislocations can have a significant and deleterious effect on the properties of semiconductors and optoelectronic materials. Dislocations in semiconductors typically are formed during crystal growth or during semiconductor device processing. Many years of research have gone into ensuring that essentially dislocation-free silicon and other crystals can be grown.

In most situations, the presence of dislocations in semiconductors is considered deleterious. For example, in gallium nitride (GaN), which is a semiconductor exhibiting a high radiative electron–hole recombination rate, the presence of dislocations has a negative effect on the optoelectronic devices by reducing the radiative recombination rate.

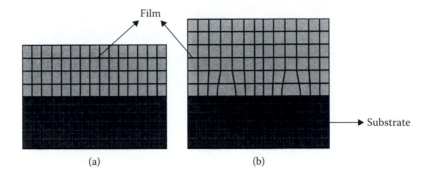

FIGURE 1.32 Illustration of a semiconductor film (a) that is coherently strained and (b) that has a lattice mismatch accommodated by misfit dislocations. (Reprinted from *Encyclopedia of Materials: Science and Technology*, Stach, E. A., and R. Hull. Dislocations in semiconductors. 2301–12, Copyright 2008, with permission from Elsevier.)

The development of LEDs and laser diodes that emit a blue or violet light was hindered by the unavailability of GaN materials with very low dislocation levels. In recent years, superior materials-processing methods for high quality GaN films have been developed, which has in turn led to the development of GaN-based blue or violet lasers.

These blue or violet laser-emitting devices have enabled the so-called Blu-ray format for high-definition (HD) optical-data storage. The Blu-ray format provides more capacity than digital video disks (DVDs). It uses a shorter wavelength (λ) of 405 nm for optically writing the information onto a disk. This wavelength is shorter than that of the typical red lasers ($\lambda = 660$ nm) that are used for authoring DVDs. For conventional compact discs (CDs), the wavelength of the laser used is longer—780 nm. The use of the shorter-wavelength GaN lasers means that a single-layer optical disk can hold 25 GB of data (about 9 hours of HD video content).

It is very difficult to grow defect-free GaN single crystals. Therefore, we typically grow GaN on another substrate, such as sapphire. This process is also used for the fabrication of many other semiconductor devices and is known as *epitaxy*. In this process, a strain is introduced at the interface if the lattice constants of the film and the substrate are not exactly the same.

In the case of GaN on sapphire (Al_2O_3), the mismatch is about 16% for sapphire. At the interface, the strain that exists due to lattice mismatch is often relieved by the formation of what are known as *misfit dislocations* (Figure 1.32). In GaN, dislocations known as *threading dislocations* originate at the interface and travel all the way to the surface. In principle, the line directions of the threading dislocations are normal to film/substrate interface. This means that a formation of the edge dislocation relives the strain, but threading dislocations also have characteristics of the screw dislocation. Dislocations in GaN, as examined by transmission electron microscopy (TEM) and high-resolution TEM (HRTEM), are shown in Figure 1.33.

In some cases, the presence of dislocations away from the active regions of electrical devices can be useful. Dislocations can attract some of the impurity atoms toward them, leaving the regions in which devices are active with fewer defects. This method of segregating impurities is known as *gettering*. Impurities in a semiconductor, such as transition metal impurities including iron and nickel, are concentrated by attracting them to defects such as precipitates and dislocations.

1.16 STACKING FAULTS AND GRAIN BOUNDARIES

Similar to point defects (that are considered zero-dimensional defects) and dislocations (one-dimensional line defects), we also find area defects (two-dimensional defects). One type of area defect is a *stacking fault*. We have seen that the FCC structure can be visualized by considering the stacking of

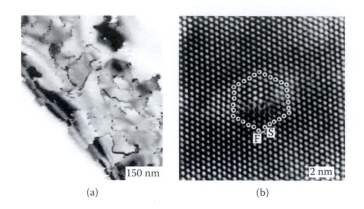

(a)   (b)

FIGURE 1.33 (a) Transmission electron microscope image of dislocations in gallium nitride deposited on sapphire (Al_2O_3). (b) The core of a dislocation as seen using high-resolution transmission electron microscopy (HRTEM): stacking faults are marked as F and S. (Courtesy of Dr. J. Narayan, North Carolina State University, Raleigh, NC.)

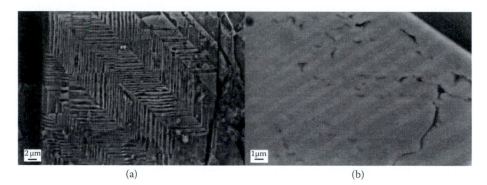

(a)   (b)

FIGURE 1.34 Ferroelectric domains in lead-free $K_{0.5}Na_{0.5}NbO_3$ ceramics. (From López-Juárez, R., et al., Lead-free ferroelectric ceramics with perovskite structure. In *Ferroelectrics-Material Aspects*, ed. M. Lallart. Rijeka, Croatia: InTech, 2011.)

atoms within the planes as ABCABCABC… (Figure 1.12). In a stacking fault, one of the planes in the expected sequence is missing. Therefore, the stacking sequence may look like ABCABABCABC…. Thus, in this stacking fault, a small region of the material shows the HCP-stacking sequence.

Other examples of area defects are domain boundaries, low-angle boundaries, and twin boundaries. A *domain* is a small region of a material in which the dielectric or magnetic polarization direction is the same. We will learn that in magnetic and ferroelectric materials, the entire material cannot be stable energetically as one large magnet or an electric dipole, respectively. To minimize the overall free energy, the material spontaneously shows the formation of multiple domains arranged in a random fashion. The formations and arrangements of domains are extremely important in determining the properties of magnetic and ferroelectric materials. A scanning electron micrograph of the domains in ferroelectric ceramics is shown in Figure 1.34.

One of the most important features of the microstructure of a polycrystalline material is the presence of grain boundaries (Figures 1.6 and 1.34). At the grain boundaries, the atomic order is disrupted. This is why grain boundaries are considered defects. Because grains are three-dimensional, the grain boundary is actually a surface in three dimensions; therefore, grain boundaries are considered three-dimensional defects.

Grain boundary is one of the most important features in terms of its effects on the properties of materials. These effects are discussed in the next section.

1.17   MICROSTRUCTURE–PROPERTY RELATIONSHIPS

1.17.1   GRAIN BOUNDARY EFFECTS

If we need to compare the electrical or other properties of two different grades of a $BaTiO_3$-based polycrystalline material, we can refer to the properties of these materials in the context of their microstructure. For example, we may conclude that finer-grained $BaTiO_3$ ceramics have finer domains and higher *dielectric constant* (*k*). We will learn in Chapter 7 that the dielectric constant is a measure of a material's ability to store an electrical charge. Grain boundary regions often also have very different electrical or magnetic properties than those of the material inside the grains.

1.17.2   GRAIN SIZE EFFECTS

As the grain size decreases, the area of the grain boundary increases. In many applications, grain boundaries are used to control the properties of materials. Most often, a polycrystalline metallic material exhibits a higher mechanical strength than that of a coarse-grained material with essentially the same composition.

The presence of grain boundaries also has a significant effect on the electrical, magnetic, and optical properties of materials. For example, the electrical resistivity of a conductor such as copper or silver typically is higher for a polycrystalline material than that for a single crystal of the same material. This is because of the increased scattering of electrons by atoms in the grain boundary regions.

The optical properties of materials are also affected by the grain boundaries and pores (holes) in polycrystalline ceramics because these contribute to the scattering of light. Al_2O_3 ceramics can be made translucent by using special additives and processing techniques that lead to larger grain size. Polycrystalline Al_2O_3 ceramics with a fine grain size are usually opaque (Figure 1.35a). This is largely because of the scattering of light from both the pores and the grain boundaries. The microstructure of larger-grain ceramics that are optically translucent is shown in Figure 1.35b. Translucent Al_2O_3 ceramics are used in many applications, such as in envelopes for high-pressure sodium vapor

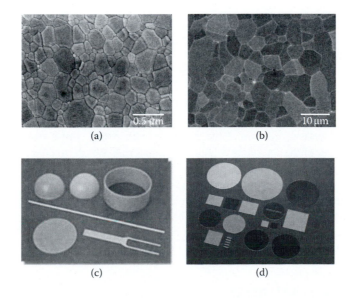

(a)   (b)

(c)   (d)

FIGURE 1.35   Effect of grain size on the optical properties of alumina ceramics. (a) Finer-sized alumina grain is opaque. (b) Polycrystalline alumina with larger grain size is translucent. (From Kim et al. 2009. *Acta Mater* 57(5):1319–26, 2009.) (c) Parts made from polycrystalline alumina with larger grain size are translucent. (Courtesy of Covalent Materials Corporation, Tokyo, Japan.) (d) Single-crystal sapphire substrates are transparent. (Courtesy of Kyocera Corporation, Kyoto, Japan.)

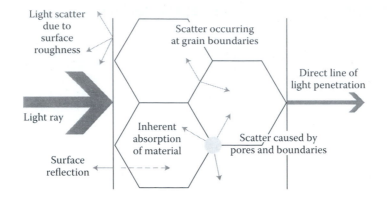

FIGURE 1.36 Processes that cause absorption and scattering of light in polycrystalline ceramics. (Adapted from Covalent Materials Corporation, Tokyo, Japan. With permission.)

lamps (Figure 1.35c). With single-crystal Al_2O_3 ceramics, the material essentially becomes transparent (Figure 1.35d) because this material has very little intrinsic absorption and there are no pores or grain boundaries to scatter light. Light scattering by the grain boundaries and the pores is schematically illustrated in Figure 1.36. In this illustration, the hexagonal regions represent the grains in a polycrystalline material. Details of the scattering are explained in the chapter on optical properties of materials.

Similar to Al_2O_3, many other transparent or translucent polycrystalline ceramic materials, such as yttrium aluminum garnet (YAG), yttrium oxide (Y_2O_3), and lead lanthanum zirconium titanate (PLZT), have also been developed for different commercial applications.

1.17.3 Microstructure-Insensitive Properties

We must emphasize two points regarding the microstructure–property relationship of materials. First, when we change the microstructure of a material, many properties can change. In the case of Al_2O_3 ceramics, as discussed before, an increase in the grain size increases optical transparency; however, decreases fracture toughness. Likewise, a change in the microstructure may result in concomitant changes in many properties.

The second point that needs to be emphasized is that not all properties of materials are sensitive to microstructure. A property that does not change with microstructure is known as a *microstructure-insensitive property*. For example, typically, the *Young's modulus* (Y) or *elastic modulus* (E) of a metallic or ceramic material will not change drastically with changes in the grain size. This is because Young's modulus, which is a measure of the difficulty with which elastic strain can be introduced, depends on the strength of the interatomic bonds in a material. When we change the microstructure, we do not change the nature of the bonds; hence, Young's modulus will not be affected. However, the so-called yield stress (σ_{YS}) of a material—a level of stress that initiates permanent or plastic deformation—typically depends strongly upon the average grain size of the material. This is because the movement of dislocations (known as slip) that causes plastic deformation is resisted by disruptions in the atomic arrangements occurring at the grain boundary regions.

We will now turn our attention to amorphous materials.

1.18 AMORPHOUS MATERIALS

In some materials, the atoms (or ions) do not exhibit an LRO. Such materials are considered *amorphous* or *noncrystalline* materials, or are simply called *glasses*. We use the term glass to refer to amorphous materials (metallic or ceramic) derived by the relatively rapid cooling of a melt. Amorphous materials typically are formed under nonequilibrium conditions during processing (e.g., relatively faster cooling of a melt or decomposition of a vapor). As a result, they tend to be thermodynamically unstable. Inorganic glasses based on silica (SiO_2), which are used to make optical fibers, and *amorphous silicon* (a-Si) are examples of amorphous materials.

1.18.1 ATOMIC ARRANGEMENTS IN AMORPHOUS MATERIALS

There exists a *short-range order* (SRO) of atoms or ions in amorphous materials. The difference between an SRO and an LRO in amorphous silicon (a-Si) and crystalline silicon (c-Si), respectively, is shown in Figure 1.37.

Note that the a-Si has a structure wherein the angles at which the silicon atom tetrahedra are connected to one another and the distances between the silicon atoms are not exactly the same throughout the structure.

a-Si is made using the so-called chemical vapor deposition (CVD) process, which involves decomposing silane (SiH_4) gas. A concentration of hydrogen atoms is also incorporated into the structure. These hydrogen atoms *pacify* some of the silicon bonds that would otherwise remain unsaturated or dangling (Figure 1.37). This is helpful for microelectronic devices based on a-Si because the unsaturated silicon bonds would make the devices electrically inactive.

Since the film deposition does not require high-temperature processing, a-Si is used in making thin-film transistors (TFTs) that are integrated onto glass surfaces. The underlying circuitry created with such a-Si-based TFTs is used to drive liquid crystal displays in personal computers, televisions, and electronic ink (e-ink)-based bistable displays found in most state-of-the-art electronic book readers (e.g., Kindle).

Another technologically important amorphous material is SiO_2-based glass. In both c-SiO_2 and a-SiO2, the silicon and oxygen ions are connected in a tetrahedral arrangement. This means that each silicon ion (Si^{4+}) is always surrounded by four oxygen ions. This is the SRO. However, in SiO_2-based glass, the angles at which these tetrahedra are interconnected vary. Thus, the distances between silicon ions in different tetrahedra also vary and LRO is not found in a-SiO_2. c-SiO_2 exhibits many equilibrium and nonequilibrium polymorphs. These have both LRO and SRO. For a given polymorph of c-SiO_2, the angles at which different silicon–oxygen tetrahedra are connected remain essentially the same.

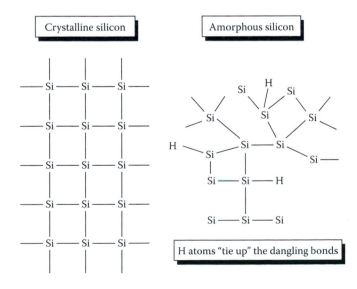

FIGURE 1.37   A schematic representation of the difference between short-range order in amorphous silicon and long-range order in crystalline silicon. (From Singh, J., *Optoelectronics: An Introduction to Materials and Devices*, McGraw Hill, New York, 1996. With permission.)

Note that a material does not have to be either crystalline or amorphous—it can exist in both forms either simultaneously or under different conditions. We often encounter materials that have both crystalline and amorphous phases. For example, in polyvinylidene fluoride (PVDF), one of the most widely used piezoelectric materials (see Chapter 9 on ferroelectrics, piezoelectrics, and pyroelectrics), the microstructure consists of crystalline regions embedded in an amorphous matrix. Most polymers are mixtures of crystalline and amorphous regions interspersed within the material. The processing or manufacturing methods used for creating a material have a significant effect on whether it is crystalline, amorphous, or a mixture of the two. The development of polymers for flexible and lightweight electronic devices is an active area of research and development known as *flexible electronics*.

1.18.2 Applications of Amorphous Materials

Several factors need to be considered before choosing an amorphous or a crystalline form of a material for a given application. For example, amorphous silica is used for making optical fibers because kilometers of continuous fibers can be drawn from molten SiO_2. In principle, we could use single-crystal SiO_2 to produce the optical fibers, because it is optically transparent. However, this is not easy or, more importantly, cost-effective. For the same reason (i.e., manufacturability and cost), a main constituent of building glass windows is also amorphous SiO_2. If a crystalline form of SiO_2 is used, it is difficult to fabricate such a flat and shiny surface.

In contrast, amorphous silica cannot be used for certain applications, such as the so-called *quartz* clocks. Only SiO_2 crystals exhibit what is described as a *piezoelectric effect*, where voltage is generated across a material when it is stressed (Sections 10.5 and 10.6). This effect is used to make quartz clocks.

The properties of a material can be very different in the amorphous and crystalline states. With some technologies, we can make use of both the amorphous and crystalline forms of a material. For example, in *phase change memory* (PCM) technology, the back-and-forth switching of a material between the amorphous and crystalline states is used for data storage. Materials based on a ternary compound of GeSbTe, known as the GST materials, are used in PCM technology. PCM technology utilizes the differences between the electrical resistivity values of the amorphous and crystalline forms of the materials.

An electrical resistor rapidly heats up the material and changes its state from crystalline to amorphous (Figure 1.38). The amorphous material, in this case, exhibits a higher resistivity than the crystalline material (Figure 1.39). These changes in electrical properties are sensed by other appropriate circuitry. The amorphous state has high resistivity and uses "1" as its code (data write).

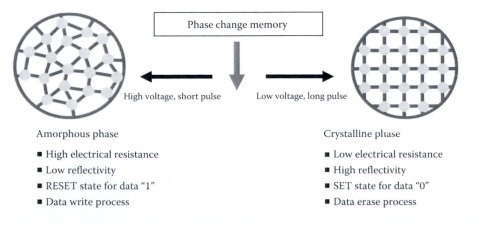

FIGURE 1.38 Phase change memory illustration. (Courtesy of Dr. Ritesh Agarwal, University of Pennsylvania, Philadelphia, PA.)

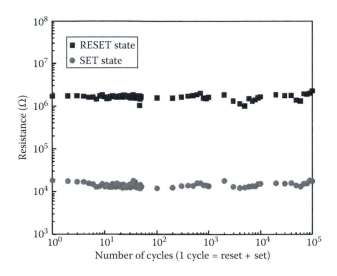

FIGURE 1.39 Resistivity of a GST-nanowire phase change memory device. (Courtesy of Dr. Ritesh Agarwal, University of Pennsylvania, Philadelphia, PA.)

The crystalline state has low resistivity and uses "0" as its code (data erase). The GST materials can be cycled rapidly and reversibly between the *reset* amorphous and *set* crystalline phases by heating and cooling. This is the basis for PCM technology.

A similar concept is used for rewritable CD and DVD media. For these technologies, the differences between the optical properties of materials in the amorphous and crystalline states are used to write and read information on the media. The amorphous phase is less reflective than the crystalline phase. This difference in the reflectivity is exploited in the optical data storage.

1.19 NANOSTRUCTURED MATERIALS

In recent years, considerable research and development has occurred with respect to *nanostructured* materials. The grain size or crystallite size of these materials is extremely small (~10 nm or less). In terms of length scales, nanostructured materials fall between those that are amorphous and crystalline. The nanostructured form causes an unusually high fraction of atoms in the material to be located at the grain boundaries (if the material is polycrystalline) or at the particle surfaces (if the material is in the form of nanoparticles). This and other phenomena (which could be quantum mechanical in nature) often lead to unique and unexpected changes in the properties of a nanostructured material.

For example, the band gap (E_g) of a semiconductor becomes larger for materials in the nanoparticle form. This increase, which results from smaller particle size, is known as *blue shift* and is easily observed by examining changes in the dispersions colors of semiconductor nanoparticles (also known as quantum dots). In addition, the availability of nanomaterials has led to the possibility of using surface effects, such as surface plasmons, because of the large surface area/volume ratio of nanomaterials. This has created new opportunities in the areas of nanophotonics and quantum computing. Also, mechanical properties of nanostructured materials are very different from their bulk counterpart. The strength of materials consisting of nanosize grains is much greater than the materials of micrometer-size grains. Also, nanostructured materials can be much more ductile than traditional bulk materials. This is because nanomaterials have the ability to suppress the formation of extended defects. Though point defects are generated in nanostructured materials, these defects can easily travel to grain boundaries, which are much closer to the location of the point defects in nanostructured materials than in traditional bulk materials. Then, the point defects of the nanostructured materials reaching the grain boundaries are annihilated because the grain boundary functions as a defect sink. Suppression of the defect generation process (or acceleration of the defect annihilation) results in unique mechanical properties (e.g., high strength and ductility) of nanostructured materials.

1.20   DEFECTS IN MATERIALS: GOOD NEWS OR BAD NEWS?

The term *defect* is somewhat of a misnomer because, when defects are present, the properties of a material actually may be improved! Thus, the term defect just refers to the fact that the atomic arrangement in a given material is disturbed and is not perfect. Amorphous materials can have a very high concentration of defects. However, the presence of point defects, line defects, area defects (domains, stacking faults), and volume defects (grain boundaries) in materials is extremely important and useful for several reasons.

The addition of antimony or boron to silicon (Figure 1.28) brought about the entire technology of ICs based on silicon. The presence of point defects also enables the conversion of insulating ceramic materials into semiconductors. There are plenty of examples where the presence of point defects is not just desirable but essential. One example is the creation of oxygen vacancies in zirconia to turn it into an ionic conductor, which allows for the transport of O^{2-} ions. In cases of metals and alloys, the presence of point defects is also critical for the enhancement of mechanical properties. For example, pure copper is too soft. The elastic modulus and yield stress of copper are improved considerably when we add other alloying elements (e.g., beryllium [Be]) to form solid solutions or precipitates.

However, the presence of defects can be detrimental in many situations. For example, in the case of solid solution- or precipitate-based strengthening of metals, electrical conductivity is lowered, although some mechanical properties, such as yield strength (σ_{YS}), are improved. The effective speed (mobility) with which electrons can move in a single-crystal silicon is far greater than that in a-Si, though highly crystalline silicon is much more difficult to fabricate than a-Si. Thus, defects can be detrimental for certain applications.

To summarize, the term defect does not mean that the material is defective. The presence of certain types of defects (i.e., imperfections in atomic or ionic arrangements) may be beneficial and, sometimes, may even be essential. However, certain types of defects have deleterious effects and must be avoided, thereby suggesting the importance of defect engineering in materials.

PROBLEMS

1.1 If the lattice constant of BCC Fe is 2.8666 Å, what is its density? Assume that the atomic mass of Fe is 55.85.

1.2 For the DC structure, show that the packing fraction is about 34%.

1.3 What is the theoretical density of Ge? Assume that the atomic radius of Ge is 122.3 pm.

1.4 What is the theoretical density of GaAs? Assume that the lattice constant of GaAs is 565 pm. The atomic masses of Ga and As are 69.7 and 74.9, respectively.

1.5 Similar to the volume-packing fractions, we can calculate the area-packing fractions for planes in a unit cell. Calculate the area-packing fractions for FCC-Ni for the planes (100) and (111). Which plane is close-packed?

1.6 Sketch the following planes and directions in a cubic unit cell: [111], [$\bar{1}$10], [011], (111), and (123).

1.7 Sketch the crystal structure of a PZT ceramic with a perovskite structure. Label the different ions clearly.

1.8 Calculate the interplanar spacing for the (100), (110), and (111) planes in Si.

1.9 Show that the Miller–Bravais indices for direction [110] would be [11$\bar{2}$0].

1.10 The equation for the formation of a Schottky defect in MgO is written as

$$\text{null} \rightarrow V''_{Mg} + V^{\cdot\cdot}_{O} \tag{1.13}$$

Write down the equation for the Schottky defect formation in Al_2O_3.

1.11 In addition to Y_2O_3, we can also add MgO to ZrO_2 to stabilize the cubic crystal structure (Equation 1.12). Write the equation that will represent the incorporation of small concentrations of MgO into ZrO_2.

1.12 Why must the Al_2O_3 ceramics used in Na (sodium) vapor lamps be transparent or translucent? How does a larger grain size help with the transparency of Al_2O_3 ceramics?

1.13 Both p-Si and a-Si are used for manufacturing solar cells. For manufacturing ICs, we use single-crystal Si. Why?

1.14 Although both graphite and diamond are based on strong covalent C–C bonds, graphite is used as a solid lubricant, whereas diamond is one of the hardest naturally occurring materials. Why is this?

1.15 What are the advantages of developing microelectronic components based on polymers?

1.16 State one example each of an application where the presence of dislocations is (a) useful and (b) deleterious.

1.17 What is the principle by which PCM technology works?

1.18 Similar to GST materials that are used in PCM, SiO_2 also exists in amorphous and crystalline forms. Compared to GST, SiO_2 is inexpensive. Will it be possible to use SiO_2 for PCM?

1.19 Why does the Blu-ray format offer much higher density for data storage in CDs and DVDs?

1.20 Conventional solar cells are made using polycrystalline silicon and offer an efficiency of about 15%. Some companies are developing solar cells based on copper indium gallium selenide (CIGS). These CIGS-based solar cells can offer efficiencies up to 40%. Why can we not just switch over to these more efficient CIGS-based solar cells?

1.21 Some companies have developed solar cells based on organic materials. The efficiency of these solar cells is about 5%. If we already have Si solar cells that operate at ~15%, why should we develop solar cells based on polymers that are not as efficient?

GLOSSARY

Allotropes: Different crystal structures exhibited by an element.

Amorphous materials: Materials in which atoms or ions do not have a long-range order.

Amorphous silicon (a-Si): An amorphous form of silicon typically obtained by the chemical vapor decomposition of silane (SiH_4) gas, which is used in thin-film transistors and photovoltaic applications.

Anions: Negatively charged ions (e.g., O^{2-}, S^{2-}).

Basis: An atom or a set of atoms associated with each lattice point. The combination of a lattice and basis defines a crystal structure.

Body-centered cubic (BCC) structure: A crystal structure in which the atoms are positioned at the corners of a unit cell with an atom at the cube center (the packing fraction is 0.68).

Bravais lattice: Any of the 14 independent arrangements of points in space.

Cations: Positively charged ions (e.g., Si^{4+}, Al^{3+}).

Close-packed structure: A crystal structure that has atoms filled in a way that achieves the highest possible packing fraction.

Coordination number (CN): The number of atoms that surround an atom on a lattice site or in an interstitial site.

Crystal structure: The geometrical arrangement in which atoms are arranged within the unit cell, described using a Bravais lattice and basis.

Crystalline material: A material in which atoms or ions are arranged in a particular three-dimensional arrangement, which repeats itself and exhibits a long-range order.

Dielectric constant (k): A measure of the ability of a material to store an electrical charge.

Diffusion: A process by which atoms, ions, or other species move, which is due to a gradient in the chemical potential (equivalent to concentration).

Domain: A small region of a ferroelectric or magnetic material in which the dielectric or magnetic polarization direction, respectively, is the same.

Dopant: Foreign atoms (e.g., boron and phosphorus) deliberately added to a semiconductor to tune its level of conductivity.

Elastic modulus (E): A measure of the difficulty with which elastic strain can be introduced into a material. This depends on the strength of the bonds within a material. Elastic modulus is the ratio of elastic stress to strain and is also known as the Young's modulus (Y).

Epitaxy: A process by which a thin film of one material is grown, typically over a single-crystal substrate of the same (homoepitaxy) or different (heteroepitaxy) composition. There is usually a good match between the lattice constants of the thin film and the substrate.

Face-centered cubic (FCC) structure: A crystal structure in which the atoms are positioned at the corners of a unit cell and also at the six face centers (the packing fraction is 0.74).

Ferroelectric substance: A dielectric material that shows spontaneous and reversible polarization.

Flexible electronics: A field of microelectronics devoted to the development of flexible electronic components or devices that is often based on polymers or thin metal foils (or on both).

Frenkel defect: A defect caused by an ion leaving its original position and entering an interstitial site.

Gettering: A method by which impurities in a semiconductor (e.g., transition metal impurities like Fe and Ni) are concentrated away from the active device regions by attracting them to defects, such as precipitates and dislocations.

Glass: An amorphous material typically obtained by the solidification of a liquid under nonequilibrium conditions. The term "glass" is often used to describe ceramic or metallic amorphous materials (or both).

Grain: A small, single-crystal region in a polycrystalline material.

Grain boundaries: The regions between the grains of a polycrystalline material.

Hexagonal close-packed (HCP) structure: A crystal structure in which atoms are arranged in a hexagonal pattern such that the maximum possible packing fraction (0.74) is obtained.

Hydrogen bond: A special form of van der Waals bonds found in materials based on polar molecules or groups (e.g., water).

Impurity: Foreign atoms introduced during the synthesis or fabrication of materials such as semiconductors. Their effect usually is not desirable.

Interstitial atoms: Atoms or ions added to a material that occupy the interstitial sites of a given crystal structure.

Interstitial site: A hole in the crystal structure that may contain atoms or ions.

Kröger–Vink notation: A notation used to indicate the point defects in materials.

Lattice: A collection of points in space. There are only 14 independent arrangements of points in space, known as the Bravais lattices.

Long-range order (LRO): Atoms or ions arranged in a particular geometric arrangement that repeats itself over relatively larger distances (~ a few micrometers up to a few centimeters). This is seen in polycrystalline and single-crystal materials.

Microstructure: A term that describes the average size and size distribution of grains, in addition to the grain boundaries and other phases (such as porosity and precipitates) and defects (e.g., dislocations).

Microstructure-insensitive property: A property that does not change substantially with alterations in the microstructure of a material (e.g., Young's modulus).

Miller–Bravais indices: A four-index system for directions and planes in the hexagonal unit cell.

Miller indices: Notation used to designate specific crystallographic directions and planes in a unit cell.

n-Type semiconductor: A semiconductor in which a majority of the charge carriers are negatively charged electrons (e.g., phosphorus-doped silicon).

Nanostructured materials: Materials with an ultrafine particle or grain size (<10 nm) that have an unusually large fraction of atoms located at the grain boundaries (for nanocrystalline materials) or surfaces (for nanoparticles), causing them to have unusual properties.

Noncrystalline materials: Same as amorphous materials—these materials exhibit no long-range order of atoms or ions.

Octahedral site: An interstitial site in which six atoms that form an octahedron surround the interstitial atom or ion—the coordination number is six.

Octant: A smaller cube obtained by dividing the unit cell into eight smaller cubes.

Organic electronics: A field of research and development based on the use of polymeric materials, such as conductors and semiconductors, for developing flexible, lightweight microelectronic devices such as transistors and solar cells.

Packing fraction: The ratio of volume or space occupied by atoms to the volume of the unit cell.

Phase: Defined as any portion of a system, including the whole, that is physically homogeneous and bounded by a surface so that it is mechanically separable from any other portion.

Phase change memory (PCM): A memory technology in which the amorphous and crystalline states of a material are used to store information as "1" or "0."

Phase diagram: A diagram indicating the different phases that can be expected in a given system of materials, assuming a condition of thermodynamic equilibrium exists.

Piezoelectric effect: The development of a voltage across a material when it is stressed.

Plastic: A polymer-based material formulated with one or many polymers and other additives (e.g., carbon black and glass fibers).

Polymorphs: Different crystal structures exhibited by a compound.

p-type semiconductor: A semiconductor in which a majority of the charge carriers are positively charged holes (e.g., boron-doped silicon).

Quantum dots: Nanocrystals of semiconductors that exhibit a change in the band gap, which in turn causes changes in their optical properties, such as color.

Radius ratio: The ratio of the radius of a cation to that of an anion.

Schottky defect: A defect in which a certain number of cations and a stoichiometrically equivalent number of anions are missing—for example, the absence of one sodium ion and one chlorine ion in NaCl.

Short-range order (SRO): Atoms and ions in amorphous and crystalline materials exhibit a short-range (up to a few Å) order in the arrangement of atoms or ions in a specific geometrical fashion.

Simple cubic (SC) structure: A crystal structure in which atoms are positioned at the corners of a unit cell (the packing fraction is 0.52).

Solid solution: A new phase of a material obtained by dissolving atoms of one element into another; this phase strengthens metallic materials and can increase conductivity.

Stacking fault: A planar defect in which one of the planes from a stacking sequence is missing.

Substitutional atoms: Atoms or ions added to a material. They occupy sites that are usually occupied by the atoms or ions of the host material.

Tetrahedral site: An interstitial site in which four atoms that form a tetrahedron surround the interstitial atom or ion; the coordination number is four.

Unit cell: The basic unit that represents an arrangement of atoms (or ions) repeating in all three dimensions.

Vacancy: The absence of an atom or an ion from its crystallographic location within a crystal structure.

van der Waals bond: A secondary bond that is present in all materials and is caused by interactions between induced dipoles. It is functionally important in materials with polar ions, atoms, or groups (e.g., water, PVC, PVDF). One special type of this order is the hydrogen bond.

Yield stress (σ_{YS}): A level of stress that initiates a permanent or plastic deformation. Yield strength depends on the grain size of materials. In general, a decrease in the grain size increases the yield strength.

Young's modulus: Also known as the elastic modulus, this is a measure of the difficulty with which elastic strain can be introduced in a material. It depends on the strength of bonds within the material and the ratio of the elastic stress to the strain.

REFERENCES

Askeland, D., and P. Fulay. 2006. *The Science and Engineering of Materials.* Washington, DC: Thomson.

Groover, M. P. 2007. *Fundamentals of Modern Manufacturing: Materials, Processes, and Systems.* New York: Wiley.

Kim, B.-N., K. Hiraga, K. Moritaa, H. Yoshida, T. Miyazaki, and Y. Kagawa. 2009. Microstructure and optical properties of transparent alumina. *Acta Mater* 57(5):1319–26.

López-Juárez, R., F. González, and M.-E. Villafuerte-Castrejón. 2011. Lead-free ferroelectric ceramics with perovskite structure. In *Ferroelectrics-Material Aspects*, ed. M. Lallart. Rijeka, Croatia: InTech.

Singh, J. 1996. *Optoelectronics: An Introduction to Materials and Devices.* New York: McGraw Hill.

Singhal, S. C. 2002. Solid oxide fuel cells for stationary, mobile, and military applications. *Solid State Ionics* 152–153:405–10.

Smart, L., and E. Moore. 1992. *Solid State Chemistry: An Introduction.* Boca Raton, FL: Chapman and Hall.

Stach, E. A., and R. Hull. 2008. Dislocations in semiconductors. In *Encyclopedia of Materials: Science and Technology,* eds. K. H. J. Buschow et al., 2301–12. Oxford: Elsevier.

2 Electrical Conduction in Metals and Alloys

KEY TOPICS

- The fundamentals of what causes a material to be a conductor of electricity
- Different types of conducting materials based on metals and alloys and their "real-world" technological applications
- The classical theory of conductivity
- The band theory of solids and its use for examining the differences among conductors, semiconductors, and insulators
- The effects of chemical composition, microstructure, and temperature on the conductivity of metals and alloys
- Real-world applications of conductive materials

2.1 INTRODUCTION

Metals are *conductors* of electricity; that is, they offer relatively little resistance to the flow of electricity. In metals, each atom donates one or more electrons and therefore becomes more stable. This process leads to a large *sea of electrons* (Figure 2.1). A high concentration of these donated electrons, which are free to move around within a metallic material, leads to a higher level of conductivity in metals.

Figure 2.2 shows one way to classify electronic materials based on the nature of their interatomic bonds. In Section 2.8, we will see that, compared to pure metals, alloys offer a higher resistance to the flow of electricity. In contrast to metals, most ceramic materials (e.g., silica [SiO_2], zirconia [ZrO_2], alumina [Al_2O_3], and silicon carbide [SiC]) exhibit strong ionic or covalent bonds (or both). Many polymers (polyethylene, polystyrene, epoxies, etc.) also primarily exhibit covalent bonds *within* the chains of atoms and van der Waals bonds *among* the chains of atoms. Therefore, ceramics and polymers are usually, but not always, electrical *insulators* and are also referred to as *dielectrics*. The use of the term insulator is sometimes preferred when emphasizing the ability of a material to withstand a strong electric field—as opposed to offering only a high electrical resistance. For example, porcelain can be described more appropriately as an insulator rather than as a dielectric.

Covalent bonds are formed by the sharing of electrons from different atoms participating in bond formation. Therefore, the valence electrons in covalent or ionic bonds are localized in dielectric and insulating materials; that is, they are trapped inside the bonds (Figure 2.1). In solids with strong ionic bonds, the electropositive elements donate their valence electrons to electronegative elements to form ionic bonds. In Section 3.2.1, we will see that many ceramics and some polymers can be made to exhibit a useful level of conductivity through deliberate modification of their composition and microstructure. The conducting ability of some materials (e.g., silicon [Si], germanium [Ge], and gallium arsenide [$GaAs$]) is in between that of insulators and conductors. These materials are known as *semiconductors* (Figure 2.2). The ranges for the values of a parameter known as the *resistivity* (ρ), discussed in Section 2.2, are also shown in Figure 2.2.

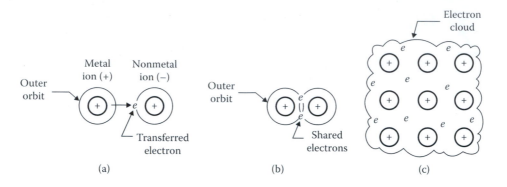

FIGURE 2.1 Illustration of (a) ionic bonds, such as those in sodium chloride and many other ceramics; (b) covalent bonds, such as those in silicon, many ceramics, and polymers; and (c) metallic bonds. (From Groover, M.P., *Fundamentals of Modern Manufacturing: Materials, Processes, and Systems*, Wiley, New York, 2007. With permission.)

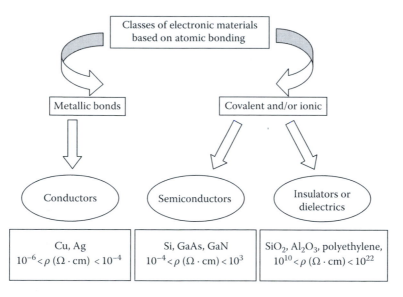

FIGURE 2.2 Classification of materials based on the nature of bonding. Typical ranges of resistivity are also shown.

Some materials, known as *superconductors*, exhibit zero electrical resistance under certain conditions (e.g., low temperatures and low magnetization). If the critical conditions for zero electric conductivity are not met, the superconductors turn to highly conductive metals. Examples of superconducting materials include elements such as mercury (Hg), intermetallic compounds such as niobium tin (Nb_3Sn), and ceramic materials such as yttrium barium copper oxide ($YBa_2Cu_3O_{7-x}$) and magnesium diboride (MgB_2). Superconductors are not shown in Figure 2.2 because the bonding in these materials is complex, and other additional mechanisms play a role in the superconductivity of these materials.

The ability of a material to conduct electricity usually depends on its temperature and its microstructure. In some instances, such as in the cases of semiconductors and insulators, a relatively higher temperature will promote a higher level of electrical conductivity because more electrons are available for the electrical conduction. In other cases, such as for essentially pure metals (which already have enough electrons for the electrical conduction), the ability to conduct electricity *decreases* with an increase in temperature because the probability of electrons colliding with lattice atoms is increased at a higher temperature.

Similarly, a material's structure has a significant effect on its ability to conduct electricity. For example, polymers such as polyethylene are usually considered to be nonconducting. However, some polymers possessing conjugated characteristics (i.e., alternation of double bonds and single bonds in a carbon chain of the polymers) exhibit a semiconducting property. An example of a polymeric material that shows a useful level of conductivity is polyaniline. The advantage of conducting polymers is that they are flexible and lightweight—unlike metals and semiconductors. Thus, novel and flexible electronic components can be developed using conducting and semiconducting polymeric materials.

An important point is that, although we can associate a general trend in conductivity with a particular class of materials, there are always exceptions. Most polymers and ceramics are dielectrics or insulators but not all. Similarly, most metals are good conductors but not all.

In some cases, it is advantageous to use conductive materials based on *composites*. Composites are materials made up of two or more materials or phases, often positioned in unique geometrical arrangements to achieve desired properties. For example, silver (Ag)-filled epoxies are used in microelectronics as conductive adhesives. These composites are prepared by dispersing fine silver powder particles in an epoxy matrix. The silver provides the electrical conductivity. Epoxy provides a means of applying a paint-like material that eventually hardens into a solid plastic.

Certain aspects of electrical conductivity in materials can be explained by considering electrons as particles and treating them using the principles of classical mechanics. However, certain other aspects related to electrical conductivity can be explained only through a quantum mechanical–based explanation in which electrons are considered waves. We begin the discussion of these topics with Ohm's law and follow this with a discussion of the classical and quantum mechanical theories of conductivity.

2.2 OHM'S LAW

The geometry of a *resistor* with resistance R is shown in Figure 2.3. A resistor is a component included in an electrical circuit to offer a predetermined value of electrical resistance.

Ohm's law is an empirical relationship that states the following:

$$V = IR \tag{2.1}$$

where V is applied voltage in volts (V), I is electrical current in amperes (A), and R is electrical resistance in ohms (Ω). In this chapter, unless stated otherwise, we will assume a direct current (DC) voltage. This is in contrast to alternating current (AC) voltage, which cycles with time.

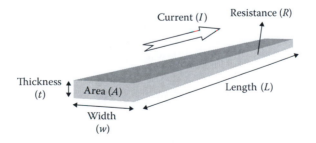

FIGURE 2.3 Geometry of a resistor used in describing Ohm's law.

The resistance (R) of an electrical resistor of length L and cross-sectional area A (Figure 2.3) is given by

$$R = \rho \frac{L}{A} \qquad (2.2)$$

The inverse of electrical resistance is known as *conductance*, which has units of Siemens or Ω^{-1}.

In Equation 2.3, ρ (rho) is defined as the *electrical resistivity*, *bulk resistivity*, or *volume resistivity*. A commonly used unit for resistivity is Ω cm. The inverse of ρ is known as *conductivity* (σ).

$$\sigma = \frac{1}{\rho} \qquad (2.3)$$

A commonly used unit for conductivity is a Siemens/centimeter (S/cm). When looking up the properties of materials or performing related calculations, it is important to check whether a value of conductivity or resistivity is being stated and what units are being used. Different sources quote values in different units. For example, resistivity values may be listed as $\Omega \cdot$ cm, $\Omega \cdot$ m, $\mu\Omega \cdot$ cm (micro-ohm-centimeter), $n\Omega \cdot$ m (nano-ohm–meter), and so on.

Most conductors obey Ohm's law: As the applied voltage is increased, the current increases linearly. The resistance does not depend on the voltage. There are, however, certain materials (e.g., zinc oxide–based formulations) and devices that, under certain conditions, do *not* obey Ohm's law. For these materials, resistance remains independent of the voltage up to a certain value. However, beyond a certain critical value of the *electric field* (E), the material undergoes an electrical breakdown, and the resistance decreases by orders of magnitudes. The electric field is defined as the voltage divided by the distance across which the voltage is applied. Materials that do not obey Ohm's law are sometimes described as variable resistors or *varistors*. Varistors are useful as current-limiting, surge-protection devices.

The magnitude of resistivity allows us to delineate the approximate boundaries among different classes of materials, as shown in Figure 2.2. Materials with a resistivity in the range of $\sim 10^{-6}$–10^{-4} $\Omega \cdot$ cm (1–10^2 $\mu\Omega \cdot$ cm) usually are considered conductors. Insulators are materials with resistivity greater than $\sim 10^{10}$ $\Omega \cdot$ cm. Resistivity values between $\sim 10^{10}$ and 10^{22} $\Omega \cdot$ cm are typical of electrically nonconducting or insulating materials. Materials that have resistivity ranging between $\sim 10^{-4}$ and 10^3 $\Omega \cdot$ cm (i.e., 10^2 10^9 $\mu\Omega \cdot$ cm) are considered semiconductors. Materials with resistivities from 10^3 to 10^{10} $\Omega \cdot$ cm are classified as *semi-insulators*. The values associated with these ranges of resistivities are not rigid.

Table 2.1 shows the typical resistivity values for some metals. The values listed in Table 2.1 are based on $T_0 = 300$ K (see Equation 2.25). The values of a parameter known as the *temperature coefficient of resistivity* (α_R; TCR) are also included. We will define α_R in Section 2.8. In Table 2.1, we see that silver is the element with the highest level of conductivity. Copper (Cu) and gold (Au) are also very good conductors of electricity. Note that the conductivity of tungsten (W) is higher than that of platinum (Pt).

The α_R values for magnetic metals are higher than those for other metals (see Section 2.8). Examples 2.1 through 2.4 illustrate some applications of the concepts discussed in this section.

Example 2.1: Resistivity and Conductivity—Units and Conversion

The resistivity of a Cu sample is 1.673 $\mu\Omega \cdot$ cm.

1. What is the value of this resistivity in $\Omega \cdot$ cm? What is the value in $n\Omega \cdot$ m?
2. The conductivity of an Ag sample is listed as 62.9×10^4 S/cm. What is the resistivity in $\mu\Omega \cdot$ cm?

TABLE 2.1

Conductivity/Resistivity Values for Selected Metals

Material	Conductivity (σ) S/cm	Resistivity (ρ) $\mu\Omega \cdot$ cm	Temperature Coefficient of Resistivity (αR) ($\Omega/\Omega \cdot$ °C)
Aluminum	37.7×10^4	2.65	4.3×10^{-3}
Beryllium	25×10^4	4.0	25×10^{-3}
Cadmium	14.6×10^4	6.83	4.2×10^{-3}
Chromium	7.75×10^4 (at 0°C)	12.9 (at 0°C)	3.0×10^{-3}
Cobalt (magnetic)	16.0×10^4	6.24	5.30×10^{-3}
Copper	59.7×10^4	1.673	4.3×10^{-3}
Gold	42.5×10^4	2.35	3.5×10^{-3}
Iridium	18.8×10^4	5.3	3.93×10^{-3}
Iron (magnetic)	10.3×10^4	9.7	6.51×10^{-3}
Lead	4.84×10^4	20.65	3.68×10^{-3}
Magnesium	22.4×10^4	4.45	3.7×10^{-3}
Mercury	1.0×10^4 (at 50°C)	98.4 (at 50°C)	0.97×10^{-3}
Molybdenum	19.2×10^4 (at 0°C)	5.2 (at 0°C)	5.3×10^{-3}
Nickel (magnetic)	14.6×10^4	6.84	6.92×10^{-3}
Palladium	9.253×10^4	10.8	3.78×10^{-3}
Platinum	10.15×10^4	9.85	3.93×10^{-3}
Rhodium	22.2×10^4	4.51	4.3×10^{-3}
Silver	62.9×10^4	1.59	4.1×10^{-3}
Tantalum	0.74×10^4	13.5	3.83×10^{-3}
Tin	0.90×10^4	11.0 (at 0°C)	3.64×10^{-3}
Titanium	2.38×10^4	42.0	3.5×10^{-3}
Tungsten	18.8×10^4	5.3	4.5×10^{-3}
Zinc	16.9×10^4	5.9	4.1×10^{-3}

Source: Webster, J.G., *Wiley Encyclopedia of Electrical and Electronics Engineering*, Vol. 4, Wiley, New York, 2002; Kasap, S.O., *Principles of Electronic Materials and Devices*, McGraw Hill, New York, 2002.

Note: Unless stated otherwise, values reported are at 300 K. For some materials, the values of the coefficient of resistivity (α_R) are shown using 300 K as the reference temperature, unless stated otherwise. To convert α_R into ppm/°C, multiply the values listed by 10^6.

Solution

1. The Cu sample resistivity in $\Omega \cdot$ cm = 1.673 $\mu\Omega \cdot$ cm × 10^{-6} $\Omega/\mu\Omega$ = 1.673 × 10^{-6} $\Omega \cdot$ cm. The Cu sample resistivity in $n\Omega \cdot$ m = 1.673 × 10^{-6} $\Omega \cdot$ cm × 10^9 $n\Omega/\Omega$ × 10^{-2} m/cm = 16.73 $n\Omega \cdot$ m.

2. The conductivity of the Ag sample = 62.9 × 10^4 S/cm. Therefore, the resistivity of the Ag sample = $(62.9 \times 10^4)^{-1}$ $\Omega \cdot$ cm = 1.589 × 10^{-6} $\Omega \cdot$ cm.
   We want the answer in $\mu\Omega \cdot$ cm; therefore, the resistivity of the Ag sample = 1.589 × 10^{-6} $\Omega \cdot$ cm × 10^6 $\mu\Omega/\Omega$ = 1.589 $\mu\Omega \cdot$ cm.

Example 2.2: Antennae for Radio Frequency–Identification Technology

Radio frequency–identification (RFID) technology is used for applications such as automatic highway toll-payment systems (e.g., EZ-Pass®), smart cards for access control, libraries, livestock tracking, and many other inventory-control applications (Figure 2.4).

A passive RFID tag comprises a conductive antenna, typically made from Cu, with a small computer chip attached to it. The computer chip holds the stored information. In one process for manufacturing antennae, we begin with a thin foil of solid Cu. A pattern is then created by etching

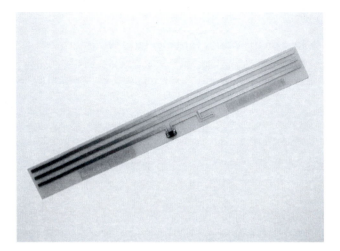

FIGURE 2.4 Typical passive radio frequency–identification (RFID) tag showing the antenna made from copper. (Courtesy of Pavel Nikitin, Intermec Technologies Corporation, Everett, WA.)

away excess Cu to form the antenna. Other methods of manufacturing antennae include using conductive inks (made from powders of Cu or Ag) or conductive polymers. These approaches offer the possibility of lowering the cost of RFID tags.

1. An antenna is made by etching solid Cu with $\sigma = 5.8 \times 10^7$ S/m. What is its electrical resistance (R)? (Assume that the antenna is in the form of a strip that is 20 cm long, 5 mm wide, and 25 μm thick.)
2. What will be the resistance of a similar antenna made by screen printing *silver paste* with $\sigma = 1.6 \times 10^6$ S/m?
3. Why do you think the conductivity of Ag ink is *lower* than that of solid Cu?

Solution

1. Let us assume that the current flows along the direction of the length. Thus, the cross-sectional area will be 5 mm × 25 μm or 0.5 cm × 25 μm × 10^{-4} cm/μm = 125 × 10^{-5} cm². Using Equation 2.2, the resistance of the Cu antenna will be:

$$R_{\text{copper}} = \rho\frac{L}{A} = \frac{20\,\text{cm}}{\left(125\times10^{-5}\,\text{cm}^2\right)\left(5.8\times10^7\,\text{S/m}\times10^{-2}\,\text{m/cm}\right)} = 0.02758\ \Omega$$

(Note the conversions of resistivity to conductivity and of meters and micrometers to centimeters.)

2. If we use Ag ink that has $\sigma = 1.6 \times 10^6$ S/m, then

$$R_{\text{silver ink}} = \rho\frac{L}{A} = \frac{20\,\text{cm}}{\left(125\times10^{-5}\,\text{cm}^2\right)\left(1.6\times10^6\,\text{S/m}\times10^{-2}\,\text{m/cm}\right)} = 1\,\Omega$$

This value of resistance is almost 36 times higher than that calculated for an antenna made by etching solid Cu.

3. An Ag conductive ink is a liquid paint-like material obtained by dispersing Ag particles in a carrier liquid. When screen-printed Ag ink is used, the electrical contact among Ag particles is not as good as that among grains of a polycrystalline solid, even after the ink "cures" subsequent to when the solvent medium is removed. This makes it harder for the current to flow from one particle to another. Thus, although *solid silver* has the highest conductivity among all metals (Table 2.1), in this case, *silver ink* has a higher resistivity than that of solid Cu.

Example 2.3: Resistance of a Long Cu Wire

Calculate the resistance of 1000 m of American wire gauge (AWG) #18 Cu wire (see Table 2.2). Assume $\sigma_{Cu} = 5.8 \times 10^7$ S/m.

Solution
From Table 2.2, for AWG #18, the wire diameter (d) is 0.040303 in (1.023696 mm).

Thus, the area of cross section $= \dfrac{\pi d^2}{4} = 8.2306 \times 10^{-7} \text{m}^2$

Therefore, the resistance of the 1000-meter-long wire will be as follows (according to Equation 2.2):

$$R = \frac{(1000 \text{ m})}{(8.2306 \times 10^{-7} \text{m}^2) \times (5.8 \times 10^7 \text{S/m})} = 20.9478 \ \Omega$$

Example 2.4: Resistance, Mass, and Length (in Feet) per Pound of Cu and Aluminum (Al) Wires

1. Calculate the mass and electrical resistance of 1000 feet of Al and Cu wires of AWG #3 (see Table 2.2).
2. What is the footage per pound of Cu and Al wire? Assume that the mass of any insulation can be ignored.
   The densities of Cu and Al are 8930 and 2700 kg/m³, respectively. Assume $\sigma_{Cu} = 5.8 \times 10^7$ and $\sigma_{Al} = 3.8 \times 10^7$ S/m.

TABLE 2.2
American Wire Gauge Diameter Conversion

American Wire Gauge Number	Conductor Diameter (inches)	Conductor Diameter (mm)
20	0.03196118	0.811814
18	0.040303	1.023696
16	0.0508214	1.290864
14	0.064084	1.627734
12	0.0808081	2.052526
10	0.10189	2.588006
8	0.128496	3.263798
6	0.16202	4.115308
5	0.18194	4.621276
4	0.20431	5.189474
3	0.22942	5.827268
2	0.25763	6.543802
1	0.2893	7.34822
0	0.32486	8.251444
00	0.3648	9.26592
000	0.4096	10.40384
0000	0.46	11.684

Source: Powerstream, http://www.powerstream.com/Wire_Size.htm

Note: The diameter stated is that of the conductor and does not include insulation dimensions.

Solution

1. For AWG gauge #3, the diameter is 0.22942 in (5.827268 mm). Thus, the area of cross section is

$$\text{Area} = \frac{\pi d^2}{4} = 2.66698 \times 10^{-5} \, \text{m}^2$$

According to Equation 2.2, the resistance of a 1000-foot Cu wire will be:

$$R_{cu} = \frac{(1000 \, \text{ft.}) \times 0.3048 \, \text{m/ft.}}{\left(2.66698 \times 10^{-5} \text{m}^2\right) \times \left(5.8 \times 10^7 \, \text{S/m}\right)} = 0.197 \, \Omega$$

Similarly, the resistance of a 1000-foot Al wire (AWG #3) will be

$$R_{Al} = \frac{(1000 \, \text{ft.}) \times 0.3048 \, \text{m/ft.}}{\left(2.66698 \times 10^{-5} \text{m}^2\right) \times \left(3.8 \times 10^7 \, \text{S/m}\right)} = 0.309 \, \Omega$$

2. The volume of 1000 feet of AWG #3 Cu wire is

Volume = $(2.66698 \times 10^{-5} \, \text{m}^2) \times 1000 \, \text{ft.} \times 0.3048 \, \text{m/ft.}$

= $8.127 \times 10^{-3} \, \text{m}^3$

The density of Cu is 8930 kg/m³; therefore, the mass of a 1000-foot Cu wire will be

= $8.127 \times 10^{-3} \, \text{m}^3 \times 8930 \, \text{kg/m}^3$

= 72.59 kg (about 160 lb.)

Thus, for a Cu AWG gauge #3 wire, the length per pound will be ~1000/160 or 6.25 feet.

Because the geometry is the same, the volume of the Al wire is the same as that of the Cu wire. The density of aluminum is 2700 kg/m³. Thus, the mass of 1000 feet of Al wire of AWG gauge #3 will be 8.129 × 10⁻³ m³ × 2700 kg/m³ = 21.948 kg or ~48.27 lb.

Therefore, for an Al wire of gauge AWG #3, the length (in feet) per pound will be ~1000/48.27 = 20.71.

2.3 SHEET RESISTANCE

In some microelectronic applications, it is customary to use a parameter known as sheet resistance (R_s) instead of resistivity (ρ). Sheet resistance is measured more easily than resistivity. As shown in Figure 2.3, let t be the thickness of the resistor. The length and width are L and w, respectively. The cross-sectional area $A = t \times w$. We can rewrite Equation 2.2 as follows:

$$R = \rho \frac{L}{t \times w} \tag{2.4}$$

In the fabrication of resistors that are used in *integrated circuits* (ICs), the designer usually works with a material that has certain values of the ratio L/w. Usually, these are chosen such that a large number of small components fit within a given area (i.e., miniaturization). Typically, the resistivity (ρ) and thickness (t) are maintained the same for resistors in a given layer of an IC. Consider a square resistor, that is, $L = w$. According to Equation 2.4, the resistance will be equal to the sheet resistance (R_s), defined as follows:

$$R_s = \frac{\rho}{t} \tag{2.5}$$

The unit for sheet resistance (R_s) is ohms (Ω) only. We write it as $\Omega/\square$ and read it as *ohms per square* to indicate that R_s represents the resistance of a square resistor. From Equations 2.4 and 2.5, we can derive an expression for R as follows:

$$R = R_s \frac{L}{w} \tag{2.6}$$

Example 2.5 illustrates the use of the concept of sheet resistance.

Example 2.5: Cu Interconnects for Integrated Circuits

Because of its high conductivity ($\sigma = 58 \times 10^4$ S/cm), Cu is widely used in modern-day ICs as a conductor material that provides connections among different components, known as interconnects. If the thickness of Cu is 80 nm, what is its sheet resistance (R_s)? If the width of this Cu conductor is 0.5 µm and the length is 300 µm, what is the resistance?

Solution

The sheet resistance can be calculated as follows:

$$R_s = \frac{\rho}{t} = \frac{1}{\sigma t}$$

$$R_s = \frac{1}{\left(58 \times 10^4 \text{S/cm}\right) \times \left(80 \times 10^{-7} \text{cm}\right)} = 0.216 \, \Omega/\square$$

From this, we can calculate the resistance as follows:

$$R = R = R_s \frac{L}{w} = 0.216 \times \frac{300 \, \mu\text{m}}{0.5 \, \mu\text{m}} = 129 \, \Omega$$

Note that, in this example, the thickness of the interconnect is very small (80 nm). In Section 2.10, we will see that, at such a low thickness, the resistivity of metals is actually higher than the "bulk" value.

Sheet resistance often is measured using what is known as the *four-point probe* or *Kelvin probe* technique (Figure 2.5). In this method, we apply an external voltage that leads to a predetermined value of current (I) between two terminals. The decrease in voltage (V) across the other two probes separated by a distance, s, is measured, and the resistivity is then calculated using the following formula:

$$\rho = 2\pi \times s \times \frac{V}{I} \tag{2.7}$$

In Equation 2.7, V is the measured voltage in volts, I is the source current in amperes, s is the spacing between probes (in cm), and ρ is the resistivity in $\Omega \cdot$ cm. Note that if $t < 5$ mm, a correction factor is needed.

The sheet resistance (R_s), measured for films and wafers using the four-point probe method, is given by Equation 2.8.

$$R_s = 4.532 \times \frac{V}{I} \times k \tag{2.8}$$

The factor 4.532 is the value of $\pi/\ln(2)$. A correction factor (k), whose value depends on the ratio of probe spacing (s) to sample diameter (d) and the ratio of wafer thickness (t) to probe spacing (s), may be needed.

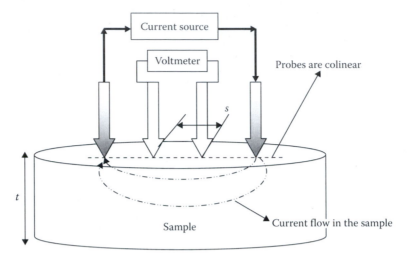

FIGURE 2.5   A schematic of the four-point probe used to measure resistivity. The spacing between probes is s, and the sample thickness is t (not to scale).

For thin films and wafers (e.g., semiconductors such as germanium [Ge] or silicon [Si]), the bulk resistivity is obtained by dividing the resistivity (ρ) by the thickness (t) and is given by Equation 2.9:

$$\rho = 4.532 \times t \times \frac{V}{I} \times k \tag{2.9}$$

When measuring the resistivity values of good conductors, the current values used are limited to ~10 mA. This avoids sample heating and limits the current density at the probe tips. For measuring high-resistivity materials, lower currents are used (~1 µA). This limits the voltage measured to ~200 mV.

Using the four-point probe technique, we can measure the spatial variations in the electrical conductivity of semiconductor crystals or other materials. For example, in the fabrication of ICs, large crystals of silicon (12-in. diameter) are used as starting materials. These crystals are then sliced into silicon wafers on which ICs are fabricated. Often, we need to measure the variation in conductivity from one wafer to another. However, in some cases, we need to measure the variation in conductivity (either deliberate or intrinsic to processing) across different parts of the same wafer. Local conductivity measurement can be achieved using the four-point probe method.

This method works well for both conductors and semiconductors. A different technique, known as the van der Pauw method, is useful for materials with high resistivity values.

2.4   CLASSICAL THEORY OF ELECTRICAL CONDUCTION

A German scientist named Paul Drude (1863–1906) proposed the classical theory of conduction in metals. In this theory, electrons are considered particles, and an atom is considered to be a nucleus surrounded by core electrons and the valence electrons (Figure 2.6). The valence electrons, which are the outermost electrons, are relatively weakly bound to the nucleus and become available for conduction. Thus, it is assumed that the electron–nucleus interaction and the electron–electron interaction are negligible for valence electrons. This leads to the formation of a sea of electrons (see Section 2.1), which is also described as *electron gas*. In the classical theory of electron conduction, the valence electrons in the sea of electrons that can freely move within the solid are a source of the electrical conductivity. The concentration of valence electrons in this sea of electrons is quite high and is comparable to the atomic density of the solid. In most solids, the atomic density ranges from $1 \times 10^{22}/\text{cm}^3$ to $1 \times 10^{23}/\text{cm}^3$. For example, for sodium (Na), there is one valence electron per atom, and each sodium atom donates this valence electron

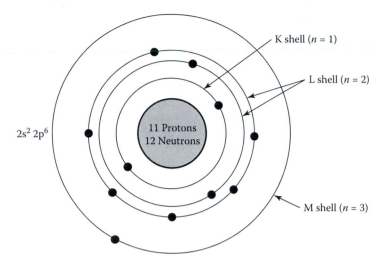

FIGURE 2.6 The structure of a sodium atom (atomic number 11), showing the nucleus surrounded by core electrons and valence electrons. (From Askeland, D., *The Science and Engineering of Materials*, 3rd ed., Thomson, Washington, DC, 1989. With permission.)

to the sea of electrons. Therefore, the free electron density of Na is the same as the atomic density of Na metal (= density ÷ [atomic mass] × [Avogadro's number]) is $2.54 \times 10^{28}/m^3$. Because of this high free electron concentration, pure metals would be expected to be good conductors of electricity.

Although classical theory provides a good start for understanding the electrical properties of solids, it does not explain several features associated with electrical conduction in materials. The limitations of classical theory in explaining electrical conduction in materials are discussed in Section 2.7. These limitations are overcome using a quantum mechanical approach, which is discussed in Section 2.13. We will now begin a more detailed discussion of the classical theory of conductivity.

2.5 DRIFT, MOBILITY, AND CONDUCTIVITY

We can rewrite Ohm's law (Equation 2.1) as follows:

$$I = \frac{V}{R} \tag{2.10}$$

Instead of the electrical current (*I*), consider the *current density* (*J*), which is defined as the current per unit area. The commonly used units for current density are A/m^2 and A/cm^2.

Another way of writing Ohm's law is by applying the relationship between the current density (*J*) and the applied electric field (*E*), as shown here:

$$J = \sigma E \tag{2.11}$$

This is known as the point form of Ohm's law. Note that *J* and *E* are vector quantities and conductivity (σ) is a tensor quantity. In this discussion, however, we treat all of these as scalar quantities. We use resistivity or conductivity instead of resistance or conductance because the resistance (*R*) depends on the geometry of the resistor, whereas the resistivity (or conductivity) is an intrinsic property of a particular material. If we have a material of a certain composition and microstructure, then the resistivity value is fixed. An exception to this occurs when we examine the properties of materials or devices at a *nanoscale* (~1–100 nm length; Section 2.10).

The distinction between resistance and resistivity is also important in situations where the resistivity of a material or a component changes across the thickness of the component. Thus, if we know the resistivity values and the geometry for different regions, we can calculate the total resistance of a component or a device.

Let us now derive the point form of Ohm's law (Equation 2.11).

The conduction electrons in metals move at very high speeds (~10^6 m/s). We can expect that, as temperature increases, the kinetic energy of electrons should also increase, causing the electrons to move at higher speeds. However, as we will see in Section 2.7, this speed is largely *independent* of the temperature. When no electric field is applied, the motions of the conduction electron are in random directions and do not result in an electrical current (Figure 2.7). This is similar to ions drifting around in an electroplating solution. A less-technical analogy is that of small schoolchildren running around without the supervision of their teacher!

Now consider a voltage V applied across the length (L) of a conductor. The electric field is given by

$$E = \frac{V}{L} \tag{2.12}$$

When we apply a DC electric field to a conductor, the electrons in the material flow from the negative to the positive terminal of the voltage supply (Figure 2.7b). This current is known as the *electron current*. As a matter of convention, we state that the current or the *conventional current* flows from the positive to the negative terminal of the battery (Figure 2.7c).

The force (F) exerted on an electron by the electric field (E) is given by

$$F = qE \tag{2.13}$$

The force results in acceleration a, given by

$$F = ma \tag{2.14}$$

Thus, the acceleration experienced by the electrons is given by

$$a = qE/m \tag{2.15}$$

Because of this externally applied driving force (i.e., the electric field), conduction electrons begin to show a net movement in the direction opposite to that of the electric field (E). This motion of charge carriers (electrons, in this case) due to an applied electric field is known as *drift*. The drift of the carriers is characterized by an average drift velocity (v_x). This movement of conduction electrons results in an electrical current density (J_x) in the positive direction "x." Because electrons are negatively charged, the electrical currents and their drift are in opposite directions (Figure 2.7c).

If an electron accelerates and then collides with an atom after time τ, it stops. From Newton's laws of motion, we know that $v = u + at$, where v is final velocity, u is initial velocity, a is acceleration, and t is time. Applying this law here, assuming the initial speed is zero, we get $v = a\tau$. Thus, the average velocity of an electron between collisions will be as follows:

$$v_{avg} = \left(\frac{1}{2}\right) a\tau$$

We should not, however, consider the average time between collisions to calculate the average velocity; we should calculate the actual velocities and then obtain the average. If this is done, we can show that the average velocity or drift velocity (v_{drift}) is given by

$$v_{drift} = v_{avg} = a\tau \tag{2.16}$$

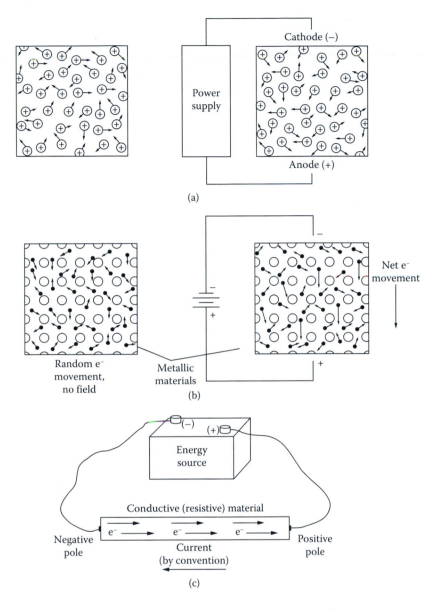

FIGURE 2.7 (a) Random motion of electrons due to thermal energy is similar to the movement of ions in an electroplating solution. (b) The drift of conduction electrons in a pure metal; note that the applied electric field (*E*) and drift are opposite, and overall, the electrons move within the material toward the anode. (c) The conventional current is said to flow from the positive to the negative terminal. The electron current (i.e., the actual motion of electrons) is in the opposite direction. (From Minges, M.L., *Electronic Materials Handbook*, Vol. 1, ASM, Materials Park, OH, 1989. With permission.)

Inserting the acceleration (Equation 2.15) into Equation 2.16, we get

$$v_{drift} = v_{avg} = \left(\frac{qE}{m} \right) \tau = \left(\frac{q\tau}{m} \right) E \tag{2.17}$$

According to Equation 2.17, the drift velocity of electrons is proportional to the applied electric field (*E*). The proportionality constant is known as the *mobility* (μ) of the charge carriers. We use the subscript "n" (to reflect the negative charge); therefore, μ_n is used to describe the mobility of electrons. The term mobility refers to the ability with which the charge carriers drift under the influence of an external electric field (E).

$$\mu_n = \frac{q\tau}{m} \tag{2.18}$$

The commonly used units for mobility are $cm^2 \cdot s^{-1} \cdot V^{-1}$ and $m^2 \cdot s^{-1} \cdot V^{-1}$.

The *mean free-path length* (λ) of conduction electrons is defined as the average distance that the electrons travel between two consecutive collision events, and it is obtained by multiplying the average speed (v_{avg}) by the time between collisions (τ).

$$\lambda = v_{avg} \times \tau \tag{2.19}$$

We can calculate the number of electrons flowing across the unit area per unit time by multiplying the drift velocity by the *carrier concentration* (n). In this case, the carriers are conduction electrons. To obtain the electrical current density, we must also multiply the carrier concentration by the charge q on each electron.

Thus, the current density (J) is given by

$$J = n \times v_{drift} \times q \tag{2.20}$$

Substituting for v_{drift} from Equation 2.17, we get

$$J = n \times \left(\frac{q\tau}{m} \right) E \times q$$

or

$$J = \left(\frac{nq^2\tau}{m} \right) E \tag{2.21}$$

This is a modified form of Ohm's law (Equation 2.11).

Therefore, in Equation 2.21, the bracketed term represents the conductivity (σ) and is given by

$$\sigma = \left(\frac{nq^2\tau}{m} \right) = \left(\frac{q\tau}{m} \right) \times n \times q = \mu_n \times n \times q$$

or

$$\sigma = \mu_n \times n \times q \tag{2.22}$$

This equation is very important because it tells us that the conductivity of a metal will be proportional to the carrier concentration (n) and the carrier mobility (μ_n) that is determined by the time between consecutive collision events. The longer the time between the scattering events or collisions (i.e., higher τ), the *higher* the mobility of the electrons will be in a metal. Note that higher mobility does not automatically guarantee a high conductivity because conductivity also depends on the carrier concentration of conduction electrons.

Thus, according to classical theory, the electrical conductivity of a metal depends on (a) the concentration of carriers transporting the electrical current (that is, how many carriers are available per unit volume of the material), (b) the mobility (μ) of the carriers, and (c) the electrical charge on each carrier. Consider a package-delivery service as an analogy. The number of packages delivered in a day depends on the following: (a) the number of delivery trucks available, (b) the speed with which the trucks can move in traffic, and (c) the number of packages each truck can carry. In some materials, the electrical current is carried by the motion of species other than electrons. We discuss these materials in Section 2.6.

2.6 ELECTRONIC AND IONIC CONDUCTORS

In semiconductors and certain types of ceramics, the motion of species other than electrons (e.g., ions) can generate an electrical current. In this case, we also need to account for their contributions to conductivity. Materials in which a significant part of the conductivity arises from the movement of ions are known as ionic conductors. An example of a ceramic that exhibits a good electrical conductivity from the ion movement is yttria (Y_2O_3)-stabilized zirconia (ZrO_2), known as YSZ. This material is used as a solid electrolyte in solid oxide fuel cells (Chapter 1). Oxygen gas sensors used in automotive and other applications are also based on YSZ. Similarly, transparent and conductive coatings based on indium–tin–oxide (ITO) are used as electrodes in touch-screen displays and solar cells. Electronic conductors are defined as materials in which electrical conduction occurs mostly due to the motions of electrons or holes. A hole is an imaginary particle that signifies the absence of an electron. Thus, metals, alloys, and semiconductors are examples of electronic conductors. If the conductivity is due to movement of ions and electrons or holes, then the material is known as a mixed conductor. We now know one more way to classify electrical conductors based on the types of charge carriers.

Note that the concept of sea of electrons is applied only to electronic conductors of metallic bonding. In other type of materials, a small portion of valence electrons participate in electronic conduction. The following examples illustrate the applications of these concepts.

Example 2.6: Carrier Concentrations in Al and Si

Calculate the concentration of conduction electrons in (a) Al and (b) Si. Assume that the densities of Al and Si are 2.7 and 2.33 g/cm³, respectively. Assuming that the electron mobility values are similar, what will be the expected concentration of electrons in terms of the values of conductivity of Al and Si? The atomic masses for Al and Si are ~27 and 28, respectively.

Solution

1. Because Al has a valence of +3, we assume that each Al atom donates three conduction electrons.

    The atomic mass of Al is 27 g; this means that the mass of 6.023×10^{23} atoms (Avogadro's number) is 27 g.

    A volume of 1 cm³ is 2.7 g of Al. The number of atoms in this volume will be

$$= \frac{6.023 \times 10^{23} \times \text{density}}{\text{atomic mass}} = \frac{6.023 \times 10^{23} \text{ atoms/mole} \times 2.7 \text{ g / cm}^3}{27 \text{ g/mole}}$$

$$= 6.023 \times 10^{22} \text{ atoms/cm}^3$$

The concentration of conduction electrons will be expected to be three times the concentration of atoms because each Al atom is assumed to donate three electrons.

Thus, the concentration of conduction electrons in Al will be

$$n_{Al} = 6.023 \times 10^{22} \frac{\text{atoms}}{\text{cm}^3} \times 3 \frac{\text{electrons}}{\text{atom}}$$

$$= 1.807 \times 10^{23} \text{ electrons/cm}^3$$

The conduction electron concentrations for Al and other metals are listed in Table 2.3.

2. Si has a valence of +4; therefore, the concentration of valence electrons (that is, the number of valence electrons in 1 cm³ of Si) will be

$$n_{Si} = \frac{6.023 \times 10^{23} \times 2.33}{28} \times 4 \frac{\text{electrons}}{\text{atom}}$$

$$n_{Si} = 2.0 \times 10^{23} \text{ electrons/cm}^3$$

TABLE 2.3

Atomic Mass, Density, Experimentally Measured Electron Mobility, and Electron Concentration for Selected High-Purity Metals

Metal	Atomic Mass (g/mol)	Density (g/cm³)	Mobility of Electrons (μ_n) (cm²/V · s)	Calculated Number of Carrier Particles (n) # (electrons/cm³)
Silver	107.868	10.5	57	5.86×10^{22}
Copper	63.546	8.92	32	8.43×10^{22} (assumes one electron per atom)
Gold	196.965	19.32	31	5.91×10^{22}
Aluminum	26.981	2.7	13	1.807×10^{23} (assumes three electrons per atom)

Source: Webster, J.G., *Wiley Encyclopedia of Electrical and Electronics Engineering*, Vol. 4, Wiley, New York, 2002; other sources.

Experimentally, we know that the electrical conductivity of Al is very high compared to that of Si. However, as we can see, the calculated concentration of valence electrons in Si is higher than that in Al. Furthermore, as we will see in Chapter 3, a very low concentration of some elements has a huge effect on the conductivity of Si. These observations show that classical theory does not explain the higher level of conductivity in some metals compared to that of Si.

Example 2.7: Calculating the Conductivity of Al from the Mobility Values

Assume that the mobility of electrons (μ_n) in Al is 13 cm²/V · s. The carrier concentration for Al is 1.807×10^{23} electrons (assuming three electrons are donated per Al atom). Calculate the expected conductivity of Al.

Solution

From Equation 2.22, the conductivity of Al will be given by the expression

$$\sigma = \left(1.807 \times 10^{23} \frac{\text{electrons}}{\text{cm}^3}\right) \times \left(13 \frac{\text{cm}^3}{\text{V} \cdot \text{s}}\right) \times \left(1.6 \times 10^{-19} \frac{\text{C}}{\text{electron}}\right)$$

$$= 37.9 \times 10^4 \text{ S/cm}$$

Because this value matches well with the data in Table 2.1, our assumption that each Al atom donates three conduction electrons appears reasonable.

Example 2.8: Carrier Concentration in Cu

The experimentally measured mobility of electrons in copper is 32 cm²/V · s. Calculate the carrier concentration (*n*) for Cu. Compare this with the values listed in Table 2.3. What does this show about the number of conduction electrons contributed per Cu atom?

Solution

We use the value of Cu conductivity (σ) provided in Table 2.1.

From Equation 2.22,

$$\sigma_{Cu} = 59.7 \times 10^4 \text{ S/cm} = (n) \times \left(32 \frac{\text{cm}^2}{\text{V} \cdot \text{s}}\right) \times \left(1.6 \times 10^{-19} \text{C}\right)$$

This gives us a concentration of conduction electrons $n = 1.16 \times 10^{23}$ electrons/cm^3.

In Table 2.3, the calculated free electron concentration for Cu is $= 8.43 \times 10^{22}$ electrons/cm^3. In arriving at this value in Table 2.3, we had assumed that each Cu atom donates one conduction electron. Because we have now estimated that the actual value of conduction electrons is 1.16×10^{23} electrons/cm^3, we know that each Cu atom must be donating more than one electron. The average number of electrons donated per Cu atom will be

$$\frac{1.16 \times 10^{23} \text{ electrons/cm}^3}{8.43 \times 10^{22} \text{ electrons/cm}^3} = 1.38$$

Thus, in Cu, the average number of electrons donated per atom is 1.38. This is expected because Cu does exhibit valences of +1 and +2.

Example 2.9: Resistivity of a Semiconductor

A particular single crystal of a semiconducting Si sample, containing a small but deliberately added level of phosphorus (P), provides conductivity with electrons as the majority carriers. Assume that the concentration of P added is 10^{18} atoms/cm^3 and that each P atom provides one electron. Ignore any other possible contributions to the conductivity of this P-doped Si sample. If $\mu_n = 700$ cm^2/V · s, what is the resistivity (ρ) of this semiconductor?

Solution
Because each P atom contributes only one conduction electron, the concentration of P atoms and that of the conduction electrons donated by P atoms are equal; therefore, $n = 10^{18}$ electrons/cm^3.

(In the next chapter, we will learn that our assumption about the concentration of conduction electrons contributed by the P atoms, as opposed to those from the Si atoms, is correct.) From Equation 2.22,

$$\sigma = \left(10^{18} \text{ electrons/cm}^3\right) \times \left(700 \frac{\text{cm}^2}{\text{V} \cdot \text{s}}\right) \times \left(1.6 \times 10^{-19} \text{C}\right)$$

Therefore, $\sigma = 112$ S/cm. Compared to metals (Table 2.1), this value of conductivity of Si containing P is still relatively small. The conductivity of essentially pure Si containing no added elements is even smaller.

We have been asked to calculate the resistivity (ρ), which is equal to $1/\sigma$. Therefore, $\rho = 8.9 \times 10^{-3} \Omega \cdot$ cm. As we will see in Chapter 3, the value of the resistivity of Si is very strongly dependent on the presence of trace levels of impurities or deliberately added elements.

2.7   RESISTIVITY OF METALLIC MATERIALS

Which factors have the greatest influence on the resistivity of metallic materials? In this section, we focus on metallic materials containing small concentrations of other elements.

We must consider the following three processes (Figure 2.8): First, atoms in a metallic material vibrate. We refer to the vibrations of atoms in a material as phonons. The conduction electrons carrying the electrical current bump into or scatter off vibrations of these atoms. The process of electron scattering of a phonon is temperature dependent. As the temperature increases, the number of electrons scattering off due to the vibrations of atoms also increases. Thus, the resistivity of a pure metal increases with increasing temperature. This increase in resistivity due to the scattering of electrons by lattice vibrations is linear. We refer to this temperature-dependent component as resistivity (ρ_T; Figure 2.9).

The second factor that affects the resistivity of metals is the scattering of electrons off defects such as vacancies, dislocations, and grain boundaries (Chapter 1). If you are unfamiliar with these terms related to the microstructure of materials, it may be worthwhile to review these concepts (see Askeland and Fulay 2006). The concentrations of different defects will depend on the

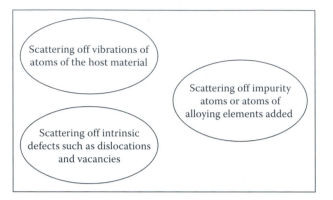

FIGURE 2.8   Schematic showing the sources of the scattering of conduction electrons in a metallic material.

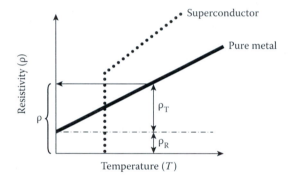

FIGURE 2.9   The temperature dependence of the conductivity of a typical metal and a superconductor.

microstructure of the material. However, we can assume that, at a given temperature or within a small temperature range, the microstructure, and therefore the concentrations of these defects (which serve as scattering centers), essentially is constant.

The third factor that influences the resistivity of metallic materials is the presence of impurities or other elements deliberately or accidentally added. Impurities cause an increase in electrical resistivity because the atoms of the added elements act as scattering centers, regardless of whether they are smaller or larger than the host atoms. Because the impurity concentrations should be independent of temperature, the effect due to the presence of impurities is generally *not* very sensitive to temperature changes.

We designate the effect of (a) microstructural defects and (b) impurities or added elements as the residual resistivity (ρ_R; Figure 2.9). These are temperature-independent effects. Resistance due to these effects will remain even if we somehow eliminate the scattering of electrons due to the vibrations of atoms. The electrical resistance of most metals does *not* become zero after cooling to extremely low temperatures (Figure 2.9).

Whenever a heat treatment or any other process causes a change in the microstructure of a material, we can expect a change in its resistivity. For example, *annealing* metals and alloys probably will increase their electrical conductivity because dislocations, a type of atomic-level defect in the arrangement of atoms, are annihilated. Similarly, if we bend an annealed copper wire or deform a metal or an alloy in some manner, the resistivity will increase. This is because of an increase in the dislocation density (see Section 2.11).

The conductivity of pure metals *decreases* with rising temperature because of the increased scattering of electrons off the phonons. Adding alloying elements or the presence of impurities causes a disruption in the periodic order of atoms and introduces a strain (see Section 2.11). This causes a sudden change in the potential energy of conduction electrons as they approach the atoms of foreign elements (the alloying elements and/or impurities). The scattering of electrons from these foreign atoms then increases, thereby causing increased resistance. Therefore, we expect the conductivity of alloys generally to be lower than that of essentially pure metals. The conductivity of alloys is controlled by *extrinsic* factors, that is, the concentration and nature of the added alloying elements (see Section 2.12). Thus, in alloys, mobility is limited by the scattering of electrons off impurity atoms. We refer to the mobility that is limited by this effect as *impurity-scattering limited drift mobility*. Conductivity (σ) in essentially *pure* metals largely is limited by the scattering of electrons from the vibrations of atoms. The value of mobility that is limited by these effects is known as *lattice-scattering limited mobility*. The conductivity of pure metals limited by lattice scattering is known as *lattice-scattering limited conductivity*.

Furthermore, because the scattering of conduction electrons is controlled by a relatively large concentration of atoms of other alloying elements, the conductivity of alloys will *not* vary significantly with temperature.

For metals, we can write the resistivity (ρ) as a sum of two components ρ_T and ρ_R:

$$\rho = \rho_T + \rho_R \tag{2.23}$$

This equation is also known as *Matthiessen's rule*, where ρ_T is the temperature-dependent part of resistivity, representing the scattering of electrons due to the vibrations of the atoms. The part ρ_R represents scattering due to the presence of impurities, added alloying elements, and microstructural defects in the arrangement of atoms within a material (Figure 2.9).

If we assume that ρ_T changes linearly with temperature, we can rewrite the dependence of resistivity (ρ) with temperature as follows:

$$\rho = AT + B \tag{2.24}$$

Instead of listing the values of A and B for different metals (Equation 2.24), we will define the TCR, designated as α_R (Section 2.1).

$$\alpha_R = \frac{1}{\rho_0}\left(\frac{\rho - \rho_0}{T - T_0}\right) \tag{2.25}$$

In Equation 2.25, ρ_0 is the resistivity at some reference temperature ($T_0 = 273$ or 300 K), and ρ is the value of resistivity at some other temperature T. Note that the value of α_R will depend on what we use as the reference temperature.

From Equation 2.25, we can show that

$$\rho = \rho_0 [1 + \alpha_R (T - T_0)] \tag{2.26}$$

Note that this can be rewritten as follows:

$$R = R_0 [1 + \alpha_R (T - T_0)] \tag{2.27}$$

The values of α_R for different metals were listed in Table 2.1. The values of α_R for some alloys are shown in Tables 2.4 and 2.5 (see Section 2.12).

The relative change in the resistivity of some metals over a wider range of temperatures is shown in Figure 2.10. Note that nickel (Ni) and iron (Fe) are magnetic, and the change in the resistivity of these materials is not linear. In addition, aluminum melts at ~657°C; thus, there may be some changes in its microstructure as it approaches higher temperatures. If the duration of a material's exposure to higher temperatures causes microstructural changes, the resistivity of the material can change. For example, in the case of metals such as aluminum and copper, annealing the material at higher temperatures will cause the annihilation of dislocations and thus cause an increase in the conductivity.

The following example explains how a change in the resistivity of platinum can be used to make precise and accurate temperature measurement devices.

TABLE 2.4

Properties of Some Heating-Element Materials

Material	Resistivity at 20°C ($\mu\Omega \cdot$ cm)	αR at 20°C ($\Omega/\Omega \cdot$ °C)	Maximum Operating Temperature (°C)	Main Applications
Nickel 80/chromium 20 (Nikorthal™-type alloys)	108	$+14 \times 10^{-3}$	1200	Furnaces, heating elements for domestic appliances
Chromium 22/aluminum 5.8/ balance iron (Kanthal™-type alloys)	145	$+3.2 \times 10^{-3}$	1400	Furnaces for heat treatment
Platinum 90/rhodium 10	18.7	–	1550	Laboratory furnaces
Platinum 60/rhodium 40	17.4	–	1800	Laboratory furnaces
Molybdenum	5.7	4.35×10^{-3}	1750	Vacuum furnaces, inert atmosphere
Tantalum	13.5	3.5×10^{-3}	2500	Vacuum furnaces
Graphite	1000	-26.6×10^{-3}	3000	Furnaces requiring nonoxidizing atmosphere
Molybdenum disilicide ($MoSi_2$)	40	1200×10^{-3}	1900	Glass industry, laboratory furnaces, ceramic processing
Lanthanum chromite ($LaCrO_3$)	2100	–	1800	Laboratory furnaces
Silicon carbide (SiC), also known as glowbar	1.1×10^5	–	1650	Industrial furnaces
Zirconia	–	–	2200	Laboratory furnaces, becomes an ionic conductor above ~1000°C

Source: Reprinted from Laughton, M.A. and Warne, D.F., *Electrical Engineer's Reference Book*, Copyright (2003), with permission from Elsevier.

TABLE 2.5

Resistivity and Temperature Coefficient Values for Selected Materials

Material	Typical Composition (wt%)	Resistivity ($\mu\Omega \cdot$ cm) at 20°C	Temperature Coefficient of Resistivity (αR) $\times 10^{-6}$ $\Omega/\Omega \cdot$ °C ($T_0 = 300$ K)
Lead (Pb)–tin (Sn) solder	Sn: 63, Pb: 37	14.7	
Brass	Cu: 60, Zn: 40	6.4	1000
	Cu: 70, Zn: 30	8.4	2000
Nichrome	Ni: 58.5, Fe: 22.5, Cr: 16, Mn: 3	100	400
Constantan	Cu: 60, Ni: 40	44.1	+2/+33
Manganin	Cu: 84, Mn: 12, Ni: 4	45	+6/–42 (12°C–100°C)
	Cu: 83, Mn: 13, Ni: 4 (wire alloy)	48.2	+15/–15 (15°C–35°C)
	Cu: 86, Mn: 10, Ni: 4 (shunt alloy)	38.3	+15/–15 (40°C–60°C)
Palladium–silver alloy	Pd: 60, Ag: 40	42–44	

Source: Webster, J. G. 2002. *Wiley Encyclopedia of Electrical and Electronics Engineering*, Vol. 4. New York: Wiley; Kasap, S. O. 2002. *Principles of Electronic Materials and Devices.* New York: McGraw Hill; Laughton, M. A. and D. F. Warne, eds. 2003. *Electrical Engineer's Reference Book.* Amsterdam: Elsevier; Askeland, D. and P. Fulay. 2006. *The Science and Engineering of Materials.* Washington, DC: Thomson; ASM International. (1990). *Properties and Selection: Nonferrous Alloys and Special Purpose Materials.* vol. 2. Materials Park, OH: ASM; Davis, J. R. 1997. *Concise Metals Engineering Data Book.* Materials Park, OH: ASM.

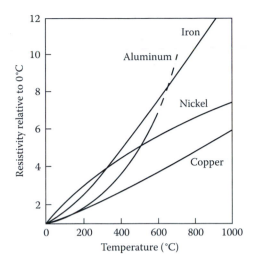

FIGURE 2.10 Relative change in resistivity with temperature (relative to 0°C) for Al, Cu, Ni, and Fe. (Reprinted from *Electrical Engineer's Reference Book*, Laughton, M. A., and D. F. Warne, Copyright 2003, with permission from Elsevier.)

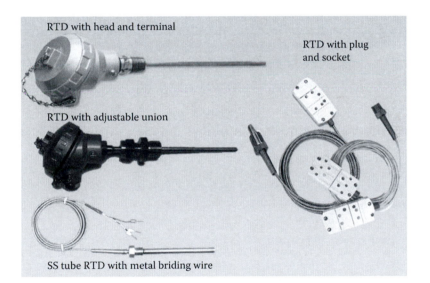

FIGURE 2.11 Commercially available resistance–temperature detector devices. (Courtesy of Omega Corporation, San Diego, CA.)

Example 2.10: Pt Resistance–Temperature Detectors

A *resistance–temperature detector* (RTD) makes use of a metal wire (e.g., that of Pt; Figure 2.11) and can operate at temperatures in the range of –200°C to +700°C. As the temperature changes, the electrical resistance changes; thus, by measuring the change in resistance, we can measure the temperature. Assume that the α_R for the Pt material used is 3.93×10^{-3} $\Omega/\Omega \cdot {}^\circ C$. The RTD wire is designed so that the resistance at 0°C is 100 Ω. (a) What is the temperature if the resistance measured is 200 Ω? (b) Why is Pt preferred for making RTDs, in comparison to W, Cu, Ag, or Au? (c) In Europe, Pt compositions with $\alpha_R = 0.00385$ $\Omega/\Omega \cdot {}^\circ C$ are often used. For $R = 200$ Ω, what is the predicted temperature if we use an RTD made from this Pt composition?

Solution

1.

$$\alpha_R = \frac{1}{\rho_0}\left(\frac{\rho - \rho_0}{T - T_0}\right)$$

$$\therefore 0.00393 = \frac{1}{R}\left(\frac{R - R_0}{T - T_0}\right) = \frac{1}{100}\left(\frac{R - 100}{T - 0}\right)$$

$$\therefore 0.393T = R - 100 \text{ or}$$

$$R = 100 + 0.393T$$

Thus, the variation in resistance is 0.393 Ω/°C.

Because the resistance measured by the RTD is 200 Ω, we rewrite this as

$$200 = 100 + 0.393T \text{ or}$$

$$T = \frac{200 - 100}{0.393} = 254.45°C$$

2. We could use Cu instead of Pt, but Cu has a lower resistivity. This means that we will have to use a longer wire. Cu will also oxidize at high temperatures. Although it can be used at high temperatures, it is brittle. Au and Ag exhibit much higher levels of conductivity, and we would need longer wires.

3. If we used a Pt composition with $\alpha_R = 0.00385$ Ω/Ω · °C, then for a wire whose resistance is 100 Ω, the change in resistance will be

$$R = 100 + 0.385T$$

$$200 = 100 + 0.385T \text{ or}$$

$$T = \frac{200 - 100}{0.385} = 259.7°C$$

Note that a seemingly small change in the TCR, from 0.00393 to 0.00385 Ω/Ω · °C, causes a considerable change in the value of the predicted temperature. In practice, the exact coefficient of the resistivity of the RTD element must be known. Also, the data for a range of temperatures must be fitted to a polynomial and not to the equation of a straight line.

2.7.1 Effect of Thermal Expansion

In Example 2.10, when calculating the resistance as a function of temperature, we ignored the change in resistance of the platinum wire due to changes in its length caused by thermal expansion. If a copper wire is heated from 25°C to 75°C, the length of the copper wire will increase. The thermal expansion coefficient for copper is (α_{Cu}) ~17 × 10^{-6}/°C. The increase in length will cause an increase in the resistance.

Note that, in some applications, this change in length—caused by a change in temperature or the presence of stress—may be important and may have to be accounted for, for example, if a type of strain gauge makes use of the change in electrical resistance to calculate the magnitude of strain.

2.8 JOULE HEATING OR I^2R LOSSES

When a material offers resistance to the flow of an electric current, the energy lost during the collisions of electrons with vibrations of atoms or other defects appears as heat. This is known as *Joule heating* or *I^2R losses*. The power dissipated by a material with resistance (R) carrying a current (I) is given by the following equation:

$$\text{Power lost} = V \times I = I^2 \times R \tag{2.28}$$

Obviously, Joule heating is not useful when the goal is to carry current with the least possible resistance. Thus, in ICs used to make computer chips, we try to minimize the I^2R losses. Heat sinks are designed and built into the computer chip packaging to conduct the heat away from the chip. The main concern regarding the use of ICs is not that the energy is wasted but that the electrical performance of components is usually reduced with increased temperatures.

In many applications, however, Joule heating can be useful. One of the most important applications of alloys such as Nichrome and Kanthal, and ceramic resistor materials is their use as heating elements (Table 2.4). This includes heating elements used in small consumer appliances, such as toaster ovens, hair dryers, and immersion water heaters. Incandescent light bulbs use tungsten filaments that produce light after being heated to high temperatures. Many materials, such as molybdenum disilicide ($MoSi_2$), are used to make heating elements for laboratory furnaces. Many of the metallic materials used as heating elements are oxidized by heating them in air. Though a dense and adherent oxide layer such as Cr_2O_3, Al_2O_3, and SiO_2 protects base metals from further oxidation, the oxide layer also increases electric resistance. Another application where Joule heating is useful is in the manufacturing of electrical fuses. In an electrical circuit, when a certain value of current is exceeded, the material in the fuse melts as a result of Joule heating. This creates an open circuit and prevents current flow.

Example 2.11: Calculation of Joule Power Losses in a Cu Bus Bar and the Cost of Electricity

A Cu *bus bar*, or a conductor used in power transmission, is 300 ft long and has a cross section of ¼ × 4 in. (a) What is the resistance of this bar at room temperature? (b) Calculate the Joule losses in kilowatts if the current is 1000 A. Assume that the current is DC. (c) What is the energy lost (in kW · h) if the current is carried for 24 hours? (d) Assuming electricity costs 12 cents/kW · h, calculate the dollar value of the energy wasted due to Joule losses on a per-year basis. Assume that the conductivity of the Cu bar is 5.8×10^7 S/m.

Solution

1.

$$R_{\text{Cu bus bar}} = \rho \frac{L}{A} = \frac{300\,\text{ft.} \times 12\,\text{in./ft.} \times 2.54\,\text{cm/in.}}{\left(\frac{1}{4} \times 4\right)\text{in.}^2 \times \left(2.54\,\text{cm/in.}\right)^2 \times \left(5.8 \times 10^7\ \text{S/m} \times 10^{-2}\,\text{m/cm}\right)}$$

$$= 2.44366 \times 10^{-3}\ \Omega$$

2. Now, we can calculate the Joule losses as follows:

$$\text{Power lost} = V \times I = I^2 \times R = (1000)^2 \times 0.00244366\ \Omega = 2443.66\ \text{W}$$

Thus, the power lost is 2.433 kW.
3. The energy lost in 24 hours will be

$$\text{Energy lost} = \text{power} \times \text{time} = 2.443\ \text{kW} \times 24\ \text{h} = 58.6\ \text{kW} \cdot \text{h}$$

As a reference, the total electric power production in the United States in 2015 was ~4×10^{12} kW · h or 4 trillion kW · h.
4. If the cost of electricity is 12 cents/kW · h, the total cost of electricity wasted due to Joule losses will be $7.03 each day. In a year, the cost of electricity wasted due to Joule losses will be 365 days × $7.03/day = $2566.

2.9 DEPENDENCE OF RESISTIVITY ON THICKNESS

So far, we have assumed that the resistivity (ρ) of a material does not change with its geometric dimensions. However, there is one exception—if the thickness of the film is comparable to the mean free-path length (λ) for the conduction electrons (see Section 2.5), then the resistivity of the material does depend on its thickness (Gupta 2003). For bulk copper (e.g., a large piece of copper or copper wire), the mean free-path length (λ) for conduction electrons is ~400 Å (40 nm). As the thickness (t) of a resistor approaches the mean free-path length (λ), the scattering of electrons from the film surface increases. In addition to surface scattering, the resistivity can also increase due to the presence of surface contaminants or due to higher resistivity phases formed as a result of surface oxidation or other chemical reactions. These effects tend to be more pronounced as the thickness of the film is in the nanoscale region. Therefore, a considerable scatter in the data concerning the resistivity of very thin films is often observed. For bulk, nearly pure, and annealed copper, the resistivity is ~1.7 $\mu\Omega \cdot$ cm (Table 2.1). For copper films with a thickness of t ~ 50 nm, the room-temperature resistivity value is ~2.5 $\mu\Omega \cdot$ cm. For copper films of 20 nm thickness, the copper resistivity is ~5 $\mu\Omega \cdot$ cm. If we include the dependence of resistivity on thickness, then Matthiessen's rule (Equation 2.23) can be modified and rewritten as follows:

$$\rho = \rho_T + \rho_R + \rho_{thickness} \tag{2.29}$$

Copper is a very important conducting material. As a result, there is a special measuring unit based on the conductivity of copper, known as the *International Annealed Copper Standard* (IACS). We define 100% IACS as a conductivity of 57.4013 × 10^6 S/m, or 57.4013 × 10^4 S/cm (or ρ = 1.74212 × 10^{-6} $\Omega \cdot$ cm). This unit is based on an annealed 1-m-long copper wire, with a cross-sectional area of 1 mm^2 and a resistance of 0.174212 Ω (Laughton and Warne 2003).

We can calculate the percentage of IACS for any material using Equation 2.30.

$$\% \text{ IACS} = \frac{1.74212}{\text{Resistivity}(\Omega \cdot m)} \times 100 \tag{2.30}$$

The conductivity of the so-called *electrolytic tough pitch* (ETP, or just TP) copper, which is 99.0% pure (~100–600 ppm oxygen), is ~100.2% IACS. The conductivity of 99.999% copper is 102.5% IACS. The IACS standard was developed in 1913 and was based on the highest-conductivity copper available at that time. Many compositions that have been developed since then have conductivity higher than 57.4013 × 10^4 S/cm. As a result, we have several high-purity copper samples on the IACS scale with conductivity greater than 100% (Table 2.6).

TABLE 2.6

Conductivity of Different Materials (at 20°C) Based on the IACS Scale

Material	% IACS Conductivity	Material	% IACS Conductivity
Annealed copper	100	Nickel	25
99.999% Copper	102.5	Iron	17
Electrolytic tough-pitch copper ~99.0%	100.2–101.5	Platinum	16
Oxygen-free high-conductivity copper	101	Tin	13
Silver	104	Lead	8
Aluminum	60		

Source: Webster, J. G. 2002. *Wiley Encyclopedia of Electrical and Electronics Engineering,* Vol. 4. New York: Wiley; Kasap, S. O. 2002. *Principles of Electronic Materials and Devices.* New York: McGraw Hill; Laughton, M. A., and D. F. Warne, eds. 2003. *Electrical Engineer's Reference Book.* Amsterdam: Elsevier; Askeland, D., and P. Fulay. 2006. *The Science and Engineering of Materials.* Washington, DC: Thomson; ASM International. (1990). *Properties and Selection: Nonferrous Alloys and Special Purpose Materials.* vol. 2. Materials Park, OH: ASM; Davis, J. R. 1997. *Concise Metals Engineering Data Book.* Materials Park, OH: ASM.

Note: IACS = International Annealed Copper Standard.

2.10 CHEMICAL COMPOSITION–MICROSTRUCTURE–CONDUCTIVITY RELATIONSHIPS IN METALS

2.10.1 INFLUENCE OF ATOMIC-LEVEL DEFECTS

We will now compare the conductivities of pure, annealed copper to that of pure, cold-worked copper. The term *cold-working* refers to a deformation process carried out at a relatively low temperature (such as bending, wire-drawing, or extrusion carried out at temperatures below the so-called recrystallization temperature). Cold-working annealed metals and alloys leads to an increase in the density of dislocations. This will cause the resistivity of the annealed metals to increase. For example, the decrease in the conductivity of sterling silver (92.5% Ag–7.5% Cu) as a function of percentage of cold work after the drawing process is shown in Figure 2.12a. The increased dislocation density caused by the cold-working process causes a simultaneous increase in the tensile strength (Figure 2.12b). These data are for a 2.3-millimeter wire that was cold-drawn after annealing. The percentage of cold work is defined as follows:

$$\text{Percentage of cold work} = \frac{A_0 - A_f}{A_0} \times 100 \tag{2.31}$$

where A_0 is the original cross-sectional area and A_f is the final cross-sectional area after cold working.

In metals and alloys, the annealing heat treatment leads to the annihilation of dislocations. Because the cold-worked pure copper has a higher dislocation density, we expect its resistivity to be higher than that of annealed pure copper.

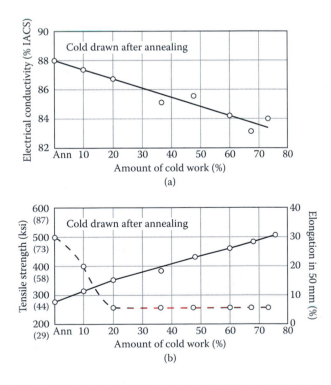

FIGURE 2.12 (a) Electrical conductivity of sterling silver (92.5% Ag–7.5% Cu), shown as %IACS, and (b) increase in tensile strength (solid line) and decrease in percentage of elongation (dotted line) as a function of % cold work. The data are for a 2.3-mm wire, which was cold-drawn after annealing. (From ASM International (1990). *Properties and Selection: Nonferrous Alloys and Special Purpose Materials*, Vol. 2, ASM, Materials Park, OH. With permission.)

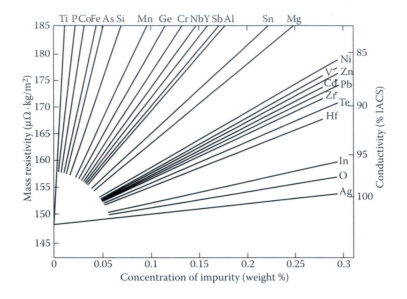

FIGURE 2.13 Effects of different impurities on the conductivity of copper. (Courtesy of Copper Development Association, New York, NY.)

2.10.2 INFLUENCE OF IMPURITIES

The conductivity of pure metals is also very sensitive to the presence of impurities. This effect does *not* come out as a prediction of the classical theory of conductivity. The change in the conductivity of copper as a function of different alloying elements is shown in Figure 2.13.

The impurities that have the most dramatic effect on the conductivity of copper are phosphorous, titanium (Ti), cobalt (Co), iron, arsenic (As), and antimony (Sb). Elements such as silver, cadmium (Cd), and zinc (Zn) produce relatively less of an increase in resistivity. You may also know that the addition of alloying elements in very small concentrations usually causes an increase in the strength of a metal. This is known as solid-solution strengthening (Chapter 1).

When alloys form, up to a certain concentration of one metal often completely dissolves in another—similar to sugar or salt dissolving in water. The resultant alloy is said to be a solid solution. For example, nickel can be added to copper to form a solid solution.

Thus, silver, cadmium, and zinc are good choices for strengthening annealed copper without causing an adverse effect on its electrical conductivity. Also note that, as shown in Figure 2.14, the effect of oxygen on the resistivity of copper is smaller than expected. This is because oxygen is usually present as a fine precipitate of copper and other oxides. Oxygen does not appear as an impurity dissolved in a solid solution.

The most widely used copper–alloy conductor is known as electrolytic tough pitch (ETP) copper, which contains about 100–650 ppm (or 1 mg/kg) oxygen. Note that "ppm" is not a scientific unit. The oxygen present in copper scavenges the dissolved hydrogen and sulfur during copper refining. Moreover, oxygen reacts with the other metal impurities that are originally dissolved in copper and causes them to precipitate as oxides. By adding an optimum concentration of oxygen, we cause an *increase* in the electrical conductivity of copper (Figure 2.14). Thus, instead of having a small level of an impurity distributed throughout the material (as in a solid solution), we can concentrate that impurity in the precipitate particles of a second phase; in this way, we can increase the conductivity without changing the nominal chemical composition. In some cases, where intricate castings need to be made or joining processes such as welding or brazing need to be used, we cannot make use of ETP copper. For these applications, *oxygen-free high-conductivity copper* (OFHC) is utilized. This copper has less than 0.001% oxygen and is more expensive. The conductivity of OFHC and

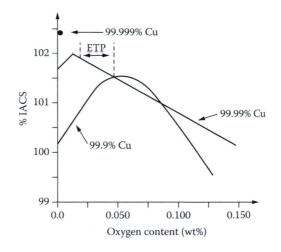

FIGURE 2.14 Effect of oxygen concentration on the conductivity of copper. (Courtesy of Copper Development Association, New York, NY.)

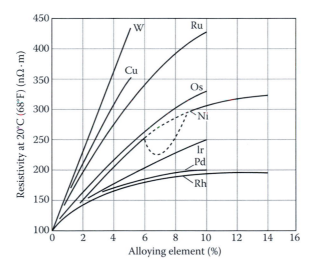

FIGURE 2.15 Effects of additions of various alloying elements on the resistivity of platinum. (From Vines, R.F. and Wise, E.M., *The Platinum Metals and Their Alloys*, International Nickel Co., New York, 1941. With permission.)

ETP copper is ~101% IACS. ETP and OFHC are the most widely used varieties of copper. Their applications include windings for motors, generators, bus bars, and so on.

The effect of the addition of different alloying elements on the resistivity of platinum is shown in Figure 2.15. Adding alloying elements also strengthens platinum; the consequent increase in strength, as represented by increased hardness, is shown in Figure 2.16.

2.11 RESISTIVITY OF METALLIC ALLOYS

Alloys are metallic materials that are characterized by the presence of more than one element. The dominant element or matrix element is what is used when referring to the alloy. For example, the term *copper alloy* means that copper is the dominant element. Other elements present in considerable concentrations are added deliberately to improve the properties of the alloy. For example, in copper–beryllium (Cu–Be) alloys, beryllium is added to increase the Young's modulus of the alloy. The electrical properties of some alloys are summarized in Table 2.5.

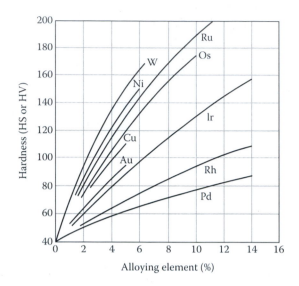

FIGURE 2.16 Increase in the hardness of platinum due to the addition of various alloying elements. (From Lampman, S.R. and Zorc, T.B., eds., *Metals Handbook: Properties and Selection: Nonferrous Alloys and Special Purpose Materials*, ASM, Materials Park, OH, 1990. With permission.)

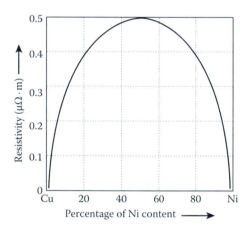

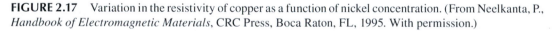

FIGURE 2.17 Variation in the resistivity of copper as a function of nickel concentration. (From Neelkanta, P., *Handbook of Electromagnetic Materials*, CRC Press, Boca Raton, FL, 1995. With permission.)

When alloys form a solid solution, we can relate the resistivity of the alloy to the resistivity of the pure metal using *Nordheim's rule*:

$$\rho_{alloy} = \rho_{matrix} + Cx(1 - x) \tag{2.32}$$

where ρ_{alloy} is the resistivity of the alloy, ρ_{matrix} is the resistivity of the base metal without any alloying elements, C is Nordheim's coefficient, and x is the *atom fraction* of the alloying element added. The first term, that is, ρ_{matrix}, accounts for the resistivity of the base or matrix material. If an alloy system forms a solid solution over a complete range of compositions (i.e., from 0% to 100% of solute), then we expect a parabolic variation in the resistivity, as predicted by Equation 2.32. This is seen in the copper–nickel system (Figure 2.17).

Alloys of platinum and palladium (Pd) are often used as electrodes in devices such as multilayer capacitors and electrical contacts. This is another example of a system in which solid solutions are

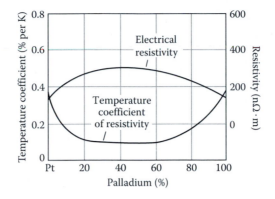

FIGURE 2.18   Electrical resistivity (in nΩ m) and temperature coefficient of resistivity (%/K) for platinum–palladium (Pt–Pd) alloys. (From Lampman, S.R. and Zorc, T.B., eds., *Metals Handbook: Properties and Selection: Nonferrous Alloys and Special Purpose Materials*, ASM, Materials Park, OH, 1990. With permission.)

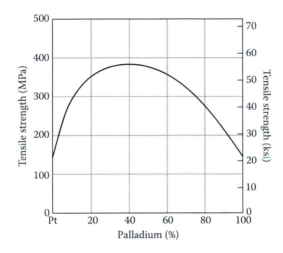

FIGURE 2.19   The change in tensile strength (in MPa on the left *y*-axis and in ksi on the right *y*-axis) of annealed platinum–palladium alloys as a function of palladium concentration. (From Lampman, S.R. and Zorc, T.B., eds., *Metals Handbook: Properties and Selection: Nonferrous Alloys and Special Purpose Materials*, ASM, Materials Park, OH, 1990. With permission.)

formed over the entire range of compositions. The electrical resistivity of platinum–palladium alloys as a function of the palladium concentration is shown in Figure 2.18.

When palladium is added to platinum or vice versa, the tensile strength increases; this effect is shown in Figure 2.19.

Note that most alloy systems do *not* show the formation of solid solutions over the entire composition ranges. Instead, different phases are often formed, or the solubility of one component into another is limited. In some alloys, heat treatments can lead to the *ordering* of atoms. Processes such as ordering, clustering, and precipitation of different phases can have a significant effect on the observed resistivity values. For example, in the copper–gold system, the formation of ordered phases leads to an *increase* in conductivity. Both composition and microstructure have a major effect on the resistivity of alloys.

Example 2.12: Resistivity of a Cu Alloy Using Nordheim's Rule

Pure Au and Ag are too soft for most applications. As a result, Cu is added as an alloying element to strengthen these metals using solid-solution strengthening. What is the resistivity of an Au alloy

containing 1.5 weight % Cu? Assume that Nordheim's coefficient (C) for Cu dissolved in Au is
450 n$\Omega \cdot$ m (Equation 2.32).

Solution

From Table 2.1, we see that the resistivity of Au is 2.35 $\mu\Omega \cdot$ cm or 23.5 n$\Omega \cdot$ m. Note that, in
Equation 2.32, the concentration of the alloy-forming element has to be expressed as an atom
fraction, which is achieved by applying the following equation:

$$x = \frac{M_{Au}w}{(1-w)M_{Cu} + wM_{Au}}$$

where x is the atom fraction of Cu, w is the weight fraction of Cu, M_{Cu} is the atomic mass of Cu,
and M_{Au} is the atomic mass of Au. In our case, the wt% of Cu is given as 1.5; thus, the weight fraction of Cu is 1.5/100 = 0.015. From the periodic table, M_{Au} = 197 g/mol, M_{Cu} = 63.5 g/mol. Thus,

$$x = \frac{197 \times 0.015}{\left[(1-0.015) \times 63.5\right] + \left[0.015 \times 197\right]} = 0.045$$

The concentration of Cu as an atom fraction is x = 0.045.
Therefore, using Nordheim's rule, we get:

$$\rho_{alloy} = Cx(1-x)$$
$$= 23.5 + 17.19 = 40.69 \sim 40.7 \text{ n}\Omega \cdot \text{m}$$
$$\rho_{alloy} = 23.5 + [400 \times 0.045(1-0.045)]$$

We can see that a small concentration of Cu increases the resistivity of Au substantially. Also,
note that Cu is actually a better conductor than Au. However, when we add Cu to Au as an alloying element, the Cu atoms disrupt the arrangement of the Au atoms. This increases the scattering
of conduction electrons in Au and causes an increase in resistivity.

As discussed in Section 2.8, we typically expect alloys to have low electrical conductivities;
moreover, the conductivity of metallic alloys does not change much with temperature (Figure 2.20).

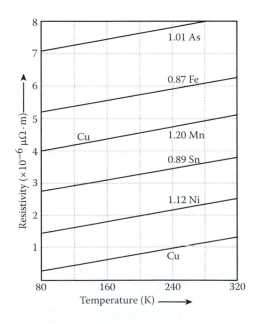

FIGURE 2.20 The resistivity of some copper alloys as a function of temperature. The data for resistivity
change of copper is also shown. (From Neelkanta, P., *Handbook of Electromagnetic Materials*, CRC Press,
Boca Raton, FL, 1995. With permission.)

The elements added to the alloys or those present as impurities in raw materials or processing may remain dissolved in the alloy and form a solid solution. Alternatively, impurities can react with one another or with other elements present to form various separate phases. Both the concentration and the manner in which the impurities are distributed in the microstructure influence the observed values of resistivity.

2.12 LIMITATIONS OF THE CLASSICAL THEORY OF CONDUCTIVITY

The classical theory based on the delocalization of valence electrons, although useful, does not explain several key features associated with the electrical conductivity. For example, in some covalent bonding materials, such as silicon or diamond (a form of carbon [C]), each atom has four valence electrons, but their free charge carrier concentration is much smaller than the valence electron concentration. Furthermore, the conductivity of Si is extremely sensitive to the presence of even very small levels of other elements (i.e., impurities) rather than the density of valence electrons. Classical theory also has a difficulty in explaining a change in the mean velocity of electrons (i.e., electron motion by thermal energy). According to classical theory, if no electric field is applied, we would expect the mean electron speed with which conduction electrons move to increase with increasing temperature, similar to the situation encountered in the kinetic theory of gases. However, it has been shown that the mean electron speed with which conduction electrons move in a material at zero electric field, is fairly constant with the temperature. This means that electrons cannot be simply treated as particles of classical physics. Moreover, classical theory cannot accurately predict the relationship between the *thermal conductivity* and the electrical conductivity of metals. If all electrons in the sea of electrons move freely and are responsible for the high electric conductivity of metals, these free electrons should also contribute to the thermal conductivity. In fact, Wiedemann and Franz experimentally found that there is a linear correlation between the thermal conductivity and the electric conductivity (i.e., good electrical conductors are good thermal conductors). However, the following quantitative studies show that the thermal conductivity predicted by classical theory is two orders of magnitude lower than the experimentally observed thermal conductivity. Unless the heat capacity of free electrons is two orders of magnitude higher than the measured value, classical theory cannot explain a correlation between the thermal conductivity and the electric conductivity.

Thus, classical theory is inadequate in explaining a number of experimental observations. This is because the interactions of a valence electron with nucleuses, other valence electrons, and impurities are ignored in the classical theory. For rigorous calculation, we cannot apply a classical particle model to describe the movement of free electrons. A more elaborate quantum mechanics–based explanation of the electron configuration can help to better understand the physics and explain the experimental observations of electron motion under the electric field or the thermal gradient.

2.13 QUANTUM MECHANICAL APPROACH TO THE ELECTRON ENERGY LEVELS IN AN ATOM

We will now turn our attention to the quantum mechanical approach to explain variations in the electrical conductivity of different solids. This approach leads to the band theory of solids. Quantum mechanics provides a powerful approach for explaining a number of features related to the conductivity of materials that are not explained by classical theory.

We will start with a short discussion on the wave-like behavior and quantized energy levels of electrons. In quantum mechanics, electrons are treated as waves rather than as particles. (The particle–wave duality will be explained further in Chapter 8, which discusses the photon as a basic element of light.) Here, we will focus on the effect of the wave-like property of materials called de Broglie waves or matter waves. According to this theory, matters exhibit wave characteristics. The wavelength (λ) of matter waves is equal to

$$\lambda = h/p = h/mv \qquad (2.33)$$

where h is the Planck constant, p is momentum, m is mass, and v is velocity. Equation 2.33 indicates that the wavelength of matter is inversely proportional to the momentum. If the mass of the matter is large, the wavelength of the matter wave is too small to detect. This is the case with a soccer ball or an apple. However, in matter with a very small mass—such as an electron—the wavelength of the matter wave is measurable and has a real physical effect on the matter's motion. This explains why the wave-like property becomes important in electrons with such a small mass.

Note that these wave characteristics make the electrons in atoms have quantized energy levels instead of continuous energy levels. Quantization means that electrons in an atom can have only certain levels of energy. Readers can imagine a guitar string that is tied at two ends. The long-lasting acoustic vibration of the string is allowed only when the standing wave is formed through the string. In other words, the boundary conditions of the string limit the wavelength of the string for the stable vibration. The wave of electrons orbiting a nucleus is analogous to the standing waves of the vibrating guitar ring in that only certain modes (energy levels) of the wave are allowed. We have briefly reviewed the basic hypothesis of quantum mechanics—the electron energy levels are quantized due to the wave-like properties of the electron.

Four kinds of quantum numbers (n, l, m, and s) are used to express the quantized energy levels of electrons and the electron configuration in atoms and solids. The principal quantum number (n) accounts for the Coulombic interactions between the positively charged nucleus and an electron. A value of $n = 1$ corresponds to the K shell, $n = 2$ corresponds to the L shell, and so on (Figure 2.6). Another quantum number is related to the angular momentum of the electrons. This is known as the orbital angular momentum quantum number (l), or azimuthal quantum number, which relates the shape of electron orbitals (e.g., sphere shape, dumbbell shape, etc.). For a given value of the principal quantum number n, there are subshells that are characterized by different values of the orbital angular momentum. In a description of the electronic configuration, a value of $l = 0$ corresponds to the letter for subshell "s." Similarly, a value of $l = 1$ corresponds to the letter for subshell "p," and so on. As an example, for an energy level of $n = 3$ (that is, the M shell), there will be 3s, 3p, and 3d subshells. Thus, different combinations of the quantum numbers n and l represent different electron energy levels.

The different allowed values of these and other quantum numbers are shown in Table 2.7. In addition to n and l, an electron has a *magnetic quantum number* (m or m_l) that represents the component of orbital angular momentum along an external magnetic field. The magnetic quantum number matters when materials are exposed to magnetic field. Finally, an electron also has a quantum number that quantizes the spin, known as *spin quantum number* (s or m_s). This quantum number becomes especially important in understanding the magnetic properties of materials (Chapter 11). According to the principles of quantum mechanics, the energy states available to electrons in atoms are discrete or quantized. There are certain energy levels that electrons within atoms are not allowed to have. For example, electron energy levels are not allowed in between the 1s and 2s levels, between the 2s and 3p levels, and so on.

A complete set of quantum numbers, that is, n, l, m, and s, describes the unique quantum state of an electron. A set of quantum numbers represents what is described as a *wave function* associated with an electron. A wave function is equivalent to an orbital or an energy level. One important principle from quantum mechanics is known as *Pauli's exclusion principle*. This principle states that no two electrons in a given system, such as an atom, can have all four identical quantum numbers. If two electrons have the same values of n, l, and m, then—according to Pauli's exclusion principle—their spins must be opposite. Such electrons are said to be *spin-paired electrons*. The possible values of quantum numbers are summarized in Table 2.7.

The maximum number of electrons allowed in a shell with a given value of n is $2n^2$. Thus, for values of $n = 1$, 2, and 3, the maximum number of electrons allowed is 2, 8, and 18, respectively. For the s, p, d, f, and g sublevels, the maximum number of states or energy levels allowed are 2, 6, 10, 14, and 18, respectively. The order in which the different energy levels are filled is as follows:

1s, 2s, 2p, 3s, 3p, 4s, 3d, 4p, 5s, 4d, 5p, 6s, 5d, 4f, 6p, 7s, 6d, 5f ….

TABLE 2.7
Quantum Numbers for Electrons

Principal quantum number (n)	n = 1, 2, 3, n = 1 is the K shell, n = 2 is the L shell, and so on. The maximum number of electrons for a shell with given n is $2n^2$.
Orbital angular momentum quantum number (l) (also known as the azimuthal quantum number)	l = 0, 1, ... (n – 1) l = 0 indicates the s subshell, l = 1 indicates the p subshell, and so on. The maximum number of electrons in the various subshells are s = 2, p = 6, d = 10, f = 14, and g = 18.
Magnetic quantum number (m or m_l)	m = $-l$, $-(l-1)$,..., 0, $(l-1)$, l
Spin (s or m_s)	s = ±½

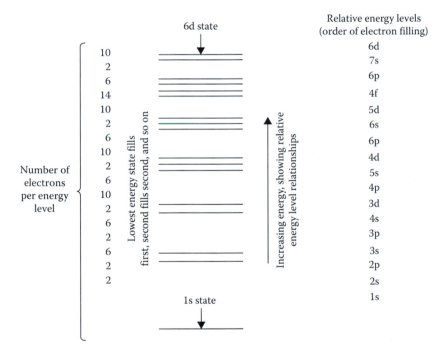

FIGURE 2.21 The order and number of electrons for different elements. (From Minges, M.L., *Electronic Materials Handbook*, Vol. 1, ASM, Materials Park, OH, 1989. With permission.)

Note that in this filling order, there are some inversions. For example, we use up the 4s level before 3d. A complete description of reasons for this is beyond the scope of this book. However, it is important to note that these inversions do occur. They play a key role in the determination of both the electronic configuration of transition elements and the magnetic properties of ceramic ferrites and iron- and nickel-based magnetic materials.

Examples 2.13 and 2.14 show how these concepts can be used to describe the electronic configuration for different elements. Figure 2.21 contains a list of the electronic configurations for different elements.

Example 2.13: Electronic Configuration for Al

Write down the electronic configuration for Al, whose atomic number (Z) is 13. Explain the meaning of the different symbols used.

Solution

For Al, $Z = 13$; this means that there are 13 electrons in one Al atom.

We start with $n = 1$. In this level, we can have $2(1)^2 = 2$ electrons. For $n = 1$, the only possible value of l is 0. In this s subshell, we can have only two electron energy levels. Thus, the first part of the configuration is $1s^2$ (read as "one s two"). Then, for $n = 2$ (or the L shell), the possible values of l are 0 and 1, that is, s and p sublevels. The total maximum number of electrons in this shell can be $2(2)^2 = 8$. Thus, when $n = 2$, for the s subshell, we can have two electrons; and for the p subshell, we can have six electrons. Thus, the configuration will read $1s^2 2s^2 2p^6$. This accounts for a total of $2 + 2 + 6 = 10$ electrons. We have only three more electrons left. We move to the $n = 3$ level (M shell). We can have up to $2(3)^2 = 18$ electrons in this level, where the possible values of l are 0, 1, and 2 or s, p, and d subshells. We start with the s subshell, place two electrons here, and then move to the p subshell and place one more electron here.

Thus, the electronic configuration for Al will be $1s^2 2s^2 2p^6 3s^2 3p^1$.

The p subshell can hold five more electrons, but these levels will remain empty because the Al atom has only 13 electrons. Similarly, the 3d and higher energy levels—such as 4s, 4p, and 4d—also will remain empty.

The electrons in the outermost shell ($n = 3$)—that is, the 3s and 3p electrons—are referred to as the valence electrons. These electrons are particularly important because they are available for both electrical conduction and chemical reactions in metallic materials.

Example 2.14: Electronic Configuration for Fe

Write down the electronic configuration of a Fe ($Z = 26$) atom.

Solution

There are 26 electrons in a Fe atom. We start with $n = 1$. This orbital can take $2(n = 1)^2$ or two electrons. The only possible value of l is 0, that is, the s subshell. Thus, the electronic configuration until this level will be $1s^2$. For $n = 2$, we can have a maximum of $2(2)^2$ or eight electrons. The possible values of l are 0 and 1; or, we can have s and p subshells. Now, the electronic configuration will read $1s^2 2s^2 2p^6$. For $n = 3$, we can have $2(3)^2$ or 18 electrons. However, we have only $26 - 2 - 8 = 16$ electrons remaining. For $n = 3$, we can have possible values of 0, 1, and 2 for l; or s, p, and d subshells. Thus, the electronic configuration will read as $1s^2 2s^2 2p^6 3s^2 3p^6 3d^8$.

However, according to the filling order stated earlier, the 4s shell will fill *before* the 3d shell. The 4s level will take two electrons. The balance of six electrons will enter the 3d shell.

Therefore, the final electronic configuration for iron will be $1s^2 2s^2 2p^6 3s^2 3p^6 4s^2 3d^6$.

The d subshell can contain a maximum of 10 electrons but actually contains only six. Thus, the d subshell is deficient by four electrons. Of the six electrons in the 3d sublevel, two are spin-paired—that is, they have all the same quantum numbers, except the spin quantum numbers (which are $+1/2$ and $-1/2$; Table 2.8).

The remaining four of the 3d electrons are *not* spin-paired. These unpaired electrons make it possible for a Fe *atom* to behave like a tiny bar magnet. This behavior contributes to making Fe a magnetic material. We will study this when we discuss magnetic materials (Chapter 9).

2.14 ELECTRONS IN A SOLID

Consider that the electron energy levels or associated spectrum in silicon that has four valence electrons per atom and 2 electrons each are in 2s orbital and 2p orbitals (Figure 2.22). Note that the energy levels of 2s orbitals and 2p orbitals are not the same. If Si atoms form the solid, how will valence electron energy levels change?

TABLE 2.8

Electron Spin States in Iron (Fe)

$n = 1$	$n = 2$		$n = 3$		$n = 4$	
1s	2s	2p	3s	3p	3d	4s
↑	↑	↑↑↑	↑	↑↑↑	↑↑↑↑↑	↑
↓	↓	↓↓↓	↓	↓↓↓	↓	↓

Source: Edwards-Shea, L., *The Essence of Solid-State Electronics*, Prentice Hall, Upper Saddle River, NJ, 1996. With permission.

Note: An arrow pointing up (↑) means $s = +1/2$, or spin up; an arrow pointing down (↓) means $s = -1/2$, or spin down. Note that the 4s level is filled before the 3d level and two of the 3d electrons are spin-paired.

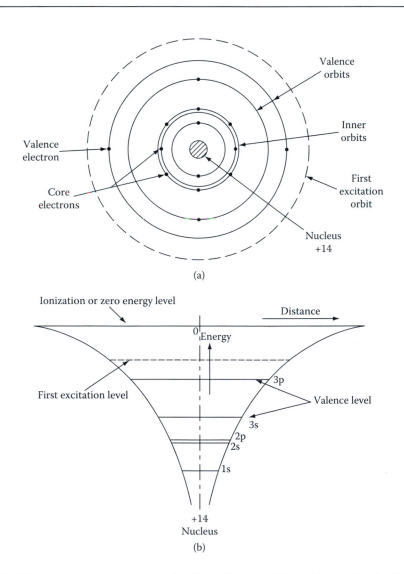

FIGURE 2.22 Electronic structure and energy levels in a Si atom: (a) The orbital model of a Si atom showing the ten core electrons ($n = 1$ and 2) and the four valence electrons ($n = 3$); (b) energy levels in the Coulombic potential of the nucleus are also shown schematically. (From Streetman, B.G. and Banerjee, S., *Solid State Electronic Devices*, 5th ed., Prentice Hall, Upper Saddle River, NJ, 2000. With permission.)

A solid material can have billions of atoms. It is important to know not just the electronic structure of individual atoms but also how different atoms interact with one another when they are in close proximity to one another. When individual atoms are brought together to form a solid material, the atomic distance decreases and the electron orbitals begin to overlap (Figure 2.23). The overlapping of the electron orbitals makes the wave characteristics of the electrons complicated. Then, the quantized and discrete electron energy levels of the atom spread out and the stable electron energy levels evolve to the band in the solid.

Figure 2.24 schematically illustrates the overlapping of energy levels of different atoms and the subsequent formation of *energy bands*, which occur in silicon. As the interatomic distance (or lattice spacing) decreases, two discrete energy levels become two separate bands. A remaining question is, "What is the physical origin of two discrete energy levels in Si atom and two energy bands in Si solid?" Apparently, the lower band and the upper band have different physical origins. We will start with two energy levels in Si atoms. The lattice structure of Si is very similar to that of diamond

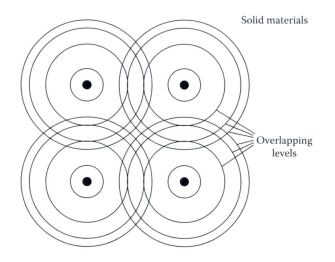

FIGURE 2.23 Overlap of the electron energy levels or wave functions as the atoms come closer. (Adapted from Edwards-Shea, L., *The Essence of Solid-State Electronics*, Prentice Hall, Upper Saddle River, NJ, 1996. With permission.)

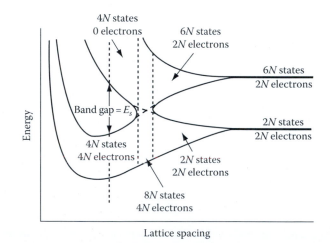

FIGURE 2.24 Formation of energy bands in silicon. (From Streetman, B.G. and Banerjee, S., *Solid State Electronic Devices*, 5th ed., Prentice Hall, Upper Saddle River, NJ, 2000. With permission.)

in which C atoms constitute a zinc blend structure (Figure 1.20). Thus, in Si lattice, four Si atoms form a face-centered cubic (FCC) structural frame. An additional four Si atoms fill four out of eight tetrahedral sites of the FCC frame and four tetrahedral sites remain empty. Because only half of the tetrahedral intersitials are filled, there are two different valence electron locations. The first type of valence electrons stay near the center of Si–Si bonds, and the Coulombic interactions between valence electrons and the nucleus consequently contribute to strengthening Si–Si bonds. This energy state of valence electrons participating in Si–Si covalent bonding is called the bound state; most valence electrons are in the bound state. The second type of electron state is known as *free state* and only a small portion of the valence electrons take the energy levels in free states. Electrons in free states are found around the empty tetrahedral sites of a Si lattice. Due to the lack of nearby Si atoms, the electrons in the free states are not bound to the atom. It is intuitively understood that the kinetic energy of valence electrons in the free state is so high that the valence electrons can easily overcome the attraction force that would have kept them bound to the specific nucleus. The energy at which an electron becomes free is referred to as $E = 0$.

As the lattice spacing decreases, two different energy levels (free states and bound states) of Si atoms evolve into two bands (Figure 2.24). In Si, the electron energy band corresponding to the bound states is called the valence band, and the electron energy band related to the free states is called the conduction band. The energy gap between the bottom of the conduction band and the top of the valence band is called band gap. This concept of the formation of energy *bands* and *band gap* in a *solid*, as opposed to the energy *levels* in individual *atoms*, is the foundation for describing the differences in the electrical properties of insulating, conducting, and semiconducting materials.

2.15 BAND STRUCTURE AND ELECTRIC CONDUCTIVITY OF SOLIDS

The band structure and electric conductivity of a solid are closely related. We will discuss a lithium atom (Li; $Z = 3$) first. The electronic configuration is $1s^2 2s^1$. This means that there are two electrons in the lowest energy level (1s level). There is only one electron in the next energy level for an atom, the 2s level. The 2s and 2p levels are collectively known as the L shell. The 2s level can accept two electrons and is thus only half-filled by a lithium atom.

Now consider a lithium crystal with N atoms. When we have N lithium atoms, there are N electrons that belong to the 2s energy band. When lithium atoms approach each other, the different 2s energy levels of the different atoms begin to overlap. This leads to the formation of a 2s band (Figure 2.25). Note that each 2s level can take two electrons; thus, for N atoms, the 2s band is capable of taking $2N$ electrons. However, there are only N 2s electrons. Thus, the 2s band is only half-full and the other half of the energy state in the 2s band is empty. Moreover, the higher energy, empty energy levels for the individual atoms also begin to overlap. Thus, the 2s band is extended and overlaps with the 2p and 3s bands.

When an electric field is applied to lithium, the electrons in the low energy–filled states gain energy, move into the empty energy states of 2*s* band, and deliver an electric charge. Therefore, solids of alkali elements, such as sodium and lithium, are very good conductors of electricity. This indicates that the valence band partially filled with valence electrons is a necessary condition for the high electric conductivity that is one of the most important physical properties of metals.

In Figure 2.25, readers may recognize that the band width for electrons of the 1s level does not increase, even though the interatomic distance decreases to the equilibrium position (marked as a dotted vertical line in Figure 2.25). Since the 1s level electrons are closer to the nucleus and the 2s band electrons shield the 1s level electrons from outside electrons, the 1s level electrons in the solid are less affected by the presence of electrons of neighboring atoms. Hence, a decrease in the interatomic spacing does not result in the significant splitting of the 1s energy level. However, if the interatomic spacing of the lithium atoms becomes much smaller than the equilibrium spacing, the 1s energy level dramatically increases, as shown in Figure 2.25. Also, note that the 1s band in the lithium is completely filled because there are two electrons in the 1s energy level of the lithium. When the band is completely filled (like the 1s band of the lithium), there are no empty energy states for the electrons to move into. Thus, electrons in the completely filled band do not contribute to electrical conduction.

We represent these energy levels of metals in the form of a band diagram (Figure 2.26). From an electrical properties viewpoint, the bands that involve the outermost or valence electrons are the most important in most cases. Thus, it is customary to show only the outermost bands in a band diagram. This outermost band consisting of the valence electrons is known as the valence band. The band that is in the bottom of the band diagram typically is almost completely filled with electrons.

Qualitative band diagrams for different types of materials are shown in Figure 2.27. Band diagrams and the magnitudes of the band gap energies provide an excellent way to classify materials into conductors, semiconductors, and insulators (Figure 2.28). As shown in Figure 2.27a, a partially filled valence band is an important feature of metals. In some metals such as magnesium (Mg),

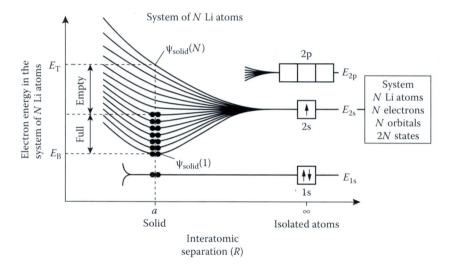

FIGURE 2.25 Formation of energy bands in lithium metal. The 2s band is only half-filled. Note that the 1s level shows very little splitting. (From Kasap, S.O., *Principles of Electronic Materials and Devices*, McGraw Hill, New York, 2002. With permission.)

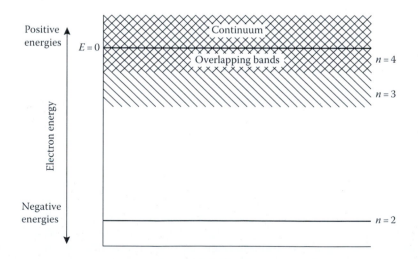

FIGURE 2.26 Schematic of a band diagram for a metal. (From Edwards-Shea, L., *The Essence of Solid-State Electronics*, Prentice Hall, Upper Saddle River, NJ, 1996. With permission.)

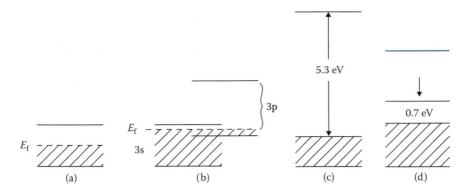

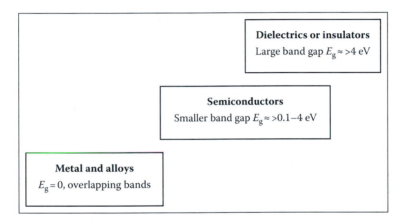

FIGURE 2.27   Band diagrams for conductors: (a) an alkali metal; (b) magnesium (Mg), a bivalent metal; (c) diamond, an insulator; and (d) germanium (Ge), a semiconductor. (From Mahajan, S. and Sree Harsha, K.S., *Principles of Growth and Processing of Semiconductors*, McGraw Hill, New York, 1998. With permission.)

FIGURE 2.28   Classification of materials based on the values of band gap (E_g) in electron volts (eV).

the valence band (3s level for Mg) is almost filled, and there is no apparent band gap. However, the empty 3p bands are overlapped with the almost filled 3s in Mg (Figure 2.27). In this band structure, when an electrical field is applied, the electrons in the almost 3s band accelerate and occupy the empty states in the 3p band and possibly higher bands. Hence, the electric conductivity of Mg is comparable to that of Na or other metals. The excitation of the valence electrons from 3s band to 3p bands explains why alkali earth elements such as Mg are classified as metal.

The allowed energy band immediately above the valence band and into which valence band electrons can move by gaining sufficient energy or momentum is known as the conduction band. Because of the overlapping of the energy levels of different atoms, the bound energy levels become almost, but not completely, continuous. In between these *energy bands*, we find regions where no electron energy states are allowed. Thus, there are *forbidden gaps* between the bands of allowed energy levels when the interatomic distance reaches the equilibrium position (Figure 2.24). If there is an energy gap between the conduction gap and the valence band in the material, the magnitude of the energy gap between the conduction and the valence bands is known as the band gap (E_g). We can now use this description of the band gap to distinguish between insulators and semiconductors. A key idea here is that when the band is completely filled, electrons cannot move to the higher energy level at a very small energy cost and carry any current; hence, conduction cannot take place. Thus, to make the electrical conduction occur in materials with the band gap (not in metals), the electrons need to acquire enough energy to jump into the empty conduction band and flow through it.

If the band gap is larger than ~4 eV, materials are classified as insulators (also called dielectrics). This is the case for diamond. Materials such as Si and Ge are grouped as semiconductors and their band gap is smaller than ~4 eV (Figures 2.27 and 2.28). The valence band of insulators is separated from the conduction band by a relatively large band gap (Figure 2.28). For example, for diamond, the band gap is 5.3 eV (Figure 2.27).

If the temperature is low, the thermal energy is not large enough to excite electrons from the valence band to the conduction. This is why insulators and semiconductors exhibit very low electric conductivity at low temperatures ($T < 100$ K). If the temperature increases, the electric conductivities of the insulators and the semiconductors become different. Even at high temperatures, the valence band of the insulator again is completely filled and the electric conductivity is still very low. For example, no appreciable electrical conduction occurs in diamond even at high temperatures. However, in the case of semiconductors such as germanium (Ge), as the temperature increases, some valence band electrons are able to jump across the smaller band gap (E_g for Ge is ~0.67 eV) and enter the conduction band. This means that the electrons locked in covalent bonds among germanium atoms can break free and jump from one bond onto another. Thus, unlike insulators, the conductivity of semiconductor materials, such as essentially pure silicon or germanium, *increases* as the temperature increases. Note that a major difference between semiconductors and metals is how a change in temperature changes the number of electrons participating in the electrical conduction (sometimes called free electrons). In metals, a free electron density is almost constant at different temperatures. In semiconductors, however, an increase in the temperature exponentially increases the free electron density. Therefore, as the temperature increases, the electric conductivity exponentially increases for semiconductors and slightly decreases for metals. Though the free electron concentration is almost constant in metals, the electron mobility is inversely proportional to temperature, and high-temperature electric conductivity is smaller than low-temperature conductivity.

When the temperature increases in the semiconductor, the electrons jump to the conduction band and the valence band becomes lack of electrons. We described this situation that *holes* are created in the valence band. As mentioned before, a hole is an imaginary positively charged particle that represents an electron missing from the valence band. When electric field is applied, holes created in the valence band can move around in the valence band similarly to the free electrons in the conduction band. Thus, the motions of both electrons and holes contribute to the conductivity of a semiconductor, and the sum of the electron conductivity and the hole conductivity equals the electrical conductivity of the semiconductor. The level of conductivity in materials such as silicon and germanium typically is lower than that of pure metals but higher than that of insulators (Figure 2.2). Therefore, we refer to materials such as germanium and silicon as semiconductors.

2.16 FERMI ENERGY AND FERMI LEVEL

Now, we discuss a new concept, the *Fermi level*, which represents the energy of an electron that is the least tightly bound to the lattice. Strictly speaking, the Fermi level at 0 K is called the Fermi energy; however, Fermi level and Fermi energy are used interchangeably in many cases. For convenience, we will use the term Fermi energy level and display it as E_F throughout this textbook. The Fermi energy level (E_F) can be explained in two different ways. First, E_F means the highest energy level that electrons occupy at $T = 0$ K. Thus, energy levels below E_F are fully taken, and those above E_F are fully empty at $T = 0$ K. Second, we can use the probability of electrons to describe E_F. When the temperature is higher than 0 K, thermal energy helps electrons move from the energy levels below E_F to the empty energy above E_F. Thus, E_F is not the highest energy level that the electrons occupy. The energy level E_F is partially empty at $T > 0$ K. Since electrons jump sequentially from the highest occupied energy levels to the lowest unoccupied energy levels, the distribution of the partially filled energy states above E_F and that of partially empty energy states below E_F is symmetric (see Figures 4.3 and 4.11). This indicates that we view E_F as the energy level that is occupied with a chance of 50% at $T > 0$ K. Other important concepts related to Fermi energy level (E_F) are work function ($q\phi$) and electron affinity ($q\chi$). Work function (ϕ) is the energy required to remove the electrons at the Fermi energy level (E_F) to the vacuum level. In addition, the energy needed to free the electron at the bottom of the conduction band to the vacuum level is called electron affinity (χ).

Now, we will consider a metal such as lithium in order to understand the new concepts added in this section. All of the 2s electrons occupy energy levels beginning from the bottom of the 2s band (E_B) to an energy level called the Fermi energy, at 0 K (E_{F0}). As noted earlier, if we define the bottom of the valence band as zero, then the highest energy level filled is the Fermi energy level (E_F), as shown in Figure 2.29(a). The Fermi energy level (E_F) of a metal is the highest energy level occupied by electrons, which is at 0 K. The Fermi energy level (E_F) of semiconductors and insulators looks a bit different from that of metals. At very low temperatures, the valence band is completely filled. As a result, there is no electrical conduction. When the temperature increases, electrons can acquire enough energy that they may be able to jump across the band gap and into the conduction band to occupy energy levels to which they are allowed access. If the electrons have any levels to move into, they would be above the valence band. Consequently, the Fermi energy level (the energy level with a 50% chance of being filled) for semiconductors and insulators will be somewhere above the top of the valence band and below the bottom of the conduction band. However, this is the band gap region into which electrons are not allowed (see Figure 2.29b)!

In the case of semiconductors, the valence band either is completely filled (at low temperatures) or is almost completely filled (at about room temperature). Thus, the probability of finding an electron in the valence band is very close to 1. The conduction band either is completely empty (at low temperatures) or is almost completely empty (at about room temperature). Thus, the probability of finding an electron in the conduction band is very small but finite. Somewhere between the top edge of the valence band and the bottom edge of the conduction band, the probability of finding an electron will be 0.5 (Figure 2.29).

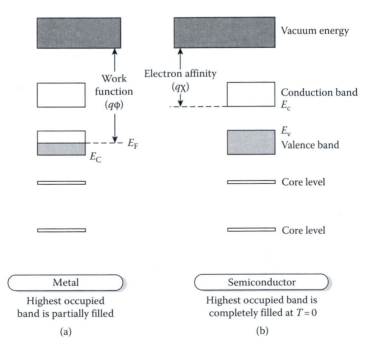

FIGURE 2.29 A schematic of band structure for (a) a typical metal and (b) a semiconductor. The work function is the energy $q\phi$. The electron affinity for a metal is shown as the energy $q\chi$. In a pure semiconductor, the Fermi energy level (E_F) is located at the center of the forbidden gap and ($E_c - E_F$) is the same as ($E_F - E_V$). (Adapted from Singh, J., *Semiconductor Devices: Basic Principles*, Wiley, New York, 2001. With permission.)

For pure semiconductors (no impurities), the Fermi energy level (E_F) is located at the center of the forbidden gap and ($E_C - E_F$) is the same as ($E_F - E_V$) (Figure 2.29). Thus, the location of E_F for semiconductors at $T > 0$ is different from that of metals (see Figure 4.3). Further details about the Fermi energy level in semiconductors will be discussed in Chapter 4 with the introduction of the Fermi–Dirac function.

The representation of E_F for a dielectric material is the same as that for a semiconductor; the only difference is that the E_g of dielectrics is larger than that of semiconductors.

The *work function* ($q\phi$) and the *electron affinity* ($q\chi$) also are schematically illustrated in Figure 2.29. The energy needed to excite an electron from E_F to the vacuum level, where it is essentially *freed from the solid*, is marked as $q\phi$ (where q is the charge on the electron). The energy required to remove an electron from the bottom of the conduction band edge to the vacuum is known as the *electron affinity* ($q\chi$) and is usually expressed in electron volts (eV). Sometimes, electron affinity is also described as χ, and its units are volts. Note that the work function and the electron affinity are different in semiconductors.

2.17 COMPARISON OF CLASSICAL THEORY AND THE QUANTUM MECHANICAL APPROACH FOR ELECTRICAL CONDUCTION

Before finishing this chapter, we will study the differences in how classical theory and the quantum mechanical approach explain electric conductivity. We can explain the electric conductivity of a semiconductor using the band structure that is derived from quantum mechanical considerations. The quantum mechanical approach assumes that a limited number of electrons are responsible for electrical conduction, even in metals. This says that only valence electrons occupying E_F or higher energy states contribute to the electric conductivity (refer to Section 3.7).

In classical theory, all electrons in the partially filled band (the valence band) contribute equally to electric conductivity. In this view, all valence electrons start moving only after an electric field is applied, and the motion of all valence electrons under an electric field is statistically the same. However, when we studied the evolution of the discrete energy levels of the band, we learned that a decrease in interatomic distance splits the energy levels in a band and that electrons occupying the different energy levels have different energies. According to quantum mechanics, electrons in the valence band of metals are not frozen even at $T = 0$ K, and the speed with which valence electrons move depends on the electron energy levels. The higher the energy level, the faster the electron motion. When an electric field is applied, fast moving electrons contribute to electric conductivity more than slow moving electrons. Therefore, the classical theory of electric conductivity needs to be modified. From the viewpoint of quantum mechanics, only a portion of valence electrons possessing higher energy levels in the valence band (in other words, valence electrons near the Fermi energy level) participate in the electrical conduction of metals, with the moving velocity much higher than the average velocity of the valence electrons.

PROBLEMS

Introduction

2.1 What are the typical ranges of resistivity for metals, plastics, and ceramics?
2.2 What is the nature of bonding among atoms for most ceramics and plastics?
2.3 Compared to metals, what is the advantage in using conducting or semiconducting plastics?
2.4 Which one of these elements shows superconductivity—Ag, Au, or Al?

Ohm's Law

2.5 What is the difference between resistance and resistivity?
2.6 Do all materials obey Ohm's law? Explain.
2.7 Calculate the resistance of an AWG #20 Cu wire one mile in length.

2.8 What is the length of an AWG #16 Cu wire whose resistance is 21 Ω?

2.9 If the wire in Problem 2.8 carries a current of 5 A, what is the current density?

2.10 Al can handle current densities of ~10^5 A/cm^2 at about 150°C (Gupta 2003). What will be the maximum current allowed in an Al wire of AWG #18 operating at 150°C?

2.11 A circuit breaker connects an AWG #0000 Cu conductor wire 300 feet in length. What is the resistance of this wire? If the wire carries 150 A, what is the decrease in voltage across this wire?

2.12 You may know that a conductor carrying an electrical current generates a magnetic field. A long wire carrying a current generates a magnetic field similar to that generated by a bar magnet. This magnet is known as an electromagnet. Consider a meter of magnetic wire AWG #2. Such wires are usually made from high-conductivity soft-drawn electrolytic Cu, and the conductor is coated with a polymer to provide insulation. What will be the electrical resistance (in ohms) of this wire?

2.13 Ground rods are used for electrical surge protection and are made from materials such as Au, Au-clad steel, or galvanized mild steel. The resistance of the actual rod itself is small; however, the soil surrounding the rod offers electrical resistance (Paschal 2001). The resistance of a ground rod is given by:

$$R = \frac{\rho}{2L\pi} \ln\left[\frac{4L}{a}\right] - 1 \qquad (2.34)$$

where R is the resistance in ohms, ρ is the resistivity of soil surrounding the ground rod (in $\Omega \cdot$ cm), L is the length of the ground rod in centimeters, and a is the diameter of the ground rod. (a) Assuming that the resistivity of a particular soil is 10^4 $\Omega \cdot$ cm, the length of the rod is 10 feet, and the diameter is 0.75 inches, what will be the resistance (R) of the ground rod in ohms? (b) Assuming that the ground rod is made from Cu, prove that the resistance of the rod itself is actually very small. (c) What will happen to the resistance of the metallic material as it corrodes over a period of many years?

2.14 The electrical resistance of pure metals increases with temperature. In many ceramics, the electrical current is carried predominantly by ions (such as oxygen ions in YSZ). Based on the data shown in Figure 2.30, calculate the electrical resistance of a 50-μm-thick YSZ element at 500°C and 800°C. Assume that the cross-sectional area is 1 cm^2 and the length is 1 m.

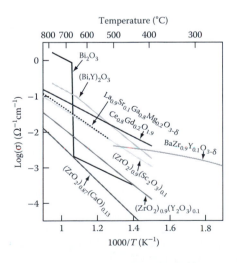

FIGURE 2.30 The conductivity of different materials used as electrolytes in the development of solid oxide fuel cells. (From Haile, S.M., *Acta Mater.*, 51, 5981–6000, 2003. With permission.)

2.15 Consider the material calcium oxide–stabilized ZrO_2 (Figure 2.30). Calculate the electrical resistance of a 50-μm-thick YSZ element. Assume that the cross-sectional area is 1 cm² and the length is 1 m at 500°C and 800°C.

2.16 If a high conductivity at temperatures above 700°C was the only consideration in selecting a material for a solid oxide fuel cell electrolyte, what material (Figure 2.30) would you choose? Besides cost, what additional factors must be considered in the selection of this material?

2.17 What is unusual about the change in resistivity as a function of temperature for bismuth oxide (Bi_2O_3)?

Classical Theory of Electrical Conduction

2.18 Calculate the mobility of the electrons in Zn in cm²/V · s. Assume that each Zn atom contributes two conduction electrons. The atomic mass of Zn is 65. The resistivity is 5.9 μΩ · cm, and the density is 7.130 g/cm³.

2.19 If the mobility of electrons in Au is 31 cm²/V · s, calculate the time between collisions (τ). Assume that the mass of electrons in Au is 9.1×10^{-31} kg. Calculate the mean free-path length (λ) of electrons in Au if the average electron speed is 10^6 m/s.

2.20 The thermal speed of electrons is about 10^6 m/s. However, the drift velocity is rather small because electrons are scattered by the vibrations of atoms. Calculate the drift velocity of electrons in Cu for an electric field of 1 V/m. Assume that the mobility of electrons in Cu is 32 cm²/V · s.

2.21 If the density of Ag is 10.5 g/cm³, what is the concentration of conduction electrons in Ag?

2.22 From the information provided in Table 2.3, calculate the expected conductivity of Ag.

2.23 Au is a face-centered cubic metal with a lattice constant of 4.080 Å. If the atomic mass of Au is 196.9655, calculate the number of conduction electrons per unit volume. Express your answer as number of electrons/cm³. Assume that each Au atom donates one conduction electron.

2.24 A semiconductor is made so that it carries electrical current primarily from the flow of electrons. If the mobility of electrons (μ_n) is 1350 cm²/V · s and the conduction electron concentration is 10^{21} cm⁻³, what is the electrical conductivity of this material?

Joule Heating

2.25 A heating element for a flat iron is rated at 1000 W. If the iron works at 220 V, what is the resistance of this heating element?

2.26 Electronic components and devices are often tested at 125°C and −55°C to check their high- and low-temperature performances. They can then be compared with the properties observed at room temperature, 25°C. For example, using 25°C as the reference temperature and +125°C as the other temperature, α_{125} can be written as follows:

$$\text{TCR}_{125} = \alpha_{R,125} = \frac{1}{\rho_{125}}\left(\frac{\rho_{125} - \rho_{25}}{125 - 25}\right) \times 10^6 \, \text{ppm/°C}$$

Write an equation to express the temperature coefficient of resistance with $T = -55$°C (note the negative sign) as the other temperature, using 25°C as the base or reference temperature (T_0).

2.27 In a circuit, the TCR_{125} value for a resistor is 100 ppm/°C. If the resistance (R) at 25°C is 1000 Ω, what is the resistance at 125°C?

2.28 Assume that the bus bar discussed in Example 2.11 is heated due to the Joule losses and now operates at 70°C. Calculate the resistance, power loss, and energy consumption for

24 hours and the total energy costs per year. Ignore the change in the length of the Cu bus bar because of thermal expansion.

2.29 Nichrome wire is used for cutting materials such as polystyrene (Styrofoam®) into different shapes, including large facades or insulation boards. (a) Calculate the length of an AWG #20 wire that needs to have a resistance of $R = 8\ \Omega$. (b) What will be the resistance of this wire if it gets heated to a temperature of 200°C? (See Tables 2.2 and 2.5.)

Resistivity of Metallic Alloys

2.30 In the nanoscale region, why does the resistivity of thin films depend upon thickness?

2.31 What elements most affect the resistivity of high-purity Cu?

2.32 Why does the addition of oxygen in limited concentrations actually *increase* the conductivity of high-purity Cu?

2.33 Why does the resistivity of pure metals increase with temperature, whereas that of alloys is relatively stable with temperature?

2.34 Nordheim's coefficient for Au dissolved in Cu is $C = 5500\ n\Omega \cdot m$. If the resistivity of Cu at 300 K is $16.73\ n\Omega \cdot m$, calculate the resistivity of an Au–Cu alloy containing 1 weight % Au.

Band Structure of Solids

2.35 What is the electronic configuration for an Mg atom ($Z = 12$)?

2.36 Draw a schematic of the band diagrams for a typical metal, a semiconductor, and an insulator.

GLOSSARY

Annealing: A heat treatment for metals and alloys in which a material is heated to a high temperature and then cooled slowly; after annealing, dislocations are annihilated, and the material exhibits a higher level of conductivity.

Azimuthal quantum number (l): See **Orbital angular momentum quantum number**.

Band diagram: A diagram showing the electron energy levels that represent the valence and conduction bands.

Band gap (E_g): The energy difference between the top of the valence band and bottom of the conduction band, which must be overcome to transfer an electron from the valence band to the conduction band.

Bulk resistivity (ρ): See **Resistivity**.

Bus bar: A conductor used in power transmission.

Carrier concentration: A concentration of species responsible for electrical conduction.

Cold-working: A process conducted at temperatures below the recrystallization temperature in which a metallic material is deformed or shaped, usually causing the resistivity of a material to increase.

Composites: Formed when two or more materials or phases are blended together, sometimes arranged in unique geometrical arrangements, to achieve desired properties.

Conductance: Inverse of resistance, whose units are Siemens, or Ω^{-1}.

Conduction band: The higher band on a band diagram, separated from the valence band by the band gap, which shows the energy levels associated with the conduction electrons.

Conductivity: A microstructure-, composition-, and temperature-dependent property that conveys the ability of a material to carry electrical current and is the inverse of resistivity.

Conductors: Materials with resistivity in the range of ~10^{-6} to $10^{-4}\ \Omega \cdot cm$ (1 to $10^2\ \mu\Omega \cdot cm$).

Conventional current: As a matter of convention, the current that flows from the positive to the negative terminal of the battery, although the electrons themselves move in the reverse direction.

Current density (J): Current per unit cross-sectional area perpendicular to the direction of the current flow.

Dielectrics: Materials that do not allow any significant current to flow through them.

Drift: Motion of charged carriers, such as electrons, holes, or ions, in response to an electric field.

Electric field (E): Voltage divided by the distance across which the voltage is being applied.

Electrolytic tough-pitch (ETP or TP) copper: High-conductivity copper containing about 100–650 ppm (or 1 mg/kg) oxygen.

Electron affinity ($q\chi$): The energy required to remove an electron from the bottom of the conduction band edge to the vacuum, sometimes also designated as χ (in volts).

Electron current: The movement of electrons from the negative to the positive terminal of the voltage supply when a DC electrical field is applied to a material.

Electronic conductors: Materials in which a dominant part of conductivity is due to the motion of electrons or holes.

Four-point probe: A setup used for making conductivity measurements, typically involving a fixed current being applied using two outer probes, with the decrease in voltage measured using two inner probes; also known as a Kelvin probe.

Hole: An imaginary positively charged particle that represents an electron missing from a bond.

I^2R losses: The heating of a material due to resistance to electrical current, with the power dissipated (for DC voltages) given by the term I^2R (see also **Joule heating**).

IACS conductivity: Conductivity of International Annealed Copper Standard; 100% IACS is defined as $\sigma = 57.4013 \times 10^6$ S/m or 57.4013×10^4 S/cm (or $\rho = 1.74212 \times 10^{-6}\ \Omega \cdot$ cm), which is based on the conductivity of an annealed 1-m-long copper wire that has a cross-sectional area of 1 mm^2 and a resistance of 0.17421 Ω.

Impurity-scattering limited drift mobility: The mobility of carrier particles in alloys, which is limited by the scattering of impurities or alloying elements and not by the thermal vibrations of the host atoms.

Indium–tin oxide (ITO): A transparent ionic conductor-based material used in touch-screen displays, solar cells, and other devices.

Insulators: Materials that neither conduct electricity (similar to dielectrics) nor easily break down electrically even in the presence of a strong electric field.

Integrated circuits (ICs): Electrical circuits typically fabricated on semiconductor substrates, such as silicon wafers, comprising resistors, transistors, and so on.

Interconnects: Conductive paths between components of an IC.

Ionic conductors: Materials in which movement of ions constitutes the major portion of the total conductivity (e.g., ITO or yttria-stabilized zirconia).

Joule heating: The heating of a material due to resistance to electrical current; for DC voltages, the power dissipated is given by I^2R (same as I^2R losses).

Kelvin probe: See **Four-point probe**.

Lattice-scattering limited conductivity: The conductivity (σ) in essentially *pure* metals; it is largely limited by the scattering of electrons by the vibrations of atoms.

Lattice-scattering limited mobility: The mobility of carriers in essentially pure metals; it is limited by scattering due to phonons and other defects in arrangements of atoms.

Magnetic quantum number (m or m_1): The electron quantum number representing the component of orbital angular momentum along an external magnetic field.

Mean free-path length (λ): The mean free-path length (λ) of conduction electrons, that is, the average distance that electrons travel before colliding again.

Mixed conductors: Materials in which conductivity occurs because of the movement of ions as well as that of electrons or holes.

Mobility (μ): The speed of carriers under the influence of a unit of external electric field (E).

Nanoscale: Length of scale between ~1 and 100 nm in which unusual effects are seen on properties of materials, devices, or structures.

Nordheim's coefficient (C): See **Nordheim's rule**.

Nordheim's rule: The resistivity of a solid-solution alloy is given by the equation

$$\rho_{\text{alloy}} = \rho_{\text{matrix}} + Cx(1-x)$$

where ρ_{alloy} is the resistivity of the alloy, ρ_{matrix} is the resistivity of the base metal without any alloying elements, C is Nordheim's coefficient, and x is the *atom fraction* of alloying element added.

Orbital angular momentum quantum number (*i*): A quantum number related to the angular momentum of the electrons (the same as azimuthal quantum number).

Oxygen-free high-conductivity copper (OFHC): A high-conductivity copper material that has less than 0.001% oxygen. It is used when electrolytic tough–pitch copper cannot be used because of potential welding or brazing problems.

Pauli's exclusion principle: No two electrons in a given system (such as an atom) can have all four quantum numbers identical.

Phonons: In pure metals, the vibrations of conduction electrons from phonons leads to a resistivity component that increases with rising temperature.

Principal quantum number (*n*): A quantum number that quantizes the electron energy, with values of $n = 1$, 2, and 3, corresponding to the K, L, and M shells, respectively.

Residual resistivity (ρ_R): The part of total resistivity arising from the effects of microstructural defects and impurities or added elements.

Resistance (*R*): The difficulty with which electrical current flows through a material. For a material with length L, cross-sectional area A, and resistivity ρ, the resistance R is given by $\rho \cdot L/A$.

Resistance-temperature detector (RTD): A temperature-measuring device based on the measurement of change in the resistivity of a metallic wire as a function of temperature.

Resistivity (ρ): The electrical resistance of a resistor with a unit length and a unit cross-sectional area. This is a microstructure- and temperature-dependent property the magnitude of which is the inverse of conductivity.

Resistor: A component included in an electrical circuit to offer a predetermined value of electrical resistance.

Semiconductors: Materials that have a resistivity ranging between 10^{-4} and $10^3\ \Omega \cdot \text{cm}$ (i.e., 10^2–$10^9\ \mu\Omega \cdot \text{cm}$).

Semi-insulators: Materials with resistivity values ranging from 10^3 to $10^{10}\ \Omega \cdot \text{cm}$.

Sheet resistance (R_s): The resistance of a square resistor of a certain resistivity and thickness.

Solid solution: A solid material in which one component (e.g., Cu) is completely dissolved in another (e.g., Ni)—similar to the complete dissolution of sugar in water.

Solid-solution strengthening: An effect in which the formation of a solid solution (i.e., the complete dissolution of one element into another) causes an increase in the yield stress. For example, a small concentration of Be in Cu increases the yield stress of Cu.

Spin-paired electrons: Electrons whose quantum numbers are identical, other than the spin quantum number, and that have spin directions opposite of each other (to satisfy Pauli's exclusion principle).

Spin quantum number (*s* or m_s): An electron quantum number that quantizes the spin; its values are $\pm\frac{1}{2}$.

Superconductors: Materials that can exhibit zero electrical resistance under certain conditions.

Temperature coefficient of resistivity (TCR): A coefficient designated as α_R and defined as

$$\alpha_R = \frac{1}{\rho_0}\left(\frac{\rho - \rho_0}{T - T_0}\right)$$

where ρ is the resistivity, T is the temperature, and ρ_0 is the resistivity at the reference temperature (T_0).

Temperature-dependent component of resistivity (ρ_T): The portion of total resistivity originating from the scattering of conduction electrons off vibrations of atoms (phonons).

Valence band: The lower band on a band diagram showing the energy levels associated with the valence electrons. This band is usually completely or nearly completely filled for metals and semiconductors.

Varistors: Materials or devices with voltage-dependent resistance.

Volume resistivity: See **Resistivity**.

Wave function: A set of quantum numbers (i.e., n, l, m, and s) that represent the wave function associated with an electron.

Work function ($q\phi$): The energy, in electron volts (eV), needed to excite an electron from E_F to the vacuum level, where it is essentially freed from the solid. Sometimes measured in ϕ (volts).

Yttria (Y_2O_3)-stabilized zirconia (ZrO_2) (YSZ): A zirconia ceramic doped with yttrium oxide; it has a cubic crystal structure and is an ionic conductor used in solid oxide fuel cells and oxygen gas sensors.

REFERENCES

Askeland, D. 1989. *The Science and Engineering of Materials*. 3rd ed. Washington, DC: Thomson.

Askeland, D., and P. Fulay. 2006. *The Science and Engineering of Materials*. Washington, DC: Thomson.

ASM International. 1990. *Properties and Selection: Nonferrous Alloys and Special Purpose Materials*. vol. 2. Materials Park, OH: ASM.

Chester, G.V., and A. Thellung. 1961. The Law of Wiedemann and Franz, *Proceedings of the Physical Society*, 77:1005–1013.

Davis, J. R. 1997. *Concise Metals Engineering Data Book*. Materials Park, OH: ASM.

Edwards-Shea, L. 1996. *The Essence of Solid-State Electronics*. Upper Saddle River, NJ: Prentice Hall.

Groover, M. P. 2007. *Fundamentals of Modern Manufacturing: Materials, Processes, and Systems*. New York: Wiley.

Gupta, T. K. 2003. *Handbook of Thick and Thin Film Hybrid Microelectronics*. New York: Wiley Interscience.

Haile, S. M. 2003. Fuel cell materials and component. *Acta Materilia* 51:5981–6000.

Kasap, S. O. 2002. *Principles of Electronic Materials and Devices*. New York: McGraw Hill.

Lampman, S. R., and T. B. Zorc, eds. 1990. *Metals Handbook: Properties and Selection: Nonferrous Alloys and Special Purpose Materials*. Materials Park, OH: ASM.

Laughton, M. A., and D. F. Warne. 2003. *Electrical Engineer's Reference Book*. Amsterdam: Elsevier.

Mahajan, S., and K. S. Sree Harsha. 1998. *Principles of Growth and Processing of Semiconductors*. New York: McGraw Hill.

Minges, M. L. 1989. *Electronic Materials Handbook. Vol. 1 of Electronic Packaging*. Materials Park, OH: ASM.

Neelkanta, P. 1995. *Handbook of Electromagnetic Materials*. Boca Raton, FL: CRC Press.

Paschal, J. M. 2001. *EC and M's Electrical Calculations Handbook*. New York: McGraw Hill.

Powerstream, http://www.powerstream.com/Wire_Size.htm.

Singh, J. 2001. *Semiconductor Devices: Basic Principles*. New York: Wiley.

Streetman, B. G., and S. Banerjee. 2000. *Solid State Electronic Devices*, 5th ed. Upper Saddle River, NJ: Prentice Hall.

Vines, R. F., and E. M. Wise. 1941. *The Platinum Metals and Their Alloys*. New York: International Nickel Co.

Webster, J. G. 2002. *Wiley Encyclopedia of Electrical and Electronics Engineering*, vol. 4. New York: Wiley.

3 Fundamentals of Semiconductor Materials

KEY TOPICS

- Origin of semiconductivity in materials
- Differences between elemental versus compound, direct versus indirect band gap, and intrinsic versus extrinsic semiconductors
- Band diagrams for n-type and p-type semiconductors
- Conductivity of semiconductors in relation to the majority and minority carrier concentrations
- Factors that affect the conductivity of semiconductors
- Device applications such as light-emitting diodes (LEDs)
- Changes in the band gap with temperature, dopant concentrations, and crystallite size (quantum dots)
- Semiconductivity in ceramic materials

3.1 INTRODUCTION

Semiconductors are defined as materials with resistivity (ρ) between ~10^{-4} and ~10^3 $\Omega \cdot$ cm. An approximate range of sensitivities exhibited by silicon (Si)-based semiconductors is shown in Figure 3.1. Elements showing semiconductivity are called *elemental semiconductors* (e.g., silicon), and the compounds that show semiconducting behavior are known as *compound semiconductors* (e.g., gallium arsenide [GaAs]). A band diagram of a typical semiconductor is shown in Figure 3.2. For most semiconductors, the band gap energy (E_g) is between ~0.1 and 4.0 eV. If the band gap is larger than 4.0 eV, we usually consider the material to be an insulator or a dielectric. In this chapter, we will learn that the composition of such dielectric materials can be altered so that they exhibit semiconductivity (see Section 3.21). As shown in Figure 3.2, the top of the valence band is known as the *valence band edge* (E_v), and the bottom of the conduction band is known as the *conduction band edge* (E_c). Recall from Chapter 2 that the band diagram shows the outermost part of the overall electron energy levels of a solid. The vertical axis shows the increasing electron energy. Thus, the magnitude of the band gap energy (E_g) is given by

$$E_g = E_c - E_v \tag{3.1}$$

3.2 INTRINSIC SEMICONDUCTORS

Let us consider the origin of the semiconducting behavior in semiconductors such as silicon. Silicon has covalent bonds; each silicon atom bonds with four other silicon atoms, which leads to the formation of a three-dimensional network of tetrahedra arranged in a diamond cubic crystal structure (Figure 3.3).

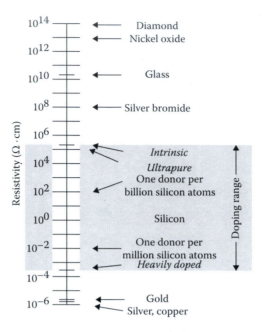

FIGURE 3.1 Approximate range of sensitivity for silicon in comparison with other materials. (From Queisser, H.J. and Haller, E.E., *Science*. 281, 945–950, 1998. With permission.)

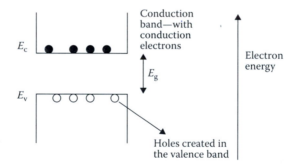

FIGURE 3.2 A band diagram for a typical semiconductor.

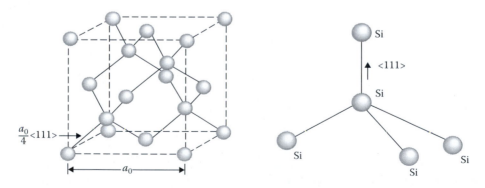

FIGURE 3.3 The diamond cubic crystal structure of silicon, showing the tetrahedral arrangement of silicon atoms. Each silicon atom is bonded to four other silicon atoms. The lattice constant is a_0. (From Mahajan, S. and Sree Harsha, K.S., *Principles of Growth and Processing of Semiconductors*, McGraw Hill, New York, 1998. With permission.)

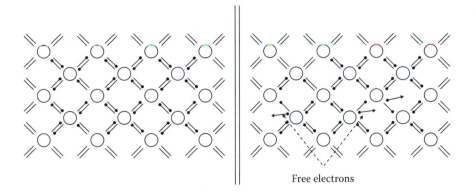

Free electrons

FIGURE 3.4 Two-dimensional representation of silicon bonding. (From Kano, K., *Semiconductor Fundamentals*, Prentice Hall, Upper Saddle River, NJ, 1997. With permission.)

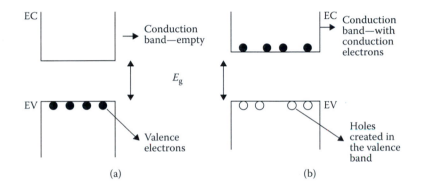

FIGURE 3.5 Typical band diagram for an intrinsic semiconductor at (a) low temperatures and (b) high temperatures.

Figure 3.4 shows a two-dimensional representation of the covalent bonds among silicon atoms. When the temperature is low (~0 K), the valence electrons shared among the silicon atoms remain in the bonds and are not available for conduction. Thus, at low temperatures, silicon *behaves* as an insulator (Figure 3.4). As the temperature increases, the electrons gain thermal energy. Note that thermally excited valence electrons have a wide distribution of energy states. A small number of these electrons gain sufficient energy to break away from the covalent bonds, which is much higher than the average energy of valance electrons, and no longer participate in the Si–Si covalent bonding. Instead, they move to the location for electrons in the *free states* (the conduction band) and travel within that band. These electrons in the conduction band (or holes in the valence band) impart semiconductivity to the material (Figure 3.4; also Section 2.14).

The electrons breaking away from the bonds (Figure 3.4) can also be shown on a band diagram (Figure 3.5). At low temperatures, the valence electrons are in the covalent bonds, that is, the *valence band* is completely filled. As the temperature increases to ~>100 K, a small fraction of electrons gain enough thermal energy to make a jump across the band gap (E_g) and into the conduction band.

When an electron breaks away from a covalent bond, it leaves behind an incompletely filled band and a *hole*; the latter is an imaginary particle that represents a missing electron from a bond (Figure 3.6). On a band diagram, a hole is an energy state left empty by an electron that moves to the conduction band. If a hole is present at Site X in one of covalent bonds, then another electron from a neighboring bond at Site Y can move into Site X. This creates a hole at

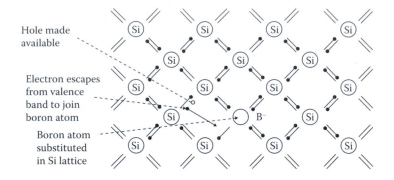

Hole made available

Electron escapes from valence band to join boron atom

Boron atom substituted in Si lattice

FIGURE 3.6 Creation of electron–hole pairs by thermal excitation. (From Kano, K., *Semiconductor Fundamentals*, Prentice Hall, Upper Saddle River, NJ, 1997. With permission.)

Site Y. Movement of an electron from Site Y to Site X is equivalent to the movement of a hole from Site X to Site Y. Thus, the movement of holes in the valence also contributes to a semiconductor's electrical conductivity.

In materials such as silicon and germanium (Ge), the band gap energy at room temperature is relatively low (E_g of Ge and Si are ~0.67 and ~1.1 eV, respectively). When we say the band gap is small, we are comparing the band gap energy with the thermal energy given by $k_B T$, where k_B is the Boltzmann's constant (8.617×10^{-5} eV/K or 1.38×10^{-23} J/K) and T is the temperature. At $T = 300$ K (~room temperature), the thermal energy $k_B T$ is ~0.026 eV.

Promoting an electron into the conduction band creates an *electron–hole pair* (EHP; Figures 3.5 and 3.6). Electrons promoted into the conduction band because of thermal energy and the resulting holes that are created in the valence band are known as *thermally generated charge carriers*.

A semiconductor in which the thermal energy is only a source of charge carrier generation is known as an *intrinsic semiconductor*. The word *intrinsic* emphasizes that no other *extrinsic* or foreign elements or compounds are present in significant enough concentrations to have any effect on the electrical properties of an intrinsic semiconductor. Appropriately, a semiconductor whose conductivity and other electrical properties are controlled by foreign elements or compounds is known as an *extrinsic semiconductor* (see Section 3.8). If the impurity concentrations and defect densities of the semiconductor are negligible, the semiconductor remains an intrinsic semiconductor. Note that, even if EHPs are created, the material still remains electrically neutral. In an intrinsic semiconductor, the concentration of the electrons available for conduction (n_i) is equal to that of the holes created (p_i).

$$n_i = p_i \tag{3.2}$$

Therefore, Figures 3.2 and 3.5 show that the electron concentration in the conduction band and the hole concentration in the valence band are equal. In this case, the carrier concentration of intrinsic semiconductors for conduction (Equation 2.22) is controlled by the temperature and the band gap (Equation 3.5).

The conductivity of the intrinsic semiconductor is given by the following equation:

$$\sigma = q \times n_i \times \mu_n + q \times p_i \times \mu_p \tag{3.3}$$

In this equation, q is the magnitude of the charge on the electron or hole (1.6×10^{-19} C), and n_i and p_i are the concentrations of electrons and holes in an intrinsic material, respectively. The terms

μ_n and μ_p are the mobilities of electrons and holes, respectively. Since $n_i = p_i$ for the intrinsic semiconductor, Equation 3.3 can be rewritten as

$$\sigma = q \times n_i \times (\mu_n + \mu_p) \tag{3.4}$$

For a given intrinsic semiconductor, the electron or hole concentrations depend mainly on the temperature.

Example 3.1 illustrates the calculation of the resistivity of an intrinsic semiconductor.

Example 3.1: Resistivity of Intrinsic Germanium

What is the resistivity (ρ) of essentially pure Ge? Assume that the mobilities of the electrons and the holes in Ge at 300 K are 3900 and 1900 cm²/V · s, respectively. Assume $T = 300$ K and $n_i = 2.5 \times 10^{13}$ cm⁻³.

Solution
We make use of Equation 3.4

$$\sigma = \left(1.6 \times 10^{-19} C\right)\left(2.5 \times 10^{13} \frac{\text{carriers}}{\text{cm}^3}\right)(3900 + 1900)\, \text{cm}^2/V \cdot s$$

Therefore,

$$\sigma = 0.0232 \text{ S/cm at } 300 \text{ K}$$

The inverse of this is the resistivity (ρ) at 300 K = 43.1 $\Omega \cdot$ cm.

3.3 TEMPERATURE DEPENDENCE OF CARRIER CONCENTRATIONS

As we can expect, at any given temperature, the larger the band gap (E_g) of a semiconductor, the lower the concentration of valence electrons (n_i) that can pass across the band gap. The relationship among the carrier concentration, the band gap, and the temperature is given by

$$n_i \propto T^{3/2} \exp\left(-\frac{E_g}{2k_B T}\right) \tag{3.5}$$

We will derive Equation 3.5 in Section 4.3 using a correlation between the electron concentration and the Fermi energy level. The exponential term dominates; hence, the plot of $\ln(n_i)$ with $1/T$ is essentially a straight line. Note that even though the increase is exponential, only a very small fraction of the *total number of valence electrons* actually gets into the conduction band. In Figure 3.7, a plot of n_i on a logarithmic scale is shown as a function of the inverse of the temperature. The slope of the line essentially is proportional to the band gap (E_g). In fact, measuring the carrier concentration as a function of temperature is one way to find out E_g of semiconductors. The band gaps of germanium, silicon, and GaAs are ~0.67, 1.1, and 1.43 eV, respectively. From Equation 3.5, we expect the concentration of free electrons at a given temperature to be the highest for germanium because it has the smallest band gap of the three. On the other hand, GaAs will have the lowest concentration of thermally generated conduction electrons because it has the largest band gap of these three materials (Figure 3.7). The band gap values and some of the other properties of semiconductors are shown in Table 3.1. The terms *direct* and *indirect* band gap semiconductors, used in Table 3.1, are defined in Section 3.5.

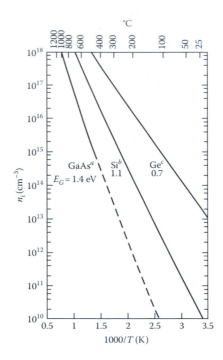

FIGURE 3.7   Intrinsic carrier concentration plotted on a log scale as a function of the inverse of temperature for germanium, silicon, and gallium arsenide. (From Grove, A.S., *Physics and Technology of Semiconductor Devices*, Wiley, New York, 1967. With permission.)

TABLE 3.1
Properties of Selected Semiconductors

Semiconductor (Band gap Type i: indirect, d: direct)	Band gap (E_g) (eV) at $T = 300$ K	Mobility of Electrons $(cm^2/V \cdot s)$ at 300 K (μ_n)	Mobility of Holes $(cm^2/V \cdot s)$ at 300 K (μ_p)	Electric Breakdown Field (V/cm)
Carbon as diamond	5.47	800	1200	10^7
Germanium (i)	0.67	3900	1900	10^5
Silicon (i)	1.12	1500	450	3×10^5
Amorphous silicon (a-Si:H) (i)	1.7–1.8	1	10^{-2}	—
SiC (α-form) (i)	2.996	400	50	2–3×10^6
GaSb (d)	0.72	5000	850	5×10^4
GaAs (d)	1.42	8500	400	4×10^5
GaP (i)	2.26	110	75	~ 5–10×10^5
InSb (d)	0.17	8000	1250	10^3
InP (d)	1.35	4600	150	5×10^5
CdTe (d)	1.56	1050	100	—
PbTe (i)	0.31	6000	4000	—

3.4   BAND STRUCTURE OF SEMICONDUCTORS

In previous sections, we learned that the energy levels are quantized and that each electron energy state is taken by one electron (Pauli's exclusion principle). What we will study here is the effect of the energy levels on the wave-like and particle-like properties of electrons.

We will start from the standpoint of the particle-like properties. In the classical theory of conductivity, electrons are considered particles. The energy (E) of a free electron can be written as:

$$E = \frac{1}{2}mv^2 \tag{3.6}$$

where m is the mass of a free electron, and v is the velocity of the electron. Here, momentum (p) is defined as mass × velocity ($p = mv$). Therefore, we can also rewrite Equation 3.6 in terms of momentum (p) and mass as follows:

$$E = \frac{p^2}{2m} \tag{3.7}$$

Note that we have also used the symbol "p" to designate the concentration of holes. The meaning of Equations 3.6 and 3.7 is that electrons taking different energy levels move at different velocity and have different momentum.

The next question is whether this difference in the velocity and the momentum influences the wave-like properties of electrons. As discussed in Section 2.13, in a quantum mechanics–based approach, an electron is considered a plane wave. According to Equation 2.33 on de Broglie waves, the wavelength of the electron wave is related to the momentum and velocity of the electron ($\lambda = h/p = h/mv$). This suggests that electrons occupying different energy levels have different electron wavelengths as well as different momentum. To quantitatively express the wave-like properties of the electron including the wavelength, a new propagation parameter, *wave vector* ($\mathbf{k}$) is introduced. The absolute magnitude of $\mathbf{k}$ and the wavelength (λ) of the electron wave are related as $|\mathbf{k}| = (2\pi)/\lambda$, which indicates that the wave vector ($\mathbf{k}$) shows the moving direction and the wavelength of the propagating electron wave.

Using quantum mechanics, we can also show that the electron momentum (p) is related not only to the velocity of the particle-like electron but also to the wave vector ($\mathbf{k}$) of the wave-like electron. If you combine Equation 2.33 ($\lambda = h/p$) and a definition of wave vector ($\mathbf{k} = 2\pi/\lambda$), you get the following relation that connects the momentum and wave vector of the electron:

$$p = \hbar\mathbf{k} \tag{3.8}$$

where $\hbar = h/2\pi = 1.054 \times 10^{-34}$ J · s, and h is the Planck's constant ($= 6.626 \times 10^{-34}$ J · s).

From Equations 3.7 and 3.8, we can write the energy of an electron as:

$$E = \frac{\hbar^2\mathbf{k}^2}{2m} \tag{3.9}$$

This shows how the wave vector (or wavelength) of the free electron wave determines the electron energy. For a free electron that does not experience any other force due to internal or external electric or magnetic fields, the relationship between its energy (E) and wave vector ($\mathbf{k}$) is a parabola with a minimum energy of zero and V_0 (Figure 3.8). The plot of electron energy as a function of the wave vector ($\mathbf{k}$) is known as the *band structure* of a material, which shows that each energy level (E) is connected to two wave vectors ($\mathbf{k}$, $-\mathbf{k}$). In Figure 3.8b, the electron energy at $k = 0$ is not zero. However, because V_0 is constant for all k values, the electron of Figure 3.8b is still considered an essentially free electron with starting energy V_0. (i.e., it is not a bound electron).

In Equations 3.6–3.9, we reviewed the relation between the energy level and the wave vector (or wavelength) for free electrons. However, truly free electrons are only in the vacuum states, and even valence electrons are still exposed to the influence of the nucleus and the neighbor electrons (recall the learnings from Section 2.12 and Section 2.17). Therefore, we need to take into account attractive and repulsive interactions that electrons meet when they travel inside materials.

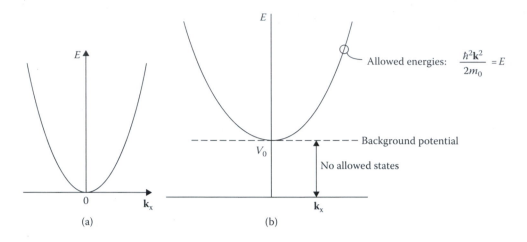

FIGURE 3.8 The band structure or energy (E) versus wave vector ($\mathbf{k}$) for (a) a free electron and (b) an essentially free electron in a band with starting energy V_0. (From Mahajan, S. and Sree Harsha K.S., *Principles of Growth and Processing of Semiconductors*, McGraw Hill, New York, 1998. With permission.)

Since the major source is the electron–nucleus interaction, electrons moving inside materials experience a periodic interaction that is determined by the atomic arrangement (i.e., crystal structure). To describe the quantum mechanical analysis of this periodic interaction, we can use a wave function that Schrödinger introduced to describe the electron wave.

$$\Psi_{(\mathbf{k_x}x)} = U(\mathbf{k_x},x)\exp(j\mathbf{k_x}x) \tag{3.10}$$

where Ψ is the wave function that is related to the probability of finding an electron. In the quantum mechanics–based approach, $\Psi_{(\mathbf{k_x}x)}\Psi_{(\mathbf{k_x}x)}^*$ is the probability of finding an electron with the certain wave vector ($\mathbf{k_x}$) at a specific location (x). In Equation 3.10, note that U is the function that accounts for the periodicity of the crystal structure, j is the imaginary number, $\mathbf{k_x}$ is the wave vector in the x-direction along which the electron (now considered a wave) is traveling, and x represents the electron location. If the electron is completely free from the interactions with the nucleus, there is no periodic potential influencing the electron motion, and $U(\mathbf{k_x},x)$ becomes a constant. Then, the wave function turns to $\Psi_{(\mathbf{k_x}x)} = A\exp(j\mathbf{k_x}x)$, where A is a constant.

In a perfect crystal, if we assume that an electron moving in a certain band has the wave vector $\mathbf{k}$ and experiences a periodic potential energy of $U(x)$, then the energy of the electron is given by:

$$E(x) = \frac{\hbar^2\mathbf{k}^2}{2m} + U(x) \tag{3.11}$$

Equation 3.11 means that electrons moving in the conduction band of the semiconductors are not completely free. Traveling electrons meet built-in electric fields that are associated with other periodically arranged atoms. Since the electron arrangements in the lattice are not the same in all directions, electrons moving in different directions experience different built-in electric fields [i.e., $U(x)$]. For example, the interatomic distance is d for electrons moving along the <100> direction of a simple cubic cell with a lattice constant d. However, if electrons move along the <111> direction in the same simple cubic cell, the interatomic distance changes to $\sqrt{3}d$.

The important consequence of the dependence of $U(x)$ on the moving direction is that electrons exposed to different $U(x)$ move around in a material as if they have different masses; this is known as the *effective mass of an electron* (m_e^* or m_n^*). The effective mass is the mass that an electron would *appear* to have in a material by responding to interactions with neighbor particles such as the nucleus and the electron. It is different from the mass of a free electron in a vacuum ($m_0 = 9.109 \times 10^{-31}$ kg). Imagine that you measure the weight of the object in air or in water. The object's weight is much reduced in water due to buoyancy, which is somewhat analogous to electron–nucleus interactions. Depending on the buoyance (analogous to the electron–nucleus interaction), the weight (analogous to the effective mass) will change. It is important to know that a change in the effective mass of an electron or hole is reflected in an E v. $\mathbf{k}$ curve. In other words, even if the wave vectors of two electron waves are the same, they may or may not possess the same energies. When two electron waves propagate along different directions, they experience different electrostatic interactions with the surrounding change and the energies of the electron waves become different. The effective mass is a parameter showing how strongly the electron or the hole is bound to the neighbor nuclei.

The concept of effective mass is illustrated in Figure 3.9. Because of the electron–nucleus interaction, the acceleration velocity of the electron by an external force such as an electric field may vary. For convenience, we will express the effective mass as a dimensionless quantity, m_e^* / m_0. This ratio is an indicator of the interactions of the electron with the atoms of the material. Similarly, holes also have an effective mass. The significance of the effective mass is as follows: Smaller effective masses for carriers (electrons or holes) mean that the carriers can move faster with less apparent inertia. This means materials with smaller apparent electron or hole masses are more useful for making faster semiconductor devices. The concept of effective mass is quantum mechanical in nature, and analogies using classical mechanics must therefore be limited in scope.

The effective masses of electrons and holes for some semiconductors are listed in Table 3.2. This table also lists the values of the *dielectric constant* (ε_r), which is a measure of the ability of a material to store a charge (see Chapter 7). The dielectric constant is defined as the ratio of the permittivity of a material (ε) and the permittivity of the free space (ε_0).

As discussed before, the E–$\mathbf{k}$ relation in Figure 3.8 depends on the effective mass (m_e^*, m_h^*) of an electron and a hole. This means that m_e^* and m_h^* can be deduced from the E–$\mathbf{k}$ curvature as follows (refer to Equation 3.9 and Figure 3.8):

$$m_e^* = \frac{\hbar^2}{\left(\dfrac{d^2 E}{d\mathbf{k}^2} \right)} \tag{3.12}$$

Since a hole is an imaginary particle that represents a missing electron in the valence band, the effective mass of the hole is the negative of the mass of the missing electron. The effective mass of holes is given by:

$$m_h^* = -\frac{\hbar^2}{\left(\dfrac{d^2 E}{d\mathbf{k}^2} \right)} \tag{3.13}$$

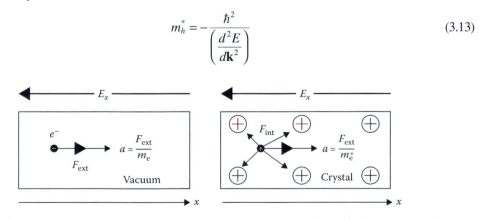

FIGURE 3.9 Illustration of the difference between the mass of an electron in a vacuum (m_0) and its effective mass m_e^*. (From Kasap, S.O., *Principles of Electronic Materials and Devices*, McGraw Hill, New York, 2002. With permission.)

TABLE 3.2

Effective Masses of Electrons and Holes and Other Properties of Different Semiconductors

Semiconductor (Band gap Type, i: indirect, d: direct)	Band gap (eV) (T = 300 K)	Mobility of Electrons at 300 K (μ_n) in cm²/V · s	Mobility of Holes at 300 K (μ_p) in cm²/V · s	Effective Mass of Electrons m_l^*/m_0 (m_l^*, m_t^*)	Effective Mass of Holes m_p^*/m_0 (m_{lh}, m_{hh})	Density (g/cm³)	Dielectric Constant (ε_r)	Lattice Constant (Å)	Melting Point (°C)
Aluminum arsenide (AlAs) (i)	2.16	1200	420	2.0	0.15, 0.76	3.60	10.9	5.66	1740
Aluminum phosphide (AlP) (i)	2.45	80	—	—	0.2, 0.63	2.4	9.8	5.46	2000
Amorphous silicon (a-Si:H)	1.7	1	10⁻²						
Cadmium telluride (CdTe) (d)	1.56	1050	100	0.1	0.37	6.2	10.2	6.482	1098
Carbon (C) as diamond	5.47	800	1200	1.4, 0.36 (at 85 K)	0.7, 2.12 (at 1.2 K)	3.51	5.7	3.566	
Gallium antimonide (GaSb) (d)	0.72	5000	850	0.042	0.06, 0.23	5.61	15.7	6.09	712
Gallium arsenide (GaAs) (d)	1.42	8500	400	0.067	0.08, 0.45	5.31	13.2	5.65	1238
Gallium phosphide (GaP) (i)	2.26	110	75	1.12, 0.22	0.14, 0.79	4.13	11.1	5.45	1467
Gallium nitride (GaN) (d)	3.47	380	—	0.19	0.60	6.1	12.2	4.5; a = 3.189; c = 5.185 (for wurtzite structure)	2530
Germanium (Ge) (i)	0.67	3900	1900	1.64, 0.082	0.04, 0.28	5.32	16	5.646	936
Indium antimonide (InSb) (d)	0.17	8000	1250	0.014	0.015, 0.4	5.78	17.7	6.474	525
Indium arsenide (InAs) (d)	0.36	~2 × 10⁴	~500	0.027	0.025, 0.41	5.68	15	6.083	1070
Indium phosphide (InP) (d)	1.35	4600	150	0.077	0.089, 0.85	4.79	12.4	5.87	
Lead telluride (PbTe) (i)	0.31	6000	4000	0.17	0.20	8.16	30	6.452	925
Silicon (Si) (i)	1.12	1500	450	0.98, 0.19	0.16, 0.49	2.33	11.8	5.430	1415
Silicon carbide SiC (α-form) (i)	2.996	400	50	0.6	1.0	3.21	10.2	3.08	2830
Zinc sulfide (ZnS) (d)	3.6	180	10	0.28	—	4.09	8.9	5.409	1650 (vaporizes)
Zinc selenide (ZnSe) (d)	2.7	600	28	0.14	0.60	5.65	9.2	5.671	1100 (vaporizes)
Zinc telluride (ZnTe) (d)	2.25	530	100	0.18	0.65	5.51	10.4	6.101	1238 (vaporizes)

Note: m_l^* and m_t^* are known as the longitudinal and translational effective masses; m_{lh}^* and m_{hh}^* are the light- and heavy-hole effective masses.

Note the negative sign in Equation 3.13 because the hole is a missing electron. Equations 3.12 and 3.13 also indicates that an electron and a hole may have the different effective mass.

The effective mass of an electron and the E–$\mathbf{k}$ curvature (Equation 3.12) is derived as follows: Since the electron can be expressed as a group of waves (we will briefly cover this concept with Figure 8.7), the electron velocity is equal to the group velocity of the associated wave (v_g), with which the boundary of the wave propagates. The group velocity is given by the following equation:

$$v_g = \frac{d}{d\mathbf{k}}(2\pi v) \tag{3.14}$$

where v is the frequency of the wave.

We rewrite Equation 3.14 as:

$$v_g = \frac{d}{d\mathbf{k}}\left(\frac{2\pi h v}{h}\right) \tag{3.15}$$

Now, we replace hv with E and $(h/2\pi)$ with $\hbar$ and we get:

$$v_g = \frac{1}{\hbar}\left(\frac{dE}{d\mathbf{k}}\right) \tag{3.16}$$

The acceleration (a) of an electron is given by:

$$a = \left(\frac{dv_g}{dt}\right) \tag{3.17}$$

Substituting v_g from Equation 3.16 into Equation 3.17, we get:

$$a = \frac{1}{\hbar}\left(\frac{d^2E}{d\mathbf{k}^2}\right)\left(\frac{d\mathbf{k}}{dt}\right) \tag{3.18}$$

Substituting $\mathbf{k}$ with $p/\hbar$, we get:

$$a = \frac{1}{\hbar^2}\left(\frac{d^2E}{d\mathbf{k}^2}\right)\left(\frac{dp}{dt}\right) \tag{3.19}$$

We rewrite Equation 3.19 as:

$$a = \frac{1}{\hbar^2}\left(\frac{d^2E}{d\mathbf{k}^2}\right)\left(\frac{d(mv)}{dt}\right) = \frac{1}{\hbar^2}\left(\frac{d^2E}{d\mathbf{k}^2}\right)F \tag{3.20}$$

From classical mechanics, we know that $F = m \times a$, or $a = F/m$. Comparing this with Equation 3.20, we obtain an expression for the effective mass of an electron as given in Equation 3.12. We can see from Equation 3.12 that the effective mass of an electron is inversely related to the E–$\mathbf{k}$ curvature. Note that different E–$\mathbf{k}$ curves in Figure 3.11 result in different effective hole mass. The larger the curvature, the smaller the effective mass.

For the conduction band of a semiconductor, we can write the relationship between the energy of an electron (E) and its wave vector $\mathbf{k}$ as follows:

$$E(\mathbf{k}) = E_c + \frac{\hbar^2 \mathbf{k}^2}{2m_e^*} \tag{3.21}$$

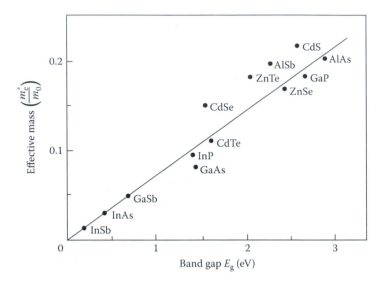

FIGURE 3.10   The relationship between electron effective mass $\left(m_e^*\right)$ and band gap (E_g). (From Singh, J: *Semiconductor Devices: Basic Principles.* 2001. Copyright Wiley-VCH Verlag GmbH & Co. KGaA. Reproduced with permission.)

where E_c is the conduction band edge energy and m_e^* is the effective mass of the electron. The effective mass of an electron depends strongly on the band gap (E_g). Since the band gap represents an energy difference between two electron states (the free state and the *bound state*), the larger band gap means a stronger electron–nucleus interaction (Section 2.14). Therefore, the smaller the value of the band gap (E_g), the smaller the value of m_e^* (Table 3.2 and Figure 3.10).

3.5   DIRECT AND INDIRECT BAND GAP SEMICONDUCTORS

The E–$\mathbf{k}$ diagrams provide one more important way to classify semiconductors. Figure 3.11 shows the band structure of a *direct band gap semiconductor.* In the E–$\mathbf{k}$ diagrams of Figure 3.11, readers need to pay attention to flipped parabolic curves at the bottom. These curves are the E–$\mathbf{k}$ diagrams of holes created in the valence band. Since electrons and holes have opposite charge signs, the shapes of the E–$\mathbf{k}$ diagrams are not the same for electrons and holes (refer to Equations 3.12 and 3.13). Note that the upper states in the E–$\mathbf{k}$ diagram correspond to higher energy levels for electrons (i.e., the farther from E_c, the higher electron energy). In contrast to electrons, holes with higher energy are found in the lower states of the valence band. In a group of semiconductors called as a direct band gap semiconductor, the maximum electron energy state in the valence band and the minimum electron energy state in the conduction band are found at the same wave vector ($\mathbf{k}$); an electron in the valence band can move into the conduction band if it has sufficient energy to cross the band gap (E_g). This process does *not* require a change in the momentum of the electron, because $\mathbf{k}$ of the moving electron does not change.

In a group of semiconductors called as an *indirect band gap semiconductor,* the valence band's maximum electron energy state and the conduction band's minimum electron energy state do not coincide at the same wave vector ($\mathbf{k}$). Hence, an electron jumping from the valence band to the conduction band needs to change its wave vector (or momentum) in addition to gaining the energy corresponding to E_g. Silicon and germanium are examples of indirect band gap semiconductors. The actual band structures of silicon, germanium, GaAs, and aluminum arsenide (AlAs) are more complex (Figure 3.12), and their analysis is beyond the scope of this book.

In many materials based on alloys of two or more semiconductors, the band structure can change from indirect to direct and vice versa. For example, GaAs is a direct band gap semiconductor.

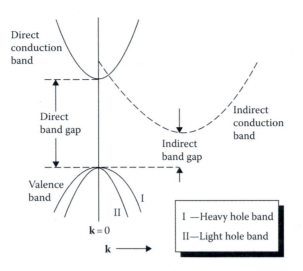

FIGURE 3.11   Band structure or E–$\mathbf{k}$ relationships for a direct and indirect band gap semiconductor. (From Singh, J: *Semiconductor Devices: Basic Principles.* 2001. Copyright Wiley-VCH Verlag GmbH & Co. KGaA. Reproduced with permission.)

When we form a solid solution with gallium phosphide (GaP; an indirect semiconductor), the band gap of $GaAs_{1-x}P_x$ remains direct up to phosphorus (P) mole fractions of ~$x = 0.45$–0.50. Beyond this (i.e., $x > 0.5$), the band gap becomes indirect until we reach GaP (Figure 3.13). Another important energy level in Figure 3.13 is associated with the nitrogen atom that is doped into GaP. The significance of the electron energy level of doped N will be discussed in Section 3.6.

3.6   APPLICATIONS OF DIRECT BAND GAP MATERIALS

Direct band gap materials exhibit strong interactions with energy in the form of light waves. Because of this, direct band gap materials are used to create *optoelectronic devices.* Electrons can move from the valence band to the conduction band by absorbing light with the energy equal to or larger than E_g. After an electron is excited into the conduction band, it falls back to the valence band by recombining with a hole. If the energy of the recombination reaction is released in the form of light, this process is known as *radiative recombination.*

$$\text{Electron} + \text{hole} \xrightarrow{\text{radiative recombination}} \text{photon}(\text{light}) \tag{3.22}$$

The radiative recombination process occurring in direct band gap materials enables the operation of light-emitting devices known as LEDs.

In certain conditions, the recombination of the electrons with the holes may not produce light. Instead, the energy takes the form of vibrations of the atoms known as phonons and appears as heat. This process is known as *nonradiative recombination.*

$$\text{Electron} + \text{hole} \xrightarrow{\text{nonradiative recombination}} \text{photon}(\text{heat}) \tag{3.23}$$

The process of radiative recombination without a change in the momentum is known as *vertical recombination* (Figure 3.14). Note that radiative recombination also occurs in some indirect band gap materials when an electron from the conduction band comes back to the valence band via a defective energy level. Nonradiative recombination (Figure 3.15a) occurs in direct band gap materials as well; however, we can minimize it by using semiconductor materials that have very few point defects, such as vacancies or interstitials, or other defects, such as dislocations. This will improve

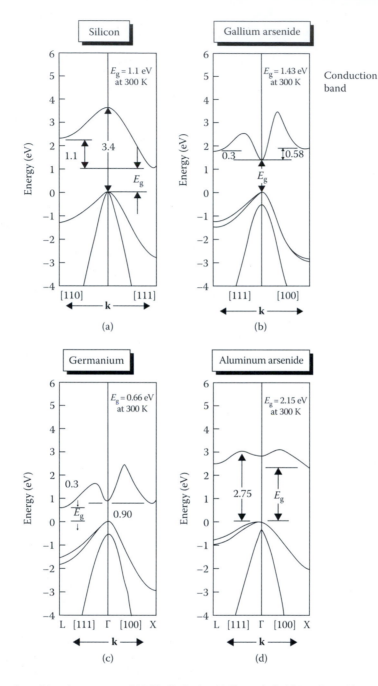

FIGURE 3.12   Actual band structures of (a) Si, (b) GaAs, (c) Ge, and (d) AlAs. (From Singh, J., *Semiconductor Devices: An Introduction*, McGraw Hill, New York, 1994. With permission.)

the LED efficiency. Nonradiative recombination also occurs at semiconductor surfaces that have incomplete or dangling bonds. Recombination dynamics is one of the factors that limits how rapidly an LED can be turned on and off.

The frequency (v) of light emitted is related to the band gap energy (E_g):

$$h v = E_g \qquad\qquad (3.24)$$

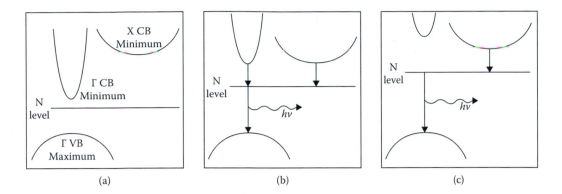

FIGURE 3.13   (a) Direct band gap of GaAs; (b) direct band gap of $GaAs_{0.5}P_{0.5}$; and (c) indirect band gap of GaP. The relative level of N dopant added for optoelectronic applications is also shown. (From Schubert, F.E., *Light-Emitting Diodes*, Cambridge University Press, Cambridge, UK, 2006. With permission.)

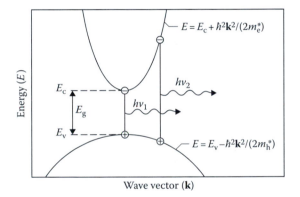

FIGURE 3.14   Vertical recombination and photon emission in a direct band gap semiconductor. (From Schubert, F.E., *Light-Emitting Diodes*, University Press Cambridge, Cambridge, UK, 2006. With permission.)

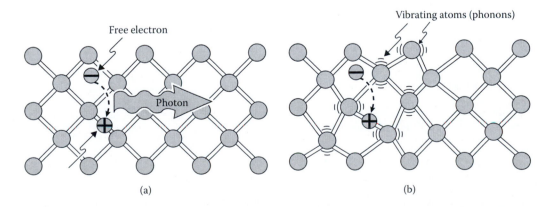

FIGURE 3.15   (a) Radiative recombination of an electron–hole pair accompanied by the emission of a photon with energy $h\nu \approx E_g$. (b) In nonradiative recombination events, the energy released during the electron–hole recombination is converted to phonons. (From Schubert, F.E., *Light-Emitting Diodes*, University Press Cambridge, Cambridge, UK, 2006. With permission.)

Now, if c is the speed of light and λ is its wavelength, then

$$c = v \times \lambda \tag{3.25}$$

Therefore, from Equations 3.24 and 3.25:

$$E_g = \frac{hc}{\lambda} \tag{3.26}$$

This relationship forms the basis for correlating the wavelength of light emitted from LEDs. Example 3.2 illustrates how Equation 3.26 can be converted into a more useful form.

Example 3.2: Wavelength of Light Emitted from LEDS

Develop a relationship between the wavelength of light (λ in micrometers) emitted from an LED and the band gap (in electron volts).

Solution
We can simplify Equation 3.26 to make it more practical for applications as follows:

$$E_g \text{ (in eV)} = \frac{\left(4.14 \times 10^{-15} \text{eV} \cdot \text{s}\right)\left(2.998 \times 10^{8} \text{m/s}\right)}{\lambda}$$

$$\lambda \text{(in m)} = \frac{12.41172 \times 10^{-7}}{E_g \text{ (in eV)}}$$

$$\therefore \lambda \text{(in μm)} = \frac{12.41172 \times 10^{-7}}{E_g \text{ (in eV)}} \text{ m} \times 10^{-6} \text{μm/m}$$

Therefore, the wavelength of light emitted from an LED is given by:

$$\lambda \text{(in μm)} = \frac{1.24}{E_g \text{ (in eV)}} \text{ μm} \tag{3.27}$$

Example 3.3 illustrates an application of optoelectronic materials.

Example 3.3: Light Emission from an LED

An LED is made using gallium nitride (GaN; $E_g = 3.47$ eV at $T = 0$ K). What is the wavelength and the color of light emitted from this semiconductor LED?

Solution
We use Equation 3.27 to calculate the wavelength:

$$\lambda \text{(in μm)} = \frac{1.24}{E_g \text{ (in eV)}} \text{ μm}$$

$$\therefore \lambda \text{(in μm)} = \frac{1.24}{3.47} \text{ μm} = 0.357 \text{ μm}$$

In nanometers, this value is 0.357 × 1000 = 357 nm, which coincides with the wavelength of UV light. We can adjust the composition of semiconductors, which causes the band gap to change. In this case, indium nitride (InN) can be alloyed with GaN to form blue lasers and LEDs that emit at λ = 470 nm. Such LEDs are found in many modern-day electronic products such as CD players, video games, and controllers. Further details of LEDs and the wavelength–color relation can be found in Chapters 6 and 8.

Indirect band gap materials such as silicon or germanium normally cannot be used for making semiconductor lasers or LEDs. This is because a radiative recombination of electrons and holes is less probable than a nonradiative recombination. In indirect band gap materials, electron–hole recombination generally results in generation of heat.

However, we cannot totally dismiss the possibility of using indirect band gap materials. In some cases, we can use an indirect band gap material for optoelectronic devices. For example, GaP is an indirect band gap material. When it is doped with nitrogen (N), a defect level is created deep in the band gap (Figure 3.13). This defect level then makes radiative recombination possible. The transition is between the N level and Γ VB (Figure 3.13c). We will learn in Section 3.8 that a *dopant* is an element or a compound deliberately added to enhance the electrical or other properties of a semiconductor. Since nitrogen has the same valence as phosphorous (i.e., +5), the doping of GaP with nitrogen is an example of *isoelectronic doping* which does not change the electron (hole) concentration of the semiconductor (Section 3.11).

The isoelectronic doping of GaP with nitrogen provides a practical application of Heisenberg's uncertainty principle. Since the electron wave function of nitrogen atoms is highly *localized* near nitrogen (small Δx), a possible range of the electron momentum broadens (large Δp). As a result, two different transitions can occur sequentially (the conduction band edge → the nitrogen impurity level → the valence and edge) (Figure 3.13). The first transition to the nitrogen impurity level is nonradiative. However, the second transition from the nitrogen level to the valence band results in emission of light (Figure 3.13c). The change in the electron momentum during the electron transition from the conduction band labeled X to the valence band labeled Γ is absorbed by the isoelectronic nitrogen atom.

Another way to modify indirect band gap characteristics is to use nanostructured or amorphous materials. For example, Si nanocrystals exhibit more direct band gap characteristics and a higher band gap than Si *bulk* due to the quantum confinement effect (or a change in the density of states). This shows that crystal structure as well as material composition are important factors that control the band structure.

3.7 MOTIONS OF ELECTRONS AND HOLES: ELECTRIC CURRENT

In this section, we will explain the motion of electrons and holes and their effects on electric conductivity using the band diagrams of electrons and holes (see Figures 3.11 and 3.12). First, let us think about the motion of an electron–hole pair of materials that are not electrically biased. When a valence electron is excited to the conduction band, the electron in the conduction band and the hole in the valence band have the same wave vector and momentum (Figure 3.16a). If their effective masses are the same, two oppositely charged particles with the same momentum move along the same direction with the same velocity. This means that the effect of the electron motion cancels out that of the hole motion and no electric current flows. Second, consider two electrons occupying the same energy in the valence band of metals (Figure 3.16b). If Electron 1 is moving with wave vector $\mathbf{k}_x$ and velocity v_x, then the effect of its motion (in terms of current generation) is nullified by the other electron (Electron 2) moving with wave vector $-\mathbf{k}_x$ and velocity $-v_x$. Since the wave vector ($\mathbf{k}$) is proportional to the momentum (p) of the electrons (Equation 3.9), the velocities of these two electrons are opposite. In metals, energy levels below E_F are fully filled. There are always two electrons that have the same energy but opposite wave vectors. Thus, in metals, the motions of all electrons in the valence band of the metal are canceled, and the net electric current is zero. Here, we see why the electron–hole pair in the semiconductors and the electron motion in the metals do not contribute to the electric conductivity if no electric field is applied. This is the basis for the following equation for the current density of a completely filled band:

$$J = (-q)\sum_{i}^{N} v_i = 0 \tag{3.28}$$

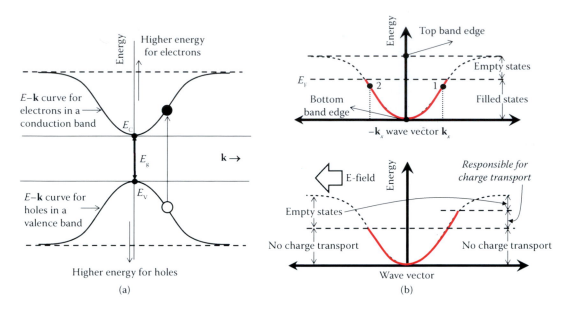

FIGURE 3.16 (a) Diagram illustrating the wave vector for an electron–hole pair in semiconductors. (b) Diagram illustrating the E–$\mathbf{k}$ curve in the valence band of metals—the wave vectors of electrons occupying the same energy level (top) and a change in E–$\mathbf{k}$ curve of materials electrically biased (bottom).

where J is the current density, q is the magnitude of the charge on the electron, and v_i is the velocity of the ith electron. Note the negative sign in Equation 3.28, which is used because of the negatively charged electron.

Then, let us examine how electric conductivity can be explained using the E–$\mathbf{k}$ curve (i.e., from the standpoint of quantum mechanics). In Section 2.17, we learned that the quantum mechanical approach considers that only electrons taking energy levels near E_F mainly contribute to electric conductivity. This view can be understood better using the schematic illustration shown at the bottom of Figure 3.16b. When the electric field is applied to the metals, the electrons moving with the wave vectors of $\mathbf{k}_x$ and $-\mathbf{k}_x$ are affected differently. If an electron with the wave vector $\mathbf{k}_x$ is accelerated (i.e., momentum increases), its counterpart with the wave vector $-\mathbf{k}_x$ is decelerated (i.e., momentum decreases) and the E–$\mathbf{k}$ curve becomes asymmetric under the electric bias. Thus, the energy levels near E_F are taken by only accelerated electrons and their motions are not canceled by decelerated electrons, which results in the appearance of a net electric current.

Now consider the valence band of a semiconductor, which has a hole created by removing the jth electron. The current density in this band will be given by the sum of all current densities minus the current density due to motion of the jth electron.

$$J = \left[(-q) \sum_i^N v_i \right] - (-q) v_j \tag{3.29}$$

Note that the first term in Equation 3.29 is zero. Thus, the current density in a valence band with one hole is given by

$$J = (+q) v_j \tag{3.30}$$

We can conclude that the current in a valence band with one hole can be described as the current by the jth electron (whose motion is uncompensated) moving with velocity $-v_j$. This current contribution is the same as that of a hole with a charge of $+q$ moving in the opposite direction, that is, with a velocity of $+v_j$. The magnitude of the wave vector associated with a missing electron is equal to the wave vector of the hole created (Figure 3.16).

An important fact to understand about electric current is that during the motion of charge carriers, the current contributions of an electron moving with a certain velocity and a hole moving with opposite velocity are the same. To understand the electrical properties of the valence band, which is almost (but not completely) filled with electrons, we consider the behavior of the holes. As an analogy, consider that we look for an empty parking spot rather than for parked cars when parking in a garage. To understand the electrical properties of the conduction band, we look at the behavior of the *electrons.*

In a band diagram or a band structure, we plot the electron energy so that it increases as the energy state goes up in the band diagram or band structure and the hole energy increases as the energy state goes down (Section 3.15). As a result, the electrons excited to the conduction band seeking the minimum energy are shown at the bottom of the conduction band on the band diagram, and the holes created in the valence band minimize their energy and are shown at the top of the valence band. The band diagrams (Figures 3.2 and 3.5), therefore, show the predominance of electrons at $E = E_c$ and that of holes at $E = E_v$.

3.8   EXTRINSIC SEMICONDUCTORS

In intrinsic semiconductors, the changes in the number of carriers are related exponentially to the changes in the temperature (Equation 3.5). This makes it very difficult to make practical and useful devices using intrinsic materials. We need semiconductors whose conductivity can be moderated by controlled changes in the composition. We would prefer that these materials show electrical properties that would not only be tunable with composition changes but that would also be stable over a wide range of temperatures (e.g., $-50°C$ to $+150°C$). In Chapter 6, we will discuss electronic devices such as transistors and diodes that can be designed and manufactured from extrinsic semiconductors. These materials show controllable variations in their conductivity due to doping and are also relatively temperature stable.

As mentioned in Section 3.2, a material in which the conductivity is largely controlled by the addition of other elements is known as an extrinsic semiconductor. In extrinsic semiconductors, the charge carriers (electrons and holes) are generated by adding *aliovalent* elements known as dopants. A dopant is an element or a compound deliberately added to a semiconductor to influence and control electrical or other properties. In particular, the electrical conductivity of semiconductors is significantly affected by the presence of foreign atoms. In contrast to dopants, *impurities* are elements or compounds that are present in semiconductors either inadvertently or because of processing limitations. For example, silicon crystals grown from melt often contain dissolved oxygen (O). This impurity comes from the quartz (SiO_2) crucibles used for the growth of silicon single crystals. Like dopants, impurities also have a profound effect on the electrical properties of semiconductors. The concentrations of impurities usually are not completely predictable, so it is difficult to estimate their effects on semiconductor properties. Thus, every effort is made in semiconductor processing to minimize the presence of impurities. In Sections 3.9 and 3.10, we describe different types of extrinsic semiconductors.

3.9   DONOR-DOPED (N-TYPE) SEMICONDUCTORS

A *donor-doped* or *n-type semiconductor* consists of an atom or an ion that provides *extra* electrons. Electrons are the *majority carriers* in these materials. Since electrons carry a negative charge, these materials are also known as n-type semiconductors.

For example, consider a silicon crystal with a low concentration of arsenic (As) forming a solid solution in which the arsenic atoms occupy the silicon sites. Each arsenic atom brings in five valence electrons. Only four of these participate in the formation of covalent bonds with other silicon atoms. The fifth electron remains bonded to the arsenic atom at low temperatures. However, as

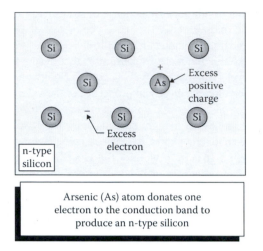

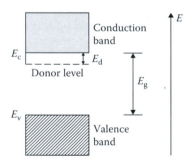

FIGURE 3.17 Illustration of arsenic-doped silicon, which is an n-type or donor-doped semiconductor. (From Singh, J.: *Semiconductor Devices: Basic Principles.* 2001. Copyright Wiley-VCH Verlag GmbH & Co. KGaA. Reproduced with permission.)

FIGURE 3.18 Band diagram for an n-type semiconductor.

the temperature increases to ~50 to 100 K, the fifth electron from arsenic dissociates itself and is available for conduction (Figure 3.17). Thus, each arsenic atom added results in one extra electron available for conduction. Note that since the dopant atoms added are neutral, the doped semiconductor remains electrically neutral and does not acquire a net negative charge. A band diagram for an n-type semiconductor is shown in Figure 3.18. In this figure, E_d demonstrates how strongly the fifth electron of As is bonded to As atoms. Note the donor energy level (E_d) in relation to the conduction band edge (E_c) (Table 3.3). Since dopant energy levels are much smaller than the band gap of semiconductors, thermal energy at room temperature is large enough to activate most dopants and to generate the charge carriers. The energy required for carrier generation is the major difference between extrinsic and intrinsic semiconductors. Examples of other n-type dopants for silicon include phosphorus (P), arsenic (As), and antimony (Sb). Section 3.21 discusses similar concepts through which we can create ceramic materials that show n-type semiconductivity.

3.10 ACCEPTOR-DOPED (p-TYPE) SEMICONDUCTORS

Another type of extrinsic semiconductor is an *acceptor-doped* or *p-type* semiconductor. The term "p-type" refers to the positive charge on holes that are the dominant charge carriers. In this type of material, the situation is opposite to that of n-type semiconductors. We add dopant atoms or ions that create a deficit of electrons. A typical example of a p-type semiconductor is silicon doped with boron (B; Figure 3.19). Each boron atom added to silicon brings in three valence electrons. Thus, the

TABLE 3.3

Energy Levels for Some Dopants in Silicon, Germanium, and GaAs

Semiconductor	Donor Dopant	Donor Energy Level (meV)	Acceptor Dopant	Acceptor Energy Level (meV)
Si	Sb	39	B	45
	P	45	Al	67
	As	54	Ga	72
			In	160
GaAs	Si	5.8	C	26
	Ge	6.0	Be	28
	Sn	6.0	Mg	28
			Si	35
Ge	Sb	9.6	B, Al	10
	P	12	Ga	11
	As	13	In	11

Source: Singh, J., *Semiconductor Devices: Basic Principles*, Wiley, New York, 2001. With permission.

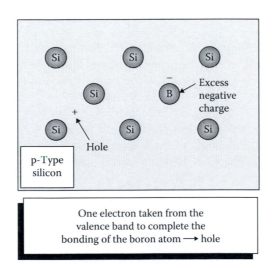

FIGURE 3.19 Illustration of a boron-doped silicon, which is a p-type or acceptor-doped semiconductor. (From Singh, J.: *Semiconductor Devices: Basic Principles*. 2001. Copyright Wiley-VCH Verlag GmbH & Co. KGaA. Reproduced with permission.)

boron atoms occupying the silicon sites have only three valence electrons that can form complete covalent bonds with three silicon atoms. The fourth boron–silicon bond is incomplete in that it is missing an electron and is thus ready to accept one. Boron-doped silicon is therefore referred to as an acceptor-doped semiconductor. Aluminum is another example of an acceptor dopant for silicon.

The missing electron in one of the boron–silicon bonds is also described as the creation of a hole. In acceptor-doped or p-type semiconductors, the majority of carriers are holes. Hence, acceptor-doped semiconductors are also known as p-type semiconductors. The band diagram for a typical p-type semiconductor is shown in Figure 3.20. Note the position of the acceptor level, E_a, which is above the valence band edge (E_v).

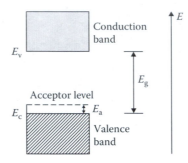

FIGURE 3.20   A band diagram for a p-type or acceptor-doped semiconductor.

3.11   AMPHOTERIC DOPANTS, COMPENSATION, AND ISOELECTRONIC DOPANTS

In some cases, the same dopant element will behave as if it is both the donor and the acceptor. Such dopants are known as *amphoteric dopants*. For example, silicon in GaAs is an amphoteric dopant. If the tetravalent silicon occupies the trivalent gallium sites, it behaves as a donor. If silicon atoms occupy the pentavalent arsenic sites, they behave as acceptors. Therefore, is silicon a donor or an acceptor dopant when added to GaAs? The answer depends on the concentration of silicon added to GaAs, the manner in which it is added, and the temperature during the addition. At low processing temperatures and lower concentrations, silicon acts as an acceptor by taking up the arsenic sites; whereas at high processing temperatures and higher concentrations, it occupies gallium sites and acts as a donor. We will learn about the ways to add dopants to semiconductors in the next chapter.

As mentioned in Section 3.6, the addition of isoelectronic dopants can be useful in some cases. For example, aluminum with a valency of 3 can be added to GaN. Thus, aluminum acts as neither an acceptor nor a donor. When isoelectronic dopants are added, they change the local electronic structure and, in turn, alter the electrical properties of the base materials. Another example of an isoelectronic dopant is germanium in silicon. Silicon–germanium (informally known as *siggy*) semiconductors have been developed. Their advantages include a higher device speed and less power consumption (Ahlgre and Dunn 2000).

In some cases, isoelectronic additions help with optical properties. As mentioned before, nitrogen-doped $GaAs_{1-x}P_x$ compositions with an indirect band gap are used for making yellow and green LEDs (Streetman and Banerjee 2000; Figure 3.13).

There are many electronic devices that contain both types of dopants in the same volume of material. For example, we may have a phosphorus-doped silicon crystal, and we add boron, an acceptor dopant, either to the entire crystal or to parts of it. The effect of the boron addition will compensate for the effect of the phosphorus doping. This is known as *compensation doping*. Note that, in order for the compensated semiconductor (either the entire crystal or a selected part of it) to behave as n-type or p-type, the net concentration of carriers created must exceed the concentrations of thermally generated carriers. A semiconductor containing both types of dopants behaves as n-type if the donor concentration (N_d) outweighs the acceptor concentration (N_a) and ($N_d - N_a$) $\gg n_i$. If a semiconductor has both types of dopants and if ($N_a - N_d$) $\gg n_i$ ($= p_i$), then the material behaves as a p-type semiconductor.

3.12   DOPANT IONIZATION

In an n-type or a p-type material, at low temperatures near 0 K, electrons and holes remain bonded to the donor or acceptor atoms, respectively. This temperature region in which the carriers remain bonded to the dopant atoms is known as the *freeze-out range* (Figure 3.21).

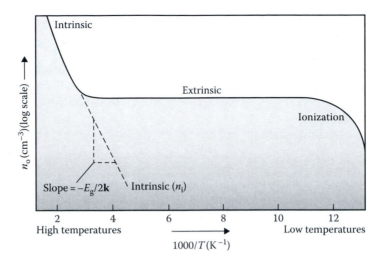

FIGURE 3.21 Carrier concentration (plotted on a log scale) as a function of the inverse of the temperature (*T*). (From Askeland, D. and Fulay P., *The Science and Engineering of Materials*, Thomson, Washington, DC, 2006. With permission.)

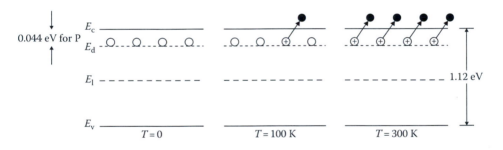

FIGURE 3.22 Dopant ionization in an n-type semiconductor. (From Kano, K., *Semiconductor Fundamentals*, Prentice Hall, Upper Saddle River, NJ, 1997. With permission.)

As the temperature increases to ~50 to 100 K, the extra electrons or holes bound to the donor and acceptor atoms become available for conduction. This process by which the carriers that have bonded with the dopant atoms dissociate themselves from the dopant atoms is known as *dopant ionization* (Figure 3.22).

As illustrated in Figure 3.22, when the fifth electron from the phosphorous atom becomes detached, we get a phosphorous ion that has an effective charge of +1. If a hole bound to a boron atom dissociates itself, we get a boron ion with an effective charge of −1. With a further increase in the temperature, the carrier concentration levels off, indicating complete ionization. This region is labeled as extrinsic in Figure 3.21 because the carrier concentration is directly linked to the dopant concentration. With a further increase in the temperature, the concentrations of thermally excited carriers (n_i and p_i) are dominant. This region is therefore labeled intrinsic because even if the material is doped, it behaves as an intrinsic semiconductor.

The energy required (E_d) to remove the fifth electron from a phosphorus atom—that is, to ionize a donor atom—is given by the following equation:

$$E_d = E_c - \frac{q^4 m_e^*}{2(4\pi\varepsilon)^2 \hbar^2} \tag{3.31}$$

In Equation 3.31, E_d is the donor energy level, E_c is the conduction band edge energy, and ε is the dielectric permittivity of the semiconductor (unit F/m). Thus, the difference $(E_d - E_c)$ tells us how far below E_c the donor energy level E_d is. This equation is an extension of Bohr's atomic model, which is used for predicting the binding energy of an electron to a hydrogen atom nucleus.

We rewrite Equation 3.31 to involve the ratio of the effective mass to the mass of electrons in a vacuum. The ratio of the permittivity of the semiconductor to the permittivity of the free space (ε_0) is its dielectric constant (ε_r).

In Example 3.4, we will show that

$$\frac{m_0 q^4}{2(4\pi\varepsilon_0)^2 \hbar^2} = 13.6\,\text{eV} \tag{3.32}$$

This is the energy with which an electron is bound to a nucleus of a hydrogen atom. Thus, from Equations 3.31 and 3.32, we get

$$E_d = E_c - 13.6 \left(\frac{m_e^*}{m_0} \right) \left(\frac{\varepsilon_0}{\varepsilon} \right)^2 \text{eV} \tag{3.33}$$

We can rewrite this as

$$E_d = E_c - 13.6 \frac{\left(\dfrac{m_e^*}{m_0} \right)}{\left(\dfrac{\varepsilon}{\varepsilon_0} \right)^2} \text{eV} \tag{3.34}$$

$$E_d = E_c - 13.6 \frac{\left(\dfrac{m_e^*}{m_0} \right)}{(\varepsilon_r)^2} \text{eV} \tag{3.35}$$

or

$$(E_d - E_c) = \frac{13.6 \left(\dfrac{m_e^*}{m_0} \right)}{(\varepsilon_r)^2} \tag{3.36}$$

Note that, in these equations, we use the effective mass of electrons (m_n^*) for direct band gap semiconductors (e.g., GaAs); whereas for indirect band gap semiconductors such as silicon, we use the *conductivity effective mass* (m_σ^*). The examples that follow illustrate the calculations of dopant levels in direct and indirect band gap materials.

The values of the dopant energy levels calculated here are approximate and use an extension of Bohr's model for the binding energy of an electron bound to the hydrogen nucleus. This is known as the hydrogen model. The actual values of the binding energy levels for different dopants used in silicon, germanium, and GaAs depend upon the specific dopant and are shown in Table 3.3. This table shows that the binding energy for most dopants is a few millielectron volts. The binding energy of dopants is significantly lower than 13.6 eV. There are two reasons for this: (1) The effective mass of the carriers is smaller than the mass of charge carriers in vacuum; and (2) the binding potential is reduced by the crystal as described by the dielectric constant (ε_r; Equation 3.36).

Shallow dopants are those whose energy levels are close to the band edge, for example, the dopants shown in Table 3.3. They reasonably follow Bohr's atomic model summarized in Equation 3.36 (Examples 3.5 and 3.7). Some atoms create defect levels that are well into the band gap ($E_d >$ 100 meV in silicon); these are known as *deep-level defects*. The levels of these dopants cannot be predicted using Bohr's atomic model (Equation 3.35). Deep-level dopants are introduced by atoms or ions that do not *fit* the semiconductor crystal; they severely distort the host lattice. Deep-level defects act as a trap for charge carriers. They can also be useful in pinning down the Fermi energy level.

Example 3.4: Value of an Expression Related to the Binding Energy for an Electron in an H Atom

Show that the term expressed in Equation 3.32 is equal to 13.6 eV. This is the energy with which an electron in a hydrogen (H) atom is bound to its nucleus.

Solution
We use $m_0 = 9.11 \times 10^{-31}$ kg, $q = 1.6 \times 10^{-19}$ C, $\varepsilon_0 = 8.85 \times 10^{-12}$ F/m, and $\hbar = 1.05 \times 10^{-34}$ J·s in Equation 3.32.

$$\frac{m_0 q^4}{2(4\pi\varepsilon_0)^2 \hbar^2} = \frac{(9.11 \times 10^{-31}\,\text{kg})(1.6 \times 10^{-19}\,\text{C})^4}{2(4 \times 3.14 \times 8.85 \times 10^{-12}\,\text{F/m})^2 \times (1.05 \times 10^{-34}\,\text{J·s})^2} = 2.189 \times 10^{-18}\,\text{J}$$

Note that 1 eV is the energy required to move the charge equal to the charge on an electron through 1 V potential difference, that is, 1 eV = 1.6 × 10^{-19} J.
Therefore, converting 2.189 × 10^{-15} J into electron volts, we get

$$(2.189 \times 10^{-18}\,\text{J}) \times \frac{1\,\text{eV}}{1.6 \times 10^{-19}\,\text{J}} = 13.6\,\text{eV}$$

Example 3.5: Donor Energy Level Calculation for Silicon-Doped Gallium Arsenide

Assuming that Si acts as a donor in a GaAs sample, calculate the position of the donor energy level (E_d) relative to the conduction band edge (E_c).

Solution
We use Equation 3.35 to calculate the relative position of the donor energy level. Since GaAs is a direct band gap semiconductor, we use the effective mass of electrons.
From Table 3.2, for GaAs $\left(m_n^* / m_0\right) = 0.067$ and the dielectric constant (ε_r) is 13.2.

$$E_d = E_c - 13.6 \frac{\left(\dfrac{m^*}{m_0}\right)}{\left(\varepsilon_r\right)^2}\,\text{eV}$$

Therefore,

$$E_d = E_c - 13.6 \frac{0.067}{\left(13.2\right)^2}\,\text{eV}$$

$$E_d = E_c - 5.22 \times 10^{-3}\,\text{eV}$$

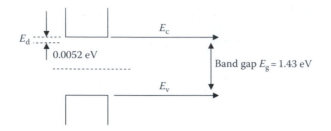

FIGURE 3.23 Illustration of the silicon donor dopant energy level in GaAs. The intrinsic Fermi energy level ($E_{F,i}$) is shown in the middle of the band gap. Diagram is not to scale (see Example 3.5).

Thus, the donor energy level for Si in GaAs is ~5.2 meV below the conduction band edge (Figure 3.23). Note that this calculation does not require the actual band gap energy value. The binding energy of many dopants is significantly lower than 13.6 eV because of the lesser effective masses and dielectric constant.

Example 3.6: Comparison of Donor Energy Level Relative to Thermal Energy

Compare the magnitude of the difference ($E_d - E_c$) calculated in Example 3.5 with the thermal energy ($k_B T$) of carriers at 300 K. (Note: Boltzmann's constant [k_B] is 8.617 × 10⁻⁵ eV/K or 1.38 × 10⁻²³ J/K.)

Solution
Since we prefer to obtain the energy in electron volts, we use the value of k_B in electron volts. The value of $k_B T$ at 300 K is ~0.026 eV.

The energy with which the donor electron of Si in GaAs is bonded is only 0.00522 eV. Thus, the thermal energy at 300 K is almost five times greater, suggesting a nearly complete donor ionization.

Example 3.7: Donor Energy Level For Phosphorus-Doped Silicon

Calculate the position of the donor energy (E_d) level relative to the conduction band edge (E_c) for P in Si. The dielectric constant (ε_r) of Si is 11.8 (Table 3.2). The ratio of the *conductivity effective mass* of electrons in Si to m_0 is 0.26.

Solution
We use the values in Equation 3.35, which have been modified to show the conductivity effective mass of the electrons in Si.

$$E_d = E_c - 13.6 \frac{\left(\dfrac{m_\sigma^*}{m_0}\right)}{\left(\varepsilon_r\right)^2} \text{ eV} \tag{3.37}$$

Note the use of the conductivity effective mass because Si is an indirect band gap material.

$$E_d = E_c - 13.6 \frac{0.26}{\left(11.8\right)^2}$$

Thus, the donor level for P in Si is ~0.025 eV or 25 meV below the conduction band edge. This is illustrated in Figure 3.24.

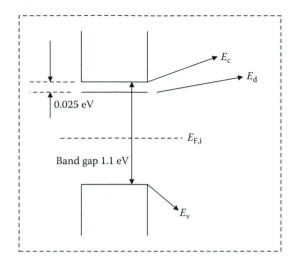

FIGURE 3.24 Position of calculated donor dopant level (E_d) in silicon relative to conduction band edge (E_c). The dotted line in the middle of the band gap is the Fermi energy level of the intrinsic semiconductor ($E_{F,i}$). Diagram is not to scale (see Example 3.7).

3.13 CONDUCTIVITY OF INTRINSIC AND EXTRINSIC SEMICONDUCTORS

The conductivity of a semiconductor is given by the following equation:

$$\sigma = q \times n \times \mu_n + q \times p \times \mu_p \tag{3.38}$$

where q is the magnitude of the charge on the electron or hole and is 1.6×10^{-19} C, n is the concentration of conduction electrons, p is the concentration of holes, and μ_n and μ_p are the mobilities of electrons and holes in the semiconductor material, respectively.

If we add one phosphorus atom to a silicon crystal and the temperature is high enough to cause donor ionization, we should get one electron in the conduction band. Let N_d be the concentration of donor atoms. Assuming that each donor dopant atom donates one electron that becomes available for conduction, we expect to get N_d number of electrons per cubic centimeter in the conduction band. Note that as the temperature increases, we expect not only donor ionization (Figure 3.22) but also some electrons in the silicon–silicon bond to break free and provide for the intrinsic or thermally generated electrons and holes (Figure 3.2).

If n_i is the concentration of thermally generated carriers, then for an n-type semiconductor, we can write the total concentration of conduction electrons (n) as

$$n = N_d + n_i \tag{3.39}$$

In an n-type semiconductor, conduction electrons generated through the doping process dominate the conductivity and are the majority carriers, that is

$$N_d \gg n_i \tag{3.40}$$

Since the concentration of conduction electrons created by doping is significantly higher than the concentration of those generated by thermal excitation (Equation 3.40), we can assume

$$n \cong N_d \tag{3.41}$$

Recall that there is an equal concentration of holes (p_i holes/cm³) corresponding to n_i electrons/cm³. In this case, we can ignore the contribution of holes and thermally excited electrons to the total conductivity because holes are *minority carriers* in n-type semiconductors. The equation for the conductivity of an n-type semiconductor can be written as

$$\sigma = q \times N_d \times \mu_n \tag{3.42}$$

The situation is analogous for the p-type semiconductor. For this type, the concentration of holes generated due to acceptor dopants is dominant ($N_a \gg p_i$). The holes are the majority carriers, and the electrons are the minority carriers.

$$p \cong N_a \tag{3.43}$$

We can ignore the contribution of thermally generated electrons and holes to conductivity, that is

$$\sigma = q \times N_a \times \mu_p \tag{3.44}$$

These concepts are illustrated in Example 3.8. Before solving examples, note that n_i and p_i of doped semiconductors are different from those of intrinsic semiconductors. Details are found in Chapter 4.

Example 3.8: Conductivity of Intrinsic Gallium Arsenide

1. What is the resistivity of intrinsic GaAs? Assume that the temperature is 300 K and $n_i = 2 \times 10^6$ electrons/cm³ (Figure 3.7).
2. How does this value compare with the conductivity of an intrinsic Ge sample? (See Example 3.1.)

Solution

1. Since this is an intrinsic semiconductor, the concentrations of electrons and holes play a role in the conductivity. From Equation 3.38,

$$\sigma = q \times n \times \mu_n + q \times p \times \mu_p$$

Table 3.1 shows that the mobility values for electrons and holes for intrinsic GaAs are 8500 and 400 cm²/V · s, respectively. The concentration of holes and electrons is the same because each electron that moves into the conduction band creates a hole in the valence band, i.e., $n_i = p_i$.
Therefore

$$\sigma = (1.6 \times 10^{-19} \text{ C})(2 \times 10^6/\text{cm}^3)(8500 + 400) \text{ cm}^2/\text{V} \cdot \text{s}$$

$$\sigma = 2.848 \times 10^{-19} \text{ S/cm}$$

The resistivity $\rho = 3.51 \times 10^8 \, \Omega \cdot \text{cm}$.
Thus, undoped intrinsic GaAs has a relatively high resistivity. This value is actually higher than $10^3 \, \Omega \cdot \text{cm}$, which generally is considered the approximate upper limit for a material to be defined as a semiconductor (see Section 3.1).

2. As seen in Example 3.1, the resistivity of intrinsic Ge was 43.1 $\Omega \cdot$ cm.
Since Ge has a smaller band gap compared to the GaAs band gap, it has a much higher concentration of electrons in the conduction band. Conversely, the mobilities of the carriers in Ge are smaller than those for GaAs (see Example 3.1). However, the effect of increased carrier concentrations dominates the effect of mobilities. As a result, intrinsic Ge ends up having a higher conductivity than GaAs.

3.14   EFFECT OF TEMPERATURE ON THE MOBILITY OF CARRIERS

Temperature has a major influence on the mobility of electrons (μ_n) and holes (μ_p). When the temperature is too low for dopant ionization, electrons and holes remain bound to the dopant atoms. In this freeze-out zone at low temperatures, the conductivity of the semiconductors is very low due to the lack of electrons in the conduction band at low temperatures. As the temperature increases, dopant ionization occurs in the extrinsic semiconductors, and electrons occupy the conduction band. In this regime, electrons traveling in the conduction band are scattered through two different mechanisms. One is scattering by ionized dopants, which mainly is related to the electrostatic attraction or repulsion between the charge carriers and the ionized dopants. The other is scattering by the host lattice atoms, which is caused by the thermal vibration of the host atoms. When the temperature increases above the freeze-out range but is not too high, the thermal energy of charge carriers is not large enough to overcome the electrostatic interactions with the ionized dopant. Also, the extent of scattering of carriers by the vibrations of atoms is small, since the host atoms do not have sufficient thermal energy. Therefore, the scattering of the carriers caused by the dopant atoms is dominant at low temperature ($T < 200$ K). The mobility of carriers limited by the scattering of the dopant or impurity atoms is known as *impurity-* or *dopant-scattering limited mobility*. As the dopant concentration increases, the impurity-scattering mechanism becomes more pronounced at low temperatures. An increase in the temperature increases the thermal energy of carriers and helps the carriers to overcome the electrostatic interaction with the ionized dopants. If the temperature reaches room temperature or higher, phonon scattering or lattice scattering (i.e., the scattering of vibrations of atoms of the host lattice atoms) begins to dominate, and the mobility decreases again. This behavior at high temperature is known as *lattice-scattering limited mobility*. Both scattering mechanisms work in extrinsic semiconductors, and highly doped semiconductors exhibit the highest mobility near room temperature where the transition from the impurity-scattering dominant regime to the lattice-scattering regime occurs.

However, in intrinsic semiconductors, only lattice scattering is available and its mobility is the highest at low temperature. An increase in the temperature keeps reducing the intrinsic semiconductor's mobility. The trends in changes in the mobility of carriers are shown in Figure 3.25.

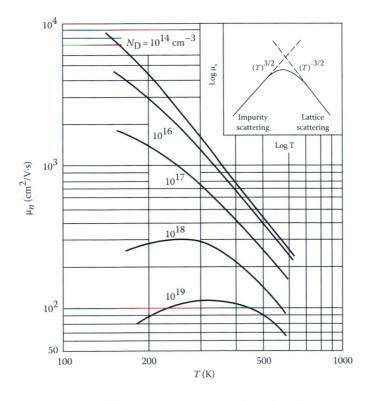

FIGURE 3.25 Electron mobility in silicon versus temperature for various donor concentrations. The inset shows the theoretical temperature dependence of electron mobility in heavily doped semiconductors. (From Sze, S.M.: *Semiconductor Devices, Physics, and Technology.* 1985. Copyright Wiley-VCH Verlag GmbH & Co. KGaA. Reproduced with permission.)

3.15   EFFECT OF DOPANT CONCENTRATION ON MOBILITY

As discussed in Section 3.14, although the dopant atoms provide carriers that contribute to conductivity, they can also act as scattering centers. This effect is important at both low temperatures and at high dopant concentrations, where the electrostatic interaction of charge carriers with the ionized dopants is significant. The variations in the electron and the hole drift mobility for silicon at different temperatures and at different dopant concentrations are shown in Figures 3.26 and 3.27. As the dopant concentration increases, the mobility decreases and

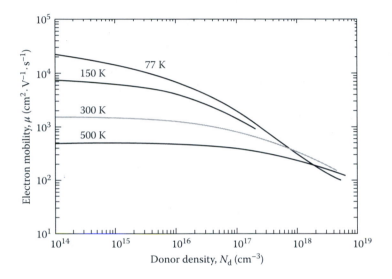

FIGURE 3.26   The variation in electron mobility for silicon at different temperatures and different donor dopant concentrations. (From Li, S.S. and Thurber W.R., *Solid State Electron.*, 20, 609–616, 1977. With permission.)

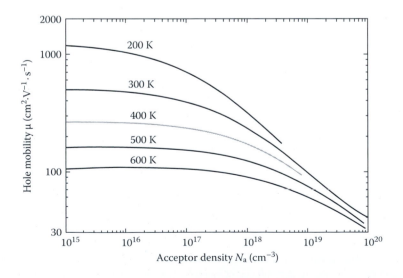

FIGURE 3.27   The variation in hole mobility for silicon at different temperatures and acceptor dopant concentrations. (From Dorkel, J.M. and Leturcq P., *Solid State Electron.*, 24(9), 821–825, 1981. With permission.)

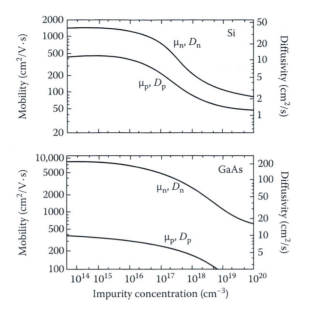

FIGURE 3.28 Mobilities and diffusivities in silicon and gallium arsenide at 300 K as a function of impurity concentration. (From Sze, S. M.: *Semiconductor Devices, Physics, and Technology.* 1985. Copyright Wiley-VCH Verlag GmbH & Co. KGaA. Reproduced with permission.)

the dependence of the mobility on temperature also decreases. Similar mobility versus dopant concentration trends are found in the direct band gap semiconductor GaAs (Figure 3.28). Note that the mobility and the diffusivity of charge carriers are linearly related in Figure 3.28. This relation is known as the Einstein relation, and its importance will be addressed in Chapter 9 (Equation 9.36).

Example 3.9 shows how to account for these changes in mobility with the dopant concentration while calculating the conductivity of extrinsic semiconductors.

Example 3.9: Resistivity of N-Type Doped Gallium Arsenide

Calculate the resistivity of donor-doped GaAs with $N_d = 10^{13}$ atoms/cm³. Assume that $T = 300$ K and that all the donors are ionized.

Solution

For GaAs at 300 K, $n_i = 2.0 \times 10^6$ electrons/cm³ (Figure 3.7).

In this case, $N_d \gg n_i$; the total concentration of conduction electrons $(n) \approx N_d = 10^{13}$ electrons/cm³. From Equation 3.38:

$$\sigma = q \times n \times \mu_n + q \times p \times \mu_p$$

We ignore the contributions of the holes because the electrons are the majority charge carriers. From Figure 3.28, for GaAs with $N_d = 10^{13}$ atoms/cm³, and $\mu_n = 8000$ cm²/V · s,

$$\sigma = (1.6 \times 10^{-19} \text{ C})(10^{13})(8000)$$

$$\therefore \sigma = 1.28 \times 10^{-3} \ \Omega^{-1} \cdot \text{cm}^{-1}$$

The resistivity (ρ) is 781 Ω · cm. This value is much lower than 3.51×10^8 Ω · cm, which is the value of resistivity for intrinsic GaAs (Example 3.8).

3.16   TEMPERATURE AND DOPANT CONCENTRATION DEPENDENCE OF CONDUCTIVITY

As shown here, the combination of dopant ionization, changes in mobility, and thermal generation of carriers using temperature (Figures 3.21 and 3.28) leads to changes in the conductivity of a semiconductor as a function of the temperature and the dopant concentration. Note that the conductivity of the intrinsic semiconductors increases with increasing temperature in spite of a decrease in the mobility. This is attributed to an increase in the electron concentration in the conduction band at a higher temperature. The change in the resistivity of silicon as a function of the dopant concentration is shown in Figure 3.29. As the dopant concentration increases, the resistivity continuously decreases. In Si with moderate doping concentration, n-type Si exhibits lower resistivity than p-type Si due to the lower electron mobility (Figure 3.28).

3.17   EFFECT OF PARTIAL DOPANT IONIZATION

For extrinsic semiconductors near room temperature, we usually assume complete dopant ionization. However, this may not always be the case. Dopant ionization is temperature-dependent (Figures 3.21 and 3.22), and its extent depends on the relative difference between the defect level and the nearest band edge. The fraction of electrons still tied to the donor atoms in an n-type semiconductor is given by

$$\frac{n_d}{n + n_d} = \frac{1}{\left\{ \dfrac{N_c}{2N_d} \exp\left(-\dfrac{(E_c - E_d)}{k_B T} \right) + 1 \right\}} \tag{3.45}$$

where n_d is the concentration of electrons still bound to donor atoms, n is the concentration of free electrons in the conduction band, $(E_c - E_d)$ is the difference between the conduction band edge and the donor energy level, N_d is the donor dopant concentration, and N_c is the effective density of states at the conduction band. N_c is a parameter indicating the maximum number of electrons that can be accommodated at the energy level $E = E_c$. Although N_c determines the upper limit of electron

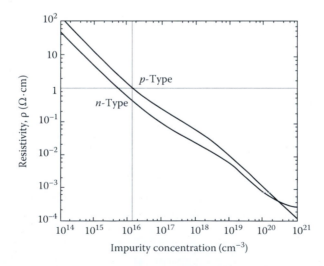

FIGURE 3.29   The change in resistivity of silicon with dopant concentration. (From Grove, A. S.: *Physics and Technology of Semiconductor Devices*. 1967. Copyright Wiley-VCH Verlag GmbH & Co. KGaA. Reproduced with permission.)

concentration at E_c, the actual number of electrons available in the conduction band depends on the distance between the donor level and the conduction band edge (i.e., donor energy level).

If the temperature is constant and the donor energy level increases, then more energy is required to release free electrons from dopants and the fraction of the electrons that remain bound to the donor atoms also increases. Similarly, as the temperature decreases, the exponential term in the denominator of Equation 3.45 decreases. This indicates that an increase in the donor energy level increases the fraction of electrons that remain bound to the donor atoms.

From a quantum mechanics base consideration, the maximum density of electrons that can stay together at the band edge (N_c) is given by:

$$N_c = 2\left(\frac{m * k_B T}{2\pi\hbar^2}\right)^{3/2} \tag{3.46}$$

Note that N_c is related to effective mass and temperature. As temperature increases, the maximum concentration of electrons that can stay at E_c (namely N_c) increases. Also, an increase in the effective mass of the carrier raises N_c. Further details about the density of states can be found in Section 4.3.

In calculating the effective density of states, we must be careful in selecting the effective mass for direct and indirect band gap materials. To calculate the density of states in the conduction band of *direct band gap materials* such as GaAs, we use the effective mass of the electron as $m*$ in Equation 3.46; that is, we substitute the value of m_e^* for $m*$.

In contrast, to calculate the density of states in the conduction band of *indirect band gap semiconductors* such as silicon, we use the *density of states effective mass* $\left(m_{dos}^*\right)$ of electrons, defined as

$$m_{dos}^* = \left[6^{2/3}\left(m_l^* m_t^{*2}\right)^{1/3}\right] \tag{3.47}$$

In Equation 3.47, m_l^* and m_t^* are known as longitudinal and transverse effective masses of electrons, respectively (Table 3.2). The values m_l^* and m_t^* for the electrons in silicon are 0.98 and 0.19, respectively. This leads us from Equation 3.47 to $m_{dos}^* = 1.08$.

A detailed discussion of these parameters is beyond the scope of this book. We will use the density of states effective masses of electrons and holes to calculate the density of states and the electron concentration in the conduction band (or the holes in the valence band), which will be discussed in Chapter 4. In addition, note that the conductivity effective mass is used in calculating the response of the carriers to the electric field and the donor energies (Example 3.7).

For p-type semiconductors, there is a relationship similar to Equation 3.44. The fraction of holes that remain bound to the acceptor atoms depends on the temperature and on the distance between the acceptor level (E_a) and the valence band edge (E_v).

$$\frac{p_a}{p + p_a} = \frac{1}{\left\{\frac{N_v}{2N_a}\exp\left(-\frac{(E_a - E_v)}{k_B T}\right) + 1\right\}} \tag{3.48}$$

where p_a is the concentration of the holes still bound to the acceptor atoms, p is the concentration of the holes in the conduction band, $(E_a - E_v)$ is the difference between the acceptor energy level and the valence band edge, and N_a is the concentration of the acceptor dopant. In Equation 3.48, N_v is the effective density of states at the valence band edge.

This value is given by the following equation if the semiconductor has a heavy-hole band and a light-hole band (e.g., for GaAs):

$$N_v = 2\left(m_{hh}^{3/2} + m_{lh}^{3/2}\right)\left(\frac{k_B T}{2\pi\hbar^2}\right)^{3/2} \tag{3.49}$$

TABLE 3.4

Effective Density of States for Silicon, Germanium, and GaAs
(T = 300 K)

Semiconductor	Conduction Band Effective Density (cm^{-3}) of States (N_c)	Valence Band Effective Density (cm^{-3}) of States (N_v)
Si	2.78×10^{19}	9.84×10^{18}
Ge	1.04×10^{19}	6.0×10^{18}
GaAs	4.45×10^{17}	7.72×10^{18}

In Equation 3.49, m_{hh} and m_{lh} are the effective masses of heavy and light holes, respectively (Table 3.2). The density of states effective mass of the holes in the valence band is given by

$$\left(m_{dos}^* \right)^{3/2} = \left(m_{hh}^{3/2} + m_{lh}^{3/2} \right) \tag{3.50}$$

For holes in silicon, $m_{hh}^* = 0.49$ and $m_{lh}^* = 0.16$, and this leads to $m_{dos}^* = 0.55$. For holes in GaAs, $m_{hh}^* = 0.45$ and $m_{lh}^* = 0.08$, and this leads to $m_{dos}^* = 0.47$.

The values of the density of states (at 300 K) for Ge, Si, and GaAs, calculated using Equations 3.48 and 3.1 and related equations, are shown in Table 3.4.

We usually assume that at or near 300 K, the dopants are completely ionized, and thus the fractions on the left-hand sides of Equations 3.45 and 3.48 are very small.

Example 3.10 shows how to account for partial dopant ionization.

Example 3.10: Partial Dopant Ionization and Conductivity

A GaAs crystal is doped with Si, which acts as a donor dopant with $N_d = 10^{15}$ atoms/cm^3.
1. Calculate the fraction of electrons that is bound to the donor atoms in GaAs.
2. What is the resistivity of this material? Assume that the donor energy (E_d) is 5.8 meV away from the conduction band edge (E_c; Example 3.5 and Table 3.3).

Solution
1. We use Equation 3.45 to obtain the fraction of electrons that is still bound to the Si dopant, assumed to be a donor for GaAs. From Table 3.3, we know that the value of ($E_c - E_F$) is 5.8 meV. Note that, at 300 K, the value of $k_B T$ is 0.026 eV.

From Table 3.4, we know that the density of states at the conduction band edge (N_c) for GaAs at 300 K is 4.45×10^{17} cm^{-3}.

$$\frac{n_d}{n + n_d} = \frac{1}{\left\{ \dfrac{4.45 \times 10^{17}\, cm^{-3}}{2 \times 10^{15}} exp\left(-\dfrac{(0.0058)}{0.026} \right) + 1 \right\}} = 0.00586$$

Thus, ~5% to 6% of the electrons remain bound to the Si atoms in this n-type GaAs sample at T = 300 K.

2. The actual number of the conduction electrons available is 0.00586×10^{15} and the conductivity is

$$\sigma = (1.6 \times 10^{-19}\text{C})(10^{13})(0.00586 \times 10^{15})(8000)$$

$$\therefore \sigma = 0.00715 \ \Omega^{-1} \cdot \text{cm}^{-1}$$

$$\therefore \rho = 139.8 \ \Omega \cdot \text{cm}^{-1}$$

3.18 EFFECT OF TEMPERATURE ON THE BAND GAP

So far, we have assumed that the band gap (E_g) of semiconductors does not change with temperature. However, the band gap of the semiconductor does change slightly with temperature because thermal expansion and contraction lead to a change in the interatomic distance. As we can see in Figure 3.30, as the interatomic spacing increases as a result of thermal expansion, the band gap of silicon will decrease slightly. This trend is seen in other semiconductors as well. Intuitively, we can think that the thermal energy decreases the electrostatic interaction between electrons and nucleus, and reduces an energy gap between the free states and bound states that are discussed in Sections 2.14–2.16. A physical meaning of the band gap is how tightly electrons are bound to the nucleus.

The changes in the band gap of semiconductors are given by the Varshni formula:

$$E_g = E_g \left(at \, T = 0\,\text{K} \right) - \frac{\alpha T^2}{T + \beta} \tag{3.51}$$

In general, the band gap of the semiconductors decreases slightly with increasing temperature, due to the thermal expansion. The *Varshni parameters* α and β for different semiconductors and their band gaps at 0 K are listed in Table 3.5 (Schubert 2006). Similarly, the band gap of a semiconductor will depend upon the type and magnitude of the stress that is applied or present. In general, a tensile stress will lead to an increase in interatomic spacing, causing a decrease in the band gap.

Examples 3.11 and 3.12 show how to account for temperature variation in the band gap.

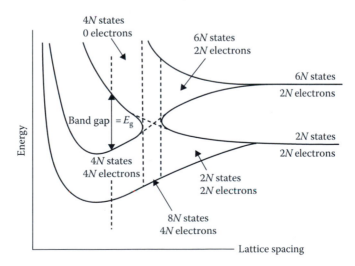

FIGURE 3.30 Energy levels in intrinsic silicon as a function of interatomic spacing. (From Streetman, B.G. and Banerjee S., *Solid State Electronic Devices*, 5th ed., Prentice Hall, Upper Saddle River, NJ, 2000. With permission.)

TABLE 3.5

Varshni Parameters for Semiconductors

Semiconductor	E_g (at 0 K)	α (10^{-4} eV/K)	β (K)
Si	1.17	4.73	636
Ge	0.744	4.77	235
GaAs	1.519	5.41	204
GaP	2.340	6.20	460
GaN	3.470	7.70	600

Source: Schubert, F.E., *Light-Emitting Diodes*, Cambridge University Press, Cambridge, UK, 2006. With permission.

Example 3.11: Variation of the Band gap with Temperature

Calculate the band gap of (1) Si and (2) GaAs at 300 K using the Varshni parameters shown in Table 3.5.

Solution

1. For Si, E_g at 0 K is 1.170 eV, α = 4.73 × 10^{-4} eV/K, and β = 636 K. Therefore:

$$E_g = E_g\left(\text{at } T = 0\,\text{K}\right) - \frac{\alpha T^2}{T + \beta}$$

$$= 1.170 - \frac{\left(4.73 \times 10^{-4}\right)\left(300\right)^2}{\left(300 + 636\right)}$$

$$= 1.124 \text{ eV}$$

2. For GaAs, the E_g at 0 K is 1.519 eV, α = 5.41 × 10^{-4} eV/K, and β = 204 K. Therefore:

$$E_g = 1.519 - \frac{\left(5.41 \times 10^{-4}\right)\left(300\right)^2}{\left(300 + 204\right)}$$

$$= 1.422 \text{ eV}$$

Thus, the band gaps of Si and GaAs at 300 K are 1.124 and 1.422 eV, respectively.

Example 3.12: Variation in the Wavelength of a Gallium Nitride LED

What is the wavelength of light emitted from a GaN LED operating at 300 K? Assume that the band gap of GaN changes according to the Varshni parameters shown in Table 3.5. How does this compare with the value calculated in Example 3.3?

Solution

Using the Varshni parameters for GaN (Table 3.5), we can show that the band gap of GaN at 300 K is 3.4 eV. For this band gap, the wavelength based on a GaN LED will be

$$\lambda\left(\text{in } \mu m\right) = \frac{1.24}{E_g\left(\text{in eV}\right)} \mu m$$

$$\therefore \lambda\left(\text{in } \mu m\right) = \frac{1.24}{3.40} \mu m = 0.365 \, \mu m = 365 \, nm$$

Note that, in most cases, we use the room-temperature values of the band gaps and *not* those at 0 K. When the room-temperature band gap values are used, we often ignore the band gap's dependence on temperature. Compared to λ = 357 nm for 0 K (Example 3.3), the wavelength of 365 nm is a little higher because of the decreased band gap.

3.19   EFFECT OF DOPANT CONCENTRATION ON THE BAND GAP

When semiconductors are heavily doped—that is, they have dopant concentrations greater than 10^{18} atoms/cm³—the average distance between the dopant atoms becomes smaller. The electron wave functions of the dopant atoms begin to overlap, which is similar to Figure 2.25. Instead of getting a single energy level associated with the dopant atoms, we get an additional band of energy levels associated with the dopant levels. Since the atomic distance between dopants is longer than the distance between the host lattice atoms, the band associated with the dopants generally is narrower than the valence band or the conduction band. If these dopant energy levels are shallow, that is, close to the band edges (E_c and E_v), the dopant band is overlapped with the main bands and the conduction band edge (i.e., the energy level at the bottom of the overlapped band) decreases. Therefore, the overall energy gap between the conduction band and the valence band becomes smaller.

This reduction in the energy gap (i.e., band gap) as a function of the dopant concentration is given by the following equation:

$$\Delta E_g (N) = -\frac{3q}{16\pi\varepsilon_r}\sqrt{\frac{q^2 N}{\varepsilon_r k_B T}} \tag{3.52}$$

where N is the dopant concentration, q is the electronic charge, ε_r is the dielectric constant, T is the temperature in K, and k_B is the Boltzmann's constant.

For silicon, the dielectric constant (ε_r) is 11.8, and Equation 3.52 is reduced to

$$\Delta E_g (N) = -22.5\sqrt{\frac{N}{10^{18}}} \text{ meV} \tag{3.53}$$

This change in the band gap for different semiconductors is shown in Figure 3.31 (Van Zeghbroeck 2004).

In some semiconductor devices, materials or parts of a material are doped so heavily that the Fermi energy level is very close to or beyond the band edge (conduction or valence). These semiconductors with such a high dopant concentration are known as *degenerate semiconductors*. The electric properties of degenerate semiconductors are closer to those of metals, although the optical

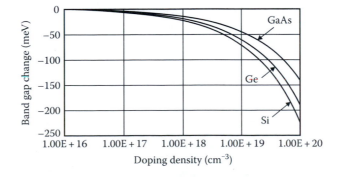

FIGURE 3.31   The relative change in the band gap of silicon as a function of dopant concentrations. (From Van Zeghbroeck, B. 2004. *Principles of Semiconductor Devices*. Available at http://ece-www.colorado.edu/~bart/book/. With permission.)

properties of the degenerate semiconductors are closer to those of the traditional semiconductors. Further discussion on E_g and E_F of the heavily doped semiconductors can be found in Section 4.9. An example of the degenerate semiconductors is tin-doped indium oxide (known as ITO). The origin of the conductivity in ITO is similar to one in silicon, which will be described in Section 3.21. The electric resistivity of ITO is as low as $2 \times 10^{-4}\ \Omega \cdot cm$, which is slightly higher than that of metals ($\sim 10^{-6}\ \Omega \cdot cm$). However, the ITO is still transparent for visible light, which is similar to wide band gap oxide semiconductors. Hence, ITO is widely used as a transparent conductor in optoelectric devices such as displays, light-emitting diodes, and emerging solar cells. Calculation of the Fermi energy level in the degenerated semiconductors is found in Section 4.9.

3.20 EFFECT OF CRYSTALLITE SIZE ON THE BAND GAP

So far, we have assumed that the materials we have considered are in a bulk form. This means that the size of the semiconductor crystals or the thickness of any thin films used in devices is significantly larger than the size of the atoms. In recent years, there has been considerable interest in the nanoparticles of semiconductors, known as *quantum dots*. These semiconductor nanocrystals, between ~2 and 10 nm, have many interesting optical and electrical properties. For example, when the size of a semiconductor crystal is reduced to a few nanometers, the band gap (E_g) becomes larger compared to the band gap of a bulk material. This is due to the fact that, as the number of atoms participating in the overlap of the wave functions decreases, so do the widths of the conduction and valence bands in the quantum dots. Thus, in direct band gap material, the wavelength of light absorbed or emitted is shorter. This is known as a *blue shift*. In the case of cadmium selenide (CdSe), a direct band gap semiconductor, larger crystals have a lesser band gap energy and appear red. As the crystallite size decreases, the band gap increases due to the so-called quantum confinement effect, and the color of the nanocrystals changes to yellow (Reed 1993).

3.21 SEMICONDUCTIVITY IN CERAMIC MATERIALS

In Chapter 2, we saw that most ceramic materials are considered to be electrical insulators. However, many ceramic materials exhibit semiconducting behavior. If the semiconductivity is due to the movement of ions, the materials are known as *ionic conductors*. We can also dope ceramics with a donor or an acceptor, similar to the way in which Si and GaAs are doped. If the predominant charge carriers are electrons or holes, these ceramics are known as *electronic conductors*.

Semiconducting compositions of ceramic barium titanate ($BaTiO_3$) are used in sensors. The insulating or nonsemiconducting formulations of $BaTiO_3$ are used for manufacturing capacitors. For $BaTiO_3$, Nb^{5+} ions (in the form of niobium oxide [Nb_2O_5]) are used as a donor dopant. In this case, Nb^{5+} ions occupy titanium ion (Ti^{4+}) sites. Since each niobium ion brings in five valence electrons (compared with four from titanium), niobium functions as a donor dopant for $BaTiO_3$ (Figure 3.32). This makes $BaTiO_3$, which is otherwise considered an insulator (E_g ~3.05 eV), behave as an n-type semiconductor. As seen in Chapter 1, we need to balance the site, mass, and electrical charge

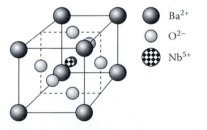

FIGURE 3.32 Donor doping of barium titanate ($BaTiO_3$) using Nb^{5+} (in the form of Nb_2O_5) ions that occupy Ti^{4+} sites results in an n-type semiconductor.

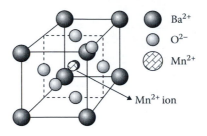

FIGURE 3.33 Illustration of acceptor-doped $BaTiO_3$, created by adding MnO. Each Mn^{2+} ion going on the titanium (Ti^{4+}) site creates two holes.

while considering the doping of compounds. This balance can be expressed using the *Kröger–Vink notation* (Section 1.14). If Y^{3+} ion is doped into the site of Ba^{2+}, the electron concentration in the conduction band of $BaTiO_3$ also increases and the ceramics exhibit the electric properties of the semiconductors.

We can create acceptor-doped or p-type $BaTiO_3$ formulations by adding manganese oxide (MnO). When we add Mn^{2+} ions in the form of MnO, they occupy the titanium (Ti^{4+}) sites (Figure 3.33). Each Mn^{2+} ion has a charge of +2 and occupies a Ti^{4+} site. This creates two holes. If Al_2O_3 is doped into $BaTiO_3$, Al occupies a Ti^{4+} site and the ceramics also show p-type semiconductivity.

Similarly, in *solid oxide fuel cells,* strontium (Sr)-doped lanthanum manganite ($LaMnO_3$) is used as a cathode material. In this material, the divalent Sr^{2+} ions occupy trivalent lanthanum ion (La^{3+}) sites. This creates a p-type semiconductor ceramic composition.

In many solid oxide fuel cells, yttria (Y_2O_3)-doped or yttria-stabilized zirconia (ZrO_2; YSZ) is used as an ionic conductor. As we studied in Example 1.8, the ionic conductivity of YSZ is due to the addition of Y_2O_3 to ZrO_2, which creates oxygen ion vacancies. To compensate for the charge imbalance of Y^{3+} taking the site of Zr^{2+}, one oxygen vacancy is formed per two Y_{Zr}^+ defects. This oxygen vacancy facilitates the diffusion of oxygen ions, thereby creating an ionic conductor. The fuel cells make use of air and fuel gases such as natural gas. They operate at relatively high temperatures (~1000°C) and generate electric power.

PROBLEMS

3.1 Si has a density of 2.33 g/cm^3. The atomic mass of Si is 28 and the valence is 4—that is, each atom has four valence electrons. Calculate the density of valence electrons in the valence band assuming all electrons are in the valence band.

3.2 The density of GaAs is 5.31 g/cm^3. The atomic masses of Ga and As are 69.7 and 74.92, respectively. Show that the density of atoms—that is, the total number of Ga and As atoms per cubic centimeter in GaAs—is 4.42×10^{22}.

3.3 At 300 K, if the concentration of electrons excited to the conduction band is 1.5×10^{10} cm^{-3}, what fraction of the total concentration of electrons is excited to the conduction band?

3.4 What is the resistivity of an intrinsic Si sample at 300 K? Assume that $T = 300$ K and the concentration of electrons in the conduction band is $n_i = 1.5 \times 10^{10}$ cm^{-3}.

3.5 What is the concentration of electrons excited to the conduction band for Si at 400 K? (See Figure 3.7.)

3.6 What is the conductivity of intrinsic Si at 400 K? Assume that the mobility is proportional to $T^{-3/2}$ in this region, and use the carrier concentrations from Figure 3.7.

3.7 Silicon carbide (SiC) is used in high-temperature, high-power, and high-frequency device applications. It exhibits various polytypes; 4H and 6H are some of the most widely used. What is the intrinsic carrier concentration for the 6H form of SiC at 666 K, if the carrier concentration changes with temperature as shown in Figure 3.34?

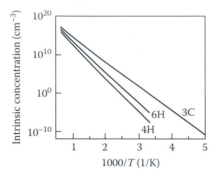

FIGURE 3.34 Changes in the intrinsic carrier concentration for SiC. (From Goldberg et al., eds.: *Properties of Advanced Semiconductor Materials GaN, AlN, SiC, BN, SiC, SiGe.* 2001. Copyright Wiley-VCH Verlag GmbH & Co. KGaA. Reproduced with permission.)

3.8 The concentration of atoms in Si is ~5 × 10^{22} atoms/cm³. If a Si crystal contains one part per billion (ppb) of Sb, what is the concentration of Sb expressed as atoms per cubic centimeter?

3.9 What is the resistivity of this Si containing 1 ppb of antimony? Assume $T = 300$ K and the same mobility values as for intrinsic Si.

3.10 Consider a Si crystal that has 1 ppb of Al as a dopant. What is the concentration of this dopant expressed as atoms per cubic centimeter?

3.11 A Si sample is doped with 10^{-4} atom % of P. Using the atomic mass of Si (28) and its density (2.33 g/cm³), show that the donor dopant concentration is 5 × 10^{22} atoms/cm³. From this, show that the conductivity of this n-type Si is ~1200 $\Omega^{-1} \cdot$ cm⁻¹.

3.12 What is the resistivity of a Si sample doped with 1 ppb of Al? Why is this resistivity higher than that for a sample doped with 1 ppb of Sb? Use your answer to justify the trends in the data shown in Figure 3.29.

3.13 Two crystals of SiC of the so-called 4H type were grown. One was grown in a N atmosphere and the other in an argon (Ar) atmosphere. Resistivity measurements showed that one of the crystals had a very low resistivity (0.007 $\Omega \cdot$ cm; Siergiej et al. 1999). Which crystal do you think this is—the one grown in a N atmosphere or in a Ar atmosphere? Explain.

3.14 The mobility of electrons in n-doped 4H-type SiC is ~1000 cm²/V · s (Siergiej et al. 1999). What is the anticipated N-doping level in an n-type doped SiC crystal that has a resistivity of 0.007 $\Omega \cdot$ cm?

3.15 What do the terms "radiative recombination" and "nonradiative recombination" mean? Explain using a sketch.

3.16 Can radiative recombination occur in indirect band gap materials? Explain.

3.17 Can nonradiative recombination occur in direct band gap materials? Explain.

3.18 An LED is made using a compound that is a solid solution between InAs and GaAs. This compound can be described as $In_xGa_{1-x}As$, where x is the mole fraction of In. If the wavelength of the LED made using this compound is 1300 nm, what is the band gap? Assume $T = 300$ K.

3.19 Amber and orange LEDs are made using AlGaInP compositions. If the wavelength of the orange LEDs is 0.6 Lmi, what is the band gap of this composition? Assume $T = 300$ K.

3.20 An LED is made using an AlGaAs composition with a band gap of 1.8 eV. What is the color of this LED? What is the wavelength in nm? Assume $T = 300$ K.

3.21 The dielectric constant of Ge is 16 (Table 3.2). The conductivity effective mass of electrons in Ge is 0.12 m_0. Calculate the difference between donor ionization energy (E_d) and conduction band edge (E_c) for Ge doped with P (see Equation 3.35).

3.22 Calculate the donor energy level position for indium arsenide (InAs), a direct band gap material. The dielectric constant is 15. The effective mass of electrons is 0.027 m_0.

3.23 What is isoelectronic doping? Explain how this can be useful for some applications.

3.24 At high dopant concentrations ($N > 10^{18}$ atoms/cm³), the band gap of a semiconductor becomes smaller. Derive a simplified equation (similar to Equation 3.53) for the change in band gap energy (in meV) for Ge. Plot the data similar to that in Figure 3.31. Notice that, for a given dopant level, the change in the band gap is the maximum for the semiconductor that has the highest dielectric constant.

3.25 Derive an expression similar to Equation 3.53 for the change in band gap as a function of dopant concentration for GaAs.

3.26 Calculate the band gap of Si, Ge, and GaAs at 200, 400, and 500 K using the Varshni parameters shown in Table 3.5.

GLOSSARY

Acceptor: An atom or ion that provides the source of holes in an extrinsic semiconductor (e.g., boron in silicon).

Acceptor-doped semiconductor (p-type): A semiconductor doped with an element that results in an electron deficit (e.g., boron-doped silicon).

Amphoteric dopants: Dopants that can act as a donor or an acceptor (e.g., silicon in GaAs).

Band gap energy (E_g): The energy difference between the conduction and valence band edges.

Band structure: The relationship between the energy (E) of a charge carrier and the wave vector (**k**).

Compensated semiconductor: A semiconductor that contains both donor and acceptor types of dopants. The semiconductor behaves as an n-type if the donor concentration ultimately outweighs the acceptor concentration and $N_d - N_a \gg n_i$; if $N_a - N_d \gg n_i$, then the material behaves as a p-type semiconductor.

Compound semiconductor: A semiconductor based on a compound (e.g., GaAs).

Conduction band edge (E_c): The energy level corresponding to the bottom of the conduction band.

Conductivity effective mass $\left(m_\sigma^*\right)$: The mass used to calculate the donor position level in indirect band gap semiconductors.

Deep-level defects: Energy level defects created deep in the band gap, usually by atoms or ions that do not easily "fit" in the crystal structure and create a severe distortion.

Degenerate semiconductors: Heavily doped semiconductors with a Fermi energy level very close to the band edge.

Density of states effective mass $\left(m_{dos}^*\right)$: Defined for electrons as $m_{dos}^* = \left[6^{2/3}\left(m_l^* m_t^{*2}\right)^{1/3}\right]$ and for holes as $\left(m_{dos}^*\right)^{3/2} = \left(m_{hh}^{3/2} + m_{lh}^{3/2}\right)$. It is used for the calculation of the density of states or for the position of the Fermi energy level.

Dielectric constant (ε_r): A measure of the ability of a material to store a charge. It is the ratio of the permittivity of a material (ε) to the permittivity of the free space (ε_0).

Direct band gap semiconductor: A material in which an electron can be transferred from the conduction band into the valence band with emission of light and without a change in its momentum.

Donor: Atoms or ions that are the source of conduction electrons in an extrinsic semiconductor (e.g., phosphorus in silicon).

Donor-doped semiconductor (n-type): A semiconductor in which the dopant atoms bring in "extra" electrons that become dissociated from the dopant atoms and provide semiconductivity.

Dopant: An element or a compound added to a semiconductor to enhance and control its electrical or other properties.

Dopant ionization: The process by which the electrons from the dopant atoms or holes from the acceptor atoms dissociate themselves from their parent atoms.

Effective density of states (N): A parameter that represents the effective allowed number of energy levels at a given band edge conduction (N_c) or (N_v).

Effective mass of an electron $\left(m_e^*\right)$: The mass that an electron would appear to have as a result of the attractive and repulsive forces an electron experiences in a crystal. It is different from the mass of an electron in a vacuum (m_0).

Electron–hole pair (EHP): The process by which an electron in the valence band is promoted to the conduction band, creating a hole in the valence band.

Elemental semiconductor: An element that shows semiconducting behavior (e.g., silicon).

Extrinsic semiconductor: A donor-doped or acceptor-doped semiconductor. The conductivity of an extrinsic semiconductor is controlled by the addition of dopants.

Freeze-out range: The low-temperature range in which the electrons and the holes remain bonded to the dopant atoms and are not available for conduction.

Hole: A missing electron in a bond that is treated as an imaginary particle with a positive charge q.

Impurities: Elements or compounds present in semiconductors either inadvertently or because of processing limitations. Their presence is usually considered undesirable.

Impurity- or dopant-scattering limited mobility: The value of mobility at low temperatures (but above the carrier freeze-out) limited by the conduction electrons scattering off of the dopant or impurity atoms.

Indirect band gap semiconductor: A semiconductor in which an electron that has been excited into the conduction band transfers into the conduction band with a change in its momentum.

Intrinsic semiconductor: A material whose properties are controlled by thermally generated carriers. It does not contain any dopants.

Isoelectronic dopants: Dopants that have the same valence as the sites they occupy (e.g., germanium in silicon or nitrogen in GaP–GaAs materials). This does not create n-type or p-type behavior.

Kröger–Vink notation: Notation used for expressing point defects in materials.

Lattice-scattering limited mobility: The limited mobility of carriers at high temperatures caused by the scattering of vibrations of atoms known as phonons.

Majority carriers: Carriers of electricity that dominate overall conductivity; for example, in an n-type semiconductor, electrons created by donor atoms would be the majority carriers.

Minority carriers: Carriers of electricity that do not dominate the overall conductivity; for example, in an n-type semiconductor, holes are the minority carriers, and in an ionic conductor, both electrons and holes are the minority carriers.

Nonradiative recombination: A process by which electrons and holes recombine, transferring the energy to vibrations of atoms or phonons. The energy appears as heat, and the process occurs in both direct and indirect band gap materials.

n-type semiconductor: A semiconductor in which the majority carriers are electrons; also known as a donor-doped semiconductor.

Optoelectronic devices: Devices such as light-emitting diodes (LEDs) that usually are based on direct band gap materials that have a strong interaction with light waves.

p-type semiconductor: A semiconductor in which the majority carriers are holes; also known as an acceptor-doped semiconductor.

Quantum dots: Nanocrystalline semiconductors whose band gap is larger than that of a bulk material.

Radiative recombination: The process by which an electron that has been excited to the conduction band falls back to the valence band, recombining with a hole and causing emission of light. It may involve a defect energy level.

Semiconductors: Elements or compounds with resistivity (ρ) between 10^{-4} and 10^3 ($\Omega \cdot cm$). This range is approximate.

Shallow-level dopants: Dopants with energy levels only a few millielectron volts from the band edge.

Solid oxide fuel cells: Fuel cells that make use of ceramic oxides, such as yttria-doped zirconia, as an electrolyte.

Thermally generated charge carriers: Electrons and holes that are generated as electrons jump from the valence band into the conduction band as a result of thermal energy.

Valence band edge (E_v): The energy level corresponding to the top of the valence band.

Varshni formula: A formula that shows the variation in the band gap with temperature (Equation 3.51).

Vertical recombination: The process of radiative recombination without a change in momentum.

Wave vector (k): In the quantum mechanics–based approach, an electron that is considered a plane wave with a propagation constant.

REFERENCES

Ahlgre, D. C., and J. Dunn. 2000. SiGe comes of age. *MicroNews-IBM* 6(1):1.

Askeland, D., and P. Fulay. 2006. *The Science and Engineering of Materials*. Washington, DC: Thomson.

Dorkel, J. M., and Ph. Leturcq. 1981. Carrier mobilities in silicon semi-empirically related to temperature, doping, and injection level. *Solid State Electron* 24(9):821–5.

Goldberg, Y., M. E. Levinshtein, and S. L. Rumyantsev, eds. 2001. *Properties of Advanced Semiconductor Materials GaN, AlN, SiC, BN, SiC, SiGe*. New York: Wiley.

Grove, A. S. 1967. *Physics and Technology of Semiconductor Devices*. New York: Wiley.

Kano, K. 1997. *Semiconductor Fundamentals*. Upper Saddle River, NJ: Prentice Hall.

Kasap, S. O. 2002. *Principles of Electronic Materials and Devices*. New York: McGraw Hill.

Levinshtein, M., S. Rumyantsev, and M. Shur, eds. 1999. *Handbook Series on Semiconductor Parameters*. London: World Scientific.

Li, S. S., and W. R. Thurber. 1977. The dopant density and temperature dependence of electron mobility and resistivity in n-type silicon. *Solid State Electron* 20(7):609–16.

Mahajan, S., and K. S. Sree Harsha. 1998. *Principles of Growth and Processing of Semiconductors*. New York: McGraw Hill.

Queisser, H. J., and E. E. Haller. 1998. Defects in semiconductors: Some fatal, some vital. *Science* 281:945–50.

Reed, M. A. 1993. Quantum dots. *Scientific American* (January):118–23.

Schubert, F. E. 2006. *Light-Emitting Diodes*. Cambridge, UK: Cambridge University Press.

Siergiej, R. R., R. C. Clark, S. Sriram, A. K. Agarwal, R. J. Bojko, A. W. Mors, V. Balakrishna, et al. 1999. Advances in SiC materials and devices: An industrial point of view. *Mater Sci Eng (B)* B61–62:9–17.

Singh, J. 1994. *Semiconductor Devices: An Introduction*. New York: McGraw-Hill.

Singh, J. 2001. *Semiconductor Devices: Basic Principles*. New York: Wiley.

Streetman, B. G., and S. Banerjee. 2000. *Solid State Electronic Devices*. Upper Saddle River, NJ: Prentice Hall.

Sze, S. M. 1985. *Semiconductor Devices, Physics, and Technology*. New York: Wiley.

Van Zeghbroeck, B. 2004. *Principles of Semiconductor Devices*. Available at http://ece-www.colorado.edu/~bart/book/.

4 Fermi Energy Levels in Semiconductors

4.1 FERMI ENERGY LEVELS IN METALS

We have seen in Chapter 1 that, in metals, the outermost band usually is nearly half-filled. Let us define the energy at the bottom of the partially filled band as $E = 0$ when $E = E_B$ (Figure 4.1).

In metals at around 0 K, the highest energy level at which electrons are present is known as the Fermi energy level (E_F). Above this level, the states are available for electrons, but no electrons occupy these states. Thus, in metals at around 0 K, the probability of finding an electron at $E > E_F$ is zero. The Fermi energy values for different metals are shown in Table 4.1. This table also shows the value of the *work function* (ϕ), which for metals refers to the energy needed to remove the outermost electron from $E = E_F$. For example, the E_F value for copper (Cu) is 7.0 eV (Table 4.1). Thus, as shown in Figure 4.1, the work function (ϕ) of copper is 4.65 eV.

From the quantum mechanical consideration, the Fermi energy level for metals at 0 K is given by

$$E_{\text{F, metal at 0 K}} = E_{\text{F,0}} = \left(\frac{\hbar^2}{2m_\text{e}} \right)\left(3\pi^2 n\right)^{2/3} \text{ (for metals)} \tag{4.1}$$

where m_e is the rest mass of the electron $\hbar = h/2\pi$, h is Planck's constant, and n is the free electron concentration in the metal. Equation 4.1 clearly shows that the metal with the higher free electron concentration (n) has higher E_F.

Example 4.1 illustrates the calculation of the Fermi energy level for a metal.

Example 4.1: Fermi Energy of a Metal at *T* = 0 K

In Cu, if we assume that each Cu atom donates one electron, the free electron concentration is 8.5×10^{22} cm^{-3}. (a) What is the value of $E_{\text{F,0}}$? (b) If the concentration of conduction electrons estimated by conductivity measurements is 1.5×10^{23} cm^{-3}, then find the number of electrons donated per Cu atom.

Solution
1. We first convert the concentration of electrons per cubic centimeter to cubic millimeters and then substitute the rest mass of electrons (9.1×10^{-30} kg) and the value of $\hbar$ as 1.05×10^{-34} J · s.

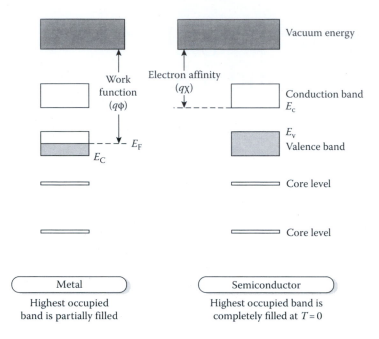

FIGURE 4.1 Fermi energy level representation for a metal and a semiconductor. In a pure semiconductor, the Fermi energy level (E_F) is located at the center of the forbidden gap and ($E_C - E_F$) is the same as ($E_F - E_V$).

TABLE 4.1

Fermi Energy Levels ($E_{F,0}$) and Work Functions (ϕ) for Different Metals

Metal	Work Function (ϕ) (eV)	Fermi Energy Level at 0 K ($E_{F,0}$)
Silver	4.26	5.5
Aluminum	4.28	11.7
Gold	5.1	5.5
Copper	4.65	7.0
Lithium	2.3	4.7
Magnesium	3.7	7.1
Sodium	2.75	3.2

Source: Kasap, S.O., *Principles of Electronic Materials and Devices*, McGraw Hill, New York, 2002.

From Equation 4.1,

$$E_{F,0} = \frac{\left(1.05\times10^{-34}\,\text{J}\cdot\text{s}\right)^2}{2\times\left(9.1\times10^{-30}\,\text{kg}\right)}\left(3\pi^2\times\left(8.5\times10^{28}\,\text{m}^{-3}\right)\right)^{2/3} = 1.135\times10^{-18}\,\text{J}$$

$$= \frac{1.135\times10^{-18}\,\text{J}}{1.6\times10^{-19}\,\text{J/eV}} = 7\,\text{eV}$$

2. From the conductivity measurements, the estimated electron concentration in Cu is ~1.15×10^{29} m^{-3}. Thus, the average number of electrons contributed by a Cu atom is larger than 8.5×10^{28}.

The number of electrons donated per Cu atom is given by

$$\frac{1.15 \times 10^{29}\,\text{m}^{-3}}{8.5 \times 10^{28}\,\text{m}^{3}} = 1.36$$

This value seems reasonable. In many compounds, Cu exhibits a valence of +1 and +2, for example Cu_2O and CuO.

The Fermi energy of metals does vary slightly with temperature and is given by

$$E_F(T) = E_{F,0}\left[1 - \frac{\pi^2}{12}\left(\frac{k_B T}{E_{F,0}}\right)^2\right] \qquad (4.2)$$

Note that the $E_{F,0}$ for metals is much higher than the $k_B T$ (Table 4.1). As a result, the E_F of metals does not change significantly with temperature. The Fermi energy level of semiconductors, especially those that are extrinsic (i.e., doped with impurities), does change appreciably with temperature (see Section 4.6).

An electron with energy $E = E_F$ represents the highest energy electron in a metal. This is the significance of the Fermi energy level in metals. If v_F is the speed of an electron with energy $E_{F,0}$, then

$$\frac{1}{2}mv_{F,0}^2 = E_{F,0} \qquad (4.3)$$

From the typical values of $E_{F,0}$ for metals, we can see that at $T = 0$ K, the highest speed of electrons with energy $E_{F,0}$ is ~10^6 m/s (see Equation 4.3). In quantum mechanical considerations, only these extremely high, fast-moving electrons near E_F deliver the electric charge. The *effective speed of electrons* (v_e) or the *root mean speed of electrons* in a metal is given by

$$\frac{1}{2}mv_e^2 \cong \frac{3}{5}E_{F,0} \qquad (4.4)$$

The effective speed of electrons (v_e) in a metal is relatively insensitive to temperature and depends on $E_{F,0}$. As mentioned in Chapter 3, however, if electrons are considered classical particles, then the speed of electrons at 0 K should be zero. This different electron speed at 0 K makes a distinction between classical theory and quantum mechanics.

4.2 FERMI ENERGY LEVELS IN SEMICONDUCTORS

We discussed Fermi Energy Level of semiconductors in Section 2.16. At low temperatures, the conduction band of semiconductors is empty, and the valence band is completely filled. Thus, the probability of finding an electron at $T = 0$ K at the valence band edge ($E = E_v$) is 1. As the temperature increases, some electrons make a transition across the band gap into the valence band. The probability of finding an electron at or above the conduction band edge ($E \geq E_c$) is zero. At higher temperatures (e.g., 300 K), the probability of finding an electron in the valence band is still close to 1, but it becomes slightly smaller than 1 near the valence band edge (E_v). In addition, there is a small (but nonzero) probability of finding an electron in the conduction band edge as well, that is, $E \geq E_c$. Thus, as we go from $E = E_v$ to $E = E_c$ and higher in the conduction band, the probability of finding an electron begins at 1 and eventually approaches zero well into the conduction band.

In general, the probability function for a system of classical particles with different energy (E) is given by the Boltzmann function:

$$f(E) = A \exp\left(\frac{-E}{k_B T}\right) \qquad (4.5)$$

where A is a constant, k_B is the Boltzmann constant, and T is the temperature. In addition to the classical Boltzmann function, electron occupation of quantized energy levels in materials is also constrained by Pauli's exclusion principle, which states that no two electrons can have the same energy level (i.e., the same set of quantum numbers). With this restriction on the occupancy of states, the probability of finding an electron occupying energy level E is given by the so-called *Fermi–Dirac distribution function*, which is defined as

$$f(E) = \frac{1}{\left[1 + \exp\left(\dfrac{E - E_F}{k_B T} \right) \right]} \tag{4.6}$$

where E is the energy of the electron, $f(E)$ is the probability of an electron occupying the state with energy E, k_B is the Boltzmann constant, T is the temperature, and E_F is the *Fermi energy level*. Thus, the Fermi–Dirac function gives the probability of filling the energy level E with an electron at T.

In Figure 4.2, readers can find that the Fermi–Dirac function becomes a step function at $T = 0$ K. If $E < E_F$, $f(E)$ is 1 at $T = 0$ K. This means that every available level below E_F is filled at $T = 0$ K. The filling of all energy states below E_F is the same for both semiconductors and metals at $T = 0$ K. Also, note that at $T = 0$ K, $f(E) = 0$ if $E > E_F$, which means that every energy level above E_F is empty.

As discussed in Section 2.16, the Fermi energy level (E_F) is the energy level for which the probability of finding an electron is 0.5 at $T > 0$ K. This feature of E_F is well reflected in Equation 4.6. Based on the band diagram of a semiconductor, E_F will be in the middle of the band gap (between $E = E_v$ and $E = E_c$), where no electrons are allowed. In other words, though the probability of filling at E_F is 50%, the actual electron density at E_F is determined by multiplying the Fermi–Dirac function with a number of electron enegy states. Since no electron energy states are allowed in the band gap of an intrinsic semiconductor, the actual number of electrons staying at E_F is zero. In intrinsic semiconductors, E_F lies at the center of the band gap (a forbidden region between E_v and E_c). In extrinsic semiconductors, E_F is closer to E_v for the p-type semiconductor and E_c for the n-type semiconductor.

At a given temperature (e.g., $T = T_1$), as E increases, $f(E)$ decreases. This indicates that the probability of finding an electron decreases, as shown in Figure 4.2.

In the regime $E > E_F$, an increase in the temperature increases $f(E)$. This means that the probability of finding an electron at higher energy levels becomes higher as the temperature increases.

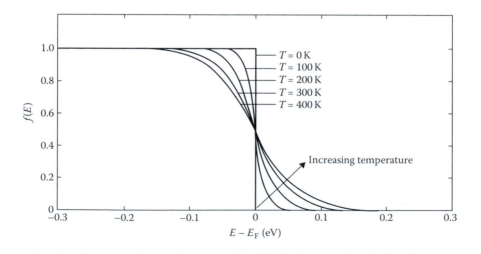

FIGURE 4.2 Fermi–Dirac function at different temperatures. E_F remains constant. (From Pierret, R.F., *Advanced Semiconductor Fundamentals*, Addison-Wesley, Reading, MA, 1987. With permission.)

In the regime $E < E_F$, an increase in the temperature decreases $f(E)$. That is, the probability of finding an electron at lower energy levels becomes smaller as the temperature increases because chances are better that the electrons have moved up to higher energy levels.

An electron creates a hole in the valence band when it is promoted to the conduction band. Because a hole can be considered to be a missing electron, the probability of finding a hole is given by

$$f(E)_{\text{hole}} = 1 - f(E) = 1 - \frac{1}{1 + \exp\left(\dfrac{E - E_F}{k_B T}\right)} \tag{4.7}$$

Another interesting situation to consider is that $(E - E_F)$ is far greater than $k_B T$ in silicon and GaAs with the band gap 1.12 and 1.42 eV for energy levels in the valence or conduction bands. If $(E - E_F) \gg k_B T$, then $\exp[(E - E_F)/k_B T] \gg 1$ and Equation 4.6 can be simplified as:

$$f(E) = \frac{1}{\left[\exp\left(\dfrac{E - E_F}{k_B T}\right)\right]} = \exp\left[-\left(\frac{E - E_F}{k_B T}\right)\right] \tag{4.8}$$

Note the negative sign in Equation 4.8. This equation is a form of the Boltzmann equation for classical particles. Thus, the *tail* of the Fermi–Dirac distribution (i.e., when E is much greater than E_F) can be described by the Boltzmann distribution equation. The importance of this is as follows: In the conduction band of a semiconductor, the number of the electron energy states available is significantly greater than the number of free electrons. Therefore, the chance of two electrons occupying the same energy level (i.e., having the same quantum numbers) is not very high. Thus, the Paul's exclusion principle ruling that two electrons cannot take the same electron energy state is not broken. We can use the Boltzmann equation (Equation 4.8) to calculate the concentration of electrons in the conduction band and the concentration of holes in the valence band (Section 4.3).

4.3 ELECTRON AND HOLE CONCENTRATIONS

The conductivity of a semiconductor depends on the carrier concentrations and their mobility. In Sections 3.14–3.16, we reviewed the effect of temperature and dopant concentration on the carrier mobility. In this section, we will learn how to calculate the carrier concentrations. If we know the probability of finding an electron at a given energy level in the conduction band, and if we know the number of the electron energy states available at a given energy level in the conduction band, we can then calculate the total number of electrons in the conduction band. The following analogy may help.

Imagine a high-rise hotel building with many guests. We want to know how many guests are in the hotel at a given time on the assumption that only one guest can stay in each room (a limitation imposed by the Pauli exclusion principle). In this case, first, we need to find how many rooms there are on each floor. Second, the probability of the room to be filled by one guest needs to be known. Third, we then calculate the total number of guests in the hotel by (i) multiplying the number of rooms and the filling probability of each floor to figure out the number of guests on each floor and (ii) adding the guest number from the bottom floor to the top floor of the hotel. The same logic can be applied to calculating the number of electrons in a conduction band.

Consider the band diagram for a typical intrinsic semiconductor (Figure 4.3). Note that the vertical axis is the electron energy (E). Here, a concept corresponding to a number of rooms available at each floor is called density of states, $N(E)$, which is a number of electron states at a specific energy level E. As $N(E)$ increases, more states are available for electrons at energy E, and a maximum number of electrons staying at E increases. A general formula of $N(E)$ is found in Equation 4.48. Note that $N(E)$ is proportional to $E^{1/2}$. A detailed calculation process to derive $N(E)$ is beyond the scope of this book.

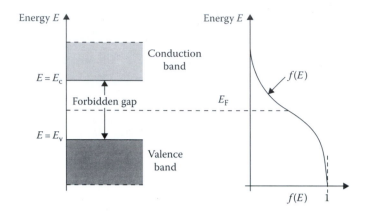

FIGURE 4.3 Band diagram and the Fermi–Dirac function for an intrinsic semiconductor at $T > 0$. (From Pierret, R.F., *Semiconductor Fundamentals Volume*, Addison-Wesley, Reading, MA, 1988. With permission.)

The probability of finding an electron at E is again calculated by the Fermi–Dirac function, $f(E)$. If the state is filled by an electron, $f(E)$ is 1. If the state is empty, $f(E)$ is 0. Then we can integrate $N(E) \times f(E)$ over the energy levels of interest.

The electron energies in the conduction band range from $E = E_c$ to $E = \infty$. At any energy level E, a product of the number of available energy levels $N(E)$ and the probability of filling the state $f(E)$ gives us the number of electrons taking the energy level E. We integrate the product $N(E) \times f(E)$ over the range $E = E_c$ to $E = \infty$ to get the total concentration of electrons in the conduction band. Strictly speaking, the top limit for the integration should be $E_c + q \times \chi$ (χ: electron affinity). When the energy of an electron becomes larger than $E_c + q \times \chi$, the electron becomes free from the attraction of the nucleus (Figure 4.1), $N(E)$ becomes zero and $N(E) \times f(E)$ equals zero. However, the function $f(E)$ approaches zero, as E tends to ∞. Therefore, we will use $E = \infty$ as the top limit for the integral for convenience.

$$n = \int_{E=E_c}^{E=\infty} N(E)f(E)dE \tag{4.9}$$

where n is the total number of electrons taking the energy levels higher than E_c. The density of states (the shaded area in the left plot of Figure 4.3) and $f(E)$ (the right plot of Figure 4.3) are used for the integral shown in Equation 4.9.

Here, we will introduce a new approach, the *effective density of states approximation*. In the semiconductor, we assume that most, if not all, electrons reside at the bottom energy level of the conduction band ($E = E_c$) to keep the sum of electron energies low. Even though the valence electrons of the semiconductor can jump to higher energy levels in the conduction band, the excited electrons still stay at the bottom of the conduction band to minimize the total energy of the system. Given the physical characteristics of the semiconductor, this assumption is reasonable. If electrons fill the energy levels above E_c, the conduction band is treated as a partially filled band, and the material cannot be treated as a semiconductor. Remember that the metal has the partially filled valence band and the semiconductor has the fully empty conduction bands. According to the quantum mechanical calculations (Section 3.17), the effective density of states at the conduction band edge $N(E_c)$ is noted as N_c, which is given by

$$N_c = 2\left(\frac{m_n^* k_B T}{2\pi\hbar^2}\right)^{3/2} \tag{4.10}$$

As explained above, in semiconductors where the excited electrons occupy only a part of the energy states at E_c, $N(E) \times f(E)$ is *not zero only at* the bottom conduction band edge ($E = E_c$). Then,

Equation 4.9 becomes $n = N(E_c) f(E_c) = N_c f(E_c)$, where $N(E_c)$ is the effective density of states at the band edge N_c. Thus, Equation 4.9 can be written as:

$$n = 2 \left(\frac{m_n^* k_B T}{2\pi\hbar^2} \right)^{3/2} f(E_c) \tag{4.11}$$

Note that because we assumed that most electrons in the conduction band reside at the bottom, we multiplied the effective density of states by the value of the probability function at $E = E_c$.

From the definition of the Fermi–Dirac function (Equation 4.6), we get

$$f(E_c) = \frac{1}{\left[1 + \exp\left(\frac{E_c - E_F}{k_B T} \right) \right]} \tag{4.12}$$

Since $(E_c - E_F) \gg k_B T$, the exponential term is much larger than 1, and hence $f(E_c)$ is written as

$$f(E) = \frac{1}{\left[\exp\left(\frac{E_c - E_F}{k_B T} \right) \right]} = \exp\left[-\left(\frac{E_c - E_F}{k_B T} \right) \right] \tag{4.13}$$

Substituting the value of $f(E_c)$ from Equation 4.13 into Equation 4.11, we get

$$n = 2 \left(\frac{m_n^* k_B T}{2\pi\hbar^2} \right)^{3/2} \exp\left[-\left(\frac{E_c - E_F}{k_B T} \right) \right] \tag{4.14}$$

By keeping the first term as N_c, n is given by:

$$n = N_c \, \exp\left[-\left(\frac{E_c - E_F}{k_B T} \right) \right] \tag{4.15}$$

The negative sign outside is shown so that the term inside is positive because E_c is above E_F. We can rewrite Equation 4.14 as follows to eliminate the negative sign in the exponential term:

$$n = 2 \left(\frac{m_n^* k_B T}{2\pi\hbar^2} \right)^{3/2} \exp\left(\frac{E_F - E_c}{k_B T} \right) \tag{4.16}$$

In Sections 4.5 and 4.6, we will see that we can calculate the concentration of electrons for intrinsic as well as extrinsic semiconductors using Equation 4.16 if we know the position of the E_F relative to E_c.

We can now derive a similar expression for calculating the concentration of holes in the valence band. Remember that the conduction band is nearly empty and has few electrons, the concentration of which we have calculated (Equation 4.16). Now, the valence band is nearly completely filled with valence electrons in the covalent bonds of the semiconductor atoms. When some of these bonds break, the electrons move to the conduction band. This creates holes in the valence band. The concentration of holes (p) in the valance band is given by

$$p = \int_{E=0}^{E=E_v} N(E)[1 - f(E)] \, dE \tag{4.17}$$

Notice that the density of states $N(E)$ is now multiplied by $1 - f(E)$, which is the probability of finding a hole. We again use the effective density of states at the valence band edge ($N(E_v)$); that is, we assume that most holes remain at the top of the valence band. Recall that the electron energy increases as the energy level moves up the band diagram, whereas the hole energy increases as the energy level goes down the band diagram. The effective density of states at $E = E_v$ is given by

$$N_v = 2\left(\frac{m_p^* k_B T}{2\pi\hbar^2}\right)^{3/2} \tag{4.18}$$

On the assumption that a total number of holes is smaller than the effective density states at E_v, holes will only stay at energy states at E_v. Then, Equation 4.17 is rewritten as

$$p = 2\left(\frac{m_p^* k_B T}{2\pi\hbar^2}\right)(1 - f(E_v)) \tag{4.19}$$

From the Fermi–Dirac function

$$1 - f(E_v) = 1 - \frac{1}{\left[\exp\left(\frac{E_v - E_F}{k_B T}\right) + 1\right]} = \frac{1 + \exp\left(\frac{E_v - E_F}{k_B T}\right) - 1}{\exp\left(\frac{E_v - E_F}{k_B T}\right) + 1} \tag{4.20}$$

$$\therefore 1 - f(E_v) = \frac{\exp\left(\frac{E_v - E_F}{k_B T}\right)}{1 + \exp\left[\frac{(E_F - E_v)}{k_B T}\right]} \tag{4.21}$$

Note that the denominator of the exponential term now contains $(E_F - E_v)$ and not $(E_v - E_F)$.

Again, if $(E_F - E_v) \gg k_B T$, then the Fermi–Dirac function simplifies to a Boltzmann function since $\exp\left[(E_F - E_v)/k_B T\right] \gg 1$. The denominator in Equation 4.21 becomes 1.

Thus, Equation 4.21 now becomes

$$1 - f(E_V) = \exp\left(\frac{E_V - E_F}{k_B T}\right)$$

$$\therefore 1 - f(E_V) = \exp\left[-\left(\frac{E_F - E_V}{k_B T}\right)\right] \tag{4.22}$$

The concentration of holes in the valence band from Equation 4.19 now becomes

$$p = 2\left(\frac{m_p^* k_B T}{2\pi\hbar^2}\right)^{3/2}\left[-\exp\left(\frac{E_F - E_v}{k_B T}\right)\right] \tag{4.23}$$

For the sake of convenience, we can represent the first term as N_v and can write Equation 4.23 as

$$p = N_v \exp-\left[\frac{E_F - E_v}{k_B T}\right] \tag{4.24}$$

The negative sign outside is shown so that the term inside is positive because E_F is above E_v. This equation gives us the concentration of holes in either an intrinsic semiconductor or an extrinsic semiconductor. If we know the difference between the valence band edge and the Fermi energy level, temperature, and the effective mass of holes, we can calculate the concentration of holes (p).

Let us multiply the equations for electron and hole concentrations. From Equations 4.15 and 4.24, we get

$$n \times p = N_c N_v \exp- \left[\frac{E_c - E_F}{k_B T} \right] \exp- \left[\frac{E_F - E_v}{k_B T} \right] = N_c N_v \exp- \left[\frac{E_c - E_v}{k_B T} \right] \qquad (4.25)$$

Since $(E_c - E_v) = E_g$, we get

$$n \times p = N_c N_v \exp \left(\frac{-E_g}{k_B T} \right) \qquad (4.26)$$

Thus, for a given semiconductor, the product of the electron concentration and the hole concentration at a given temperature is constant. This is known as the *law of mass action for semiconductors*. It is similar to the ideal gas law: pressure × volume is a constant. For example, for silicon (Si) ($E_g \approx 1.1$ eV at 300 K), the value of $n \times p$ is constant, whether we have intrinsic, n-type, or p-type silicon. For an n-type silicon crystal, created by doping with antimony (Sb), the electron concentration in the conduction band is higher compared to n_i. The increase in the concentration of electrons in an n-type semiconductor is compensated by a decrease in the concentration of holes in the valence band. Similarly, for a p-type semiconductor, holes are the majority carriers, that is, $p \gg n$. To maintain a constant value of $n \times p$, the concentration of electrons (n) decreases.

4.4 FERMI ENERGY LEVELS IN INTRINSIC SEMICONDUCTORS

For an intrinsic semiconductor, $n_i = p_i$. Applying Equations 4.15 and 4.24 to an intrinsic semiconductor, we get

$$N_C \exp \left[- \left(\frac{E_C - E_{F,i}}{k_B T} \right) \right] = N_V \exp \left[- \left(\frac{E_{F,i} - E_V}{k_B T} \right) \right] \qquad (4.27)$$

In this equation, $E_{F,i}$ is the Fermi energy level of an intrinsic semiconductor. Taking a natural logarithm of both sides, we get

$$E_{F,i} = \frac{1}{2} \left(E_c + E_v \right) + \frac{1}{2} k_B T \ln \left(\frac{N_v}{N_c} \right) \qquad (4.28)$$

Substituting for N_c and N_v from Equations 4.10 and 4.18, respectively, we get

$$E_{F,i} = \frac{1}{2} \left(E_c + E_v \right) + \frac{3}{4} k_B T \ln \left(\frac{m_p^*}{m_n^*} \right) \qquad (4.29)$$

The first term in Equation 4.29 represents the middle of the band gap. Thus, the Fermi energy level of an intrinsic material ($E_{F,i}$) is very close to the center energy level of the band gap (E_{midgap}), as shown in Figure 4.4:

$$E_{F,i} - E_{midgap} = \frac{3}{4} k_B T \ln \left(\frac{m_p^*}{m_n^*} \right) \qquad (4.30)$$

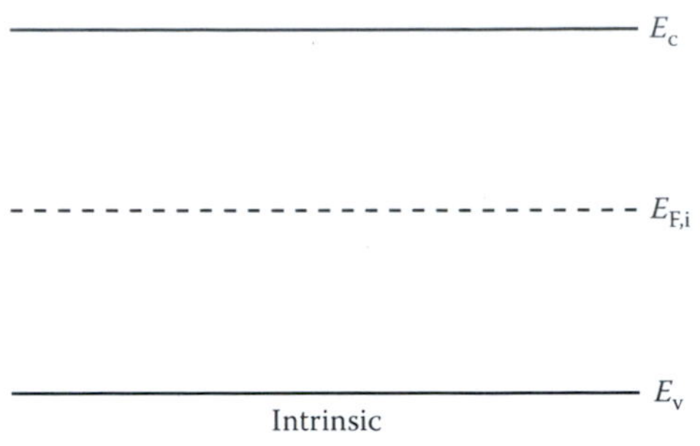

$$\text{————————————————}\ E_c$$

$$\text{- -}\ E_{F,i}$$

$$\text{————————————————}\ E_v$$

Intrinsic

FIGURE 4.4 Fermi energy level position for a typical intrinsic semiconductor. (From Pierret, R.F., *Advanced Semiconductor Fundamentals*, Addison-Wesley, Reading, MA, 1987. With permission.)

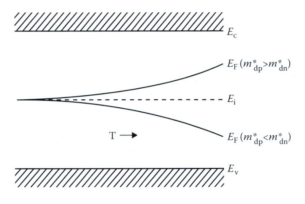

FIGURE 4.5 Variation of $E_{F,i}$ with the temperature. When the effective mass of holes is higher, the Fermi energy level is in the upper half of the band gap. (With kind permission from Springer Science+Business Media: *Semiconductor Physical Electronics*, 2006, Li, S. S.)

For intrinsic semiconductors with a larger effective mass of holes, N_v is higher than N_c. Thus, the intrinsic Fermi energy level ($E_{F,i}$) moves away from the valence band and is therefore slightly above the middle of the band gap (E_{midgap}), as shown in Figure 4.5. For intrinsic materials with a larger effective electron mass, the $E_{F,i}$ is slightly below the middle of the band gap (E_{midgap}). The effective masses also change with the temperature, as does the band gap (E_g). The variation of $E_{F,i}$ as given by Equation 4.30, is shown in Figure 4.5. As temperature increases, a deviation of $E_{F,i}$ from E_{midgap} increases. Note that the dependence of E_g on temperature, as given by the Varshni parameters, is not considered in Figure 4.5.

The slight deviation that causes the $E_{F,i}$ to shift from the middle of the band gap as a result of the difference between the effective mass of holes and electrons is calculated in Example 4.2.

Example 4.2: Intrinsic Fermi Energy Level for Si

Calculate the locations of the Fermi energy level for intrinsic Si located relative to the middle of the band gap. Assume that the effective masses for electrons and holes in Si are 1.08 and 0.56, respectively.

Solution
From Equation 4.30, we get

$$E_{F,i} - E_{midgap} = \frac{3}{4}(0.026 \text{ eV})\ln\left(\frac{0.56}{1.08}\right) = -12.8 \text{ meV}$$

Thus, the intrinsic Fermi energy level for Si is 12.8 meV below the middle of the band gap. Compared to 550 meV (~half of the band gap [E_g] of ~1.1 eV), this deviation is much smaller.

4.5   CARRIER CONCENTRATIONS IN INTRINSIC SEMICONDUCTORS

For intrinsic semiconductors, we can derive the electron and the hole concentrations using Equation 4.16 and Equation 4.23, respectively, as follows:

$$n_i = N_C \exp\left[-\left(\frac{E_C - E_{F,i}}{k_B T}\right)\right] \tag{4.31}$$

$$p_i = N_V \exp\left[-\left(\frac{E_{F,i} - E_V}{k_B T}\right)\right] \tag{4.32}$$

As mentioned previously, the negative sign is included so that the terms inside the brackets are positive because E_c is above $E_{F,i}$ and $E_{F,i}$ is above E_v. In Equations 4.31 and 4.32, n_i and p_i are the electron and hole concentrations, the subscript i indicates the intrinsic semiconductor, $E_{F,i}$ is the Fermi energy level for an intrinsic semiconductor, and N_c and N_v are the effective density of states for the conduction and valence band edges, respectively.

We now multiply Equation 4.31 by Equation 4.32 to get

$$n_i \times p_i = N_c N_v \exp-\left[\frac{E_c - E_{F,i}}{k_B T}\right] \exp-\left[\frac{E_{F,i} - E_v}{k_B T}\right] = N_c N_v \exp-\left[\frac{E_c - E_v}{k_B T}\right]$$

or since $E_c - E_v = E_g$, the band gap energy

$$n_i \times p_i = N_C N_V \exp\left[-\left(\frac{E_g}{k_B T}\right)\right] \tag{4.33}$$

Comparison of Equation 4.26 and Equation 4.33 shows that the $n \times p$ value is a constant for a given semiconductor at a given temperature. It does not matter whether the semiconductor is extrinsic or intrinsic.

We have now shown that the $n \times p$ value remains constant for a given semiconductor at a given temperature:

$$n \times p = n_i \times p_i \tag{4.34}$$

This is a very important equation since it allows us to make a connection between the properties of intrinsic and extrinsic semiconductors.

Another way to rewrite Equation 4.33 based on the relation of $n_i = p_i$, is as follows:

$$n_i = \sqrt{N_C N_V} \exp\left[-\left(\frac{E_g}{2k_B T}\right)\right] \tag{4.35}$$

We can see quantitatively from Equation 4.35 that the concentration of thermally generated carriers changes with the temperature.

We discussed the values of n_i and p_i for different semiconductors in Chapter 3. The intrinsic carrier concentrations are strongly temperature-dependent and are inversely related to the band gap (E_g).

An increase in temperature increases the intrinsic carrier concentration and an increase in the band gap decreases the intrinsic carrier concentration. The values of electron or hole intrinsic concentrations for silicon, germanium (Ge), and gallium arsenide (GaAs) are shown in Figure 4.6.

As we can see in Figure 4.6, the value of n_i for Si at 300 K is ~1.5×10^{10} electrons/cm^3.

The band gap of the semiconductors (E_g) also changes with temperature. This dependence is given by the Varshni parameters. For silicon, the change in densities of states N_c and N_v (Equations 4.10 and 4.18) and the variation in the band gap (E_g) with a temperature in the range 200–500 K are given by the following simplified equations (Green 1990; Bullis and Huff 1994):

$$N_c = 2.86 \times 10^{19} \left(\frac{T}{300} \right)^{1.58} \tag{4.36}$$

$$N_v = 3.10 \times 10^{19} \left(\frac{T}{300} \right)^{1.82} \tag{4.37}$$

The Varshni formula that describes the temperature dependence of the band gap for silicon is

$$E_g \text{ (in eV)} = 1.206 - 2.73 \times 10^{-4}\, T \tag{4.38}$$

In Equations 4.36 through 4.38, the density of states is expressed as states per cubic centimeter, and E_g is expressed in electron volts. Note that these Varshni parameters are for the temperature range 200–500 K and are slightly different from the values of 1.206 and 2.73×10^{-4} for a wider temperature range, as discussed in Chapter 3.

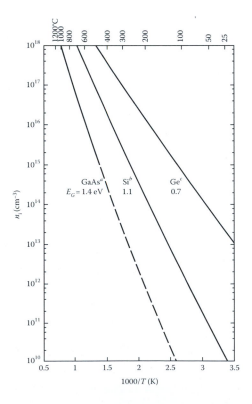

FIGURE 4.6 Intrinsic carrier concentrations for gallium arsenide, silicon, and germanium. (From Grove, A.S.: *Physics and Technology of Semiconductor Devices*. 1967. Copyright Wiley-VCH Verlag GmbH & Co. KGaA. Reproduced with permission.)

TABLE 4.2

Density of States (N_c, N_v), Band gap (E_g), and Intrinsic Carrier Concentration (n_i) Values for Si at Different Temperatures

T (K)	N_c (cm^{-3})	N_v (cm^{-3})	E_g (eV)	n_i (cm^{-3})
200	1.52×10^{19}	1.43×10^{19}	1.1483	5.03×10^{4}
250	2.15×10^{19}	2.20×10^{19}	1.1367	7.59×10^{4}
300	2.86×10^{19}	3.10×10^{19}	1.1242	1.07×10^{10}
350	3.65×10^{19}	4.13×10^{19}	1.1104	3.92×10^{11}
400	4.51×10^{19}	5.26×10^{19}	1.0968	6.00×10^{12}
450	5.43×10^{19}	6.49×10^{19}	1.0832	5.11×10^{13}
500	6.46×10^{19}	7.81×10^{19}	1.0695	2.89×10^{14}

For silicon, a more precise equation for the change in n_i is

$$n_i = 1.4514 \times 10^{20} \exp\left(\frac{-6997.4}{T}\right)\left(\frac{T}{300}\right)^{1.715} \tag{4.39}$$

The values of N_c, N_v, E_g, and n_i from Equations 4.36 through 4.39, calculated as a function of the temperature, are shown in Table 4.2. The value of n_i for silicon at 300 K is 1.07×10^{10} electrons/cm³, whereas the normally quoted value is ~1.5×10^{10} electrons/cm³. The use of these higher n_i values leads to the prediction of a higher current in silicon-based devices.

4.6 FERMI ENERGY LEVELS IN N-TYPE AND P-TYPE SEMICONDUCTORS

In Section 4.4, we saw that the Fermi energy level of an intrinsic semiconductor lies close to the middle of the band gap. Now we will find the location of E_F for extrinsic materials. Since $n \times p$ is constant for a given semiconductor at a given temperature,

$$n \times p = n_i \times p_i \tag{4.40}$$

Since the intrinsic semiconductor has the same number of electrons and holes (i.e., $n_i = p_i$), we can rewrite Equation 4.40 as

$$n \times p = n_i^2 = p_i^2 \tag{4.41}$$

Recall from Equation 4.15 that

$$n = N_C \exp\left[-\left(\frac{E_c - E_F}{k_B T}\right)\right]$$

Substitute for N_c from Equation 4.31 and rewrite Equation 4.15. Then, you will get

$$\therefore n = n_i \exp\left(\frac{E_F - E_{F,i}}{k_B T}\right) \tag{4.42}$$

This is also an important equation because it relates the conduction band electron concentration in an *extrinsic semiconductor* to E_F of the semiconductor in both *extrinsic intrinsic* forms. Since the product $n \times p$ is a constant, we can also calculate the concentration of holes in an n-type semiconductor.

Similarly, in a p-type semiconductor, the concentration of holes (p) is related to the density of states available in the valence band and to the Fermi energy level by Equation 4.24:

$$p = N_V \exp\left[-\left(\frac{E_F - E_V}{k_B T}\right)\right]$$

We substitute the value of N_v from Equation 4.32:

$$p_i = N_V \exp\left[-\left(\frac{E_{F,i} - E_V}{k_B T}\right)\right]$$

Therefore,

$$p = p_i \exp\left[-\left(\frac{E_F - E_{F,i}}{k_B T}\right)\right] \tag{4.43}$$

To remove the negative sign, and since $n_i = p_i$, we rewrite Equation 4.43 as

$$p = n_i \exp\left[-\left(\frac{E_{F,i} - E_F}{k_B T}\right)\right] \tag{4.44}$$

Note that in Equation 4.44, $p \approx N_a$. Thus, we can calculate the position of the Fermi energy level relative to the center of the band gap. As discussed in Section 4.4, the Fermi energy level for an intrinsic semiconductor ($E_{F,i}$) lies close to the center of the band gap (Figures 4.4 and 4.5).

To summarize, we now have a set of two important equations that allows us to relate the dopant concentration of an extrinsic semiconductor (n-type or p-type) to their Fermi energy levels.

$$n = n_i \exp\left(\frac{E_F - E_{F,i}}{k_B T}\right)$$

$$p = n_i \exp\left(\frac{E_{F,i} - E_F}{k_B T}\right) \tag{4.45}$$

From Equation 4.45, we can conclude that, as the doping level increases, the Fermi energy level of an n-type semiconductor will move closer to the conduction band edge. *The Fermi energy level of n-type semiconductors lies in the upper half of the band gap.* Similarly, as the doping level increases, the Fermi energy level of a p-type semiconductor will move closer to the valence band edge. *The Fermi energy level of p-type semiconductors lies in the lower half of the band gap.* Figure 4.7 shows the variation of E_F of Si as a function of the dopant concentrations (Pierret 1988). Note that this was calculated assuming $n_i = 10^{10}$ cm^{-3}.

4.7   FERMI ENERGY AS A FUNCTION OF THE TEMPERATURE

For an intrinsic semiconductor, the dependence of Fermi energy level on temperature is marginal. If the effective mass ratio between the electron and the hole is 1 regardless of temperature, $E_{F,i}$ stays at the center between E_c and E_v (Equations 4.28–4.30). However, because of the change in the effective masses of holes and electrons with temperature (Figure 4.5), the $E_{F,i}$ varies slightly with temperature.

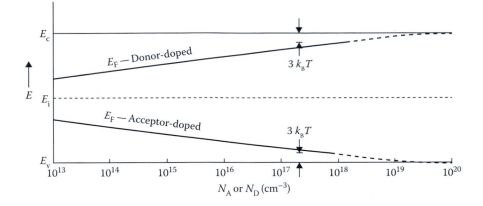

FIGURE 4.7 Variation in E_F as a function of the dopant concentration of silicon at 300 K. (Note: It assumes $n_i = 10^{10}$ cm^{-3}.) (From Pierret, R.F., *Semiconductor Fundamentals Volume*, Addison-Wesley, Reading, MA, 1988. With permission.)

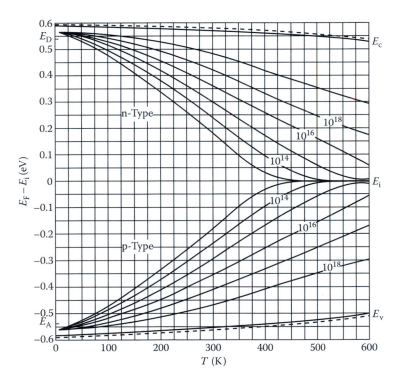

FIGURE 4.8 Position of E_F of silicon for different temperatures and dopant concentrations. (From Pierret, R.F., *Advanced Semiconductor Fundamentals*, Addison-Wesley, Reading, MA, 1987. With permission.)

When an extrinsic semiconductor is at very low temperatures, a carrier freeze-out occurs, causing it to behave as an intrinsic semiconductor. Thus, at very low temperatures, E_F is equal to $E_{F,i}$. As the temperature increases, donor ionization occurs. The Fermi energy level then moves to the location given by Equation 4.45. At higher temperatures, the thermally generated carriers (i.e., the intrinsic effect) begin to dominate. The semiconductor behaves as if it is intrinsic, and the E_F goes back to close to the center of the band gap, that is, near the $E_{F,i}$ (Figure 4.8).

As we will see in Section 4.9, when the dopant concentration is very high, the Fermi energy level moves very close (~ within $3 k_B T$) to a band edge. Such materials are called *degenerate semiconductors*. We cannot use Equation 4.45 to calculate the position of E_F for these materials.

The calculation of E_F for degenerate semiconductors is discussed in Section 4.9. The calculation of the location of E_F relative to $E_{F,i}$ for *nondegenerate semiconductors* is shown in Examples 4.3 and 4.4.

Example 4.3: Fermi Energy Level in n-Type Si

A Si crystal is doped with 10^{16}/cm³ Sb atoms. At 300 K:

1. What is the concentration of electrons (n)?
2. What is the concentration of holes (p) in this n-type Si?
3. Where is the Fermi energy level (E_F) for this material relative to the Fermi energy level for intrinsic Si (E_i)?

Solution

1. The donor dopant (Sb) concentration (N_d) is 10^{16} atoms/cm³. The intrinsic electron concentration (n_i) in Si at 300 K (Figure 4.6) is 1.5×10^{10} electrons/cm³ (Figure 4.6). Since $N_d \gg n_i$, we have assured $n \approx N_d$. We assume that the complete dopant ionization occurs at 300 K; therefore, the electron concentration (n) $\approx N_d = 10^{16}$ electrons/cm³.
2. From Equation 4.41,

$$n \times p = n_i^2 = p_i^2$$

Therefore,

$$10^{16} \times p = (1.5 \times 10^{10})^2 = 2.25 \times 10^{20}$$

Thus, the concentration of holes in this n-type semiconductor is $p = 2.25 \times 10^4$ holes/cm³. When compared to the electron concentration of 10^{16} (100,000 trillion), there are only 22,500 holes/cm³. Therefore, holes are considered a minority carrier in an n-type semiconductor. We had seen this concept in Chapter 3 but had not calculated the actual concentration of minority carriers.

3. Now, we calculate the relative position of E_F relative to $E_{F,i}$ using Equation 4.42:

$$n = n_i \exp\left(\frac{E_F - E_{F,i}}{k_B T} \right)$$

or

$$E_F - E_{F,i} = k_B T \ln\left(\frac{n}{n_i} \right) \tag{4.46}$$

Therefore,

$$E_F - E_{F,i} = (0.026 \text{ eV}) \ln\left(\frac{10^{16}}{1.5 \times 10^{10}} \right) = 0.349 \text{ eV}$$

Note that we used $k_B T = 0.026$ eV.

Thus, the Fermi energy level for this n-type semiconductor is 0.349 eV *above* $E_{F,i}$. The band diagram for this n-type Si is shown in Figure 4.9. In this diagram, the Fermi energy level is designated as $E_{F,n}$.

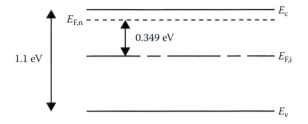

FIGURE 4.9   A band diagram and the Fermi energy level location for n-type silicon (see Example 4.3).

Example 4.4: Fermi Energy Level in B-Doped Si

A Si crystal is doped with $10^{17}/cm^3$ boron (B) atoms. At 300 K:

1. What is the concentration of holes (p)?
2. What is the concentration of electrons (n) in this p-type Si?
3. Where is the Fermi energy level (E_F) for this material relative to the Fermi energy level for intrinsic Si (E_i)?

Solution

1. The acceptor dopant (B) concentration (N_a) is 10^{17} atoms/cm³. We assume that complete dopant ionization occurs at 300 K. Therefore, the hole concentration (p) ≈ N_a = 10^{17} holes/cm³.
2. From Equation 4.41,

$$n \times p = n_i^2 = p_i^2$$

   Therefore,

$$n \times 10^{17} = (1.5 \times 10^{10})^2 = 2.25 \times 10^{20}$$

   Thus, the concentration of electrons in this p-type semiconductor is 2.25×10^3 electrons/cm³. Therefore, electrons are considered a minority carrier in a p-type semiconductor.
3. Now, we calculate the relative position of E_F for this p-type semiconductor relative to the $E_{F,i}$ using Equation 4.44:

$$p = n_i \exp\left(\frac{E_{F,i} - E_F}{k_B T} \right)$$

or

$$E_{F,i} - E_F = k_B T \ln\left(\frac{p}{ni} \right) \tag{4.47}$$

Therefore,

$$E_{F,i} - E_F = \left(0.026 \ eV\right)\ln\left(\frac{10^{17}}{1.5 \times 10^{10}} \right) = 0.407 \ eV$$

or
$E_F = E_{F,i} - 0.407$ eV
   Thus, the Fermi energy level (E_F) for this p-type semiconductor is 0.407 eV *below* the $E_{F,i}$. The band diagram for this n-type Si is shown in Figure 4.10.

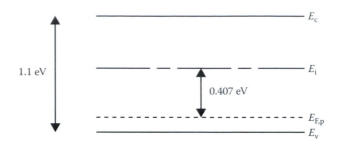

FIGURE 4.10   A band diagram for boron-doped silicon (see Example 4.4).

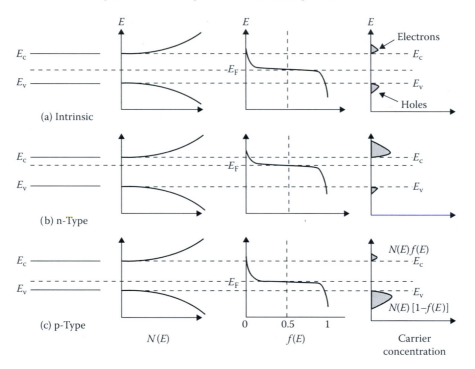

FIGURE 4.11   The Fermi energy level, the Fermi–Dirac distribution function, and the related band diagram for (a) an intrinsic, (b) an n-type, and (c) a p-type semiconductor. (From Streetman, B.G. and Banerjee S., *Solid State Electronic Devices*, Prentice Hall, Upper Saddle River, NJ, 2000. With permission.)

4.8   FERMI ENERGY POSITIONS AND THE FERMI–DIRAC DISTRIBUTION

It is clear from Equation 4.45 that the E_F of a semiconductor moves closer to the band edge as the doping level increases. This change in E_F is also reflected in the Fermi–Dirac distribution function (Figure 4.11). In the intrinsic semiconductor, $E_{F,i}$ is near the center between E_c and E_v, depending on the ratio of N_v/N_c or m_p^*/m_e^*, and the inversion center of the Fermi–Dirac function is at $E_{F,i}$. Therefore, the electron distribution in the conduction band (Equation 4.9) and the hole distribution in the valence band (Equation 4.17) are almost symmetric (Figure 4.11a). In addition, note that the density of states $N(E)$ in Figure 4.11 is proportional to $E^{1/2}$ and is given by

$$N(E) = \frac{V}{4\pi^2}\left(\frac{2m}{\hbar^2}\right)^{3/2} E^{1/2} \tag{4.48}$$

where V is total volume that electrons can occupy.

For n-type semiconductors (Figure 4.11b), the E_F is closer to the conduction band edge, which means that the Fermi–Dirac function $f(E)$ is also pushed up on the diagram. Though $N(E)$ does not depend on the doping type and concentration, a change in the Fermi–Dirac function alters the electron concentration at E ($n(E)$) that is expressed as $n(E) = N(E) \times f(E)$. Figure 4.11b shows that the high electron concentration in the conduction band and low hole concentration in the valence band can be explained using a shift of $f(E)$ toward E_c.

Similarly, for a p-type semiconductor (Figure 4.11c), the Fermi energy level is closer to the valence band edge; thus, the Fermi energy function is pushed down. Note that for p-type semiconductors, the value of $f(E_v)$ is less than 1 because there are holes in the valence band edge. The value of $f(E_v)$ (i.e., the probability of finding an electron) is slightly less than 1, and this $(1 - f(E_v))$ is the probability of finding a hole in the valence band. From Figure 4.11, we learn how a shift in E_F modifies $f(E)$ and $n(E)$, although there is no change in $N(E)$.

4.9   DEGENERATE OR HEAVILY DOPED SEMICONDUCTORS

As mentioned in Section 4.6, with very high levels of doping, the Fermi energy level of a semiconductor comes very close to the band edge. When $|E_F - E_c|$ or $|E_F - E_v|$ is $\leq 3\, k_B T$, the semiconductor is known as a degenerate semiconductor (Figure 4.12). The position of the Fermi energy level for degenerate semiconductors cannot be calculated using Equation 4.45.

In degenerate semiconductors, the dopant concentration is high, and the distance between dopant atoms is very small. As we discussed in Chapter 3 (Equation 3.53 and Figure 3.31), there no longer is a discrete energy level (E_d or E_a) at such high dopant concentrations. Instead, we get a band of energies near the band edge, and the band gap effectively becomes narrow. If the doping levels are very high, then the Fermi energy level of an n-type material actually lies in the conduction band. The Fermi energy level of a p-type semiconductor lies in the valence band (Figure 4.13). Tunnel diodes can be made by using such degenerate semiconductors. In these devices, a phenomenon known as electron tunneling occurs. For silicon at 300 K, a donor dopant concentration, (N_d) $\approx \geq 1.6 \times 10^{18}$ atoms/cm³, and an acceptor dopant concentration, (N_a) $\approx \geq 9.1 \times 10^{17}$ atoms/cm³, result in a degenerate semiconductor (Pierret 1988). Terms such as *highly* or *heavily* doped and "n⁺–material" or "p⁺–material" are also used to describe degenerate semiconductors.

We use the *Joyce–Dixon approximation* (Equation 4.49) to calculate E_F for degenerate semiconductors:

$$E_F = E_c + k_B T \left[\ln \frac{n}{N_c} + \frac{1}{\sqrt{8}} \frac{n}{N_c} \right] \text{(for n-type)}$$

$$\text{(4.49)}$$

$$E_F = E_v - k_B T \left[\ln \frac{p}{N_v} + \frac{1}{\sqrt{8}} \frac{p}{N_v} \right] \text{(for p-type)}$$

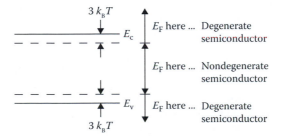

FIGURE 4.12 Definition of degenerate and nondegenerate semiconductors. (From Pierret, R.F., *Semiconductor Fundamentals*, Addison-Wesley, Reading, MA, 1988. With permission.)

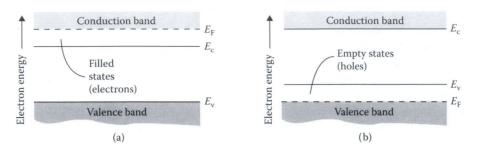

FIGURE 4.13   Fermi energy levels for (a) degenerate n-type and (b) degenerate p-type semiconductors. (From Neaman, D., *An Introduction to Semiconductor Devices*, McGraw Hill, New York, 2006. With permission.)

Equation 4.49 shows that E_F can be higher than E_c (or lower than E_v) when the carrier concentration (n or p) is larger than the effective density of states at the band edges (N_c or N_v). This indicates that electrons (or holes) can partially fill the conduction band (or the valence band) in the degenerated semiconductors, which is observed in the metals. In the degenerate semiconductors, an increase in the doping level increases E_F and decreases E_g. Hence, as discussed in Chapter 3, the degenerated semiconductors are sometimes considered to exhibit the electric properties of metals. Example 4.5 illustrates the use of the Joyce–Dixon approximation for the E_F calculation.

Example 4.5: Fermi Energy Level for a Degenerate Semiconductor

A GaAs n-type sample is doped so that the concentration of electrons is 10^{17}. What is the value of the E_F for this sample relative to E_c? Assume that $T = 300$ K and that the semiconductor is degenerate.

Solution
The density of states for GaAs at 300 K is 4.45×10^{17} cm^{-3} and $k_B T = 0.026$ eV.

$$E_F - E_C = \left(0.026\,\text{eV}\right)\left[\ln\frac{10^{17}}{4.45\times10^{17}} + \frac{1}{\sqrt{8}}\left[\frac{10^{17}}{4.45\times10^{17}}\right]\right]$$

Therefore, $(E_F - E_c)$ is -0.037 eV and E_F is 0.037 eV below E_c. Note that this value is within $3\,k_B T$ of E_c.

4.10   FERMI ENERGY LEVELS ACROSS MATERIALS AND INTERFACES

We have discussed in detail how to calculate the location of E_F for intrinsic and extrinsic semiconductors. The significance of the location of E_F will become clearer in Chapter 5 when we discuss the formation of the *p-n junction*, *the metal-semiconductor junction*, and devices based on the same. When electrically different materials (e.g., a p-type semiconductor and a metal) are brought together, or when electrically different interfaces are created within the same material, the equilibrium Fermi energy level remains invariant through the contact. There is no discontinuity or gradient that can exist in the equilibrium Fermi energy level. As an example, consider two materials, A and B, with Fermi energy levels $E_{F,A}$ and $E_{F,B}$, respectively (Figure 4.14). It does not matter whether these materials are the same or different semiconductors or whether they are n-type or p-type. Materials A and B can also be metals.

We usually create a p-type material in a crystal that is uniformly donor-doped or vice versa, or we deposit a metal on a semiconductor crystal. When we bring Materials A and B with different Fermi energy levels together (Figure 4.14), electrons from Material A (which has a higher Fermi energy level and an overall higher energy) begin to flow to Material B (Figure 4.14c), until the electrons in both materials have the same energy distribution or the same E_F in both Materials A and B (Figure 4.14d). The flow of electrons from Material A to Material B will be proportional to the

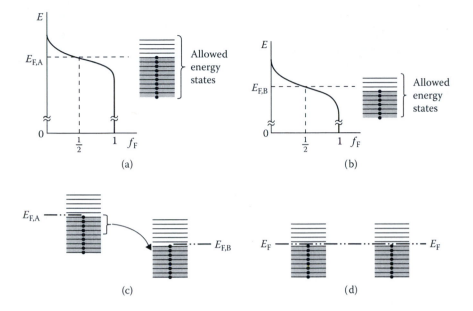

FIGURE 4.14 Materials A and B with different Fermi energy levels before and after making a contact. (a) Material A in thermal equilibrium, (b) Material B in thermal equilibrium, (c) Materials A and B at the instant they are placed in contact, and (d) Materials A and B in contact at thermal equilibrium. (From Neaman, D., *An Introduction to Semiconductor Devices*, McGraw Hill, New York, 2006. With permission.)

number of energy states filled with electrons in Material A and to the number of empty energy states available in Material B.

The rate of transfer of electrons from Material A to Material B at the energy level E is proportional to $[N_A(E) f_A(E)] \times [N_B(E) (1 - f_B(E))]$ (i.e., the electron population density of Material A at $E \times$ the hole population density of Material B at E). Similarly, the rate of transfer of electrons from Material B to Material A at the same energy level E is proportional to $[N_B(E) f_B(E)] \times [N_A(E)(1 - f_A(E))]$.

When equilibrium is reached following the contact between the materials with different Fermi energy levels, there is no net current flow. At equilibrium, the rate of the transfer of electrons from Material A to Material B is equal to that of the transfer of electrons from Material B to Material A. Therefore,

$$[N_A(E) f_A (E)] \times [N_B(E)(1 - f_B(E))] = [N_B(E) f_B(E)] \times [N_A(E)(1 - f_A(E))]$$

This simplifies to

$$f_A(E) = f_B(E)$$

or

$$1 + \exp\left(\frac{E - E_{F,A}}{k_B T} \right) = 1 + \exp\left(\frac{E - E_{F,B}}{k_B T} \right)$$

This is true only if $E_{F,A} = E_{F,B}$.

Therefore, the Fermi energy level remains constant across Materials A and B, or

$$\frac{dE_F}{dx} = 0 \qquad\qquad (4.50)$$

In this discussion, if Materials A and B were n-type and p-type semiconductors, respectively, then we would create a p-n junction by bringing these materials together. In Chapter 5, we will discuss the use of this important concept, called *invariance of Fermi energy*, in formation of p-n junctions.

PROBLEMS

4.1 A metal has an electron concentration of 10^{22} atoms/cm³. What is the value of the Fermi energy level at 0 K? (Use Equation 4.1.)

4.2 What is the maximum speed of an electron in Cu at 0 K? Does this speed increase significantly with temperature? Explain.

4.3 What is the effective speed of electrons in Cu at 0 K? Does this speed depend strongly on the temperature? Explain.

4.4 What is the E_F at $T = 300$ K for the metal discussed in Problem 4.1?

4.5 How is the Fermi energy level of a semiconductor different from both the donor and acceptor levels? Show using an illustration.

4.6 A semiconductor has $E_F = 0.26$ eV below the conduction band edge. What is the probability that the state 0.026 eV ($k_B T$) above the conduction band edge is occupied by an electron? What type of semiconductor is this material—n-type or p-type?

4.7 An energy level is located at 0.3 eV above the E_F. Calculate the probability of occupancy for this state at 0, 300, and 600 K.

4.8 What is the probability that an energy state 0.4 eV below E_F is empty for the temperatures of 0, 300, and 600 K?

4.9 Show that the Fermi–Dirac function is symmetric around E_F, that is, that the probability of a state ΔE above E_F is the same as the probability of finding a hole at a state E below E_F.

4.10 The reduced effective density of states mass for electrons and holes in GaAs are ~0.067 and 0.50. Use Equation 4.30 to calculate the exact position of the Fermi energy level for intrinsic GaAs at 300 K. How is this location different from that for Ge at 300 K?

4.11 Use Equation 4.15 to express the concentration of electrons (n) in the conduction band. Then, use a similar equation to express the concentration of electrons n_d in the donor state E_d and density of states N_d.

4.12 The reduced effective density of states mass for electrons and holes in Ge are ~0.56 and 0.40 at 300 K. Use Equation 4.30 to calculate the exact position of the Fermi energy level for intrinsic Ge at 300 K.

4.13 Use Equation 4.39 to calculate the intrinsic electron concentration in Si at 350 and 400 K.

4.14 What is the resistivity of the intrinsic Si sample discussed in Problem 4.13? Use the mobility values from Table 3.1 in Chapter 3, and note that the mobility does depend on the temperature.

4.15 Derive an equation equivalent to Equation 4.39 that describes the variation in the concentration of thermally excited electrons in Ge.

4.16 Derive an equation equivalent to Equation 4.39 that describes the variation in the concentration of thermally excited electrons in GaAs.

4.17 A Si sample is doped so that $n = 2 \times 10^5$ cm⁻³. What is the hole concentration for this material? Is this an n-type or a p-type Si? Assume that $T = 300$ K.

4.18 What is the conductivity (σ) of the sample described in Problem 4.17?

4.19 A Si crystal is doped with 10^{15} cm⁻³ of P atoms. Calculate the E_F of this crystal relative to the center of the band gap, which can be assumed as the location of $E_{F,i}$.

4.20 A GaAs crystal is doped such that the E_F is 0.2 eV above the E_v. Is this an n-type or a p-type GaAs? What are the electron and hole concentrations in this material? Assume that $E_g = 1.43$ eV and $T = 300$ K.

4.21 What is the resistivity (ρ) at 300 K of the sample discussed in Problem 4.20? Use the appropriate mobility values from Table 3.1 in Chapter 3.

4.22 A Si crystal is first uniformly doped with a donor dopant, and then the process of doping the entire crystal with an acceptor is started. Sketch the variation of the E_F of this semiconductor, starting out as an n-type material as a function of the increasing acceptor concentration. Where is the E_F when the donor and acceptor concentrations are equal?

4.23 A Si crystal is doped with 10^{16} cm^{-3} of Sb atoms (see Example 4.3), and then the entire crystal is doped with 3×10^{17} atoms/cm^3 of B. What is the new location of the Fermi energy level for this crystal containing both B and Sb?

4.24 What is the resistivity of Si (at 300 K) discussed in Problem 4.23 after Sb-doping and B-doping? Use the appropriate values of mobility and consider the dependence of mobility on the dopant concentration.

4.25 As stated in Section 4.9, for Si at 300 K, a donor dopant concentration (N_d) $\approx \geq 1.6 \times 10^{18}$ atoms/cm^3 or an acceptor dopant concentration (N_a) $\approx \geq 9.1 \times 10^{17}$ atoms/cm^3 result in a degenerate semiconductor. Verify that this is correct using the definition of a degenerate semiconductor.

4.26 What is the minimum donor dopant concentration for GaAs at 300 K, at which we consider it degenerate? (Hint: The solution for $[E_F - E_c]$ must be less than 3 $k_B T$.)

4.27 A sample of a GaAs crystal is doped so that $N_d = 10^{17}$ donor atoms/cm^3 (see Example 4.5). Calculate the Fermi energy position using Equation 4.45, in which we assume that the sample is nondegenerate. Compare the result with that used in Example 4.5, where we used the Joyce–Dixon approximation.

GLOSSARY

Degenerate semiconductors: Semiconductors that are heavily doped so that the Fermi energy moves very close (within 3 $k_B T$) to a band edge.

Effective density of states approximation: The assumption that most electrons or holes reside at the bottom of the conduction band or the top of the valence band, respectively. Thus, N_c or N_v can be used in calculations of the concentrations of electrons and holes in the conduction and valence bands, respectively.

Effective speed of electrons (v_e): or the **root mean speed of electrons** in a metal is given by $(1/2)mv_e^2 \cong (3/5)E_{F,0}$.

Fermi–Dirac distribution function: The function that gives the probability of occupancy of a state by an electron and is defined as

$$f(E) = \frac{1}{\left[1 + \exp\left(\dfrac{E - E_F}{k_B T}\right)\right]}$$

Fermi energy (E_F): The energy level at which the probability of occupancy of a state is 0.5.

Invariance of Fermi energy: When materials with different Fermi energy levels are brought in contact with each other under equilibrium conditions, the Fermi energy level remains continuous across the interface between the different materials.

Joyce–Dixon approximation: An equation used for calculating the Fermi energy level for degenerate semiconductors.

Law of mass action for semiconductors: For any given semiconductor, the value of $n \times p$ is a constant at a given temperature.

p-n junction: The electrical junction between a p-type and an n-type semiconductor. This is the basis for many devices, including diodes and transistors.

Work function ($q\phi$): The energy required to remove an electron from E_F to a vacuum.

REFERENCES

Bullis, W. M., and H. R. Huff. 1994. Silicon for microelectronics. In *Encyclopedia of Advanced Materials,* vol. 4, eds. D. Bloor, S. Mahajan, M. C. Flemings, and R. J. Brook. Oxford, UK: Pergamon Press.

Green, M. A. 1990. Intrinsic concentrations, effective density of states, and effective mass in silicon. *J Appl Phys* 67:2944–54.

Grove, A. S. 1967. *Physics and Technology of Semiconductor Devices.* New York: Wiley.

Kasap, S. O. 2002. *Principles of Electronic Materials and Devices.* New York: McGraw Hill.

Li, S. S. 2006. *Semiconductor Physical Electronics.* New York: Springer.

Neaman, D. 2006. *An Introduction to Semiconductor Devices.* New York: McGraw Hill.

Pierret, R. F. 1987. *Advanced Semiconductor Fundamentals.* Reading, MA: Addison-Wesley.

Pierret, R. F. 1988. *Semiconductor Fundamentals.* Reading, MA: Addison-Wesley.

Streetman, B. G., and S. Banerjee. 2000. *Solid State Electronic Devices.* Upper Saddle River, NJ: Prentice Hall.

5 Semiconductor p-n Junctions

KEY TOPICS

- Formation of a p-n junction
- Concept of built-in potential
- Band diagram for a p-n junction
- Diffusion and drift currents in a p-n junction
- What makes the p-n junction useful?
- Current–voltage (*I–V*) curve for a p-n junction
- Some devices based on p-n junctions

5.1 FORMATION OF A p-n JUNCTION

Consider a hypothetical experiment in which we put together two crystals of a semiconductor and form a contact: one is an n-type and the other is a p-type. A schematic illustration on the change at the contact is shown in Figure 5.1.

When we bring the two semiconductors together to form a p-n junction, electrons start flowing from the n-side to the p-side to minimize the concentration gradient, which is called the diffusion current. This will continue until a new common Fermi energy level (E_F) is established through the contact (Section 4.10). Since electrons are negatively charged, the direction of the current associated with their motion is opposite to the direction of their actual motion. The direction of the current associated with the diffusion of electrons is from the p-side to the n-side of the p-n junction (Figure 5.2).

On the other hand, a p-doped material has a much higher concentration of holes on the p-side than on the n-side. Thus, when a p-n junction is formed, the holes diffuse from the p-side to the n-side. Since holes are positively charged particles, the diffusion current induced due to the hole movement is in the same direction as the hole diffusion, and hole diffusion current and electron diffusion current flow along the same direction (Figure 5.2).

Note that a concentration gradient of dopant atoms exists because there is a significant concentration of donor atoms on the n-side and almost no donor atoms on the p-side. However, there is no diffusion of donor atoms or ions because their mobility is very low at or near room temperature. Thus, here we only consider the diffusion of holes and electrons.

Since diffusion is concentration gradient driven, we expect that the flow of electrons and holes induced will continue until the concentrations of electrons and holes become equal on both sides of the p-n junction. For example, if we join a crystal of copper (Cu) and a crystal of nickel (Ni) and then heat this to a high temperature (e.g., 500°C) to promote diffusion (Figure 5.3), after a few hours, the copper atoms will diffuse into the nickel crystal and vice versa. This process of interdiffusion will continue until the concentrations of copper and nickel atoms are equal on both sides of the original interface.

We will now discuss the difference between the interdiffusion of copper and nickel atoms (Figure 5.3) and the diffusion of holes and electrons in the formation of a p-n junction (Figures 5.1 and 5.2). When electrons in the n-type material begin to diffuse out into the p-type material, they leave behind positively charged donor ions. For example, when a donor such as phosphorus (P) is added as a neutral atom, it donates an electron and becomes ionized, turning into a P^{1+} ion.

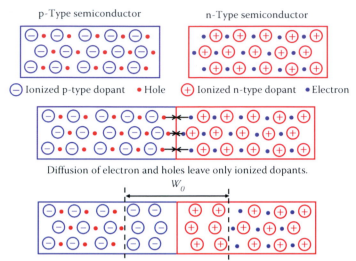

FIGURE 5.1 Schematics on the diffusion of charge carriers and the formation of the depletion region after the p-type semiconductor makes a junction with the n-type semiconductor.

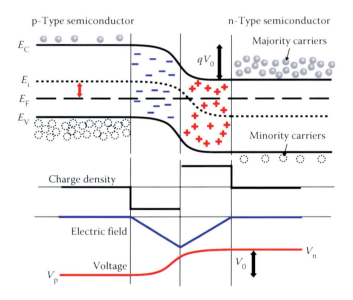

FIGURE 5.2 The directions of particle motion for the diffusion and drift of electrons and their corresponding currents for a p-n junction under equilibrium. (From Edwards-Shea, L., *The Essence of Solid State Electronics*, Prentice Hall, Upper Saddle River, NJ, 1996. With permission.)

Similarly, when holes from the p-side begin to diffuse onto the n-side of the p-n junction, they leave behind negatively charged acceptor ions (e.g., boron; B^{1-} ion).

Thus, as electrons diffuse from the n-type into the p-material, they leave behind a region of positively charged donor ions, shown with + signs in Figure 5.1. Similarly, as holes diffuse from the p-side to the n-side, they create a region comprised of negatively charged acceptor ions. Thus, at the p-n junction, there is an area known as the *depletion region,* so named because it is depleted of electrons and holes. It is also called the *space-charge region* or the *space-charge layer,* which refers to the positive and negative charges present in this region (Figure 5.4).

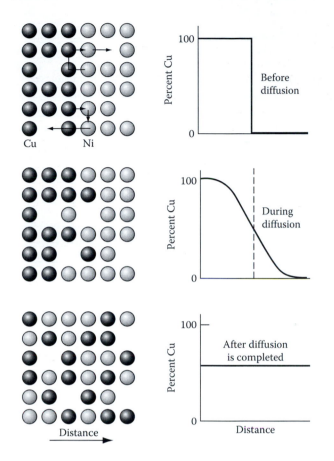

FIGURE 5.3 Illustration of the interdiffusion of copper and nickel atoms. (From Askeland, D. and Fulay P., *The Science and Engineering of Materials*, Thomson, Washington, DC, 2006. With permission.)

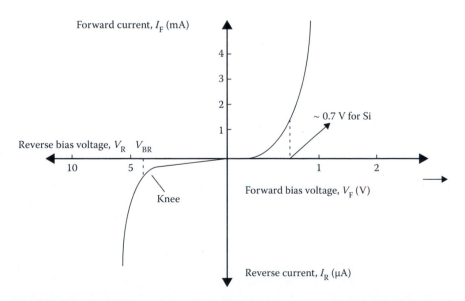

FIGURE 5.4 *I–V* curve for a silicon-based p-n junction. Breakdown and knee voltage are shown. Note the current and voltage scales for forward and reverse bias are different.

Across the width of the depletion region (w_0), the electric charge is not zero (negative in p-type semiconductor and positive in n-type semiconductor). An *internal electric field* (E) is set up because of the presence of positively charged donor ions and negatively charged acceptor ions (Figure 5.4). This electric field (E) is directed from the positively charged donor ions to the negatively charged acceptor ions, and the *built-in* potential is generated across the depletion region (Figure 5.3). If an electron tries to diffuse from the n-side to the p-side, the electron begins to *see* the acceptor ions and experiences the *built-in* potential. The motion of such electrons toward the p-side is opposed by the negatively charged acceptor ions $\left(N_a^- \right)$ (or the *built-in* potential). Similarly, if a hole tries to diffuse from the p-side to the n-side, it begins to experience the repelling force of the positively charged donor ions $\left(N_d^+ \right)$. We will use the symbols N_d and N_a for the concentrations of donor and acceptor atoms or ions, respectively. Thus, the built-in electric field stops the diffusion of electrons from the n-side to the p-side and the diffusion of holes from the p-side to the n-side.

The voltage corresponding to the internal electric field is known as the *contact potential*, which is the same as the *built-in potential* (V_0), and is expressed in volts. This potential difference can appear when a p-n junction between different materials is formed. As we can see from Figure 5.3, the built-in electric field is directed toward the $-x$-direction. In the depletion region, the electric field is related to the potential as follows:

$$E(x) = -\frac{dV(x)}{dx} \tag{5.1}$$

The electrostatic potential (V) is higher on the n-side, which is a part of the depletion layer with a net positive charge. The contact potential (V_0) is the difference between V_n and V_p, where V_n and V_p are the electrostatic potentials in the n- and p-neutral regions, respectively. Thus,

$$V_0 = V_n - V_p \tag{5.2}$$

We plot the electron energy, which is related to the electric potential by a factor of $-q$, on the band diagram. Therefore, $E_{c,n}$, $E_{F,n}$, and $E_{v,n}$ (that is, different energy levels on the n-side) appear lower on the p-n junction band diagram than the corresponding levels on the p-side.

The development of the contact potential, also known as the *diffusion potential*, makes the p-n junction interesting and useful for device applications. This *built-in* potential prevents further diffusion of electrons (from the n- to the p-side) and holes (from the p- to the n-side) and maintains the width of the depletion region at the equilibrium value (W_0). If no external bias is applied, the equilibrium is made, and there is no net electric current flowing through the junction. We will see in Section 5.9 that, by applying a *forward bias* (that is, by connecting the positive terminal of an external voltage supply to the p-side), both the built-in potential and the depletion junction width can be decreased. If this happens, the electric current flows through the p-n junction as the diffusion of electrons and holes resumes. Conversely, if we apply a *reverse bias* by connecting the negative terminal of an external voltage supply to the p-side, then the reverse bias adds to the built-in potential barrier (see Section 5.10). Thus, the p-n junction will be able to carry very little diffusion current. However, there still is a net electric current even under the reverse bias state. Let us take a look at the charge carrier transport through the junction more carefully in the next section.

5.2 DRIFT AND DIFFUSION OF CARRIERS

A p-n junction is a *tunable device* and is used to create diodes, transistors, and so on. The explanation in Section 5.1, which is only based on the diffusion, is a bit intuitive and not very rigorous. To understand the nature of the p-n junction, we have to classify a driving force of the carrier transport at the junction into drift as well as the diffusion. Strictly speaking, the diffusion current is not zero at the equilibrium state. Instead, since the drift and the diffusion cause two currents flowing in parallel but opposite directions, the net electric current becomes zero.

The built-in or internal electric field (E) created in the p-n junction stops further diffusion of electrons and holes. It also plays another important role by setting up drift currents (Figures 5.2 and 5.4). The term *drift* refers to the motion of charge carriers under the influence of an internal or external electric field. When *electrons* thermally generated in the p-side (where they are minority carriers) experience the internal electric field at the junction, the minority carriers are driven toward the positively charged region of donor ions. The internal electric field causes electrons on the p-side to drift toward the n-side (Figure 5.2). Similarly, thermally generated *holes* in the n-side drift toward the negatively charged space-charge region of the p-side. Due to a difference in the polarity, these drifts of the charge carriers cause electric currents (called drift currents) to flow along the same direction. The total drift current in a p-n junction is known as the *generation current* (see Section 5.11). It is an important part of the current–voltage (I–V) characteristics of a p-n junction.

Note that the directions of the diffusion of electrons and the drift of electrons are opposite. Due to the electron diffusion, the current is from the p-side to the n-side. The electron drift current is from the n-side to the p-side. The directions of the diffusion of majority carriers and the drift of minority carriers on the n-side and p-side are shown in Figures 5.2 and 5.4.

For a p-n junction at equilibrium (Figure 5.2), the drift and diffusion current densities (J) cancel out because a p-n junction at the equilibrium carries no electrical current.

Therefore,

$$J_p(\text{diffusion}) + J_p(\text{drift}) = 0 \tag{5.3}$$

and

$$J_n(\text{diffusion}) + J_n(\text{drift}) = 0 \tag{5.4}$$

Subscripts p and n refer to the motion of holes and electrons, respectively.

5.3 CONSTRUCTING THE BAND DIAGRAM FOR A p-n JUNCTION

In Chapter 4, we learned how to calculate the relative position of the Fermi energy level for a semiconductor (Examples 4.2 and 4.3) and the invariance of the Fermi energy level through the contact (Section 4.10). We will now draw the band diagram for a p-n junction using this information and graphically calculate the value of qV_0 and, hence, V_0. This is illustrated in Example 5.1.

Example 5.1: Estimation of Contact Potential from the Band Diagram

Consider a p-n junction in Si. Assume that the n- and p-sides have a dopant concentration of $N_d = 10^{16}$ and $N_a = 10^{17}$ atoms/cm³, respectively. (a) Calculate the Fermi energy position for the n-type and p-type semiconductors, and draw the band diagrams for the n-type and p-type semiconductors before they are joined. (b) Draw the band diagram for this p-n junction and estimate its contact potential (V_0). Assume that $T = 300$ K, $n_i = 1.5 \times 10^{10}$ electrons/cm³, and $E_g = 1.1$ eV.

Solution
1. As we have seen in Chapter 4,

$$n = n_i \exp\left(\frac{E_F - E_{F,i}}{k_B T}\right) \tag{5.5}$$

Assuming a complete donor ionization, that is, $n \approx N_d = 10^{16}$ atoms/cm³, we get

$$E_{F,n} - E_{F,i} = (0.026 \text{ eV}) \ln\left(\frac{10^{16}}{1.5 \times 10^{10}}\right) = 0.349 \text{ eV}$$

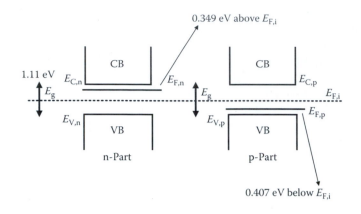

FIGURE 5.5 Band diagrams for n-type and p-type regions (see Example 5.1).

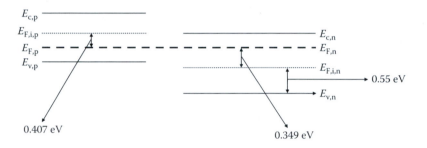

FIGURE 5.6 Summary of steps to be followed to draw a p-n junction band diagram.

Thus, the Fermi energy level for this n-type semiconductor is 0.349 eV above $E_{F,i}$. This is shown in Figure 5.5.

Similarly, for the p-side,

$$E_{F,i} - E_F = k_B T \ln\left(\frac{p}{n_i}\right) \tag{5.6}$$

Substituting $N_a = 10^{17}$ atoms/cm$^3 = p$ and the value of n_i,

$$E_{F,i} - E_{F,p} = (0.026 \text{ eV}) \ln\left(\frac{10^{17}}{1.5 \times 10^{10}}\right) = 0.407 \text{ eV}$$

or

$$E_{F,p} = E_{F,i} - 0.407 \text{ eV}$$

2. Now, we draw the band diagram for the p-n junction (Figure 5.6). First, we recognize that a common Fermi energy level is established after the n-type and p-type semiconductors are joined. Therefore, we draw a horizontal line that represents this common E_F (Figure 5.6). On the n-side of this junction, but far away from the boundary of the interface between the two materials, the common Fermi energy level E_F is 0.349 eV above the $E_{F,i,n}$. Using a suitable scale, we draw a line 0.349 eV below $E_{F,i,n}$ and label it $E_{F,n}$. This represents the intrinsic Fermi energy level position on the n-side. Since this is almost at the middle of the band gap, we draw the conduction band edge $E_{c,n}$ at 0.55 eV (1/2 of E_g) above $E_{F,i,n}$. We next draw $E_{v,n}$ 0.55 eV below the $E_{F,i,n}$.

On the p-side of this junction, but far away from the boundary of the interface between the two materials, the $E_{F,p}$ is 0.407 eV below the dotted $E_{F,i,p}$ line. Thus, we draw

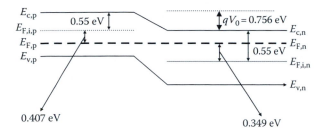

FIGURE 5.7 A band diagram for the p-n junction in Example 5.1.

a line 0.407 eV above the common energy level $E_{F,p}$ and label it as $E_{F,i,p}$. This represents the intrinsic Fermi energy level position on the p-side. We then draw the conduction band edge on the p-side E_{cp} at 0.55 eV (1/2 of E_g) above $E_{F,i,p}$. We next draw $E_{v,p}$ 0.55 eV below the $E_{F,i,p}$ line. Note that this automatically puts the band edges on the p-side higher than their counterparts on the n-side.

The steps needed to complete the p-n junction band diagram are summarized in Figure 5.6. We now join the conduction band edges and valence band edges on both sides as shown in Figure 5.7. We will see in Section 5.7 that the variation of electron energy across the depletion region is parabolic.

From this band diagram, we can see that

$$qV_0 = \text{height of the energy barrier} = (E_{c,p} - E_{c,n}) \tag{5.7}$$

Looking at the p-side of the diagram,

$$E_{c,p} = 0.407 \text{ eV} + 0.55 \text{ eV} \tag{5.8}$$

Looking at the n-side of the diagram,

$$E_{c,n} = 0.55 \text{ eV} - 0.349 \text{ eV} \tag{5.9}$$

Therefore, subtracting Equation 5.9 from Equation 5.8, we get qV_0 = the height of the energy barrier setup = $E_{c,p} - E_{c,n}$ = 0.407 eV + 0.55 eV − 0.55 eV + 0.349 eV = 0.756 eV. Thus, the value of the contact potential for this p-n junction (V_0) is 0.756 V (Figure 5.7). Note that the contact potential is $V_n - V_p$ (Figure 5.1).

5.4 CALCULATION OF CONTACT POTENTIAL

If the doping level on the n-side increases, then $E_{F,n}$ will be farther away from $E_{F,i,n}$ (Figure 5.6) and the E_{cn} and $E_{v,n}$ will move down to center themselves around $E_{F,i,n}$. If nothing changes on the p-side, then the value of qV_0 will be expected to increase. Similarly, if the doping level on the p-side is increased, then $E_{F,i,p}$ will move up relative to $E_{F,p}$ (Figure 5.5). Then, $E_{v,p}$ and E_{cp} will also move up to center around $E_{F,i,p}$. If nothing changes on the n-side, qV_0 will increase.

We will now derive an equation that quantitatively describes how the value of V_0 changes with the dopant concentrations on the n- and p-sides of the p-n junction.

According to Fick's first law of diffusion, the flux of diffusing species is proportional to the negative of the concentration gradient. The negative sign means that there will be a movement of species from a region of higher concentration to a region of lower concentration. In this case, we want to calculate the *electrical charge* flowing per unit time not just the number of electrons or holes. Therefore, we will multiply the flux of the species by q, the magnitude of the charge of an electron or a hole.

Thus, the diffusion current density (J) due to the motion of the holes is

$$J_{p,\text{diffusion}} = -q \times D_p \times \frac{dp(x)}{dx} \tag{5.10}$$

We have taken the direction from p to n as the positive x-direction (Figure 5.2). In this equation, D_p is the diffusion coefficient for holes.

The drift current, which is due to the movement of the holes from the n-side—where they are the minority carriers—to the p-side, is given by

$$J_{p,\text{drift}} = q \times \mu_p \times p(x) \times E(x) \tag{5.11}$$

In Equation 5.11, $p(x)$ is the hole concentration along the x-direction, and $E(x)$ is the built-in electric field. We know the hole concentration on both the p-side and the n-side, where the holes are minority carriers.

The current induced by the diffusion of majority carriers and the current caused by the drift of minority carriers are in opposite directions and cancel each other out (Equation 5.3). Therefore, we get

$$q \times D_p \times \frac{dp(x)}{dx} = q \times \mu_p \times p(x)E(x) \tag{5.12}$$

Simplifying,

$$\frac{dp(x)}{dx} \frac{1}{p(x)} = \frac{\mu_p}{D_p} E(x) \tag{5.13}$$

Since we want to calculate the value of V_0, the contact potential, we change the electric field (E) to electrostatic potential (V) by substituting for $E(x)$ from Equation 5.1 into Equation 5.13:

$$\frac{dp(x)}{dx} \frac{1}{p(x)} = -\left[\frac{\mu_p}{D_p}\right] \frac{dV(x)}{dx} \tag{5.14}$$

Using the so-called *Einstein relation* (not derived here) and applying it to holes,

$$\frac{\mu_p}{D_p} = \frac{q}{k_B T} \tag{5.15}$$

From Equations 5.14 and 5.15, we get

$$\left[\frac{dp(x)}{dx}\right]\left[\frac{1}{p(x)}\right] = \left[\frac{q}{k_B T}\right]\frac{dV(x)}{dx} \tag{5.16}$$

We know the concentrations of holes on both the p-side and the n-side in the neutral regions, that is, the regions away from the p-n junction that do not have any built-up net charge. We assume a one-dimensional model; that is, we assume that the carriers will diffuse and drift along the x-direction (+ or −) only. We now integrate Equation 5.16 from the p-side to the n-side:

$$-\frac{q}{k_B T} \int_{V_p}^{V_n} dV = \int_{p_p}^{p_n} \frac{dp}{p} \tag{5.17}$$

In Equation 5.17, V_p and V_n are the electrostatic potentials on the p-side and n-side of the neutral regions, where there is no built-up net charge. The hole concentrations in the neutral regions on the p-side and the n-side are p_p and p_n, respectively. The subscripts indicate the side of the p-n junction. Simplifying Equation 5.17,

$$-\frac{q}{k_B T}(V_n - V_p) = \ln(p_n) - \ln(p_p) = \ln\left(\frac{p_n}{p_p}\right) \tag{5.18}$$

Note that the potential difference $V_n - V_p$ is V_0, the contact potential.

$$-\left(\frac{q}{k_B T}\right)V_0 = \ln\left(\frac{p_n}{p_p}\right) \tag{5.19}$$

Eliminating the negative sign, we get

$$V_0 = \frac{k_B T}{q}\ln\left(\frac{p_p}{p_n}\right) \tag{5.20}$$

For a *step junction*, we move abruptly from the p-side, with N_a acceptors per cubic centimeter, to the n-side with N_d donors per cubic centimeter. We rewrite Equation 5.20 as

$$V_0 = \frac{k_B T}{q}\ln\left(\frac{N_a N_d}{n_i^2}\right) \tag{5.21}$$

To get Equation 5.21, we used $p_p = N_a$ and $p_n \cdot N_d = n_i^2$. This form is useful in calculating the contact potential (V_0) associated with a p-n junction.

We can rewrite Equation 5.20 as

$$\frac{p_p}{p_n} = \exp\left(\frac{qV_0}{k_B T}\right) \tag{5.22}$$

Since $p_p \times n_p = p_n \times n_n = n^2$, we can write

$$\frac{p_p}{p_n} = \frac{n_n}{n_p} = \exp\left(\frac{qV_0}{k_B T}\right) \tag{5.23}$$

We can see from Equation 5.23 that, if the dopant concentration on the p-side increases, then qV_0 will increase. We also saw this in the calculation of V_0 from the p-n junction band diagram in Section 5.3 and Figure 5.7.

From Equation 5.22, we substitute for p_p and p_n in terms of the density of states (N) and the difference between E_F relative to the valence band edge on each side:

$$\frac{p_p}{p_n} = \frac{N_v \exp-\left(\dfrac{E_{F,p} - E_{v,p}}{k_B T}\right)}{N_v \exp-\left(\dfrac{E_{F,n} - E_{v,n}}{k_B T}\right)} = \exp\left(\frac{qV_0}{k_B T}\right) \tag{5.24}$$

In Equation 5.24, the additional subscript for E refers to the side of the junction. Thus, $E_{F,p}$ is the Fermi energy level on the p-side, $E_{v,n}$ is the valence band edge on the n-side, and so on. Rearranging Equation 5.24,

$$\exp\left(\frac{qV_0}{k_B T}\right) = \frac{\exp-\left(\dfrac{E_{F,p} - E_{v,p}}{k_B T}\right)}{\exp-\left(\dfrac{E_{F,n} - E_{v,n}}{k_B T}\right)}$$

Note that the Fermi energy is constant for a p-n junction under equilibrium; that is, $E_{F,n} - E_{F,p} = 0$, so we get

$$\exp\left(\frac{qV_0}{k_B T}\right) = \exp\left(\frac{E_{F,p} - E_{v,p}}{k_B T}\right)\exp\left(\frac{E_{v,p} - E_{v,n}}{k_B T}\right) = \exp\frac{E_{v,p} - E_{v,n}}{k_B T}$$

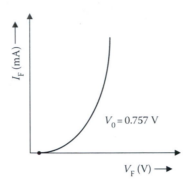

FIGURE 5.8   The *I–V* curve for the p-n junction in Example 5.2, with a forward bias.

Therefore,

$$qV_0 = E_{v,n} - E_{F,p} \tag{5.25}$$

The contact potential barrier energy (qV_0) is the difference between the valence and conduction band edges on each side of the p-n junction. This was shown in Figure 5.7, the band diagram for a p-n junction. Example 5.2 examines the application of these equations.

Example 5.2: Calculation of Contact Potential (V_0)

A step junction in Si is such that the n- and p-sides have dopant concentrations of $N_d = 10^{16}$ and $N_a = 10^{17}$ atoms/cm³, respectively (see Example 5.1). Calculate the contact potential (V_0), assuming that $T = 300$ K and $n_i = 1.5 \times 10^{10}$ electrons/cm³.

Solution
From Equation 5.21,

$$qV_0 = \frac{k_B T}{q}\ln\left(\frac{N_a N_d}{n_i^2}\right) = (0.026 \text{ eV})\ln\left(\frac{10^{17}\text{atoms/cm}^3 \times 10^{16}\text{atoms/cm}^3}{(1.5 \times 10^{10}\text{electrons/cm}^3)}\right)$$

$$qV_0 = 0.757 \text{ eV}$$

Thus, the value of the potential energy barrier set up by the built-in electric field for this p-n junction is 0.757 eV and the corresponding voltage (V_0) is 0.757 V, which is the same as in Example 5.1.

Using forward bias (the "p-side to positive," a useful mnemonic), we overcome this built-in potential, and the diffusion current starts to flow again. This assumes that the voltage drop is very small in the neutral regions, where there is no space charge accumulated. For a forward $V_F > 0.757$ V, electrical current (V_F) of the p-n junction begins to increase dramatically (Figure 5.8).

5.5   SPACE CHARGE AT THE P-N JUNCTION

The depletion region shown in Figure 5.4 has an extremely small concentration of electrons or holes, and ionized dopants in the depletion region are not electrically compensated. The charge density for the depletion region on the n-side of the junction is $q \times N_d$, where N_d is the concentration of donor ions. Assume that the cross-sectional area of the p-n junction is A and the width of the depletion region on the n-side is $x_{n,0}$. The subscript n refers to the n-side, and the subscript 0 stands for a p-n

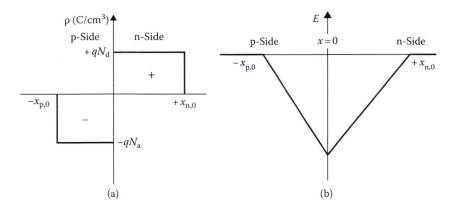

FIGURE 5.9 (a) Space charge density and (b) electric field variation at a p-n junction. (From Neaman, D., *An Introduction to Semiconductor Devices*, McGraw Hill, New York, 2006. With permission.)

junction under equilibrium; that is, no external voltage or bias is applied. Thus, the volume of the depletion region on the n-side is $A \times x_{n,0}$. In this region, the total positive electrical charge is

$$Q_+ = q \times A \times x_{n,0} \times N_d \tag{5.26}$$

Similarly, if $x_{p,0}$ is the width of the penetration of the depletion region into the p-side and the cross-sectional area is A, then the magnitude of the negative charge on the p-side of the depletion region is

$$Q = q \times A \times x_{p,0} \times N_a \tag{5.27}$$

The donor and acceptor ion charges accumulated on the n- and p-sides must be equal for an electrical neutrality of the entire p-n junction, that is, $Q_+ = Q_-$ or

$$q \times A \times x_{n,0} \times N_d = q \times A \times x_{p0} \times N_a \tag{5.28}$$

Therefore,

$$\frac{x_{n,0}}{N_a} = \frac{x_{p,0}}{N_d} \tag{5.29}$$

The width of penetration of the depletion region varies *inversely* to the dopant concentration. For example, if the p-side of the semiconductor is doped more heavily than the n-side (as in Example 5.1), the penetration of the depletion region on the p-side will be smaller. If one side is heavily doped, it needs a lesser volume of that material to compensate for the charge on the other side (Figure 5.9). If one side is lightly doped (in this case, the n-side), then we need more volume of that material to compensate for the charge on the other side. This means a higher penetration depth in the one-dimensional model.

5.6 ELECTRIC FIELD VARIATION ACROSS THE DEPLETION REGION

Using *Poisson's equation,* we can calculate the electric field and the electron energy variation across the depletion region. According to the one-dimensional form of this equation, the gradient of the electric field is related to the charge density and the dielectric permittivity of the material (ε):

$$\frac{d^2\phi}{dx^2} = -\frac{\rho(x)}{\varepsilon} = -\frac{dE(x)}{dx} \tag{5.30}$$

where ϕ is the electric potential, ρ is the charge density, ϵ is the dielectric permittivity of the material, E is the electric field, and x is the distance. Note that $\epsilon = \epsilon_0 \times \epsilon_r$, where ϵ_r is the dielectric constant and ϵ_0 is the permittivity of the free space. The dielectric constant (ϵ_r) is a measure of the ability of a dielectric material to store an electrical charge.

We assume that the doping on the n- and p-sides is uniform, and hence the charge density is as shown in Figure 5.9a.

Applying Poisson's equation to the n-side region in the depletion zone ($x = 0$ to $x = x_{n,0}$),

$$\frac{dE}{dx} = \frac{qN_d}{\epsilon}\left(\text{for } 0 \le x \le x_{n,0}\right) \qquad (5.31)$$

We can obtain the electric field (E) variation by integrating Equation 5.31.

$$E = \int \frac{qN_d}{\epsilon} dx = \frac{qN_d}{\epsilon} x + C_1 \qquad (5.32)$$

Note that when $x = x_{n,0}$, $E = 0$, that is, outside the depletion layer, there is no net built-up charge and the electric field becomes zero. Using this condition in Equation 5.32, we get

$$0 = \frac{qN_d}{\epsilon} x_{n,0} + C_1$$

$$\therefore C_1 = \frac{qN_d}{\epsilon} x_n \qquad (5.33)$$

Using this value of the constant of integration (C_1) in Equation 5.32, the electric field variation across the n-side of the depletion region is given by

$$E(x) = -\left(\frac{qN_d}{\epsilon}\right)(x_{n,0} - x)\ 0 \le x \le x_{n,0} \qquad (5.34)$$

The maximum in the electric field occurs at $x = 0$. Its magnitude is given by

$$|E_{max}| = \left(\frac{qN_d}{\epsilon}\right)x_{n,0} \qquad (5.35)$$

Similarly, we can show that on the p-side of the junction, in the depletion region,

$$E(x) = \left(\frac{qN_a}{\epsilon}\right)(x_{p,0} + x) - x_{p,0} \le x \le 0 \qquad (5.36)$$

The variation in the electric field across the depletion layer width (w), that is, from $-x_{p0}$ to x_{n0}, is shown in Figure 5.9.

As we can see from Poisson's equation, the electric field variation is in the form of a straight line for a uniform charge density. With a linear variation in the electric field, the electric potential will have a parabolic change.

5.7 VARIATION OF ELECTRIC POTENTIAL

We use Poisson's equation to compute the electrostatic potential (ϕ) across the p-n junction. We can obtain the electrostatic potential by integrating the electric field across the distance over which the potential appears.

$$\phi(x) = -\int E(x)\,dx \qquad (5.37)$$

For the p-side of the depletion region, substituting for $E(x)$ from Equation 5.36 into Equation 5.37, we get

$$\phi(x) = -\int \left(\frac{-qN_a}{\epsilon} \right)(x_{p,0} + x)\,dx \tag{5.38}$$

Therefore,

$$\phi(x) = \frac{qN_a}{\epsilon}\left(\frac{x^2}{2} + x_{p,0}\,x \right) + C_2 \tag{5.39}$$

Since the electrostatic potential is zero at $x = -x_{p,0}$, we can calculate the integration constant C_2 in Equation 5.39 as follows:

$$C_2 = \frac{-qN_a}{2\epsilon} x_{p,0}^2 \tag{5.40}$$

Substituting the value of C_2 in Equation 5.39, we get

$$\phi(x) = \frac{qN_a}{2\epsilon}(x + x_{p,0})\ (-x_{p,0} \le x \le 0) \tag{5.41}$$

We will now calculate the electrostatic potential (ϕ) on the n-side of the depletion layer by integrating the electric field from Equation 5.34 as follows:

$$\phi(x) = -\int \left(-\frac{qN_d}{\epsilon} \right)(x_{n,0} - x)\,dx \tag{5.42}$$

Therefore,

$$\phi(x) = \left(\frac{qN_d}{\epsilon} \right)\left(x_{n,0}x - \frac{x^2}{2} \right) + C_3 \tag{5.43}$$

The electrostatic potential is continuous across the p-n junction, that is, at $x = 0$, the value of potential (ϕ) can also be calculated using Equation 5.41 and is equal to the value given by Equation 5.43. Therefore, we evaluate the integration constant C_3 using the following equation:

$$C_3 = \left(\frac{qN_a}{2\epsilon} \right)x_{p,0}^2 \tag{5.44}$$

Substituting this value of C_3 into Equation 5.43, we get

$$\phi(x) = \left(\frac{qN_d}{\epsilon} \right)\left(x_{n,0}x - \frac{x^2}{2} \right) + \frac{qN_a}{2\epsilon} x_{p,0}^2\ (0 \le x \le x_{n,0}) \tag{5.45}$$

This variation in the electrostatic potential across the p-n junction is shown in Figure 5.10 (Neaman 2006).

The value of the contact potential (V_0) can be calculated by evaluating this equation at $x = x_{n,0}$. Therefore,

$$V_0 = \phi(x = x_{n,0}) = \frac{q}{2\epsilon}\left[N_d x_{n,0}^2 + N_a x_{p,0}^2 \right] \tag{5.46}$$

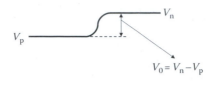

FIGURE 5.10 Electrostatic potential variation across the p-n junction.

Note that the electrostatic potential (ϕ; unit is volts) and the electron energy (unit is electron volts) shown on the band diagram are related by the following equation:

$$\text{Electron energy} = -q \times \phi \tag{5.47}$$

The electrostatic potential is higher on the n-side (Figure 5.9), which means that the electron energy is lower on the n-side. One way to visualize the higher electrostatic potential on the n-side is to note that this is the side of the p-n junction with positively charged donor ions left behind (Figure 5.1).

5.8   WIDTH OF THE DEPLETION REGION AND PENETRATION DEPTHS

The semiconductor p-n junction is electrically neutral as a whole (Equations 5.28 and 5.29).
Substituting for x_p from Equation 5.29 into Equation 5.46 and solving for x_n, we get

$$x_n = \left[\left(\frac{2\varepsilon V_0}{q} \right) \left(\frac{N_a}{N_d} \right) \left(\frac{1}{N_a + N_d} \right) \right]^{1/2} \tag{5.48}$$

Note that we dropped the subscript 0 in $x_{n,0}$, and that the p-n junction is not biased.
Similarly, we can solve for x_p by substituting for x_n from Equation 5.29 into Equation 5.46:

$$x_p = \left[\left(\frac{2\varepsilon V_0}{q} \right) \left(\frac{N_d}{N_a} \right) \left(\frac{1}{N_a + N_d} \right) \right]^{1/2} \tag{5.49}$$

Now, the width of the depletion layer is

$$w = x_p + x_n \tag{5.50}$$

Substituting for x_p and x_n,

$$w = \left[\left(\frac{2\varepsilon V_0}{q} \right) \left(\frac{N_a + N_d}{N_a N_d} \right) \right]^{1/2} \tag{5.51}$$

Thus, for a p-n junction with known doping levels, we can calculate the built-in potential (V_0) using Equation 5.21 or 5.46. We can calculate the depletion layer width using Equation 5.51, and the maximum electric field in a p-n junction at $x = 0$ using Equation 5.35. Examples 5.3 and 5.4 illustrate the calculation of the contact potential (V_0), the depletion layer width (w), and the maximum electric field.

Example 5.3: Calculation of the Depletion Region Width

A step junction in Si is such that the n- and p-sides have a uniform dopant concentration of $N_d = 10^{16}$ and $N_a = 10^{17}$ atoms/cm³, respectively (see Examples 5.1 and 5.2). (a) On which side will the depletion layer penetration be smaller? Why? (b) Calculate the penetration depths on the n-side and p-side. (c) Calculate the total depletion layer width (w).

Solution

1. For this p-n junction, the depletion layer penetration depth on the p-side will be smaller because of the relatively higher acceptor dopant concentration.
2. From Equations 5.48 and 5.49,

$$X_n = \left[\left(\frac{2 \times 11.8 \times 8.85 \times 10^{-14}\,\text{F/cm} \times 0.757\,\text{V}}{1.6 \times 10^{-19}\,\text{C}} \right) \left(\frac{10^{17}\,\text{atoms/cm}^3}{10^{16}\,\text{atoms/cm}^3} \right) \left(\frac{1}{10^{17} + 10^{16}} \right) \right]^{1/2}$$

$X_n = 2.98 \times 10^{-5}$ cm or 298 nm

The space charge width on the p-side (X_p) is given by

$$X_p = \left[\left(\frac{2 \times 11.8 \times 8.85 \times 10^{-14}\,\text{F/cm} \times 0.755\,\text{V}}{1.6 \times 10^{-19}\,\text{C}} \right) \left(\frac{10^{16}\,\text{atoms/cm}^3}{10^{17}\,\text{atoms/cm}^3} \right) \left(\frac{1}{10^{17} + 10^{16}} \right) \right]^{1/2}$$

$X_p = 2.98 \times 10^{-6}$ cm or 0.0 298 μm or 29.8 nm

Because of the higher dopant concentration, the penetration depth on the p-side (X_p) is smaller.
3. The total width of the depletion layer (w) will be = 298 + 29.8 = 327.8 nm.

Example 5.4: Calculation of the Maximum Electric Field

A p-n junction in Si has $N_d = 10^{16}$ atoms/cm³ and the acceptor doping level is $N_a = 5 \times 10^{17}$ cm⁻³.

1. Calculate the built-in potential (V_0) and (b) the maximum electric field in the depletion region.

Solution

1. From Equation 5.21,

$$qV_0 = k_B T \ln\left(\frac{N_a N_d}{n_i^2} \right) = (0.026\ \text{eV}) \ln\left(\frac{5 \times 10^{17}\,\text{atoms/cm}^3 \times 10^{16}\,\text{atoms/cm}^3}{(1.5 \times 10^{10}\,\text{electrons/cm}^3)^2} \right) = 0.795\ \text{eV}$$

The built-in voltage (V_0) is 0.795 V.
2. The maximum electric field occurs at $x = 0$, the metallurgical boundary of the p-n junction.
   From Equation 5.35, when $x = 0$, we get

$$E(X = 0) = -\frac{\left(1.6 \times 10^{-19}\,\text{C}\right)\left(10^{16}\,\text{atoms/cm}^3\right) X_n}{8.85 \times 10^{-14}\,\text{F/cm} \times 11.8} \tag{5.52}$$

We calculate X_n value from Equation 5.48:

$$X_n = \left[\left(\frac{2\varepsilon V_0}{q}\right)\left(\frac{N_a}{N_d}\right)\left(\frac{1}{N_a + N_d}\right)\right]^{1/2}$$

$$X_n = \left[\left(\frac{2 \times 11.8 \times 8.85 \times 10^{-14}\,\text{F/cm} \times 0.795\,\text{V}}{1.6 \times 10^{-19}\,\text{C}}\right)\left(\frac{5 \times 10^{17}\,\text{atoms/cm}^3}{10^{16}\,\text{atoms/cm}^3}\right)\left(\frac{1}{5 \times 10^{17} + 10^{16}}\right)\right]^{1/2}$$

$$X_n = 3.19 \times 10^{-5}\,\text{cm}\quad \text{or}\quad 0.319\,\mu\text{m}$$

Therefore, the magnitude of the electric field is

$$E_{max}(X = 0) = \left|\frac{1.6 \times 10^{-19} \times 10^{16}}{11.8 \times 8.85 \times 10^{-14}}\right|(3.19 \times 10^{-5}\,\text{cm}) = 4.91 \times 10^4\,\text{V/cm}$$

$$E_{max} = 4.91 \times 10^4\,\text{V/cm}$$

This is a fairly large electric field at ~50,000 V/cm. The depletion region has almost no free electrons or holes; therefore, there is very little drift current despite this large built-in electric field.

5.9 DIFFUSION CURRENTS IN A FORWARD-BIASED P-N JUNCTION

The real utility of the p-n junction is that its ability to conduct can be changed with doping and by application of an external voltage. If the positive terminal of the power supply is connected to the p-side and the negative terminal is connected to the n-side, it is called *forward biased*.

In a forward-biased p-n junction, the electrostatic potential at the p-side is increased and the applied electric field opposes the existing internal electric field at the junction (Figure 5.11a). This means that the total height of the potential barrier working as a diffusion barrier will be reduced (Figure 5.12b).

Electrons, the majority carriers on the n-side, can now diffuse more easily to the p-side. In the p-region, these injected electrons move toward the positive terminal, where they are collected. While making their way through the p-region, some of the electrons will recombine with the holes. The positive terminal of the power supply compensates for the holes lost as a result of the recombination. The current due to the electron diffusion is thus maintained by the electrons from the n-side. The negative terminal of the battery provides these electrons.

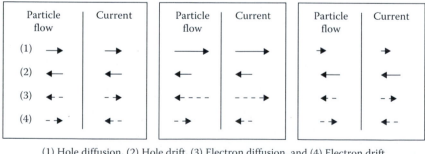

(1) Hole diffusion, (2) Hole drift, (3) Electron diffusion, and (4) Electron drift

(a)   (b)   (c)

FIGURE 5.11 The directions of particle motion for the diffusion and drift of electrons and their corresponding currents for the p-n junction with (a) zero bias, (b) forward bias, and (c) reverse bias. (From Askeland, D. and Fulay P., *The Science and Engineering of Materials*, Thomson, Washington, DC, 2006. With permission.)

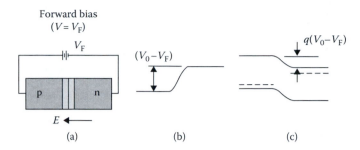

FIGURE 5.12 (a) The bias direction for a forward-biased p-n junction, (b) electrostatic potential variation, and (c) a band diagram. (From Streetman, B.G. and Banerjee S., *Solid State Electronic Devices*, 5th ed., Prentice Hall, Upper Saddle River, NJ, 2000. With permission.)

Similarly, we can see on the electrostatic potential diagram (Figure 5.12c) that the energy barrier for hole diffusion is reduced. This means that more holes, the majority carriers on the p-side, diffuse into the n-side. The holes injected in the n-side diffuse toward the negative terminal of the power supply. Some holes will end up recombining with the electrons on the n-side. The negative terminal of the power supply replaces the electrons lost to the recombination. The current is sustained as more holes continue to diffuse over to the n-side (Figure 5.11b). The positive terminal of the power supply provides these holes.

Note that the forward bias does not change the drift current (Figures 5.11a and 5.11b).The motion of the majority carriers produces a *forward current* (I_F). Under a sufficient forward bias, the total diffusion current increases and the p-n junction begins to conduct. The forward bias also reduces the depletion layer width (w) because the external field decreases the electric field in the depletion layer (Figure 5.12b). We can obtain the new values of the penetration depths and the depletion layer width by substituting V_0 by ($V_0 - V_f$) in Equations 5.48, 5.49, and 5.51 for these quantities.

This is due to the fact that the minority carrier concentration near the ends of the depletion region is increased. For example, since more electrons diffuse from the n- to the p-side under the forward bias, the electron concentration in the neutral p-region (i.e., the minority carrier concentration) near the depletion region increases. When the forward bias is applied, this local increase in the minority carrier concentration near the ends of the depletion region compensates for a decrease in the contact potential and maintains the drift current as it was.

5.10 DRIFT CURRENT IN REVERSE-BIASED p-n JUNCTION

Consider a reverse-biased p-n junction. This means that we apply the n-side of the junction to the positive terminal of an external direct current (DC) voltage supply (Figure 5.13).

Applying a reverse bias, that is, connecting the n-side (which has the positive space-charge region) to the positive terminal, causes the electrostatic potential on the n-side (V_n) to increase (Figure 5.13b). An increase in the potential, in turn, increases the length of the depletion region at the p-n junction.

Recall that electron energy is related to electrostatic potential by $-q$. Thus, with a reverse bias, the Fermi energy level on the n-side is decreased. The band diagram for a reverse-biased p-n junction is shown in Figure 5.13c.

The internal electric field and the external electric field follow the same direction, from the n-side to the p-side. Consequently, the depletion layer width (w) for a reverse-biased junction is larger than that for a p-n junction with no bias (Figure 5.13a).

We can calculate the width of the depletion region (w) and the penetration depths in the p- and n-regions of a p-n junction by substituting ($V_R + V_0$) for V_0 in Equations 5.48, 5.49, and 5.51 derived in Section 5.8.

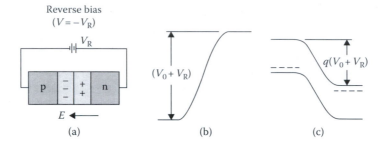

FIGURE 5.13 (a) A reverse-biased p-n junction showing the directions of built-in and applied electric fields, (b) the variation in the electrostatic bias potential that can be calculated, and (c) a band diagram. (From Streetman, B.G. and Banerjee S., *Solid State Electronic Devices*, 5th ed., Prentice Hall, Upper Saddle River, NJ, 2000. With permission.)

Therefore, for a reverse-biased p-n junction,

$$x_n = \left[\left(\frac{2\varepsilon(V_0 + V_R)}{q} \right) \left(\frac{N_a}{N_d} \right) \left(\frac{1}{N_a + N_d} \right) \right]^{1/2} \tag{5.53}$$

$$x_p = \left[\left(\frac{2\varepsilon(V_0 + V_R)}{q} \right) \left(\frac{N_d}{N_a} \right) \left(\frac{1}{N_a + N_d} \right) \right]^{1/2} \tag{5.54}$$

The depletion layer width under the reverse bias can be calculated by adding x_n and x_p values or by using the following equation:

$$w = \left[\left(\frac{2\varepsilon(V_0 + V_R)}{q} \right) \left(\frac{N_a + N_d}{N_a + N_d} \right) \right]^{1/2} \tag{5.55}$$

In the reverse-biased junction, electrons attempting to diffuse from the n-side to the p-side under a concentration gradient encounter a larger energy barrier of height $q(V_0 + V_R)$, which can be visualized easily on a band diagram (Figure 5.13c). The magnitude of the diffusion current due to the diffusion of electrons from the n-side to the p-side becomes very small. Similarly, holes attempting to diffuse from the p-side to the n-side, again under a concentration gradient, encounter a larger potential barrier that can be easily visualized on the electrostatic potential diagram (Figure 5.10b). Because of the increased energy barrier, the magnitude of the total diffusion current due to the movement of holes and electrons is negligible because of the increased energy barrier (Figure 5.13c).

Although the diffusion current becomes negligible, the drift of the minority carriers does not change in the reverse-biased state (Figure 5.13c). Therefore, there is a net current in a reverse-biased p-n junction due to the drift current. Since the drift current does not depend on the magnitude of the applied voltage, the electric current is constant in the reverse-bias state.

Consider the drift current of the minority carriers through the depletion region. In a p-n junction at room temperature, there are thermally created electron–hole pairs (EHPs) in both the neutral regions and in the depletion regions of the p-n junction. The thermal generation in the neutral region is a major source of minority carriers in the *neutral region*. Once the minority carriers are generated, they first diffuse to the depletion region under a concentration gradient. For example, there are thermally generated holes in the n-side neutral region $(x > x_n)$. However, the concentration of holes near the space-charge region is essentially zero at $x = x_n$. Thus, the holes flow from the neutral region to the boundary of the depletion layer due to the concentration gradient. These holes then drift toward the n-side under the influence of the electric field present in the depletion region. The *hole diffusion length* (L_p) is the distance that thermally generated holes on the neutral n-side can travel from the transition region to the n-side through the electric field in the depletion region. Similarly, the *electron diffusion length* (L_n) is the average distance that an electron can travel before recombining with a hole. Carriers generated within the length of the diffusion distance will successfully pass to the depletion layer without recombination. Holes or electrons generated at a distance larger than the diffusion distance (for that carrier) will most likely recombine and thus will not participate in the drift through the depletion region. All thermally generated carriers therefore do not end up contributing to the drift current, which is so called *generation current* or *reverse current*.

The first source for the drift current is induced by the diffusion of thermally generated *minority carriers* in the neutral region followed by their drift across the depletion region. This drift current of the minority carriers generated in the outside of the depletion region is written as I_{Shockley} and is calculated by the Shockley equation:

$$I_{\text{Shockley}} = \left[\left(\frac{qD_p}{L_p N_d} \right) + \left(\frac{qD_n}{L_n N_a} \right) \right] n_i^2 \tag{5.56}$$

In the Shockley equation, q is the magnitude of the charge on the electron or hole, D is the diffusion coefficient, L is the diffusion length, and subscripts p and n refer to the holes and electrons, respectively. Recall from Chapter 4 that the intrinsic carrier concentration (n_i) increases exponentially with the temperature and is inversely related to the band gap (E_g).

The second source for the drift current is from the thermally generated carriers *inside* the depletion region. The internal electric field separates these carriers, and they drift toward the neutral regions. This aspect of the reverse current is given by

$$I_{\text{generation,depletion}} = \frac{qWn_i}{\tau g} \tag{5.57}$$

In Equation 5.57, τ_g is the *mean thermal generation time*. This is the average time needed to thermally create an EHP. The total current due to the drift of thermally generated minority carriers, whether created in the neutral region or in the space-charge region, is called the generation current (I_0).

Thus, the total generation current (I_0) is given by combining Equations 5.56 and 5.57:

$$I_0 = \left[\left(\frac{qD_p}{L_p N_d} \right) + \left(\frac{qD_n}{L_n N_a} \right) \right] n_i^2 + \frac{qW_i}{\tau_g} n_i \tag{5.58}$$

From Equation 5.58, we can see that either the first term with n_i^2 n or the second, thermal-generation term with n_i controls the generation current. In Figure 5.14, the reverse current for a germanium (Ge) diode is shown. In this case, we show a *dark current* to avoid any current due to the photogeneration of EHPs.

At low temperatures, the diffusion of the minority carriers from the neutral region to the depletion region is not significant, and the total generation current is controlled (<240 K) by the second term, thermal generation within the depletion region. At lower temperatures, the slope of the line is proportional to $E_g/2$ because $n_i \propto \exp(-E_g/2kT)$. At higher temperatures, the diffusion term dominates, and the current is controlled by the n_i^2 term. The slope of the line shown in Figure 5.14 is nearly proportional to the band gap of the semiconductor at high temperatures.

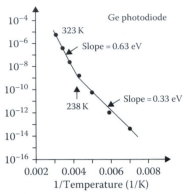

FIGURE 5.14 Generation current (I_0) for a germanium diode with $1/T$. The effect of photogenerated carriers is excluded (i.e., this is a dark current). (From Kasap, S.O., *Principles of Electronic Materials and Devices*, McGraw Hill, New York, 2006. With permission.)

5.11 OVERALL *I–V* CHARACTERSTICS IN A p-n JUNCTION

The generation current or *total drift current* (I_0) is *independent* of the applied bias. Although the forward current (I_F), which is due to the diffusion of majority carriers, is strongly affected by the type and amount of bias, the magnitude of the generation current (I_0) is independent of the applied voltage (Figure 5.11).

The generation current depends on the temperature because the carrier concentration due to thermal excitation and the diffusion of the minority carriers toward the depletion region also depend on the temperature. Similarly, the absorption of light energy leads to the creation of EHPs. This also causes an increase in the generation current.

Thus, the total current (I) flowing through the p-n junction is given by

$$I = I_{\text{generation}} + I_{\text{diff}} \qquad (5.59)$$

When there is no applied voltage (Figures 5.11a and 5.15a), the total drift current and the current due to the diffusion of majority carriers are equal and opposite, that is,

$$|I_{\text{diff}}| = |I_{\text{generation}}| \text{ for } V = 0 \qquad (5.60)$$

When the p-n junction is reverse-biased (Figures 5.13c and 5.15c), the diffusion of majority carriers is almost negligible, and the total current is equal to the generation current.

$$I \cong I_{\text{generation}} \text{ for reverse bias} \qquad (5.61)$$

Therefore, the generation current (I_0) is also known as the *reverse-bias saturation current* (I_s). The generation current usually is in the range of 10^{-14}–10^{-12} A. The magnitude of the generation current depends on its doping levels, the size of the cross-sectional area of the junction, temperature, and so on (Equation 5.58). It does not depend on the reverse-applied voltage.

For a p-n junction under a forward bias (Figures 5.12b and 5.15b), the diffusion current ($I_{\text{diffusion}}$) changes exponentially with the applied voltage and is given by

$$I_{\text{diffusion}} = I_0 \left(\frac{qV}{kT} \right) \qquad (5.62)$$

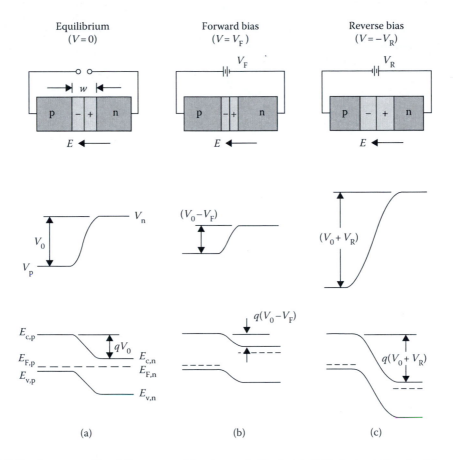

FIGURE 5.15 Summary of band diagrams and the changes in the electric field, electrostatic potential, and depletion layer width of a p-n junction under (a) zero, (b) forward, and (c) reverse bias. (From Streetman, B.G. and Banerjee S., *Solid State Electronic Devices*, 5th ed., Prentice Hall, Upper Saddle River, NJ, 2000. With permission.)

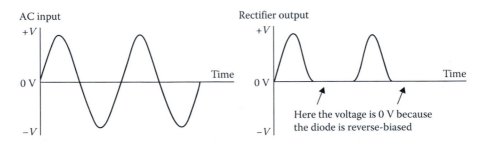

FIGURE 5.16 Half-wave rectification using a diode.

Thus, the total current under a forward bias is given by the difference between I_{diff} and $I_{\text{generation}}$:

$$I = I_0 \left[\exp\left(\frac{qV}{k_{\text{B}}T} \right) - 1 \right] \tag{5.63}$$

Equation 5.63 is also known as the *ideal diode equation*. The current–voltage (*I–V*) curve for a p-n junction is shown in Figure 5.16.

If the applied bias (V) is large compared to the magnitude of (k_BT/q) or the temperature is very low, then the exponential term will be much larger than 1, and Equation 5.63 can be rewritten as

$$I \approx I_0 \exp\left(\frac{V}{(q/k_BT)}\right) \tag{5.64}$$

At room temperature $(T = 300 \text{ K})$, the value of (q/k_BT) is ~0.026 V, hence

$$I \approx I_0 \exp\left(\frac{V(\text{in V})}{(0.026 \text{ V})}\right) \text{at } T = 300 \text{ K} \tag{5.65}$$

The p-n junction behaves like a one-way electrical valve. It functions as a rectifier by allowing current to flow through under a forward bias but not under a reverse bias.

A summary of p-n junction band diagrams and changes in the electric field in the depletion layer width, as well as electrostatic potential variation under zero, forward, and reverse bias, is shown in Figure 5.15.

5.12   DIODE BASED ON A P-N JUNCTION

The electrical symbol for a diode is shown in Figure 5.17. A diode can serve several useful functions—including rectification, which means the filtering of the electric current with a specific polarity. For an alternating current voltage (AC) input, the diode produces the output voltage shown in Figure 5.16.

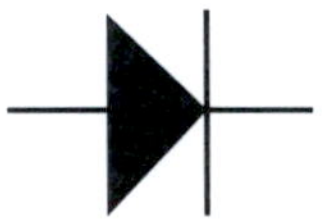

FIGURE 5.17   Symbol for a p-n junction diode in an electrical circuit.

Two diodes can be used to create a full-wave rectification of AC voltage. The p-n junction is also used as the basic building block for solar cells and devices known as transistors. Millions of transistors are connected to form miniature circuits, which are then used to create semiconductor chips that are used in computers and other electronic equipment.

An actual I–V characteristic curve for a p-n junction is shown in Figure 5.18. Note that, under a forward bias, the current (I) is in milliamperes (mA). Under a reverse bias, the current is in microamperes (μA). When the forward-bias voltage is less than the built-in potential V_0, a small current is set up across the p-n junction. This built-in potential of the p-n junction is called *knee voltage*. If the dopant concentrations are high, the knee voltage is proportional to the band gap of the semiconductors (Figure 5.19). As the forward-biased voltage becomes larger than the knee voltage, the current also increases dramatically. Note that this increase is nonlinear; that is, it does not follow Ohm's law (Figure 5.18).

For moderately doped silicon (Si), the knee voltage is ~0.7 V, which is close to the built-in potential (V_0) of silicon p-n junctions, as shown in Example 5.1. For Ge with the band gap smaller than Si, the typical knee voltage of the p-n junctions is ~0.3 V. Since the band gap of GaAs is larger than that of Si, the typical knee voltage of GaAs p-n junctions is more than 1.0 V. A schematic of I–V curves expected for p-n junctions for different semiconductors is shown in Figure 5.19.

As the temperature increases from 25°C to 25 + ΔT, the knee voltage decreases slightly. At an elevated temperature, the forward current (I_F) at any given voltage is higher than that at 25°C. As expected, more minority carriers are generated on both sides of the p-n junction at higher temperatures. Thus, there is a very slight increase in the reverse-bias current at higher temperatures. The sketch in Figure 5.18 is not to scale and exaggerates this difference in the change in the magnitude of the generation current. Note that the abrupt increase in the reverse current at $V = V_{BR}$ in Figure 5.18 is not related to the drift current. This is called the *breakdown* of the diode and will be explained in Section 5.13.

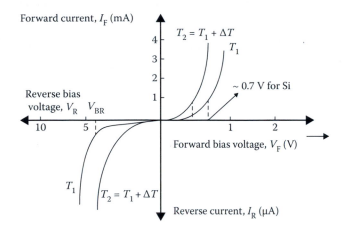

FIGURE 5.18 The *I–V* curves for representative silicon-based p-n junctions. The effect of increased temperature on the lowering of breakdown and knee voltage is shown. Note the current scale for forward and reverse bias is in milli- and microamperes, respectively. Similarly, the voltage scales for forward and reverse bias are different. (From Floyd, T., *Electric Circuit Fundamentals*, 7th ed., Prentice Hall, Upper Saddle River, NJ, 2006. With permission.)

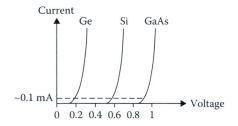

FIGURE 5.19 Schematic of *I–V* curves for different semiconductors showing different expected knee voltages (current axis not to scale). (From Kasap, S.O., *Principles of Electronic Materials and Devices*, McGraw Hill, New York, 2002. With permission.)

Example 5.5: Current in a Si P-N Junction Diode

The built-in potential (V_0) of a Si p-n junction diode is 0.8 V. Assume that the reverse-bias saturation current (I_0) is 10^{-13} amperes (A). A forward bias of 0.5 V is then applied.

1. What is the new height of the potential barrier seen on the electrostatic potential diagram?
2. What is the current flowing through this diode with a forward bias of 0.5 V?
3. What is the current if the forward bias changes to 0.6 V?

Compare the different current values with each other.

Solution
1. When the applied voltage is 0.5 V, the new height of the barrier seen on the electrostatic potential will be ($V_0 - V_F$) = (0.8 − 0.5) = 0.3 V (Figure 5.14b).
2. Since the forward bias applied (V = 0.5 V) is much larger than kT/q (0.0259 V), that is, $qV \gg kT$, the exponential term is much larger than 1, and we use Equation 5.65. Therefore,

$$I_F \approx \left(10^{-13} A\right) \exp\left(\frac{0.5\ V}{0.0259\ V}\right) = 2.42 \times 10^{-5} A$$

This is about 24 μA.

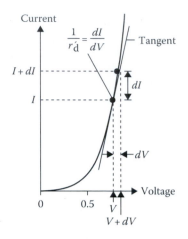

FIGURE 5.20 A graphical illustration of the dynamic resistance of a p-n junction. (From Kasap, S.O., *Principles of Electronic Materials and Devices*, McGraw Hill, New York, 2002. With permission.)

3. When the forward bias is 0.6 V, the current is given by

$$I_F \approx \left(10^{-13}\,\text{A}\right)\exp\left(\frac{0.6\,\text{V}}{0.0259\,\text{V}}\right) = 1.15 \times 10^{-3}\,\text{A}$$

This current of about 1 mA is about 40 times greater than the current when the bias of 0.5 V was applied. This shows that the effective resistance of the p-n junction drops by ~40 times with a bias increase of only 0.1 V.

The p-n junction under the forward bias has a *dynamic resistance* (r'_d) that changes with the applied voltage (Figure 5.20). Most materials (e.g., metals, alloys and semiconductors) show a resistance that is constant and not voltage-dependent.

The dynamic resistance of a p-n junction can be calculated from the slope of the I–V curve at a given value of voltage:

$$r'_d = \frac{dV_{\text{applied}}}{dI_F} \tag{5.66}$$

A p-n junction circuit uses a resistor (R_L), known as the limiting resistor, to limit the total current and to protect the p-n junction from damage due to excessive current. When such a resistor is used in series with the diode, the forward current is given by

$$I_F = \frac{(V_{\text{applied}} - V_{\text{knee}})}{R_L} \tag{5.67}$$

If a limiting resistor is not used, the p-n junction can be damaged by Joule heating.

If the dynamic resistance of the p-n junction is accounted for, then the forward current is given by

$$I_F = \frac{(V_{\text{applied}} - V_{\text{knee}})}{(R_L + r'_d)} \tag{5.68}$$

The application of this equation is illustrated in Example 5.6.

Example 5.6: Forward Current in a Diode

A Si p-n junction functions as a diode and is connected to a resistor, $R_L = 1000\ \Omega$. Assume that the knee voltage is 0.7 V.

1. What is the forward current if a 4.5 V forward bias is applied?
2. What is the forward current, assuming that the diode offers a dynamic resistance of 10 Ω at the selected I_F value?

Solution
1. From Equation 5.67,

$$I_F = \frac{(4.5 - 0.7)}{1000\ \Omega} = 3.8\ \text{mA}$$

2. When the dynamic resistance of the diode (r_d') must be accounted for, we add that resistance to the R_{limit}.
   From Equation 5.68,

$$I_F = \frac{(4.5 - 0.7)}{(1000 + 10)\ \Omega} = 3.762\ \text{mA}$$

As can be expected by adding the dynamic resistance, the magnitude of I_F decreases.

5.13   REVERSE-BIAS BREAKDOWN

A reverse bias is sometimes so high that it causes a *reverse-bias dielectric breakdown* of the p-n junction (Figure 5.21). When this happens, the current flowing through the p-n junction increases rapidly. Critical breakdown voltages (V_{BR}) for silicon diodes generally start at ~60 V. The term breakdown is a bit misleading; if the p-n junction breaks down electrically, this does not mean that the device is permanently damaged. The *I–V* relation in Figure 5.21 is reversible. The reversible breakdown of a p-n junction, which should be distinguished from a broken or defective p-n junction, is useful in some applications.

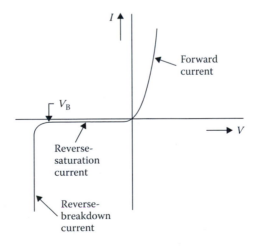

FIGURE 5.21   The *I–V* characteristic of a p-n junction including a reverse-bias breakdown. (From Mahajan, S. and Sree Harsha K.S., *Principles of Growth and Processing of Semiconductors*, McGraw Hill, New York, 1998. With permission.)

A limiting resistor (R_L) is used to restrict the flow of current during a reverse-bias breakdown. If such a current-limiting resistor is not used, the p-n junction will be damaged because of overheating caused by the excessive current. This will occur in both forward and reverse bias.

The reverse-bias breakdown can occur via two mechanisms. In the *avalanche breakdown* mechanism, a high-energy conduction electron drifting through large electric potential at the junction collides with a silicon–silicon bond and knocks off a valence electron from that bond, which makes this electron free. Consequently, a number of the free electrons are doubled. This proliferation process of the free electron under large reverse-bias is known as *impact ionization* (Figure 5.22).

From a band diagram viewpoint, the impact ionization process sends an electron from the valence band to the conduction band, creating a hole in the valence band. The first high-energy electron continues, knocking off another electron from one more silicon–silicon covalent bond and continuing the process. The impact ionization process is shown on the band diagram in Figure 5.23.

In the mechanism known as an avalanche breakdown, multiple collisions from a single high-energy electron can create many EHPs.

A p-n junction with one side heavily doped is known as a one-sided junction. If the p-side is heavily doped, we refer to this as a p$^+$-n junction. The depletion layer of a p$^+$-n is mainly on the n-side. The charge density distribution for this junction is shown in Figure 5.24. When this junction breaks down electrically, the breakdown will occur on the n-side at a location where the electric field is at a maximum.

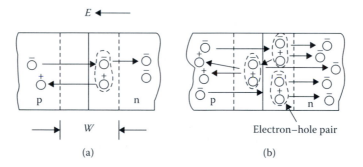

FIGURE 5.22 Illustration of the impact ionization process. (a) A single ionizing collision with a host lattice atom by an incoming electron and (b) multiplication of carriers due to multiple collisions. (From Mahajan, S. and Sree Harsha K.S., *Principles of Growth and Processing of Semiconductors*, McGraw Hill, New York, 1998. With permission.)

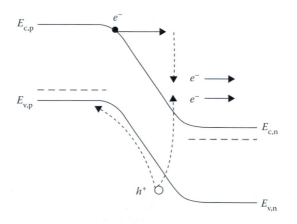

FIGURE 5.23 Impact ionization process illustrated on a band diagram. (From Streetman, B.G. and Banerjee S., *Solid State Electronic Devices*, Prentice Hall, Upper Saddle River, NJ, 2000. With permission.)

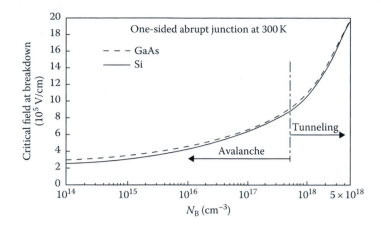

FIGURE 5.24 Critical field at the breakdown for one-sided abrupt p-n junctions in silicon and GaAs (at $T = 300$ K). (From Mahajan, S. and Sree Harsha K.S., *Principles of Growth and Processing of Semiconductors*, McGraw Hill, New York, 1998. With permission.)

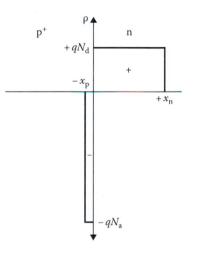

FIGURE 5.25 Schematic of the charge density for a p⁺-n junction. (From Neaman, D., *An Introduction to Semiconductor Devices*, McGraw Hill, New York, 2006. With permission.)

Figure 5.24 shows the critical breakdown field as a function of background doping for a *one-sided p-n junction,* in which one of the sides is very heavily doped. If the p-side is heavily doped, the junction is referred to as p⁺-n; if the n-side is heavily doped, the junction is shown as p-n⁺. The charge density expected for a p⁺-n junction is shown in Figure 5.25.

The breakdown voltage (V_{BR}) for a p⁺-n junction is given by

$$V_{BD} = \frac{\varepsilon E_{critical}^2}{2qN_d} \; V_{BR} = \frac{\varepsilon E_{critical}^2}{2qN_d} \tag{5.69}$$

where N_d is the donor dopant concentration and $E_{critical}$ is the critical field that causes breakdown (Singh 2001). The breakdown field for a p⁺-n junction will depend upon the level of doping. This change for the silicon and gallium arsenide (GaAs) p-n junctions is shown in Figure 5.24.

Example 5.7 shows the calculation of the breakdown voltage by the avalanche breakdown mechanism.

Example 5.7: Breakdown Voltage for a P-N Si Diode

In a Si diode, $N_a = 10^{19}$ atoms/cm³ and $N_d = 5 \times 10^{15}$ atoms/cm³. If the critical breakdown field is 4×10^5 V/cm, what is the breakdown voltage for this diode at $T = 300$ K?

Solution

The p-side is heavily doped; therefore, the breakdown will occur in the depletion layer that extends to the n-side. We apply Equation 5.69:

$$V_{br} = \frac{(11.8)(8.85 \times 10^{-14} \text{ F/cm})(4 \times 10^5 \text{ V/cm})^2}{2(1.6 \times 10^{-19} \text{ C})(5 \times 10^{15} \text{ atoms/cm}^3)} = 103.5 \text{ V}$$

For diodes that have a higher breakdown field, the breakdown voltage will be higher. Thus, such diodes as those made from silicon carbide (SiC), which has a high breakdown voltage, are useful for high-temperature, high-power applications.

As the doping level increases, another mechanism of breakdown can be applied. In the *Zener tunneling mechanism*, electrons tunnel across the p-n junction instead of climbing over the energy barrier. This is commonly seen in heavily doped p-n junctions. The breakdown occurs across very thin depletion regions and thus typically occurs at low voltages (~0.1–5 V). Heavier doping means a smaller depletion region width (Equations 5.48 and 5.49). With Zener tunneling, electrons from the p-side valence band can tunnel across into the conduction band on the n-side because the reverse bias pushes down the E_c on the n-side and aligns it with the valence band edge (E_v) on the p-side. The Zener tunneling mechanism is also known as the Zener effect or band-to-band tunneling (Figure 5.26).

The probability of Zener tunneling (T) is given by

$$T \approx \exp\left(-\frac{4\sqrt{2m_e^*}\,E_g^{3/2}}{2lq\hbar E} \right) \tag{5.70}$$

where m_e^* is the reduced effective mass of the electron, E_g is the band gap, and E is the electric field (Singh 2001). The Zener breakdown is important in heavily doped junctions and in narrow-bandgap materials. The value of probability (T) generally needs to be ~10^{-6} for tunneling to initiate this breakdown process. Example 5.8 illustrates the calculation of the tunneling probability.

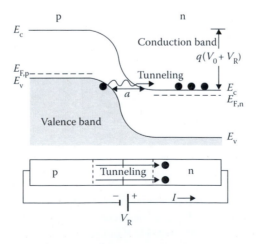

FIGURE 5.26 Zener effect showing the breakdown of a heavily doped p-n junction. (From Kasap, S.O., *Principles of Electronic Materials and Devices*, McGraw Hill, New York, 2002. With permission.)

Example 5.8: Zener Tunneling in InAs

Calculate the probability of tunneling in indium arsenide (InAs) if the applied electric field (E) is 3×10^5 V/cm. Assure that the m_e^* for electrons in InAs is 0.02 m_0^*. The band gap of InAs is 0.4 eV.

Solution
From Equation 5.70, the probability of tunneling is

$$T = \exp\left(\frac{-4\sqrt{2(0.02 \times 9.1 \times 10^{-31}\,\text{kg})}(0.4\,\text{eV} \times 1.6 \times 10^{-19}\,\text{J/eV})^{3/2}}{3(1.6 \times 10^{-19}\,\text{C})1.05 \times 10^{-34}\,\text{J·s})(3 \times 10^{4}\,\text{V/cm} \times 10^{2}\,\text{cm/m})} \right)$$

Note the use of SI units, including the conversion of the electric field from V/cm to V/m.
$T = 2.82 \times 10^{-4}$, which is greater than 10^{-6}. Therefore, Zener tunneling is important for this value of the electric field in InAs.

5.14 ZENER DIODES

The breakdown by either the Zener or the avalanche mechanism is the basis for the *Zener diode* (Figure 5.27a), a device that is typically operated under a reverse bias. The *I–V* curve for a Zener diode, including the reverse-breakdown region, is shown in Figure 5.27b. The normal operating region for a regular p-n junction diode is also shown in this figure for comparison (Figure 5.27c).

Since the Zener diode has an avalanche or Zener breakdown after a certain applied voltage is reached, the voltage across this diode remains essentially constant. Therefore, Zener diodes are used as voltage regulators (Example 5.9). Under a forward bias, the Zener diode will function like a regular diode, with the typical knee voltage of ~0.7 V for silicon diodes.

The voltage at which a Zener diode will begin to conduct current under a reverse bias is known as the *Zener voltage*. Diodes can be designed to have a Zener voltage ranging from a few volts to a few hundred volts.

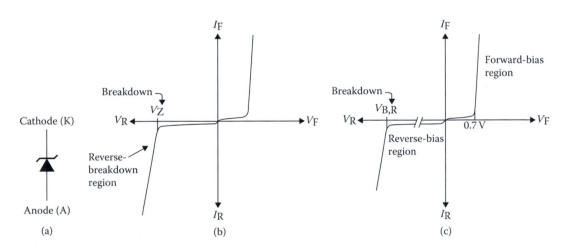

FIGURE 5.27 (a) Symbol for a Zener diode. The anode (+) is on the left and cathode (−) is on the right. The positive terminal of the power supply is connected to the cathode of the Zener diode to operate under a reverse bias. (b) Typical *I–V* curve for a Zener diode. (c) Normal operating region for a regular p-n junction diode. (From Floyd, T., *Electronics Fundamentals: Circuits, Devices, and Applications*, 4th ed., Prentice Hall, Upper Saddle River, NJ, 1998. With permission.)

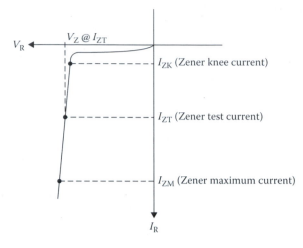

FIGURE 5.28 Close-up of a reverse-bias portion of the *I–V* curve for a Zener diode. (From Floyd, T., *Electronics Fundamentals: Circuits, Devices, and Applications*, 4th ed., Prentice Hall, Upper Saddle River, NJ, 1998. With permission.)

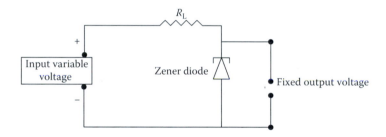

FIGURE 5.29 A Zener diode as a voltage regulator (see Example 5.9).

In practice, the Zener *I–V* curve is not completely vertical (Figure 5.28). We define ΔV_Z as the change in voltage across the reverse-biased Zener diode as the current changes from a Zener test current (I_{ZT}) to a higher value, up to the Zener maximum current (I_{ZM}). The Zener diode impedance or resistance (Z_Z) is given by

$$Z_z = \frac{\Delta V_z}{\Delta I_z} \tag{5.71}$$

In Equation 5.71, $\Delta I_Z = I_{ZM} - I_{ZT}$.

Example 5.9 shows how a Zener diode under a reverse bias functions as a voltage regulator.

Example 5.9: Zener Diode Voltage Regulator

Consider a voltage regulator supply connected to a 200 Ω current-limiting resistance (R_l) and a reverse-biased Zener diode. The knee current (I_{ZK}; Figure 5.28) is 0.2 mA, and the maximum current (I_{ZM}) is 100 mA. Assume that the voltage across the Zener diode is always 10 V. What is the range of input voltages that can be regulated using this Zener diode?

Solution

The job of this circuit (Figure 5.29) is to provide a 10 V constant output even though the input voltage of the variable power supply will change. We need to find the range of voltages that can be controlled using this Zener diode and the 200 Ω resistor.

If the applied voltage is too low, the Zener diode will not conduct. When the applied voltage is just above the Zener voltage, the current is close to the knee current of 0.2 mA. This current flows through the Zener diode and the 200 Ω resistor. The voltage drop across the resistor is given by Ohm's law:

$$V_{resistor} = I \times R = I_{ZK} \times R = 0.2 \text{ mA} \times 200 \ \Omega = 40 \text{ mV}$$

The voltage drop across the Zener diode is always 10 V. The total input voltage will be 10 V + 0.04 V = 10.04 V. This is the minimum input voltage that can be controlled. If the applied voltage is lower than this, then the Zener diode will not conduct.

If we now increase the voltage beyond 10.04 V, the Zener diode will conduct a current. The maximum current this diode is designed for is 100 mA. The voltage drop across the resistor for this current is given by

$$V_{resistor} = I \times R = I_{ZM} \times R = 100 \text{ mA} \times 200 \ \Omega = 20 \text{ V}$$

Once again, the voltage drop across the Zener diode is only 10 V. Under these conditions, the total voltage must be 20 + 10 = 30 V.

This combination of a Zener diode and a current-limiting resistor would provide a constant 10 V output even as the input voltage changes between 10.04 and 30 V. If the voltage exceeds 30 V, it exceeds the maximum allowed current of 100 mA and will damage the Zener diode.

PROBLEMS

5.1 A p-n junction is formed in Ge so that the donor and acceptor concentrations are 10^{16} and 10^{17} atoms/cm^3, respectively. Ge has a small band gap ($E_g \sim 0.67$ eV). Based on this, do you expect the built-in potential to be smaller or greater than Si ($E_g \sim 1.1$ eV)? What is the built-in potential (V_0) for this p-n junction? Use the intrinsic carrier concentrations for Ge from Figure 3.7 and assume $T = 300$ K.

5.2 A p-n junction is formed in GaAs so that the donor and acceptor concentrations are 10^{16} and 10^{17} atoms/cm^3, respectively. Show that the built-in potential for this p-n junction is 1.22 V. Use the intrinsic carrier concentrations for GaAs from Figure 3.7 and assume that $T = 300$ K.

5.3 A p-n junction is such that the acceptor and donor dopant levels are 0.16 eV from the nearest band edge. The doping level on both sides is 10^{16} atoms/cm^3. What is the built-in potential for this junction?

5.4 A Si p-n junction at 300 K is P-doped on one side such that $N_d = 5 \times 10^{16}$ atoms/cm^3. On the acceptor side, the B dopant level is 10^{16} atoms/cm^3. Calculate the contact potential, maximum electric field, width of the depletion region, and the penetration depths on each side. It may be easier to set up a spreadsheet to solve this and other problems.

5.5 For the junction described in Example 5.3, calculate the maximum electric field in V/cm.

5.6 For the p-n junction discussed in Equation 5.4, calculate the value of the penetration depth on the p-side and the width of the depletion layer (w).

5.7 What is the forward current in a diode as discussed in Example 5.6, if the forward bias is 0.4 and 0.7 V?

5.8 What is the breakdown field for a Si p$^+$-n junction where the $N_d = 5 \times 10^{16}$ atoms/cm^3? Assume that the p-side is very heavily doped and the breakdown therefore occurs on the n-side. Use the data in Figure 5.25.

5.9 What is the breakdown voltage for the diode discussed in Example 5.8, if $N_d = 2 \times 10^{17}$ atoms/cm^3?

5.10 What is the breakdown voltage for a diode in a diamond p-n junction? Assume that the breakdown field is 10^7 V/cm and $N_d = 10^{16}$ atoms/cm^3. How does this value compare with the breakdown voltage for a similar diode made from Si?

5.11 What donor doping level will be needed for a Si p$^+$-n diode if the breakdown voltage is 30 V? Use the data in Figure 5.25 and assume that the p-side is heavily doped so that the breakdown occurs on the n-side.

5.12 What is the minimum voltage that can be regulated using a Zener diode with a knee current (I_{ZK}) of 2 mA and a V_Z of 15 V?

5.13 True or false: The Zener diode operates on a breakdown mechanism by avalanche or tunneling. Explain.

GLOSSARY

Avalanche mechanism: A high-voltage breakdown mechanism occurring in diodes in which a conduction electron with high energy scatters a valence electron and transfers it into the conduction band, creating a hole in the valence band. This is different from the Zener mechanism, which occurs at lower voltages.

Band-to-band tunneling: A tunneling mechanism in which electrons from the valence band of the p-side tunnel flow into the conduction band on the n-side. See also **Zener tunneling**.

Built-in electric field: See **Internal electric field**.

Built-in potential (V_0): See **Contact potential**.

Contact potential (V_0): This is the potential difference that appears when a junction is formed between different materials (also known as the built-in potential). On the band diagram, this is expressed as energy qV_0. We need to apply a voltage greater than V_0 or provide energy greater than qV_0 to cause the p-n junction to conduct.

Dark current: A current measured for a nonilluminated p-n junction in order to avoid any current due to the photogeneration of electron–hole pairs, as opposed to thermal generation.

Depletion region: A region at the electrical interface between the n- and p-regions of a p-n region that is depleted of charge carriers, that is, electrons and holes. A built-in electric field exists over this region.

Diffusion: The motion of electrons, holes, and ions from a region of high concentration and chemical potential to a region of low concentration.

Diffusion potential: See **Contact potential**.

Drift: The motion of charge carriers under the influence of an internal or external electric field.

Dynamic resistance (r_d'): The voltage-dependent resistance of a p-n junction under a forward bias when the applied voltage is less than the voltage near the knee of the I–V curve.

Einstein relation: An equation that describes the relationship between the mobility of species and their diffusion coefficient.

$$\frac{\mu}{D} = \frac{q}{k_B T}$$

Electron diffusion length (L_n): The average distance an electron can diffuse before recombining with a hole.

Forward bias: A voltage applied to a p-n junction in which the positive terminal of an external voltage supply is connected to the p-side of the p-n junction. This bias can overcome the built-in electric field and cause the p-n junction to conduct.

Forward current (I_F): The current in a p-n junction under a forward bias.

Generation current (I_0): The total current due to the drift of thermally generated carriers under the influence of an electric field in the depletion region. These go from the n-side to the p-side.

Hole diffusion length (L_p): The average distance a hole can diffuse before recombining with an electron.

Ideal diode equation: The equation that describes the I–V characteristics of an ideal p-n junction (Equation 5.63):

$$I = I_0 \left[\exp\left(\frac{qV}{k_B} \right) - 1 \right]$$

Impact ionization: A process in which a high-energy electron collides with other electrons, breaking bonds in a material and causing one of the valence electrons to move into the conduction band, which creates a hole. This process eventually leads to an avalanche breakdown in a p-n junction.

Internal electric field: The electric field developed by positively charged donor ions, which are left behind as the electrons diffuse from the n-side to the p-side, and negatively charged acceptor ions, which are left behind as the holes diffuse from the p-side to the n-side.

Knee voltage: The voltage on the I–V curve for a diode, at which the forward current begins to increase exponentially. This value is very close to the built-in potential (V_0) for the p-n junction.

One-sided p-n junction: A p-n junction in which one side is very heavily doped. This side is indicated with a + superscript (e.g., a p^+-n junction is where the p-side is very heavily doped).

Poisson's equation: An equation relating the gradient of an electric field to the charge density and the dielectric permittivity of the material (ε), given by Equation 5.30:

$$\frac{d^2\phi(x)}{dx^2} = \frac{\rho(x)}{\varepsilon} = \frac{dE(x)}{dx}$$

Reverse bias: A voltage applied to a p-n junction so that the negative terminal of the external voltage supply is connected to the p-side of the p-n junction. This reverse bias adds to the built-in electric field and makes the p-n junction nonconducting.

Reverse-bias dielectric breakdown: A high value of reverse-bias applied voltage, which causes a high level of current in the p-n junction.

Reverse-bias saturation current (I_s): A current resulting from the drift of thermally generated carriers in a p-n junction, which is the same as the generation current (I_0). It is *independent* of the applied voltage.

Shockley equation: An equation that describes one source of the generation current in a p-n junction originating from the diffusion of minority carriers that are generated in the neutral regions of the p-n junction (Equation 5.56):

$$I_{Shockley} = \left[\left(\frac{qD_p}{L_p N_d} \right) + \left(\frac{qD_n}{L_n N_a} \right) \right] n_i^2$$

Space-charge layer: See **Depletion region**.

Space-charge region: See **Depletion region**.

Step junction: A p-n junction in which the transition from the p-side to the n-side is abrupt.

Zener breakdown mechanism: A low-voltage breakdown mechanism in which a p-n junction breaks down when electrons tunnel across the p-n junction instead of climbing over the energy barrier. Electrons from the p-side valence band tunnel across into the conduction band on the n-side because the reverse bias pushes down the E_c on the n-side and aligns it with the valence band edge (E_v) on the p-side. This is seen more often in heavily doped junctions with smaller band gaps.

Zener diode: A diode based on a Zener or avalanche breakdown occurring in a p-n junction.

Zener tunneling: The tunneling seen in the Zener breakdown mechanism, also known as band-to-band tunneling.

Zener voltage: The voltage at which a Zener diode begins to carry current under a reverse bias (typical values ~1–200 V).

REFERENCES

Askeland, D., and P. Fulay. 2006. *The Science and Engineering of Materials*. Washington, DC: Thomson.

Edwards-Shea, L. 1996. The *Essence of Solid State Electronics*. Upper Saddle River, NJ: Prentice Hall.

Floyd, T. 1998. *Electronics Fundamentals: Circuits, Devices, and Applications,* 4th ed. Upper Saddle River, NJ: Prentice Hall.

Floyd, T. 2006. *Electric Circuit Fundamentals,* 7th ed. Upper Saddle River, NJ: Prentice Hall.

Kasap, S. O. 2002. *Principles of Electronic Materials and Devices*. New York: McGraw Hill.

Kasap, S. O. 2006. *Principles of Electronic Materials and Devices*. New York: McGraw Hill.

Mahajan, S., and K. S. Sree Harsha. 1998. *Principles of Growth and Processing of Semiconductors*. New York: McGraw Hill.

Neaman, D. 2006. *An Introduction to Semiconductor Devices*. New York: McGraw Hill.

Singh, J. 2001. *Semiconductor Devices: Basic Principles*. New York: Wiley.

Streetman, B. G., and S. Banerjee. 2000. *Solid State Electronic Devices*, 5th ed. Upper Saddle River, NJ: Prentice Hall.

6 Semiconductor Devices

KEY TOPICS

- Metal–semiconductor junctions
- Schottky and ohmic contacts
- Solar cells
- Light-emitting diodes (LEDs)
- The operation of a bipolar junction transistor (BJT)
- Principle of the field-effect transistor (FET)

In Chapter 6, we will discuss several semiconductor devices utilizing a metal–semiconductor junction, semiconductor–semiconductor junction, and semiconductor–dielectric interface. The devices of interest are Schottky/ohmic contacts, light-emitting diodes (LEDs), and several *transistors*. The solar cell is another very important device based on the p-n junction of the semiconductor, and we will briefly study it in this chapter after we discuss the optical properties of materials. In addition to our coverage in this chapter, the optical aspect of LEDs will also be reviewed in Chapter 8. This combined knowledge regarding semiconductors and optical materials will help readers better understand solar cells and LEDs more effectively.

6.1 METAL–SEMICONDUCTOR CONTACTS

In semiconductor device fabrication, metals are often deposited as electrodes onto semiconductor surfaces. The deposition of a metallic material onto a semiconductor results in an *ohmic contact* or in a *Schottky contact*. The ohmic contact between a metallic material and a semiconductor is such that there is no energy barrier to block the flow of carriers in either direction at the metal–semiconductor junction. In some cases, however, the contact between a metal and a semiconductor is rectifying; this type of contact is known as the Schottky contact.

To better understand the origin of Schottky and ohmic contacts, recall that the *work function* of a material is the energy (E) required to remove an electron from its Fermi energy level (E_F) and set it free (Figure 6.1). Note that, although there are electrons at $E = E_F$ for a metal, there are no electrons at $E = E_F$ for a semiconductor.

Therefore, we define the *electron affinity of a semiconductor* ($q\chi_S$) as the energy needed to remove an electron from the conduction band and set it free (Figure 6.1). It is *not* the lowest energy required to remove an electron because when an electron is removed from the conduction band, another electron must move from the valence band to the conduction band, which requires additional energy.

For semiconductors, we usually refer to the values of electron affinity and not the work function because electrons are present in the conduction band but not at $E = E_F$. The work function and electron affinity values for different metals and semiconductors are shown in Table 6.1. The values of the electron affinity and the work function are often expressed as potential in volts (V) or as energy in electron volts.

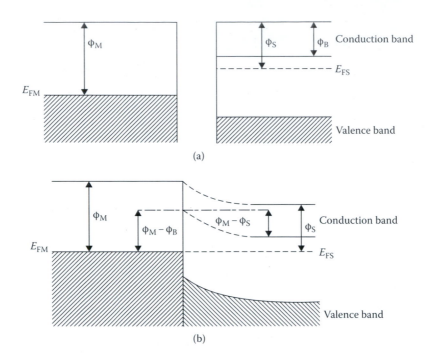

(a)

(b)

FIGURE 6.1   Energy diagrams for a junction between a metal and an n-type semiconductor ($\phi_M > \phi_S$) (a) before contact and (b) after contact, when the Fermi levels agree ($E_{FM} = E_{FS}$). (From Solymar, L. and Walsh D., *Electrical Properties of Materials*, 6th ed., 1998, by permission of Oxford University Press.)

TABLE 6.1
Some Metal Work Function ($q\phi_M$) and Semiconductor Electron Affinity ($q\chi_s$) Values

Metal	Work Function ($q\phi_M$) (eV)	Semiconductor	Electron Affinity ($q\chi_s$) (eV)
Silver (Ag)	4.26	Germanium (Ge)	4.13
Aluminum (Al)	4.28	Silicon (Si)	4.01
Gold (Au)	5.1	Gallium arsenide (GaAs)	4.07
Chromium (Cr)	4.5	Aluminum arsenide (AlAs)	3.5
Molybdenum (Mo)	4.6		
Nickel (Ni)	5.15		
Palladium (Pd)	5.12		
Titanium (Ti)	4.33		
Tungsten (W)	4.55		

6.2   SCHOTTKY CONTACTS

A Schottky contact is a rectifying contact between a metal and a semiconductor. It can function as a diode, known as the *Schottky diode* (Figure 6.2).

The Schottky contact is formed if $\phi_M < \phi_S$ for a p-type semiconductor or $\phi_M > \phi_S$ for an n-type semiconductor. The subscripts "M" and "S" stand for metal and semiconductor, respectively.

Anode

Cathode

FIGURE 6.2   Symbol for a Schottky diode.

6.2.1 BAND DIAGRAMS

Consider a metal–semiconductor system in which the work function of the metal (ϕ_M) is greater than that for an n-type semiconductor (ϕ_S; Figure 6.1a).

We will construct a band diagram for this metal–semiconductor junction using the principle of the invariance of Fermi energy. We will follow steps similar to those used in Chapter 5 for creating a band diagram for a p-n junction. At this metal–n-type semiconductor junction, where $\phi_M > \chi_S$, electrons flow from the higher-energy states of the semiconductor conduction band to the lower-energy states of the metal. This creates a positively charged depletion region in the n-type semiconductor. A negative surface charge builds up on the metal. Since the free electron concentration in a metal is very high, this charge is within an atomic distance from the surface. A built-in electric field is directed from the n-type semiconductor to the metal. The associated built-in potential (V_0) prevents the further flow of electrons from the n-type semiconductor to the metal (Figure 6.3).

As we can see in Figure 6.3, there is also a barrier ($q\phi_B$) to the flow of electrons from the metal to the n-type semiconductor:

$$q\phi_B = q\phi_M - q\chi_s \tag{6.1}$$

This barrier, called the *Schottky barrier* ($q\phi_B$), prevents any further injection of electrons into the n-type semiconductor from the metal. When a metal is deposited onto a semiconductor, we may think that the electrons from the metal flow easily into the semiconductor. However, this is not always the case.

The magnitude of the built-in potential (qV_0) for transferring an electron from the conduction band of the semiconductor to the Fermi energy level metal is given by

$$qV_0 = q\phi_B - qV_n \tag{6.2}$$

where qV_n is the energy difference between E_c and E_F of the n-type semiconductor (Figure 6.3).

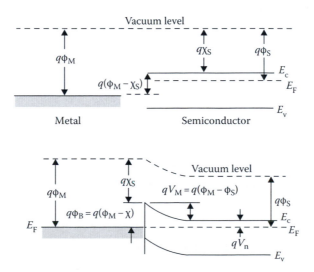

FIGURE 6.3 Formation of a Schottky contact between a metal and an n-type semiconductor ($q\phi_M > q\phi_S$). (From Sze, S.M.: *Semiconductor Devices, Physics, and Technology*, 2nd ed. 2002. Copyright Wiley-VCH Verlag GmbH & Co. KGaA. Reproduced with permission.)

Substituting for the Schottky barrier from Equation 6.1 ($q\phi_B$) as ($q\phi_M - q\chi_S$) in Equation 6.2, we get

$$qV_0 = q\phi_M - q\chi_S - qV_n$$

$$qV_0 = q\phi_M - q\chi_S - (q\phi_S - q\chi_S)$$

$$qV_0 = q\phi_M - q\phi_S$$

Thus, the built-in potential barrier (qV_0) is equal to the difference between the two work functions of the materials forming the Schottky contact.

$$qV_0 = q\phi_M - q\phi_S \tag{6.3}$$

6.2.2 SURFACE PINNING OF THE FERMI ENERGY LEVEL

It may seem from Equation 6.1 that, for different metals deposited on a given semiconductor (e.g., silicon [Si]), the barrier height ($q\phi_B$) changes with the work function of the metal ($q\phi_M$). However, in practice, the Schottky barrier height does not change appreciably for different metals deposited onto a given semiconductor (Table 6.2). As shown in Table 6.2, the Schottky barrier height of ~0.8 eV essentially is independent of the metal deposited for n-type silicon. For n-type gallium arsenide (GaAs), the Schottky barrier height is ~0.9 eV.

Many metals deposited on silicon are thermodynamically unstable, so that when the semiconductor–metal contact is exposed to high temperatures during processing, the metals react with the silicon and form a *silicide* intermetallic compound. For example, platinum (Pt) reacts with silicon and forms platinum silicide (PtSi). This lowers the Schottky barrier by ~0.06 eV, from 0.90 for platinum to 0.84 eV for PtSi. Tantalum silicide (TaSi$_2$) and titanium silicide (TiSi$_2$) are the preferred materials for forming Schottky contacts in silicon semiconductor processing. The values of the Schottky barrier for some silicides are listed in Table 6.2.

TABLE 6.2
Schottky Barrier Heights (in Electron Volts) for Metals and Alloys on Different Semiconductors

Metal/Alloy	n-Si	p-Si	n-GaAs
Aluminum (Al)	0.72	0.8	0.80
Titanium (Ti)	0.5	0.61	
Chromium (Cr)	0.61		
Tungsten (W)	0.67		
Nickel (Ni)	0.61		
Molybdenum (Mo)	0.68		
Gold (Au)	0.79	0.25	0.9–0.95
Silver (Ag)			0.88
Platinum (Pt)			0.86–0.94
Platinum silicide (PtSi)	0.85	0.2	
Nickel silicide (NiSi)	0.7	0.45	
Nickel silicide (NiSi$_2$)	0.70		
Tantalum silicide (TaSi$_2$)	0.59		
Titanium silicide (TiSi$_2$)	0.60		

The Schottky barrier is independent of the metal or alloy used to create it because of the *surface pinning* of the semiconductor's Fermi energy level in the interface region (Figure 6.4).

At the semiconductor surface or at its interface with another metal or material, additional energy levels are introduced into the otherwise forbidden band gap. The physical interface between the semiconductor and its surroundings (another metal, surrounding atmosphere, and so on) is not perfect; there are dangling or incomplete bonds, which means the atoms at the surface or interface are not fully coordinated with the other atoms. For example, inside a single crystal of silicon, an atom of silicon should be coordinated with four other silicon atoms in a tetrahedral fashion. However, this is not the case for silicon atoms at the surfaces or interfaces with other metals or materials. The interface between a metal and a semiconductor is not sharp at an atomistic level. There is a very small region at the interface in which it is unclear whether the material is a metal or a semiconductor. Nanoscale oxide particles, intermetallic compounds, and so on may be present at this interface. These surface atoms have incomplete bonds and other imperfections and introduce a large number of energy states or available defect-related energy levels into the interface region of the semiconductor's band gap (Figure 6.4).

Thus, although theoretically there are no energy levels allowed in the band gap, some energy levels occur in the band gap near the surface or interface region in the bulk of the semiconductor. These states effectively *pin* the Fermi energy level of the semiconductor in the interface region.

Surface pinning means that the Fermi energy level in the surface region of the semiconductor is at a fixed level (ϕ_0) that does not change with the addition or removal of electrons in the rest of the semiconductor (via doping). This also means that the position of the semiconductor's conduction band edge is fixed in the interfacial region. Thus, regardless of the metal deposited, the Fermi energy level of the metal must align with the pinned Fermi energy level (ϕ_0; Figure 6.4). When the Fermi energy level is pinned, the Schottky barrier height ($q\phi_B$) for the injection of electrons from a metal into the semiconductor is given by

$$q\phi_B = E_g - q\phi_0 \tag{6.4}$$

The Schottky barrier of different metals deposited on a given semiconductor essentially is constant (Table 6.2).

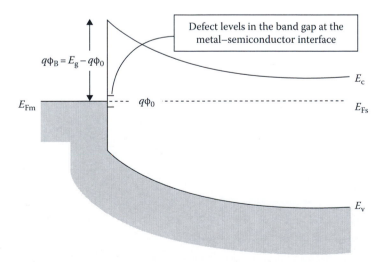

FIGURE 6.4 Pinned Fermi energy level ($q\phi_0$) in semiconductors. (From Singh, J.: *Semiconductor Devices: Basic Principles*. 2001. Copyright Wiley-VCH Verlag GmbH & Co. KGaA. Reproduced with permission.)

6.2.3 Current–Voltage Characteristics for Schottky Contacts

We now consider the ideal current–voltage (I–V) curve for a Schottky contact between an n-type semiconductor and a metal, such that $q\phi_M > q\chi_S$. When there is no applied bias, some electrons on the semiconductor side will have a high enough energy to overcome the built-in potential barrier (qV_0) and flow onto the metal side. This process of *thermionic emission* creates the thermionic current. If the n-type semiconductor is heavily doped, the depletion layer is thin, and it is possible for electrons to tunnel from the n-side semiconductor to the metal. Thermionic emission and tunneling both create a flow of electrons from the semiconductor to the metal. The resultant *conventional current* (I_{MS}) is directed from the metal to the semiconductor (Figure 6.5). This current is balanced by the conventional current resulting from the flow of electrons from the metal to the semiconductor (I_{SM}). These currents cancel each other out in a Schottky contact under equilibrium.

A forward bias is applied to this Schottky contact by connecting the positive terminal of a power supply to the metal side. The applied voltage is opposite the internal field and is directed from the n-type semiconductor to the metal. The potential barrier for the flow of electrons from the semiconductor to the metal is reduced from qV_0 to $q(V_0 - V_F)$. There is therefore an increased flow of electrons from the semiconductor to the metal. This means that the current directed from the metal to the n-type semiconductor (I_{MS}) increases. This is represented in Figure 6.5b by the relatively thicker and longer arrow for I_{MS}.

Note that since the Schottky barrier does not change much for a given semiconductor (Table 6.2), the *current due to the motion of electrons* from the metal to the semiconductor (I_{SM}) does not change. Thus, the value of I_{SM} does not change under an applied bias. The net result is that, under a forward bias, the overall conventional current flow from the metal to the semiconductor (I_{SM}) increases.

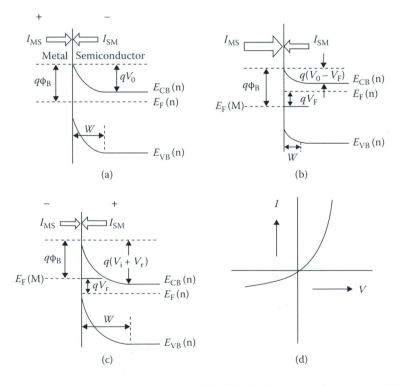

FIGURE 6.5 Current flows in a Schottky contact: (a) no bias; (b) forward bias; (c) reverse bias; and (d) the *I–V* curve. (From Mahajan, S. and Sree Harsha K.S., *Principles of Growth and Processing of Semiconductors*, McGraw Hill, New York, 1998. With permission.)

Under a reverse bias—that is, when the metal side is connected to the negative terminal of a DC power supply, the potential barrier for the flow of electrons from the semiconductor toward the metal increases from qV_0 to $q(V_0 + V_r)$. This causes the resultant I_{MS} to decrease, whereas the value of I_{SM} remains unchanged (Figure 6.5c). The resultant current–voltage ($I–V$) curve for a Schottky diode is shown in Figure 6.5d.

We have considered a Schottky contact formed between an n-type semiconductor and a metal (Figure 6.6a and b). This type of rectifying contact can occur between a p-type semiconductor and a metal when $\phi_M < \phi_S$ (Figure 6.6).

The current through a Schottky diode is given by the following equation:

$$I = I_S\left[\exp\left(\frac{qV}{\eta k_B T}\right) - 1\right] \tag{6.5}$$

where I_S is the reverse saturation current, V is the applied bias, and η is the ideality factor for a Schottky diode. The value of η is between 1 and 2 and is closer to 1 for a Schottky diode. The reverse-bias saturation current (I_S) is given by

$$I_s = A \times \left(\frac{m^* q k_B^2}{2\pi^2 \hbar^3}\right) \times T^2 \times \exp\left(-\left\{\frac{q\phi_B}{k_B T}\right\}\right) \tag{6.6}$$

where A is the area through which the Schottky current flows.

The term $\left(\dfrac{m^* q k_B^2}{2\pi^2 \hbar^3}\right)$ is known as the *effective Richardson constant* (R^*).

$$R^* = \left(\frac{m^* q k_B^2}{2\pi^2 \hbar^3}\right) = 120\left(\frac{m^*}{m_0}\right) \mathrm{A\,cm^{-2}K^{-2}} \tag{6.7}$$

where m^* is the carrier effective mass and m_0 is the carrier rest mass.

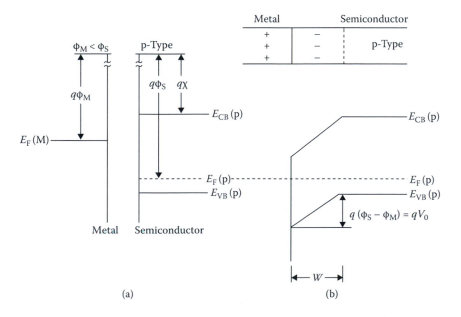

FIGURE 6.6 Energy-band diagrams of a metal/p-type semiconductor contact with $\phi_M < \phi_S$: (a) two materials isolated from each other and (b) at thermal equilibrium after the contact is made. (From Mahajan, S. and Sree Harsha K.S., *Principles of Growth and Processing of Semiconductors*, McGraw Hill, New York, 1998. With permission.)

TABLE 6.3

Effective Richardson Constants for Semiconductors

Carriers and Semiconductor	Effective Richardson Constant (R^*) ($A \cdot cm^{-2} \cdot K^{-2}$)
Electrons in silicon (Si)	110
Holes in silicon (Si)	32
Electrons in GaAs	8
Holes in GaAs	74

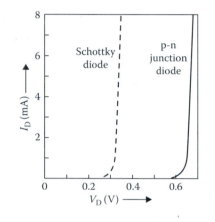

FIGURE 6.7   Comparison of the I–V curves for a p-n junction and a Schottky diode. (From Neaman, D., *An Introduction to Semiconductor Devices*, McGraw Hill, New York, 2006. With permission.)

Instead of the saturation current (I_S), we can write Equation 6.7 in the form of a saturation current density (J_S) as

$$J_S = \left(\frac{m^* q k_B^2}{2\pi^2 \hbar^3} \right) \times T^2 \times \exp\left(-\left\{ \frac{q\phi_B}{k_B T} \right\} \right) \tag{6.8}$$

The values of the effective Richardson's constant (R^*), predicted from Equation 6.7, are high. Values from more detailed calculations are shown in Table 6.3.

The Schottky diode I–V curve (Figure 6.5) appears very similar to that for a p-n junction diode. However, there are important differences in the magnitude of the currents, which are illustrated in Example 6.1 and Figure 6.7.

Example 6.1: Schottky Diode Current

A Schottky diode made from a Si and tungsten (W) junction has a saturation current density (J_s) of 10^{-11} A/cm². (a) What is the current density if the applied forward bias (V) is 0.3 V? (b) If the cross-sectional area of this diode is 5×10^{-4} cm², what is the value of the current (I) in mA?

Solution

1. We rewrite Equation 6.5 as

$$J = J_S \left[\exp\left(\frac{qV}{\eta k_B T} \right) - 1 \right] \tag{6.9}$$

We assume that under a forward bias of 0.3 V, the exponential term is much larger than 1. This is because (k_BT/q) ~0.026 V at 300 K. We also assume $\eta = 1$. Therefore,

$$J = J_S\left[\exp\left(\frac{V}{\left(\frac{k_BT}{q}\right)}\right)\right] = 10^{-11}\text{A/cm}^2 \times \exp\left(\frac{0.3\text{ V}}{0.026\text{ V}}\right)$$

$$J = 10.25 \text{ A/cm}^2$$

2. The magnitude of the current is I = 10.25 A/cm^2 × (5 × 10^{-4} cm^2) = 5.12 mA. This is a large forward current. A typical Si p-n junction diode, with a built-in potential of ~0.7 V, hardly carries any current for a forward voltage of 0.3 V.

Schottky diodes have a lower voltage drop compared to p-n junction diodes because they are made from a metal–semiconductor junction not a p-n junction. A forward current of ~1 mA can be achieved for forward voltages as small as ~0.1–0.4 V.

6.2.4 ADVANTAGES OF SCHOTTKY DIODES

The Schottky diode is considered a *majority device*, that is, minority carriers do not play an important role. Therefore, Schottky diodes exhibit faster switching times. The lower forward voltage means that a Schottky diode does not dissipate as much power. Because of these advantages, Schottky diodes are used in high-speed computer circuits.

The junction capacitance of a Schottky diode is lower compared to that of a typical p-n junction based on the same semiconductor. Since the reverse saturation current is high, the voltage and current ratings of a Schottky diode for a forward bias are lower.

Silicon Schottky diodes have relatively smaller breakdown voltages. The silicon diodes work well up to a breakdown voltage of ~100 V. The resistance of the diode increases significantly in silicon diodes that have larger breakdown voltages. For any given forward voltage drop, the value of the current decreases significantly with higher breakdown voltage diodes. This limits the use of silicon Schottky diodes to rectify relatively lower voltages (Figure 6.8).

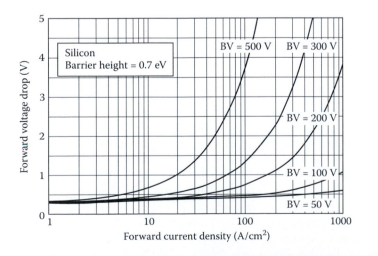

FIGURE 6.8 Calculated forward *V–I* characteristics for silicon Schottky diodes. BV = breakdown voltage. (From Baliga, B.J., *Silicon Carbide Power Devices*, World Scientific, Singapore, 2005. With permission.)

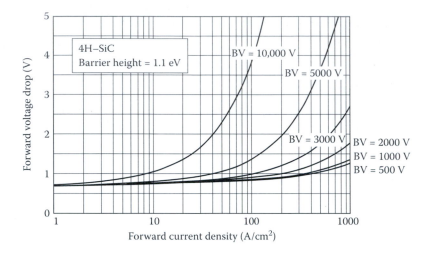

FIGURE 6.9 Calculated forward *V–I* characteristics for silicon carbide Schottky diodes. BV = breakdown voltage. (From Baliga, B.J., *Silicon Carbide Power Devices*, World Scientific, Singapore, 2005. With permission.)

We can use semiconductors with higher breakdown voltages to control currents and voltages in high-powered electronics. For example, there is considerable interest in using silicon carbide (SiC) Schottky diodes. The forward current characteristics for a form of silicon carbide (known as 4H-SiC) are shown in Figure 6.9.

Compared to silicon diodes, these diodes have a higher Schottky barrier height of ~1.1 eV. They can also be designed for higher breakdown voltages and still offer a lower resistance. Thus, compared to silicon, Schottky diodes made using SiC are better suited for high-power applications that involve higher voltages and currents.

6.3   OHMIC CONTACTS

Ohmic contacts are necessary in many semiconductor devices, such as in solar cells and transistors. Ohmic contacts are formed when $\phi_M < \phi_S$ for an n-type semiconductor or $\phi_M > \phi_S$ for a p-type semiconductor. These contacts are nonrectifying, meaning that there is no energy barrier to block current flow in either direction. The use of the word *ohmic* does not, however, mean that the resistance of the contact is constant with voltage.

6.3.1   BAND DIAGRAM

The band diagrams for a metal–semiconductor junction forming an ohmic contact are shown in Figure 6.10.

Ohmic contacts can be formed on semiconductor surfaces using different strategies such as those shown in Figure 6.11. The usual approach is to make tunneling possible using a Schottky barrier of a lower height and then doping to reduce the depletion layer width.

Aluminum (Al) metallization is often used to create an ohmic contact on silicon. As we can see from Table 6.2, aluminum can form a Schottky contact on silicon by itself. However, when an evaporated aluminum film on p-type silicon is heated to ~450–550°C, the aluminum diffuses into the silicon and creates a heavily doped (p^+) layer at the interface between the aluminum and p-type silicon. Silicon can also diffuse out into aluminum and form an aluminum–silicon alloy. The contact between the p^+ layer of silicon and the aluminum–silicon alloy is ohmic in nature. Similarly, gold (Au) containing a small concentration of antimony (Sb), when deposited on n-type silicon, can

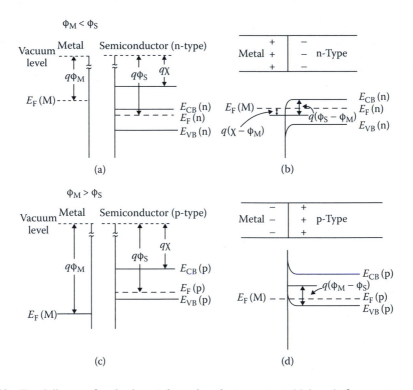

FIGURE 6.10   Band diagram for ohmic metal–semiconductor contact: (a) $\phi_M < \phi_S$ for an n-type semiconductor, (b) the corresponding equilibrium band diagram for the junction, (c) $\phi_M > \phi_S$ for a p-type semiconductor, and (d) the corresponding band diagram for the junction. (From Mahajan, S. and Sree Harsha K.S., *Principles of Growth and Processing of Semiconductors*, McGraw Hill, New York, 1998. With permission.)

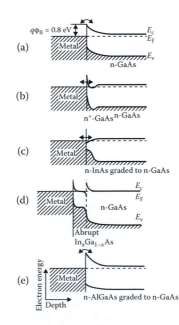

FIGURE 6.11   Strategies for creating ohmic contacts on n-GaAs: (a) Thermionic emission barrier; (b) field emission (tunneling) barrier; (c) graded composition low-resistance barrier; (d) heterojunction contact reduced barrier; and (e) graded composition enhanced barrier. (Adapted from Mahajan, S. and Sree Harsha K.S., *Principles of Growth and Processing of Semiconductors*, McGraw Hill, New York, 1998.)

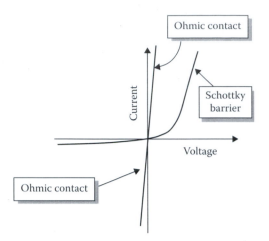

FIGURE 6.12 A comparison of the *I–V* curves for an ohmic and a Schottky contact. (From Singh, J.: *Semiconductor Devices: Basic Principles*. 2001. Copyright Wiley-VCH Verlag GmbH & Co. KGaA. Reproduced with permission.)

create an ohmic contact by diffusing some of the antimony into the n-type silicon and creating an n⁺ layer at the interface. Electrons can tunnel through this barrier and create an ohmic contact. The current–voltage (*I–V*) characteristics of an ohmic contact and a Schottky contact are compared in Figure 6.12.

6.4 SOLAR CELLS

A *solar cell* is a p-n junction–based device that generates an electric voltage or current upon optical illumination. In this section, the basic characteristics of the solar cells are addressed. We will study detailed operating principles and advanced device design concepts of solar cells in Chapter 9 after we discuss optical materials. Solar cells are useful in alternative energy technologies that are *greener* and utilize resources such as energy from the sun and wind. The field of research and development converting light energy into electricity is known as *photovoltaics*.

The sun emits most of its energy in the wavelength of 0.2 to 4 μm. Solar cells are made using semiconductors that can absorb energy in this wavelength spectrum. The band gap of the semiconductors used is $h\nu > E_g$, where ν is the frequency of light. *Absorptivity* is the ability of the semiconductor to absorb solar radiation. This also is important to solar cells. The semiconductors used in solar cells include crystalline silicon, amorphous silicon (a:Si:H), polycrystalline silicon, silicon ribbons, nanocrystalline silicon, and GaAs. Polycrystalline silicon is the most widely used because it costs less than single-crystal silicon, the second most widely used (Green 2003). Amorphous silicon is attractive because it can be used for deposition on large areas. Other compound semiconductors such as *copper indium diselenide* ($CuInSe_2$, CIS) and cadmium telluride (CdTe) can also be used, and they provide a very high absorption of incident light. A schematic of the structure of a solar cell is shown in Figure 6.13.

Usually, the top layer in a solar cell is an n-type material and the bottom layer is a p-type semiconductor. There are other elements in the solar cell structure, such as an antireflective coating (Figure 6.13). The coating helps capture as much of the light energy incident on the solar cell as possible by minimizing reflection losses. Similarly, in addition to the p-n junction, electrical ohmic contacts are required for operating an electrical circuit. The bottom contact is made using metals such as aluminum or molybdenum. The top contact is in the form of metal grids or transparent

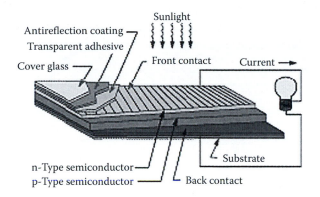

FIGURE 6.13 Schematic of the structure of a solar cell. (Courtesy of Solar Energy Technology Program, U.S. Department of Energy, Washington, DC.)

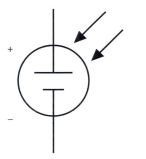

FIGURE 6.14 Symbol for a solar cell in an electrical circuit.

conductive oxides such as indium tin oxide (ITO) so that light can still get through to the p-n junction. The symbol for a solar cell in an electrical circuit is shown in Figure 6.14.

Experimental data for a silicon solar cell I–V curve is shown in Figure 6.15. From this I–V curve, we can characterize the major performance parameters of the solar cell (maximum output current, maximum output voltage, and power conversion efficiency). Note that the open-circuit voltage is slightly less than 0.7 V and the contact potential (V_0) for silicon is ~0.7 V. If the incident photons in solar cells have an energy ($h\nu$) less than the band gap energy (E_g), then no electron–hole pairs are produced and the incident energy is wasted—that is, it is not used for conversion into electrical energy. Similarly, if the incident photon energy ($h\nu$) is too high compared to E_g, then electron–hole pairs are created, and the difference ($h\nu$–E_g) will appear as heat. This will also make the solar cell inefficient. Thus, the efficiency of the solar cell conversion is maximized by better matching the semiconductor band gap with the solar spectrum. Details about the relation between the band gap of the semiconductor and the theoretical efficiency of the solar cell can be found in Chapter 9.

Note that both direct and indirect band gap semiconductors can be used in solar cells. This is different from the requirement of LED semiconductors. Since electrons and holes recombine to produce photons, LEDs are built on direct band gap semiconductors. In some ways, the solar cell and the LED make use of opposite effects. Currently, the best power conversion efficiency of a single-junction GaAs solar cell is 29.1%; however, it is more expensive than silicon. The state-of-the-art silicon solar cell has a power conversion efficiency of 25.0%, which is lower than the GaAs solar cell. Thin film solar cells of direct band gap semiconductors ($CuInSe_2$, CdTe) offer the highest conversion efficiencies—slightly more than 21%. More complex solar cells based on multiple layers of different semiconductors offer higher efficiencies (up to 40%) because as the band gap varies across the thickness of the solar cell, we can capture photons of different energies. However, the cost of these solar cells is relatively high because they require a complicated fabrication process. There also is a growing interest in using the nanoparticles of semiconductors and organic materials for solar cell applications.

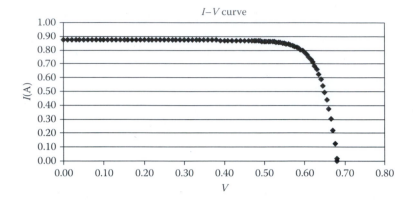

FIGURE 6.15 Experimental *I–V* curve for a silicon solar cell. (Courtesy of Professor Martin Green, University of New South Wales, Australia.)

6.5 LIGHT-EMITTING DIODES

An LED normally uses a direct band gap semiconductor in which the process of electron–hole recombination results in *spontaneous emission* of light. The symbol for an LED is shown in Figure 6.16. In this section, we will focus on the electrical aspect of LEDs. The optical aspect of LEDs and further information about using the semiconductor junction for a laser diode are discussed in Chapter 8.

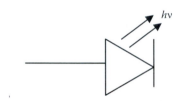

FIGURE 6.16 Symbol for an LED.

The first practical LED was reported by Holonyak at General Electric in 1962. He built red light emitting LEDs using III–V compound semiconductors. Later, multicolor emitting LEDs were fabricated, which was followed by the invention of organic light-emitting diodes (OLEDs).

6.5.1 Operating Principle

The underlying operating principle of an LED has already been mentioned in Chapter 5 in a discussion of indirect and direct band gap semiconductors as well as the processes of radiative and nonradiative electron–hole recombination (Figure 6.17). The recombination dynamics—how rapidly the electrons and holes can recombine and produce light—determine the speed at which an LED can be turned on and off. The speed with which an LED can be turned on and off is important in some fiber-optic applications.

Most LEDs are made using direct band gap materials. However, as has been noted in Chapter 5, it is possible to use indirect band gap materials to make LEDs, provided a defect energy level can be introduced by doping (see Section 6.5.3). This defect level provides an intermediate state in which radiative recombination is possible.

LEDs are constructed as forward-biased p-n junctions made most often from direct band gap semiconductors (Figure 6.18). When the p-n junction forming an LED is forward-biased, the electrons with a high energy can overcome the built-in potential barrier and arrive at the p-side of the junction. These minority carriers then combine with the holes (i.e., the majority carriers) on the p-side.

Similarly, some of the holes from the p-side of the junction also make it across the depletion layer onto the n-side and recombine with the electrons there. These recombination processes result in the emission of light in direct band gap materials (Figure 6.18). For indirect band gap materials

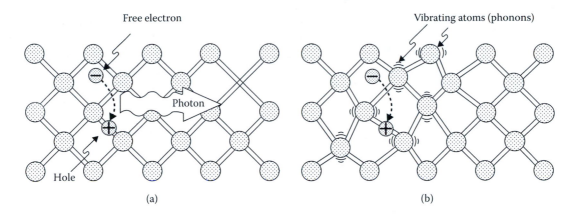

FIGURE 6.17 Illustration of radiative and nonradiative recombination: (a) Radiative recombination of an electron–hole pair accompanied by the emission of a photon with energy $h\nu = E_g$. (b) In nonradiative recombination events, the energy released during the electron–hole recombination is converted to phonons. (From Schubert, E.F., *Light-Emitting Diodes*, Cambridge University Press, Cambridge, UK, 2006. With permission.)

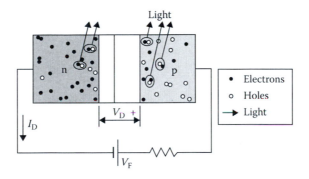

FIGURE 6.18 LED operating principle.

such as silicon or germanium, the recombination process also leads to the generation of heat. Thus, no LEDs are made using silicon or germanium. Note that even in direct band gap materials, some nonradiative recombination occurs and has a negative effect on LED efficiency.

6.5.2 LED MATERIALS

The light emitted from an LED may be in the visible, infrared, or ultraviolet ranges. LEDs that emit in the visible range are widely used in displays and many other consumer applications. Figure 6.19 shows some of the materials used for making LEDs.

The response of the eye to different wavelengths corresponding to various colors is also shown in this figure. The wavelength of light emitted (λ) is related to the band gap of the semiconductor (E_g) by the following equation:

$$\lambda = \frac{h\nu}{E_g} = \frac{1.24}{E_g(\text{in eV})}\ \mu m \tag{6.10}$$

LEDs that emit in the infrared range are used in fiber-optic systems. Gallium nitride (GaN)–based LEDs are relatively new and emit in the near-ultraviolet and blue range. The band gap of GaN is ~3.4 eV. GaN is alloyed with other nitrides (e.g., indium nitride and aluminum nitride) to obtain blue, amber, or green LEDs (Figure 6.20).

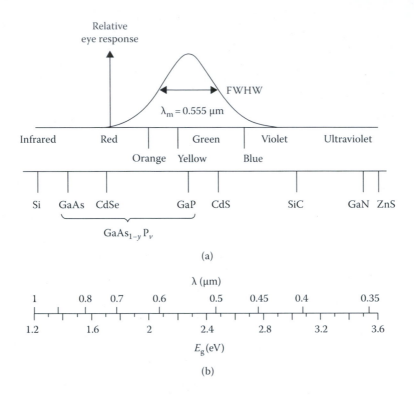

(a)

(b)

FIGURE 6.19 Semiconductors used in LEDs relative to (a) color wavelengths and relative eye response and (b) wavelength emitted. (From Sze, S.M.: *Semiconductor Devices, Physics and Technology*, 2nd ed. 2002. Copyright Wiley-VCH Verlag GmbH & Co. KGaA. Reproduced with permission.)

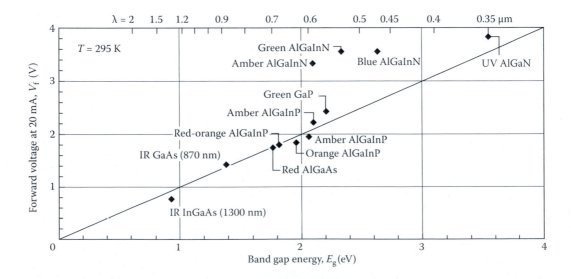

FIGURE 6.20 Band gaps of LEDs emitting different visible colors of light. IR = infrared. (From Schubert, E.F., *Light-Emitting Diodes*, Cambridge University Press, Cambridge, UK, 2006. With permission.)

Thus, the band gap of semiconductors used for making LEDs can be intentionally varied by forming solid solutions among multiple semiconductors. This band gap engineering approach leads to variations in the color of the light emitted. For example, GaAs is a direct band gap material with a band gap of 1.43 eV and a corresponding wavelength of 870 nm in the infrared range. However, it can be alloyed with aluminum arsenide (AlAs) and other materials to form LEDs of different colors.

The quality of materials used in the manufacture of durable LEDs is extremely important from the viewpoint of the presence of atomic level defects such as vacancies and dislocations. For example, as pointed out by Nakamura et al. (1995), scientists knew for a long time that a blue or ultraviolet LED could be made using GaN-based materials. However, until recently, the available materials had too many defects—such as certain types of dislocations—that prevented the creation of a blue LED. Thus, defects such as dislocations should be minimized in LED device materials.

Organic light-emitting diodes (OLEDs) or *polymer light-emitting diodes* (PLEDs) have recently been developed in which there is an electron-injecting cathode instead of an n-layer. There also is a hole-injecting anode. This technology has considerable promise as a more energy-efficient alternative to lighting by conventional sources such as incandescent or fluorescent lights. Several companies have recently introduced high-quality displays based on OLEDs.

6.5.3 LEDs Based on Indirect Band Gap Materials

Gallium phosphide (GaP) has an indirect band gap of 2.26 eV. When solid solutions of GaAs, a direct band gap material, are formed with GaP, the band gap of the solid solution $GaAs_{1-x}P_x$ is direct until $x = 0.45$ (Figure 6.21a). In fact, the most commonly used composition in this system is $GaAs_{0.6}P_{0.4}$, which emits in the red region (with $E_g \sim 1.9$ eV).

As the mole fraction x of phosphorus (P) in the $GaAs_{1-x}P_x$ increases, the materials become an indirect band gap. These materials can be doped with a dopant such as nitrogen (N) to make LEDs that emit yellow to green wavelengths. When nitrogen, an isoelectronic dopant, is added to these materials, it substitutes for some of the phosphorus atoms at their locations. Although the valence of phosphorus and nitrogen is the same (+5), the positive nucleus of a nitrogen atom is less shielded than the positive nucleus of a phosphorus atom. Isoelectronic impurities such as nitrogen atoms tend to bind the conduction electrons more tightly, that is, the electron wave function is highly localized. According to the Heisenberg uncertainty principle because the location of the electron is more fixed (i.e., Δx is small), its momentum is spread out (i.e., Δx is large). Hence, there are two possible

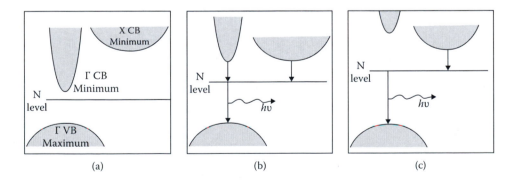

FIGURE 6.21   Schematic of the band diagram for (a) $GaAs_{1-x}P_x$ ($x < 0.45$) with a direct band gap; (b) nitrogen-doped indirect band gap $GaAs_{1-x}P_x$ ($x > 0.45$); and (c) transitions in direct band gap GaAs, crossover GaAsP, and indirect band gap GaP. (From Schubert, E.F., *Light-Emitting Diodes*, Cambridge University Press, Cambridge, UK, 2006. With permission.)

vertical transitions via the isoelectronic nitrogen defect level. One of these is a radiative transition, in which the nitrogen atoms introduce a defect energy level near the conduction band of GaP or $GaAs_{1-x}P_x$. The change in momentum that occurs when an electron makes the transition from the indirect conduction band (labeled X) to the valence band (labeled Γ) is absorbed by the impurity atom (Figure 6.21c). This enables the emission of light with an energy that is slightly less than that for the band gap of GaP or $GaAs_{1-x}P_x$.

6.5.4 LED Emission Spectral Ranges

We usually think of an LED as a device that emits the light of one wavelength. In practice, the light emitted from an LED is *not* all at one wavelength corresponding to the band gap (E_g). The reason for this spectral width is that carriers in semiconductors have a range of energies. Thus, some electrons with an energy higher than E_c show a vertical recombination with holes in the valence band without a change in carrier momentum. For these electrons, which recombine with holes at a higher energy, the frequency of light emitted is also higher (shown as $h\nu_2$; Figure 6.22).

The maximum emission intensity occurs at

$$E = E_g + (1/2)(k_B T) \tag{6.11}$$

At half mask, the full width of the luminescence intensity as a function of energy is 1.8 $k_B T$. Thus, the width of the energy is given by

$$\Delta E = 1.8\, k_B T \tag{6.12}$$

The spectral width ($\Delta\lambda$) is given by

$$\Delta\lambda = \frac{1.8\, kT\lambda^2}{hc}$$

where λ is the wavelength of light emitted corresponding to the band gap energy (E_g). For example, a GaAs LED emitting at 870 nm has a spectral width ($\Delta\lambda$) of ~28 nm, that is, the wavelengths range

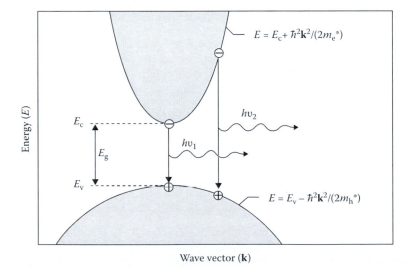

FIGURE 6.22 Vertical recombination of electrons and holes with energy $E = E_g$ and higher. (From Schubert, E.F., *Light-Emitting Diodes*, Cambridge University Press, Cambridge, UK, 2006. With permission.)

from 870 nm ($E = E_g = 1.43$ eV) to $870 - 28$ nm $= 842$ nm. Note that higher electron energies mean that the wavelength of light emitted is smaller. Thus, an LED emits wavelengths in the range of λ to $(\lambda - \Delta\lambda)$. Compared to the human eye's sensitivity, this spectral range is small; hence, for all practical purposes, the color of a given LED appears monochromatic.

Although spectral purity is not a major concern in display applications, it is very important for some applications in fiber-optic systems. In these applications, LEDs cannot be used to send signals over longer distances because the signal spreads; that is, different wavelengths start at the same point but arrive at the destination at different times. In such applications, devices known as *laser diodes* must be used. In a laser diode, photons of a given energy and wave vector are absorbed by the semiconductor, causing the generation of electrons and holes. The electrons and holes recombine and release *stimulated radiation*, an emission of coherent photons; that is, the photons emitted are in phase with the incident photons (i.e., with the same energy and wave vector). Blue laser diodes based on GaN offer a higher resolution for writing a digital video disk (DVD) and are used in the latest DVDs. This laser diode also forms the basis of the so-called Blu-ray disks and disk drives.

6.5.5 *I–V* CURVE FOR LEDS

Because LEDs are usually operated as forward-biased p-n junctions, their *I–V* curves are similar to the curves we have seen for regular silicon-based p-n junctions. The *I–V* curves for different LED materials are compared to those for silicon in Figure 6.23.

As we can expect, the knee of these various curves appears at a voltage close to the built-in potential for these different materials. Thus, for germanium, the knee occurs around 0.4 V. For silicon, which cannot be used to make LEDs because of its indirect band gap, the knee occurs at ~0.7 V. This is close to the built-in potential for a silicon p-n junction.

The typical forward voltage required to drive an LED is about 1.2–3.2 V. This is much greater than that for a silicon-based diode. Similarly, the reverse-bias breakdown voltage of a typical LED is ~3 to 10 V. This is much smaller than that for a silicon diode.

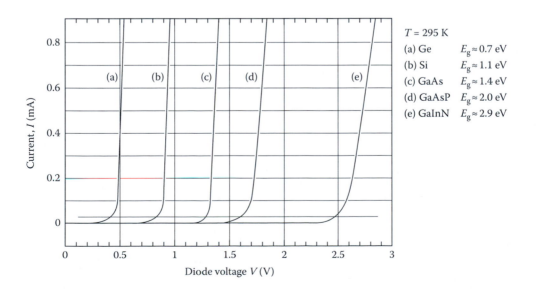

FIGURE 6.23 Room-temperature *I–V* curves for different LED materials and silicon. (From Schubert, E.F., *Light-Emitting Diodes*, Cambridge University Press, Cambridge, UK, 2006. With permission.)

A typical LED circuit uses a series resistance (R_s) to limit the forward current and to prevent any damage to the LED because of excessive current and associated heat. This series resistance (R_s) is separate from the internal resistance associated with the LED itself. The use of a series resistance is discussed in Example 6.2.

Example 6.2: LED Circuits with Series Resistance

An LED is driven by a maximum voltage supply of 8 V. For the driving voltage across an LED of 1.8–2.0 V, what should be the series resistance (R_s)? Assume that the LED current is 16 mA.

Solution

In this example, the maximum voltage for the circuit shown in Figure 6.24 is 8 V. This voltage appears between the resistor (R_s) and the LED. We assume that the resistance of the LED itself—that is, the smaller resistance of the neutral parts of the p-n junction—is small (~5 Ω) and can be ignored.

Supplied voltage is divided to LED and series resistance. Therefore, forward current (I_F) is given by

$$I_F = \frac{V_{Supply,max} - V_{F,min}}{R_S}$$

$$\therefore 16 \times 10^{-3}\,A = \frac{(8-1.8)\,V}{R_S}$$

$$\therefore R_S = \frac{7.2\,V}{16 \times 10^{-3}\,A} = 387.5\,\Omega$$

This can be rounded to 390 Ω. Therefore, we connect a 390-Ω resistor in the series to ensure that the current in this LED does not exceed 16 mA.

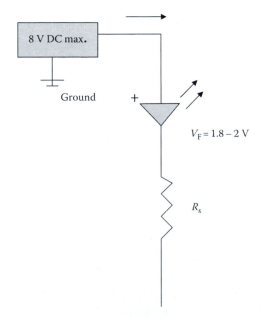

FIGURE 6.24 LED circuit showing a series resistance (R_S).

6.5.6 LED EFFICIENCY

The efficiency of an LED depends on what happens inside the active region in which the recombination of holes and electrons occurs—that is, the efficiency of the spontaneous emission process. The efficiency of an LED device also depends on what happens to the light that is emitted.

The photons that are emitted can be lost due to reabsorption into the semiconductor, which leads to regeneration of electron–hole pairs. Some photons are reflected at the p-n junction–air interface and some are lost (i.e., they do not come out of the LED structure) because of total internal reflection. To minimize the losses by reflection at the semiconductor–air interface, the LED is encapsulated in a dielectric dome.

In advanced LED structures, the two sides of the junction are made from different materials (Figure 6.25). An LED with p- and n-sides made from different semiconductors is known as a *heterojunction* LED. In these LEDs, an active region is sandwiched between two layers of different semiconductors. The compositions are selected such that the emitted photons are not absorbed by the top or bottom layer. As a result, these heterojunction LEDs are more efficient than *homojunction* LEDs (Figure 6.26), in which both sides of the p-n junction are made from the same semiconductor.

As a light source, LEDs are more efficient than conventional light sources such as the incandescent lamp (Figure 6.27).

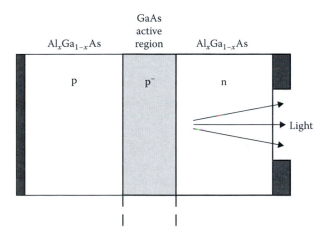

FIGURE 6.25 Schematic of a heterojunction LED. (From Singh, J., *Optoelectronics: An Introduction to Materials and Devices*, McGraw Hill, New York, 1996. With permission.)

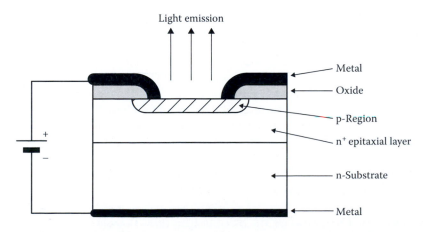

FIGURE 6.26 Schematic of a homojunction LED. (From Sparkes, J.J., *Semiconductor Devices*, Chapman and Hall, London, 1994. With permission.)

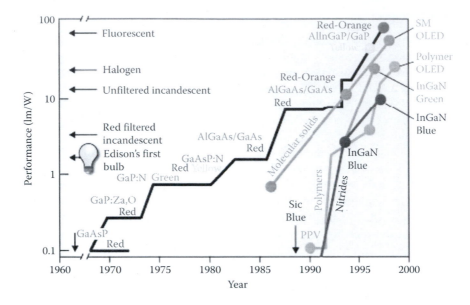

FIGURE 6.27 The performance of LEDs and OLEDs compared to incandescent and fluorescent lamps. (From Bergh, A., et al., *Physics Today,* 54, 12, 2001. With permission.)

6.5.7 LED Packaging

In addition to the p-n junction, an LED used as a light source has other parts. Almost all LEDs are mounted into a package that has two electrical leads. The anode lead is usually longer. The assembly also has a reflector with a hemispherical epoxy encapsulant (Figure 6.28).

Some high-powered LEDs are mounted on a heat sink and use a silicone encapsulant and a plastic cover (Figure 6.29). Packaging of OLEDs used in flat-panel displays and as light panels is also a very important area of research and development.

6.6 BIPOLAR JUNCTION TRANSISTOR

The term transistor is an abbreviation of transfer resistor, a device whose electrical resistance can be changed by an external voltage. Therefore, the transistor can amplify voltage and/or current of the input signal. On December 16, 1947, Bardeen, Brattain, and Shockley reported a device known as the point contact transistor. They received the Nobel Prize in Physics in 1956 for the discovery of the transistor effect.

Transistors are an indispensable component of modern-day integrated circuits (ICs). The two major types of transistors are the *bipolar junction transistor* (BJT) and the *field-effect transistor* (FET). A type of FET known as a *metal oxide semiconductor field-effect transistor* (MOSFET) is the most widely used transistor in ICs (see Section 6.11). Millions of MOSFETs are integrated onto semiconductor chips. Transistors play a vital role in running most computers and other state-of-the-art electronic equipment. In this section, we discuss the BJT. We will discuss the FET in Section 6.7.

The term *bipolar* is used to describe the BJT because this particular type of transistor uses both electrons and holes in its operation. In the BJT, the *transistor action* involves controlling the voltage at one terminal by applying voltages at two other terminals.

6.6.1 Principles of Operation of the Bipolar Junction Transistor

The BJT has two p-n junctions that are connected back-to-back (Figure 6.30). In the BJT, there are three regions: an *emitter*, a *base*, and a *collector* (Figure 6.30a). In the so-called *npn transistor*

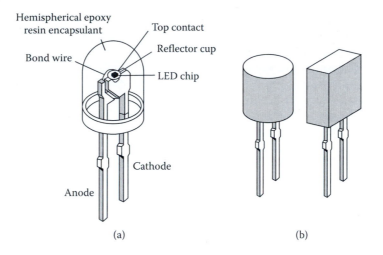

(a)

(b)

FIGURE 6.28 LED packaging: (a) LED with hemispherical encapsulant and (b) LEDs with cylindrical and rectangular encapsulant. (From Schubert, E.F., *Light-Emitting Diodes*, Cambridge University Press, Cambridge, UK, 2006. With permission.)

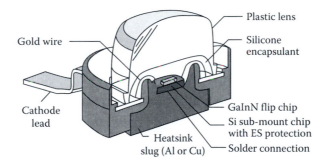

FIGURE 6.29 Packaging for high-powered LEDs showing the heat sink. (From Schubert, E.F., *Light-Emitting Diodes*, Cambridge University Press, Cambridge, UK, 2006. With permission.)

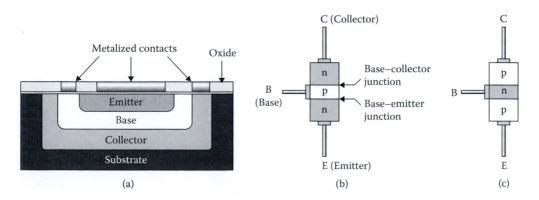

(a)

(b)

(c)

FIGURE 6.30 Schematic of bipolar junction transistors: (a) basic epitaxial planar structure; (b) npn; and (c) pnp. (From Floyd, T.L., *Electronics Fundamentals: Circuits, Devices, and Applications*, 4th ed., Chapman & Hall, Boca Raton, FL, 1998. With permission.)

(Figure 6.30b), the emitter is the most heavily donor-doped n-type (n^{++}). The base is a p-type region that is relatively thin, moderately doped (p^+), and sandwiched between two n-regions. The collector is a lightly doped n-type region. A similar logic applies to how the different regions are arranged in a *pnp transistor* (Figure 6.30c; see Section 6.6.8). Doping levels are $\sim 10^{19}$, 10^{17}, and 10^{15} cm^{-3} for the emitter, base, and collector regions, respectively. The npn transistors are more popular than pnp transistors because the former's electrons have a higher mobility.

As we will discuss in Section 6.6.2, the terms emitter, base, and collector originate from the transistor operation. The emitter region is the main source of injecting or emitting carriers, hence the name emitter. Carriers are emitted by the emitter and flow through the base and into the collector region. The original transistor reported in 1947 used a germanium (Ge) crystal as a mechanical base. This is how the term base came into being.

The symbols for npn and pnp transistors and their schematics showing different terminals and junctions are shown in Figure 6.31.

The BJT is a three-terminal, two-junction device. The direction of the arrow on the transistor symbols shows the direction of conventional current flow under *active mode* (see Section 6.6.6).

Thus, for the npn transistor in active mode, the *conventional current* flows from the collector to the base and then to the emitter. For the npn transistor, a majority of the electrons injected from the emitter flow into the base and then into the collector.

Similarly, for a pnp transistor in active mode, the conventional current flows from the emitter to the base to the collector (Figure 6.31b). In this case, the positively charged holes are injected from the emitter into the base and then into the collector.

6.6.2 Bipolar Junction Transistor Action

Consider the current flows for the npn transistor in the forward active mode (Figure 6.32). In this mode, the base–emitter junction is forward-biased. The base–collector junction is reverse-biased. This is considered the *active mode* or *forward active mode* for the transistor. The band diagram for an npn transistor under these biases is shown in Figure 6.33. The band diagram for an npn transistor without bias is also shown for comparison.

The resultant current flows for an npn transistor in an active mode are shown in Figure 6.33.

In the *common base* (CB) *configuration*, the base electrode is common to the emitter and collector circuits. The transistor can also be connected in other configurations such as *common emitter* (CE) and *common collector* (CC; Figure 6.34).

6.6.3 Current Flows in an npn Transistor

To better understand the following discussion, the reader should review Sections 5.10 and 5.11 concerning current flows in a forward-biased p-n junction.

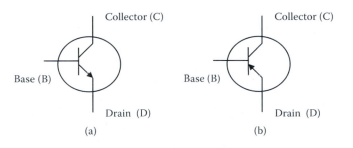

FIGURE 6.31 Symbols for bipolar junction transistors: (a) npn and (b) pnp. (From Floyd, T.L., *Electronics Fundamentals: Circuits, Devices, and Applications*, 4th ed., Chapman & Hall, Boca Raton, FL, 1998. With permission.)

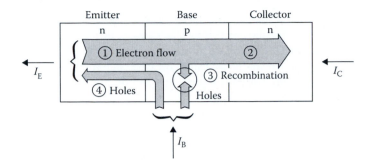

FIGURE 6.32 Common base configuration of current flows for an npn transistor in forward active mode. (From Neamen, D., *An Introduction to Semiconductor Devices*, McGraw Hill, New York, 2006. With permission.)

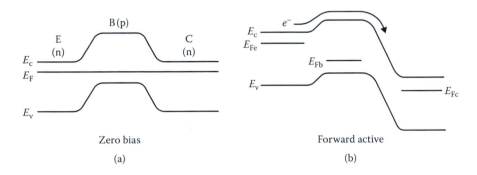

FIGURE 6.33 Band diagram for an npn transistor (a) without bias and (b) in active mode. (From Neamen, D., *An Introduction to Semiconductor Devices*, McGraw Hill, New York, 2006. With permission.)

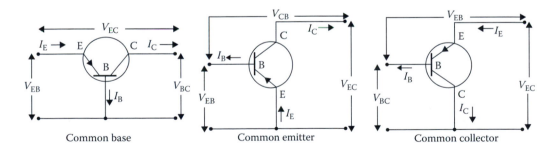

FIGURE 6.34 Common base (CB), common emitter (CE), and common collector (CC) configurations for an npn BJT. (From Singh, J.: *Semiconductor Devices: Basic Principles*. 2001. Copyright Wiley-VCH Verlag GmbH & Co. KGaA. Reproduced with permission.)

Consider the current flows in the CB configuration and in active mode (Figure 6.32):

1. The base–emitter junction is forward-biased. Electrons from the emitter region are injected into the base region. This flow of negatively charged carriers sets up one part of the emitter current ($I_{E,1}$) from the base toward the emitter because electrons have a negative charge.
2. The base–collector junction is under a reverse bias. Thus, at the collector–base interface, there is a very low concentration of electrons.

3. Therefore, on one side of the base, at the emitter–base junction, there is a significant injection of electrons into the base. On the other side of the base, at the collector–base interface, there is a very low electron concentration. There is a significant concentration gradient of electrons across the base, so that electrons diffuse across the base region into the collector region.

4. Once the electrons enter the collector region, they are driven by the internal electric field in the base–collector space-charge region. Note that this field is directed from the n-type collector toward the p^+ base region. This means that electrons are attracted toward the positively charged depletion layer on the n-side.

5. To summarize, for an npn transistor in an active mode, electrons are injected from the n^{++} region into the base region and continue on to the collector. Recombination of electrons occurs as they travel through the p-type base region.

6. One of the goals in the design of npn transistors is to transfer as much of the current from the emitter into the collector. This means that recombination occurring in the base region must be minimized. One way is to minimize the width of the base region (i.e., the p-region, in this case). A narrow base region is desirable because it minimizes the recombination of holes and electrons. The length of the neutral p-type material in the base must be much smaller than the diffusion length of electrons (L_n).

7. The number of electrons that flows into the collector is controlled by the injection of electrons in the base. The injection of electrons depends on the voltage applied to the base–emitter voltage (V_{BE}). This is the transistor action. In a BJT, the current at one terminal is controlled by the voltage at the other two terminals. In this case, the output—that is, the collector current (I_C)—is controlled by the input voltage across the base and the emitter (V_{BE}).

8. Note that the transistor action is possible only if the two p-n junctions connected back to back are *interacting p-n junctions*. This means that junctions are designed so that the carriers injected from forward biasing of the emitter–base junction flow into the base–collector junction. For example, if the base region is very wide, then the injected carriers simply recombine and no transistor action results. This is the difference between a transistor and two diodes (p-n junctions) connected back-to-back.

9. In terms of magnitude, there are secondary current flows in addition to this flow of electrons from the emitter to the collector. They are very important in terms of the use of transistors as amplifiers.

10. In an npn transistor, some of the electrons are injected from the emitter into the p-type base region and recombine with the holes. The base contact connected to a power source supplies the replacement of these holes. This is one component of the base current ($I_{B,1}$).

11. For the forward-biased emitter–base junction, holes diffuse from the p-doped base region to the emitter n^{++} region. This is the second component of the base current ($I_{B,2}$) and also makes up the second component of the emitter current ($I_{E,2}$). Since the base is lightly doped compared to the emitter, the current caused by the diffusion of holes is relatively small.

12. There is also a small reverse-bias current associated with the base–collector junction.

6.6.4   TRANSISTOR CURRENTS AND PARAMETERS

6.6.4.1   Collector Current

Due to the injection of electrons from the emitter into the base region and their journey into the collector region, the collector current (I_C) in an npn BJT is related to the base–emitter voltage (V_{BE}) by the following equation. This is also one part of the emitter current ($I_{E,1}$).

$$I_{E,1} = IC = I_{S,1} \exp\left(\frac{V_{BE}}{V_t}\right) \tag{6.13}$$

Equation 6.13 describes the transistor action. We control the current at the collector (I_C) by controlling the base–emitter voltage (V_{BE}).

6.6.4.2 Emitter Current

The total emitter current (I_E) in the npn transistor has two components. The first component ($I_{E,1}$) is due to the injection of electrons from the emitter into the base, which is equal to the collector current (I_C; Equation 6.13). The second component ($I_{E,2}$) is due to the diffusion of holes from the p-type base into the n-type emitter. Note that this current, $I_{E,2}$, is part of the emitter current only. It does not become part of the collector current. The expression for this current ($I_{E,2}$) is

$$I_{E,2} = I_{S,2} \exp\left(\frac{V_{BE}}{V_t} \right) \tag{6.14}$$

where $I_{S,2}$ involves the minority-carrier (i.e., the hole, in this case) parameters in the emitter.

Thus, the total emitter current (I_E) is given by

$$I_E = I_{E,1} + I_{E,2} = I_C + I_{E,2} = I_{SE} \exp\left(\frac{V_{BE}}{V_t} \right) \tag{6.15}$$

where I_{SE} represents the sum of $I_{S,1}$ (Equation 6.13) and $I_{S,2}$ (Equation 6.14).

The ratio of the collector current (I_C) to the emitter current (I_E) is known as the *common base current gain* (α).

$$\alpha = \frac{I_C}{I_E} \tag{6.16}$$

Note that the $I_{E,2}$ component is not a part of the collector current. Therefore, I_C is always smaller than I_E. Thus, the common base current gain (α) is smaller than 1 and should be closer to 1.

6.6.4.3 Base Current

For the npn transistor in a forward active mode, some of the electrons injected from the emitter into the base recombine with the holes in the p-region. These holes must be replaced by a flow of positive charge into the base. This is the first component to the base current, ($I_{B,1}$). This part of the current is proportional to the rate at which the holes are recombining with the electrons being injected into the base region. This, in turn, is related to the V_{BE}, the magnitude of the forward bias. The other part of the base current ($I_{B,2}$ or $I_{E,2}$) in an npn transistor is due to the diffusion of holes from the p-type base into the n-type emitter.

There are two components of the base current (I_B), both of which are proportional to $\exp(V_{BE}/V_t)$. The ratio of the collector current (I_C) to the base current (I_B) is known as the base-to-collector *current amplification factor* (β).

$$\beta = \frac{I_C}{I_B} \tag{6.17}$$

This parameter is also known as the *common emitter current gain* (β). The base current (I_B) is usually small, and the value of β is >100. The directions of the conventional current flow in npn and pnp transistors are summarized in Figure 6.35. The relationship between transistor parameters is shown in Example 6.3.

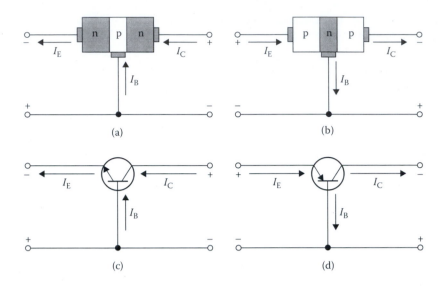

FIGURE 6.35 Directions of conventional current flows in bipolar junction transistors: (a) npn, (b) pnp, (c) npn, and (d) pnp. (From Floyd, T.L., *Electronics Fundamentals: Circuits, Devices, and Applications*, 4th ed., Chapman & Hall, Boca Raton, FL, 1998. With permission.)

Example 6.3: The Relationship between Transistor Parameters α and β

Show that the common base current gain (α) and common emitter current gain (β) are related by the following equation:

$$\beta = \frac{\alpha}{1-\alpha} \tag{6.18}$$

Solution

The emitter current (I_E) is equal to the sum of collector and base currents:

$$I_E = I_C + I_B \tag{6.19}$$

Divide both sides of this equation by I_C:

$$I_E/I_C = 1 + (I_B/I_C)$$

The common base current gain $\alpha = I_C/I_E$ and the common emitter gain (β) is I_C/I_B. Therefore, we get

$$(1/\alpha) = 1 + (1/\beta)$$

or

$$\beta = \frac{\alpha}{1-\alpha}$$

Thus, the closer the value of α to 1 (i.e., the closer a transistor is to getting most of the emitter current to the collector), the higher the value of β.

6.6.5 ROLE OF BASE CURRENT

The base current (I_B) is small in magnitude, but it is controllable and therefore plays a very important role in the functioning of a BJT. We have focused on showing that the collector current (I_C) can

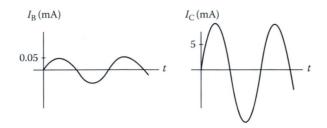

FIGURE 6.36 Schematic of the amplification achieved in a common emitter transistor circuit. Small changes in the base current (I_B) cause large changes in the collector current (I_C). (From Streetman, B.G. and Banerjee S., *Solid State Electronic Devices*, 5th ed., Prentice-Hall, Englewood Cliffs, NJ, 2000. With permission.)

be restricted by controlling the emitter current (I_E) through V_{BE}. Small variations in the base current (I_B) can lead to large changes in I_C. This is how a BJT can be used as a *current amplifier*. A similar amplification effect is seen in situations involving alternating current (AC), as shown in Figure 6.36.

Consider an npn transistor in an active mode. Assume that the supply of holes available to compensate for the holes lost in recombination is now limited. If the electron injection from the emitter continues into the base region, the negative charge builds up in the base, which causes reduction of the forward bias for the emitter–base junction. This in turn creates a loss of electron injection and thus reduces the collector current (I_C). By controlling the base current (I_B), we can control the collector current (I_C). This situation is similar to the way the smallest step in a chemical reaction controls its overall kinetics, or how the weakest link in a structure determines its overall strength. Therefore, the base current (I_B) is sometimes known as the *controlling current*, while the collector current (I_C) or the emitter current (I_E) is known as the *controlled current*.

6.6.6 Transistor Operating Modes

The BJT is similar to a two-way valve that can be turned on in either direction and can deliver either a desired level of flow or no flow at all. We have considered a situation in which the emitter–base junction was forward-biased and the collector–base junction was reverse-biased. This combination of biases results in an active mode for transistor operation. In this active region, we use the base current to control the collector current. We can also use the emitter–base voltage to control the collector current.

If both the emitter–base and collector–base junctions are forward-biased, the transistor is said to be in *saturation mode*. A small biasing voltage results in a large current; the transistor is in the *on* state. If both junctions are reverse-biased, the transistor is in *cutoff mode*, in which no current flows. This is the *off* or *cutoff* state of the transistor. When the roles of the emitter and the collector are switched—that is, when the collector–base is forward-biased and the emitter–base is reverse-biased—the transistor is in *inverted* or *inverse active mode*. The polarities of the two junctions that describe the different modes of operation for a BJT are summarized in Figure 6.37.

6.6.7 Current–Voltage Characteristics of the Bipolar Junction Transistor

Consider the BJT in CB configuration (Figure 6.34). The *I–V* characteristics of a BJT in this configuration are shown by plotting the collector current (I_C) as a function of reverse-biased V_{BC} on the base–collector junction for various fixed values of I_E (Figure 6.38). Note the negative sign associated with V_{BC} on the x-axis.

For a BJT in CB configuration, when the transistor is in a cutoff mode, the emitter current (I_E) is nearly zero (a fraction of a microampere). This is shown as I_{CBO} in Figure 6.39. The emitter current

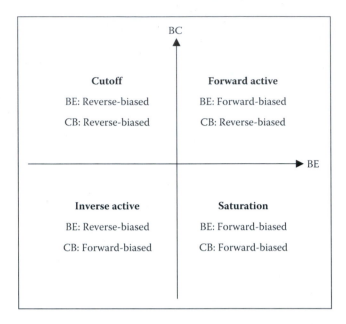

FIGURE 6.37 Different modes for operating a bipolar junction transistor.

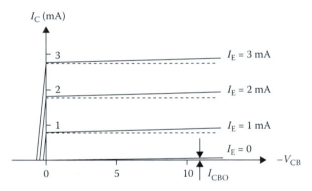

FIGURE 6.38 *I–V* characteristics of a bipolar junction transistor in common base configuration. (From Kasap, S.O., *Principles of Electronic Materials and Devices*, McGraw Hill, New York, 2002. With permission.)

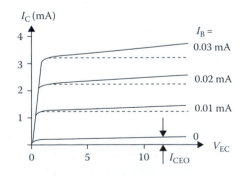

FIGURE 6.39 *I–V* characteristics of a bipolar junction transistor in a common emitter configuration. (From Kasap, S.O., *Principles of Electronic Materials and Devices*, McGraw Hill, New York, 2002. With permission.)

increases exponentially with base–emitter voltage (V_{BE}; Equation 6.15). The collector current (I_C) increases with increasing emitter current (I_E). Once the emitter current (I_E) is finite, the collector current is not zero for $V_{BC} = 0$. The collector current (I_C) is independent of the base–collector voltage (V_{BC}).

In Figure 6.38, we have shown the I_C to be nearly constant with V_{BC}. The slight increase in the emitter current with a reverse-biased base–collector voltage (V_{BC}), shown by the solid lines, is called the *early effect*. As V_{BC} increases, the width of the depletion layer (W) associated with the base–collector junction increases. This means that the width of the neutral part of the base region through which the electrons must travel to get to the collector becomes shorter, and the number of electrons lost to the recombination with holes is reduced. This in turn causes a slight increase in I_C as a function of the increasing reverse bias (V_{BC}).

If the polarity of a base–collector junction is changed so that it is forward-biased, then the collector current is the difference between the forward currents associated with the two junctions because the two forward currents subtract from each other.

The collector current (I_C) for an npn BJT in a CE configuration is shown as a function of the emitter–collector voltage (V_{EC}) in Figure 6.39.

In Example 6.4, we show the actual calculations of the transistor voltages and currents in a CE circuit.

6.6.8 CURRENT FLOWS IN A pnp TRANSISTOR

The current flows in a pnp transistor are similar in concept to those seen in Section 6.6.7 for an npn transistor (Figure 6.38). The doping levels follow a similar pattern; that is, the emitter has the highest level of doping (p++), the n-type base is moderately doped, and the p-type collector is lightly doped.

In this case, when the emitter–base junction is forward-biased, the emitter region injects holes into the base region. These holes flow through the relatively small, n-type base region and are then swept up by the internal electric field at the base–collector junction, ultimately arriving at the collector. Similar to an npn transistor, there is a current due to the diffusion of electrons from the base to the emitter region, shown with a filled arrow in Figure 6.40. The resultant current is directed from the emitter to the base because the electrons are negatively charged (Figure 6.40). There is also a

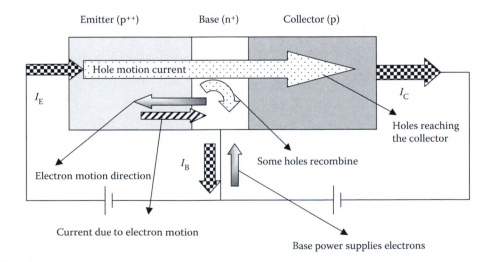

FIGURE 6.40 Summary of current and carrier flows in a pnp transistor in active mode.

recombination of injected holes and electrons in the base region. The electrons lost to the recombination are made up for by the base power supply. Similar to the npn transistor, there is a base current
(I_B). A small reverse-biased current also exists at the base–collector junction. The particle motions
and currents associated with this are not shown in Figure 6.40.

The relationships among the base, emitter, and collector currents, as well as transistor parameters α and β, are the same as those defined in Equations 6.16 and 6.17 for the npn transistor.

6.6.9 Applications of Bipolar Junction Transistors

The BJT can be used as an on-and-off switching device. In this application, the devices switch
between the cutoff and saturation regions (Figure 6.41).

A transistor can be used as current amplifier (Figure 6.36). The current and voltage analysis for
a BJT can be performed as shown in Example 6.4.

Example 6.4: Current and Voltage Analysis for a Bipolar Junction Transistor

Consider the transistor circuit shown in Figure 6.42. The value of resistance R_C connected
between the collector and its power supply is 200 Ω. The resistance (R_B) connected between
the base and the power supply is 15 kΩ. (a) Is this a pnp or npn transistor? (b) In what configuration is the transistor connected? (c) In what mode is the transistor? (d) What are the voltages
across the different terminals (V_{CB}, V_{CE}, and V_{BE})? What are the transistor currents I_C, I_E, and I_B?
Assume that $\beta = 200$, $V_{BB} = 5$ V, and $V_{CC} = 20$ V. Calculate the voltage drops across the resistors R_B and R_C.

Solution
1. The arrow on the transistor symbol in Figure 6.42 indicates the direction of the conventional current flow. In this transistor, the conventional current flows from the collector →
   base → emitter, so the electrons flow from the emitter → base → collector. Therefore,
   this is an npn transistor (Figure 6.31a).
2. In the circuit diagram shown in Figure 6.42, we can see that this is a common emitter configuration, in which the emitter junction is common to the base and collector
   circuits.
3. The base–emitter junction is forward-biased because the p-type base is connected
   to the positive of the voltage supply (V_{BB}). The collector–emitter junction is reverse-
   biased because the n-type emitter is connected to the positive of the power supply (V_{CC}).
   Therefore, this transistor is in forward active mode.

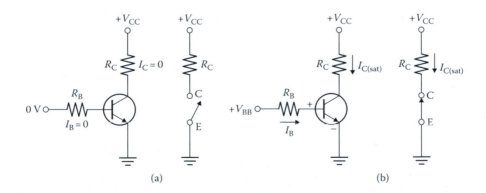

(a)   (b)

FIGURE 6.41 A transistor operating as a switch: (a) cutoff, open switch and (b) saturation, closed switch.
(From Floyd, T.L., *Electronics Fundamentals: Circuits, Devices, and Applications*, 4th ed., Chapman & Hall,
Boca Raton, FL, 1998. With permission.)

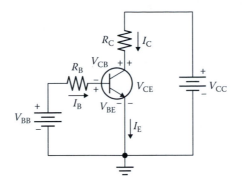

FIGURE 6.42 A transistor circuit for current and voltage analysis. (From Floyd, T.L., *Electric Circuit Fundamentals*, 7th ed., Chapman & Hall, Boca Raton, FL, 2006. With permission.)

4. Since the base–emitter junction is forward-biased, the voltage drop at this junction is ~0.7 V, similar to most other forward-biased Si p-n junctions. Thus, $V_{BE} = 0.7$ V, and the voltage (V_{RB}) across the resistor (R_B) is

$$V_{RB} = V_{BB} - V_{BE}$$

$$(V_{RB} = 5 - 0.7 = 4.3 \text{ V})$$

By examining the base–emitter side of the circuit and applying Kirchhoff's law, which states that the algebraic sum of voltage drops around a closed loop is zero, the current I_B is given by

$$V_{BB} = (I_B \times R_B) + V_{BE}$$

Therefore,

$$I_B = \frac{V_{BB} - V_{BE}}{R_B}$$

$$\therefore I_B = \frac{(5 - 0.7) \text{ V}}{15,000 \text{ }\Omega} = 287 \text{ }\mu\text{A}$$

$$I_B = 0.287 \text{ mA}$$

Since the current gain (β) is 200, from Equation 6.17, we get

$$\beta = \frac{I_C}{I_B}$$

$$\therefore I_C = 200 \times 0.287 = 57.3 \text{ mA}$$

Since the emitter current $I_E = I_C + I_B$, we get

$$I_E = 57.3 \text{ mA} + 0.287 \text{ mA} = 57.587 \text{ mA}$$

We will now look at the collector side of the circuit (Figure 6.42). The voltage drop (V_{RC}) across the resistor R_C is given by

$$V_{RC} = I_C \times R_C = 57.3 \times 10^{-3} \text{ A} \times 200 \text{ }\Omega = 11.46 \text{ V}$$

Since the voltage V_{CC} is 20 V and the voltage across the resistor R_C is 11.46 V, the voltage V_{CE} is

$$V_{CE} = V_{CC} - V_{RC} = 20.0 - 11.46 = 8.54 \text{ V}$$

If we complete two electrical paths around the transistor, one from the collector to the base directly and the other from the collector to the emitter via the base, then the following relationship will be true:

The collector base voltage $V_{CB} = V_{CE} - V_{BE}$.
Therefore, $V_{CB} = 8.54 - 0.7 = 7.84$ V.

To summarize, the base current for this circuit $I_B = 0.287$ mA, and the collector current I_C is 200 times larger at 57.3 mA. The emitter current (I_E) is the sum of these two currents: $I_E = 57.587$ mA.

The base–emitter junction is forward-biased, so $V_{BE} = 0.7$ V (assumed for a typical p-n junction). The values of V_{CE} and V_{CB} are 8.54 and 7.84 V, respectively.

6.7 FIELD-EFFECT TRANSISTORS

The concept of the FET is relatively simple and was proposed in 1926 by Lilienfeld (Figure 6.43) before the discovery of the BJT. An FET is a device whose electrical resistance is controlled by the voltage.

The discovery of BJT was serendipitous, in that the research originally aimed at developing the FET led to the discovery of the BJT. Although the concept of the FET existed for several years, the quality of materials and especially interfaces required to achieve the effect did not exist. Therefore, the initial FETs were based on the BJT. Today, however, millions of types of FETs (known as MOSFETs) are routinely integrated onto computer chips (see Section 6.11).

In an FET, the *source* and *drain* regions are separated by the *channel* region. A *gate* is located between the source and the drain. The conductivity of the channel region is affected by the applied electric field at the gate. This is known as the *field effect*, hence the name field-effect transistor. One of the main distinctions between the FET and the BJT is that in an FET, the current mainly is carried by one type of carrier (electrons or holes). Therefore, an FET is a *unipolar device*, whereas a BJT is a bipolar device.

The gate of an FET controls the flow of carriers between the source and the drain and must be electrically isolated from the channel so that it can influence the flow of carriers without having to draw any significant amount of current flowing from the source to the drain through the channel region.

6.8 TYPES OF FIELD-EFFECT TRANSISTORS

The manner in which the gate isolation is achieved defines the different types of FETs (Figure 6.44). As shown in Figure 6.44, there are two main strategies for gate isolation. We have

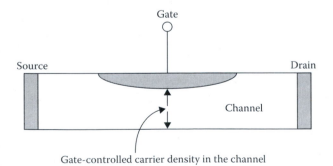

FIGURE 6.43 Illustration of the underlying principle for a field-effect transistor. (From Singh, J.: *Semiconductor Devices: Basic Principles*. 2001. Copyright Wiley-VCH Verlag GmbH & Co. KGaA. Reproduced with permission.)

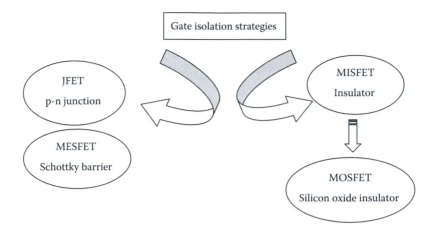

FIGURE 6.44 Different types of field-effect transistors based on different ways of isolating the gate region from the channel.

seen both of the basic effects associated with these strategies in the current flows at a p-n junction. The first strategy relies on changes in the width (W) of the depletion layer in the channel region as a function of biasing voltage applied at the gate electrode. The second strategy relies on increasing or decreasing the conductivity of the channel region by creating a higher or lower carrier concentration.

In the first strategy, we create a reverse-biased p-n junction or a Schottky contact. If a reverse-biased p-n junction is used for gate isolation, we refer to the device as a *junction field-effect transistor* (JFET). If a Schottky contact is used for gate isolation, then we refer to the device as a *metal semiconductor field-effect transistor* (MESFET). These devices are used mainly with III–V semiconductors such as GaAs and indium phosphide (InP). In JFET and MESFET, we use a doped semiconductor to provide free carriers (electrons or holes). When we apply a bias to the gate, we change the depletion layer width in the channel region width (W). This in turn changes the flow of the current between the source and the drain for a given voltage bias applied between the source and the drain. We can then turn the JFET or MESFET on and off.

In the second strategy for gate isolation, we use an insulator deposited on the gate region. We deposit a metal contact onto the insulator. The device is known as a *metal insulator semiconductor field-effect transistor* (MISFET). The most widely used version of the MISFET is silicon oxide (SiO_2) as a *gate dielectric*. This device is known as the *metal oxide semiconductor field-effect transistor* (MOSFET).

There are two variations on MOSFET operation. In the first, the channel region is not conducting; that is, the transistor is in the off state. When we apply a voltage to the gate, we create a small region with a high carrier concentration underneath the gate insulator. After a voltage is applied between the source and drain regions, a current begins to flow through the channel, turning on the transistor. This mode is known as the *enhancement-mode MOSFET*.

In the second variation, the channel region already has a high conductivity; that is, the transistor is in the on state. When we apply a reverse bias to the gate, we deplete the carrier concentration in the channel region underneath the gate. If we now apply a voltage between the source and the drain, the transistor will not conduct because carriers are absent in the channel region, and thus the transistor is turned off. This way of operating a MOSFET is known as the *depletion-mode MOSFET*.

We will first discuss the operation of MESFET in Section 6.9, and then MISFET and MOSFET in Sections 6.10 and 6.11, respectively.

6.9 MESFET *I–V* CHARACTERISTICS

6.9.1 MESFET WITH NO BIAS

A MESFET uses a change in the width of the depletion layer (W) as a means of controlling the current flow from the source to the drain.

For a MESFET with no bias applied to the gate (V_{GS}; Figure 6.45a), when a small bias is applied between the source and the drain (V_{DS}), a small amount of current flows to the drain (I_D). The magnitude of this current is given by (V_{DS}/R), where R is the resistance of the channel region. Thus, initially, there is an ohmic region in that I_D increases with the increasing drain bias (V_{DS}). As the voltage V_{DS} increases further, the depletion layer width also increases (Figure 6.45). This causes the cross-sectional area of the channel current to decrease. This area is labeled as the conducting channel in Figure 6.45b. The channel resistance increases. The net result is that the channel current decreases. Therefore, with an additional drain bias (V_{DS}), the drain current (I_D) increases but at a slower rate. A typical *I–V* characteristic curve for a MESFET, labeled $V_{GS} = 0$, shows this trend (Figure 6.46).

As the drain bias voltage increases even more, the depletion layer width also increases. This is shown as the shaded area under the gate region in Figure 6.45b. Since the increase in the voltage from 0 (at the source) to V_{DS} (at the drain) is not uniform across the channel, the width of the depletion layer actually is higher on the drain side. The increase of V_{DS} can be so high that it can actually pinch off the channel and saturate the drain current. If the applied drain bias (V_{DS}) increases further, the device will break down at $V = V_B$ (Figure 6.46).

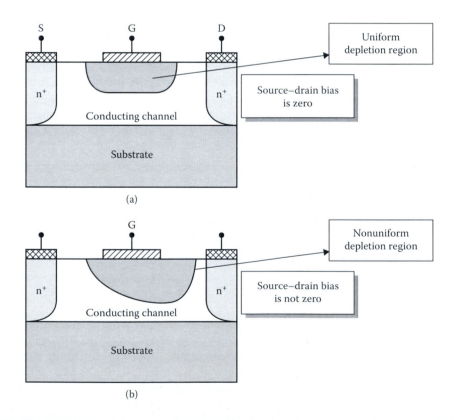

FIGURE 6.45 Changes in the width of the depletion layer with (a) zero bias and (b) non-zero bias. (From Singh, J.: *Semiconductor Devices: Basic Principles*. 2001. Copyright Wiley-VCH Verlag GmbH & Co. KGaA. Reproduced with permission.)

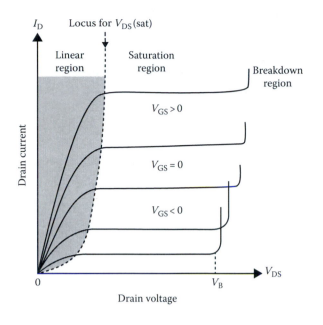

FIGURE 6.46 Typical *I–V* characteristics of a MESFET. (From Singh, J., *Semiconductor Devices: Basic Principles*, Wiley, New York, 2001. With permission.)

6.9.2 MESFET with a Gate Bias

When the gate bias (V_{GS}) is > 0 (Figure 6.46), the depletion layer width of the channel region decreases and, overall, the value of the drain current (I_D) increases for a given value of drain bias V_{DS}. If the gate bias is negative (V_{GS} < 0), the depletion layer width increases, and the overall I_D decreases for a given value of drain bias. The channel region pinch-off then occurs at lower voltages (Figure 6.46).

6.10 METAL INSULATOR SEMICONDUCTOR FIELD-EFFECT TRANSISTORS

The second strategy for gate isolation involves the use of an electrical insulator, that is, a dielectric, nonconducting material, with a relatively large band gap (Figure 6.44) that is known as the gate dielectric. To apply a voltage to the gate, a metal or metal-like material is deposited onto the gate dielectric. This device, in which a metal is deposited onto an insulator to form an FET, is known as the MISFET. One of the best dielectrics for silicon is SiO_2. A MISFET with silica as the gate dielectric is known as a MOSFET. A high conductivity polycrystalline silicon known as *polysilicon* (*poly-Si* or *polysil*), is used as a gate electrode instead of a metal in the semiconductor processing of a MOSFET. The device is still traditionally referred to as a MOSFET.

A thorough analysis and discussion of the operating principles for different FETs and the analysis of related circuits is beyond the scope of this book. Please consult introductory electrical engineering textbooks on microelectronics for a detailed discussion of these topics. In Section 6.11, we present an overview of the MOSFET, which is the most commonly used transistor.

6.11 METAL OXIDE SEMICONDUCTOR FIELD-EFFECT TRANSISTORS

6.11.1 MOSFET in Integrated Circuits

The MOSFET is the most important transistor device in ICs. Millions of transistors are integrated into a very small area of a silicon chip. The number of transistors on a computer chip has grown exponentially due to significant advances in the science and technology of silicon microelectronic device processing (Figure 6.47).

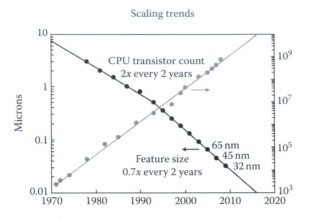

FIGURE 6.47 Transistor dimension scale to improve performance, reduce power, and reduce cost per transistor. (Courtesy of Intel Corporation, US.)

The chart shown in Figure 6.47 is consistent with *Moore's law* (named for Roger Moore, cofounder of Intel Corporation), which states that the number of active devices on a semiconductor chip doubles every 18 months.

6.11.2 Role of Materials in MOSFET

The structure of a conventional MOSFET is shown in Figure 6.48a. The current MOSFET design uses silica (SiO_2) as the gate dielectric and polysilicon as the conductor that forms the gate electrode.

Since the goal and trend in microelectronics has been to make transistors as small as possible (Figure 6.47), the thickness of the SiO_2 used as a gate dielectric has been getting smaller. The SiO_2 layer in state-of-the-art transistors currently is only ~1.2 nm (12 Å; Figure 6.49). The thinness of the gate dielectric layer provides a serious challenge in scaling down the overall size of transistors in the future. Very thin oxide layers such as SiO_2 can break down electrically and start to *leak* current due to the fact that the oxide layer experiences a very high electric field.

Researchers are therefore developing new gate dielectrics that insulate at even very small thicknesses. In 2007, the Intel Corporation reported a new hafnium oxide (HfO_2)–based insulator with a dielectric constant (k or ε_r) of ~25 (Figure 6.48b). The dielectric constant of SiO_2 is ~3.8. These higher dielectric constant materials, known as *high-k gate dielectrics*, allow the use of higher gate insulator thicknesses. The capacitance of a capacitor is directly proportional to its dielectric constant and inversely proportional to its thickness. Thus, higher-k gate dielectrics can be used with a higher insulator thickness, instead of using lower-k materials with smaller thickness.

The higher thickness of the gate dielectric in turn helps reduce the current leakage, and thus, less energy is wasted as heat, resulting in a higher energy efficiency. Therefore, a computer can run for a longer time using the same battery power. Similarly, researchers in the semiconductor industry are also developing new alloys to replace polysilicon gate electrodes (Figure 6.48) because polysilicon is not compatible with HfO_2-based oxide.

6.11.3 NMOS, PMOS, and CMOS Devices

The term *MOS* simply means a metal oxide semiconductor such as SiO_2 on silicon. MOSFETs can be designed so that the current from the source to the drain is carried by electrons or holes; these devices are known as NMOS (or *n-channel*) and PMOS (or *p-channel*), respectively. Since the mobility of holes is less, the PMOS transistor occupies more area than a typical NMOS transistor.

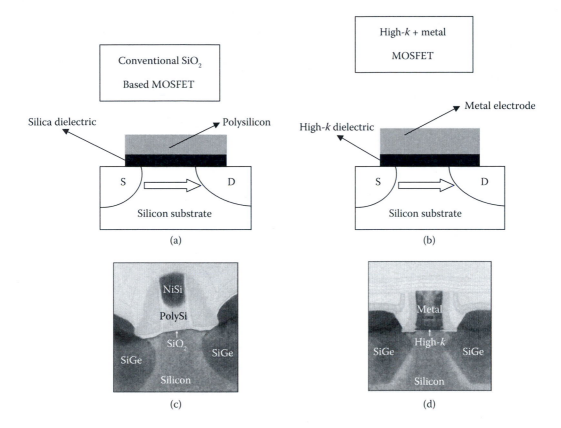

FIGURE 6.48 Structure of a conventional MOSFET based on (a) silicon oxide and polysilicon; (b) new high-k dielectrics and conductors; (c) silica gate dielectric with polysilicon gate electrode (65 nm transistor); and (d) hafnium-based dielectric with metal gate electrode (45 nm transistor). (Courtesy of Intel Corporation, US.)

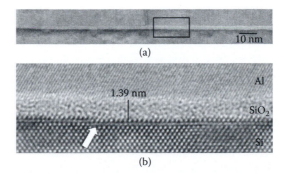

FIGURE 6.49 A transmission electron microscope image of a transistor cross-section showing the gate oxide layer (1.2 nm); (a) low magnification image, (b) high magnification image of an area marked with a rectangle in (a). (Courtesy of Intel Corporation, US.)

To create an NMOS transistor, we start with a p-type substrate (e.g., silicon; Figure 6.50). We then create n-type source and drain regions, shown as n$^+$. The gate is isolated from the channel using SiO_2, and an n-type polysilicon is used to create the gate electrode. Note that the channel region is a p-type semiconductor.

When a positive voltage is applied to the gate, a channel of electrons is created between the source and the drain. This MOSFET is known as an n-channel MOSFET. The regions outside the

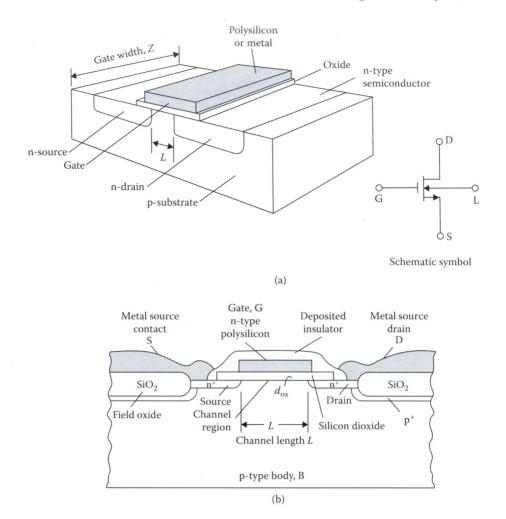

FIGURE 6.50 Schematic of an NMOS device: (a) structure and (b) cross-sectional view. (From Singh, J., *Semiconductor Devices: Basic Principles*, Wiley, New York, 2001. With permission.)

transistor channel are heavily doped with acceptors, shown as p$^+$, to electrically isolate the drain and source regions from the substrate. This ability to electrically isolate devices is extremely important in integrating a large number of transistors into the small area of a silicon chip. When the gate voltage is sufficient with respect to the substrate or source (V_{GS}), there is a flowing drain current (I_D) shown in the *transfer characteristics* of an NMOS transistor (Figure 6.51a). These are different from the so-called *output characteristics*, which refer to the change in the drain current (I_D) as a function of source-to-drain voltage (V_{GS}) for a fixed-gate bias.

We can also create a device that is complementary to the NMOS transistor. In a PMOS transistor, there is an n-type substrate. The source and the drain are p$^+$ type. The gate region has an insulator with an electrode. When a *negative* voltage is applied to the gate, a region of n-type substrate under the gate develops a positive charge. The channel is then p-type, and the device is known as a p-channel or PMOS transistor. When a sufficiently large, negative gate voltage is applied, the PMOS transistor shows a drain current. This appears in the transfer characteristics of the PMOS transistor in enhancement mode (Figure 6.51b).

We can also operate the NMOS and PMOS devices in depletion mode (see Section 6.11.6). For an NMOS in depletion mode, we can turn off the channel conductivity by applying a negative gate

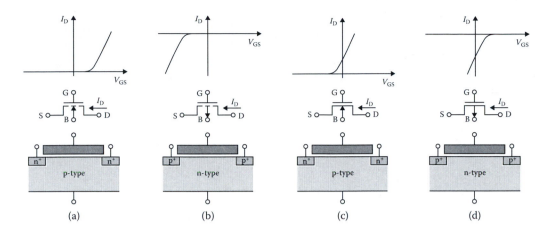

FIGURE 6.51 Transfer characteristics for the enhancement and depletion modes of an NMOS (a) and (c) and a PMOS (b) and (d). (From Dimitrijev, S., *Principles of Semiconductor Devices*, 2006, by permission of Oxford University Press.)

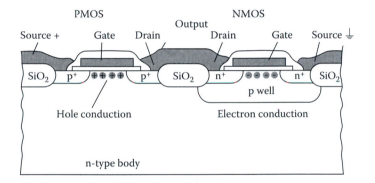

FIGURE 6.52 An illustration of a CMOS device. (From Singh, J., *Semiconductor Devices: Basic Principles*, Wiley, New York, 2001. With permission.)

voltage (Figure 6.51c). For a PMOS in depletion mode, we turn off the transistor by applying a positive voltage to the gate to deplete carriers in the channel (Figure 6.51d).

We can integrate the NMOS and PMOS devices onto the same substrate. This is known as a complementary MOSFET, or CMOS. CMOS devices consume relatively less power than NMOS and PMOS and are used in many applications of microelectronics (Figure 6.52).

6.11.4 ENHANCEMENT-MODE MOSFET

MOSFET can be operated in two modes (see Section 6.8). In enhancement-mode MOSFET, the conductivity of the channel region increases when a voltage is applied to the gate region and a current begins to flow from the source to the drain. This is called the on state of the transistor. Enhancement-mode MOSFET can be NMOS (n-channel) or PMOS (p-channel; Figures 6.53 and 6.51a and b).

When no voltage is applied to the gate and the resistance of the channel region is too high, no current flows through from the source to the drain. This is called the "off" state of the transistor. Typical voltage–current characteristics are shown in Figure 6.54. This shows the increase in the drain current (I_D) as a function of drain bias with respect to the source (V_{DS}). We assume that the

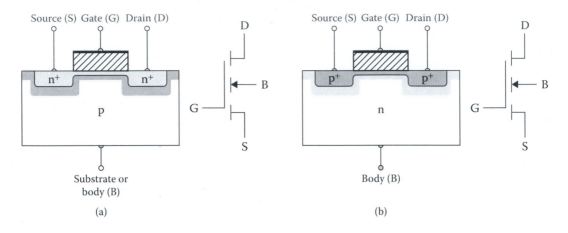

FIGURE 6.53 Schematic of (a) NMOS (n-channel) and (b) PMOS (p-channel) enhancement-mode MOSFETs. (From Neaman, D., *An Introduction to Semiconductor Devices*, McGraw Hill, New York, 2006. With permission.)

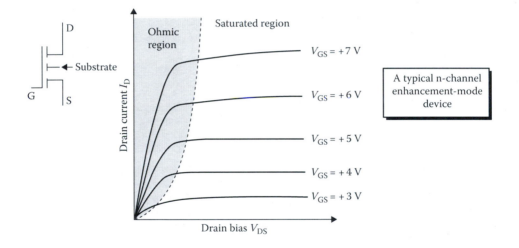

FIGURE 6.54 Typical *I–V* curves for MOSFETs. (From Singh, J., *Semiconductor Devices: Basic Principles*, Wiley, New York, 2001. With permission.)

source and the substrate or the body are connected, but this is not always the case. If a voltage is applied between the body or the substrate and the source, this is known as the body effect and can be accounted for when estimating I_D.

In enhancement-mode MOSFET, applying a bias between the gate and the substrate (V_{GS}) increases the conductivity of the channel region. The mechanism through which this occurs is very different from that in MESFET, which makes use of changes in the width of the depletion layer (see Section 6.9).

6.11.5 MECHANISM FOR ENHANCEMENT MOSFET

Consider an n-channel MOSFET, that is, an NMOS made from a p-type semiconductor (Figure 6.50). The gate dielectric forms a parallel plate capacitor. The dielectric of a capacitor normally is sandwiched between two electrodes. However, in this case, one side of the capacitor is a conductor (polysilicon or some other metal), and there is a semiconductor underneath the gate dielectric. When a positive bias is applied to the gate electrode, negative charges are induced on the other side, that is, the dielectric–semiconductor interface. At a certain voltage called the *threshold voltage* (V_{TH}), a very thin layer of the original p-type semiconductor begins to behave like an n-type semiconductor. This is known as an *inversion layer*. When a small drain bias (V_{DS}) is applied, electrons can then flow from the source to the drain (Figure 6.54).

If the drain bias (V_{DS}) increases even more, then the voltage between the gate and the inversion layer near the drain decreases as does the concentration of electrons near the drain region. This can continue with increasing V_{DS}. Eventually, this reaches a point where the voltage difference between the gate and the channel becomes so small that an inversion layer cannot be maintained. This will pinch off the channel region because there usually are very few electrons left near the drain region. This process causes the drain current (I_D) value to saturate (Figure 6.54).

If the gate bias with respect to the substrate (V_{GS}) increases, then the concentration of electrons present in the inversion layer increases. This causes an increase in the drain current when a small bias is applied to the drain (Figure 6.54).

The *I–V* characteristics for an enhancement-type MOSFET (Figure 6.54) and those for a MESFET look similar (Figure 6.46). However, the fundamental underlying phenomena responsible for their operations are very different. For a MESFET, we change the depletion layer width (W) of the channel region as a function of the gate bias (see Section 6.9). For an enhancement-type MOSFET (either NMOS or PMOS), we create an inversion layer in the channel region by inducing a charge at the insulator–semiconductor interface.

The most common MOSFET used is the n-channel or an NMOS in enhancement mode.

6.11.6 DEPLETION-MODE MOSFET

In a depletion-mode MOSFET, a MOSFET is created such that it is normally in an on state without any bias to the gate. This is a *ready-to-go* state that is achieved, for example, by creating a highly doped n-type region that forms a conductive channel between the source and the drain. In this case, the substrate is p-type. The free carriers in the n-type conductive channel are from donor doping. Similarly, we can create a p-channel MOSFET and operate it under the depletion mode. The structures of a depletion n-channel and p-channel MOSFET are shown in Figure 6.55.

When a negative voltage is applied to the gate of an NMOS, the conductivity of the channel region decreases because the carriers are depleted from the region. This switches the transistor into its off state, shown in the transfer characteristics (Figure 6.51c). When a positive gate voltage is

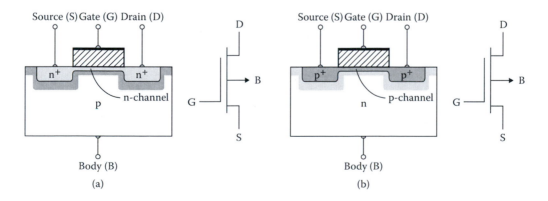

FIGURE 6.55 Schematic of depletion-mode (a) n-channel and (b) p-channel MOSFET. (From Neaman, D., *An Introduction to Semiconductor Devices*, McGraw Hill, New York, 2006. With permission.)

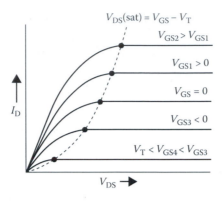

FIGURE 6.56 Characteristic *I–V* curves for an n-channel MOSFET in depletion mode. (From Neaman, D., *An Introduction to Semiconductor Devices*, McGraw Hill, New York, 2006. With permission.)

applied to a PMOS in depletion mode, the transistor is switched off (Figure 6.51d). The *I–V* curves for an NMOS depletion-layer MOSFET are shown in Figure 6.56.

Thus, if we consider the threshold voltage to be the gate voltage that is required to either just form the channel or to just deplete it in the case of depletion-mode MOSFET, then the gate voltage is positive for the enhancement-mode NMOS and is negative for depletion-mode NMOS (Figure 6.51a and c). For PMOS, the threshold voltage is negative for the enhancement type and positive for the depletion type (Figure 6.51b and d).

PROBLEMS

6.1 Define the terms "work function" and "electron affinity." Why do we use the term *work function* for metals and *electron affinity* for semiconductors?

6.2 Sketch the band diagrams for a Schottky contact and an ohmic contact from an n-type semiconductor in contact with a metal.

6.3 What is the principle by which a solar cell operates?

6.4 A Si p-n junction has an open-circuit voltage of 0.6 V. What value of J_L is required to produce this open-circuit voltage? Assume $T = 300$, $n_i = 1.5 \times 10^{10}$ cm^{-3}; $N_a = 10^{18}$ and $N_d = 10^{16}$ atoms/cm^3; $L_n = 10$ μm and $L_p = 25$ μm; and $D_p = 20$ and $D_n = 10$ cm^2/s.

6.5 Use the diffusion coefficient values in the previous example to calculate the carrier lifetimes.

6.6 Show that the maximum possible open-circuit voltage for a p-n junction in Si with $N_a = 10^{17}$ and $N_d = 3 \times 10^{16}$ atoms/cm^3 is 0.782 V.

6.7 If the photocurrent density J_L is 150 mA/cm^2, what is the open-circuit voltage for the p-n junction in Example 6.2?

6.8 Consider a Si p-n junction solar cell with electron and hole lifetimes of 3×10^{-7} s and 10^{-7} s, respectively. If the acceptor and donor doping levels are 3×10^{17} and 10^{16} atoms/cm^3, respectively, what is the open-circuit voltage (V_{OC})? Assume $D_p = 20$, $D_n = 10$ cm^2/s, and $J_L = 10$ mA/cm^2.

6.9 A solar cell with an area of 1 cm^2 has I_L of 15 mA. If the saturation current (I_S) is 3×10^{-11} A, (a) calculate the open-circuit voltage (V_{OC}) and (b) calculate the short-circuit current in milliamperes. (c) If the fill factor is 0.7, what is the power extracted for each solar cell in milliwatts?

6.10 Why is GaAs better suited than Si for making efficient solar cells? Why, then, is it not widely used?

6.11 Do solar cells require indirect or direct band gap semiconductors? Explain.

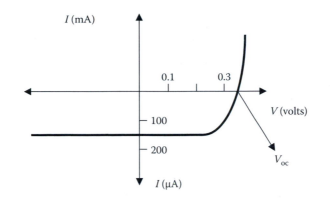

FIGURE 6.57 *I–V* curve for a solar cell. (From Edwards-Shea, L., *The Essence of Solid-State Electronics*, Prentice Hall, Englewood Cliffs, NJ, 1996. With permission.)

6.12 The *I–V* curve for a solar cell is shown in Figure 6.57. Calculate the fill factor for this solar cell. (Hint: Draw the largest possible rectangle that fits inside the given *I–V* curve.)

6.13 For the *I–V* curve shown in Figure 6.20, what is the open-circuit voltage (V_{oc}) for this solar cell? Show that the fill factor for this solar cell is about 0.8.

6.14 What is the basic principle of an LED? How is it different from that of a solar cell?

6.15 What is the difference between spontaneous and stimulated radiation?

6.16 Do LEDs always require a direct band gap semiconductor?

6.17 What mechanisms cause LEDs' overall efficiency to be lower?

6.18 What techniques can one use to produce LEDs that emit a "white" light?

6.19 In Example 6.3, what is the current through the circuit with a resistance of 390 Ω that is used in a series? Assume that the voltage driving the LED changes to 2.0 V.

6.20 In Example 6.3, what is the current through the circuit with a resistance of 800 Ω that is used in a series? Assume that the voltage driving the LED changes is 1.8 V.

6.21 What is the minimum drive voltage for a blue LED that emits at 470 nm?

6.22 Show that the minimum drive voltage for an LED that emits at 1550 nm is 0.8 V.

6.23 Why is the BJT considered a bipolar device?

6.24 An engineer uses a Si crystal to create two p-n junctions connected back-to-back. However, he or she finds that this device does not function as a transistor. Explain why.

6.25 A transistor has a base current of 50 µA. If the collector current is 3.7 mA, what is the value of emitter current (I_E)? What are the values of transistor parameters α and β?

6.26 In Example 6.4, what is the value of α?

6.27 In the circuit shown in Example 6.4 (Figure 6.42), assume that the values of R_B = 10 kΩ, R_C = 100 Ω, V_{BB} = 5 V, and V_{CC} = 10 V. Assume that the current gain is 100. Calculate the transistor currents and voltages across different junctions.

6.28 In the circuit shown in Example 6.4 (Figure 6.42), assume that the values of R_B = 50 kΩ, R_C = 500 Ω, V_{BB} = 5 V, and V_{CC} = 10 V. Assume that the current gain (β) is 90. Calculate the transistor currents and voltages across different junctions.

6.29 What does the term "field effect" mean?

6.30 Explain how the different gate-isolation strategies used for JFET and MESFET differ from those used for MOSFET.

6.31 What does enhancement-mode MOSFET mean? How is this different from depletion-mode MOSFET?

6.32 Explain the need for developing new high-*k* gate dielectrics for MOSFETs.

GLOSSARY

Active mode: An npn transistor in which the emitter–base junction is forward 0 biased and the base–collector junction is reverse 0 biased (Figure 6.37) is an example of a transistor in active mode.

Base: The moderately doped middle region sandwiched between the collector and the emitter of a transistor.

Base current: A flow of positive charge into the base; controls the collector and emitter currents.

Base-to-collector current amplification factor (β): The ratio of the collector current to the base current $\beta = I_C/I_B$. Typical values are $\sim>100$, also known as common emitter current gain.

Bipolar device: Any microelectronic device the function of which depends on the motions of both electrons and holes (e.g., BJT).

Bipolar junction transistor (BJT): A device with three terminals (emitter, base, and collector) based on two interacting p-n junctions connected back-to-back. The current at the collector can be controlled by regulating the voltage between the two other terminals, the emitter and the base. The most commonly used BJT is the npn type because of the higher mobility of the electrons.

CMOS: A MOSFET in which both NMOS and PMOS transistors are integrated. The acronym stands for complementary metal oxide silicon. The term FET is implied.

Collector: The lightly doped region in a transistor that ultimately receives the charge carriers emitted by the emitter.

Common base (CB) configuration: A BJT configuration in which the base electrode is common to the emitter and collector circuits. The transistor can be connected in other configurations as well (e.g., common emitter).

Common base current gain (α): The ratio of collector current to the emitter current: $\alpha = I_C/I_E$. Maximum value possible is 1.

Common emitter (CE) configuration: A BJT configuration in which the emitter electrode is common to the base and collector circuits.

Common emitter current gain (β): The ratio of the collector current to the base current $\beta = I_C/I_B$. Typical values are $\sim>100$, also known as the base-to-collector current amplification factor.

Controlled current: A collector or emitter current, which is controlled by the base current.

Controlling current: See **Base current**.

Copper indium diselenide ($CuInSe_2$): A compound semiconductor material for high-efficiency solar cells. Also known as CIS.

Current amplifier: The use of a BJT in which a small change in the base current (I_B) causes large changes in the collector current (I_C).

Cut-off mode: A mode of an npn operation in which both the emitter–base and base–collector junctions are reverse-biased.

Depletion mode: A MOSFET that is normally in the "on" state without bias. This is achieved by creating a highly doped region forming a conductive channel. When a negative bias is applied to the gate region, the conductivity of the channel region decreases and can turn off the device (Figure 6.55).

Early effect: In a BJT with a common base configuration, when V_{BC} reverse bias increases, the width of the depletion layer (W) associated with the base–collector junction gets larger. This means the width of the base region through carrier diffusion gets a little shorter, causing a slight increase in I_C as a function of V_{BC}.

Effective Richardson constant: The term $\left(\dfrac{m^* q k_B^2}{2\pi^2 \hbar^3} \right)$ that is related to the reverse saturation current due to thermionic emissions in a Schottky contact.

Electron affinity of a semiconductor ($q\chi_S$): The energy required to remove an electron from the conduction band and set it free.

Emitter: The heavily doped region in a transistor that injects minority carriers into the base.

Enhancement mode: The mode in which a potential forward or reverse bias is applied to the gate region of a MOSFET that normally is turned off; the conductivity of the channel region increases and a current can begin to flow from the source to the drain (Figure 6.53).

Fill factor (F_f): The ratio of $V_m \times I_m$ to $V_{oc} \times I_{oc}$. Most solar cells have a fill factor of ~0.7.

Field effect: The effect of an applied electric field on the conductivity of the channel region.

Field-effect transistor (FET): A unipolar device based on the control of charge carrier flow between a source and a drain, using voltage applied to a region known as the gate. The applied electric field changes the conductivity of the channel region, hence the name field-effect transistor.

Forward active mode: See **Active mode**.

Gate: The region in an FET between the source and the drain.

Gate dielectric: A nonconducting, dielectric material that isolates the gate electrode from the channel region. Usually the gate dielectric is silicon oxide (SiO_2), although new gate dielectrics with higher dielectric constant are being developed.

Heterojunction: A p-n junction in which the n- and p-sides are made using two different semiconductors.

High-k gate dielectrics: Recently developed materials used as gate insulators (e.g., hafnium oxide with k ~25). These materials allow for a higher thickness of the dielectric compared to silica, which leads to smaller leakage of current and more efficient computer chips.

Homojunction: A p-n junction in which the n- and p-sides are made using the same semiconductor.

Interacting p-n junctions: Two p-n junctions connected back-to-back and designed such that transistor action is possible. This means that the carriers injected from forward biasing of the emitter–base junction flow into the base–collector junction. If the base region is too long, injected carriers simply recombine and there is no transistor action.

Inverse active or inverted mode: A mode of npn transistor operation in which the roles of emitter and collector are reversed—the emitter–base is reverse-biased and the collector–base is forward-biased.

Inversion layer: In a MOSEFT, a thin layer underneath the gate that exhibits semiconductivity opposite to that of the rest of the substrate when a voltage is applied. In a p-type substrate, it becomes enriched with electrons and behaves as an n-type.

Junction field-effect transistor (JFET): An FET in which a reverse-biased p-n junction is used to isolate the gate region from the channel current flow. This is used mainly for III–V semiconductors such as GaAs.

Laser diode: In a laser diode, electrons and holes are generated when photons of a given energy and wave vector are absorbed by the semiconductor. The electrons and holes recombine and cause a stimulated emission of coherent photons; that is, the photons emitted are in phase with the incident photons that have the same energy and wave vector.

Majority device: A device such as a Schottky diode in which currents are generated primarily through the motions of majority carriers. The motion of minority carriers does not play a primary role.

Metal insulator semiconductor field-effect transistor (MISFET): An FET in which a dielectric material is used to isolate the gate region from the channel current flow. A metal electrode forming an ohmic contact with the dielectric allows application of the gate voltage.

Metal oxide semiconductor field-effect transistor (MOSFET): An FET, usually based on silicon, in which SiO_2 is used as a dielectric for isolating the gate region from the channel current flow. A relatively high-conductivity polysilicon (in this context, a metal) forms the electrode that allows application of the gate voltage.

Metal semiconductor field-effect transistor (MESFET): An FET in which a metal that forms a Schottky contact with the semiconductor is used to isolate the gate region from the channel current flow. This is used mainly for III–V semiconductors such as GaAs.

Moore's law: Named after Roger Moore, cofounder of Intel Corporation, this law states that the number of active devices on a semiconductor chip doubles every 18 months.

MOS: A metal oxide semiconductor, for example, silicon oxide on silicon. It is usually used to indicate a MOSFET whose operation can be explained using an MOS capacitor.

n-channel: The conduction from the source to the drain in the channel (which is p-type) occurs by electrons; hence, the channel is known as an n-channel or an NMOS device.

NMOS: A MOSFET in which electrons carry the channel current from source to drain.

npn transistor: A BJT with a p-type base sandwiched between the n-type emitter and the source.

Ohmic contact: A contact between a metallic material and a semiconductor such that there is no barrier to block the flow of current at the junction. The contact resistance may or may not follow Ohm's law. This contact is seen if $\phi_M < \phi_S$ for an n-type semiconductor or if $\phi_M > \phi_S$ for a p-type semiconductor.

Open-circuit voltage (V_{oc}): The maximum possible voltage generated from a solar cell.

Optical generation: The generation of carriers (electrons and holes) as a result of the absorption of light energy in a semiconductor or a p-n junction. This process forms the basis for solar cell operation.

Organic light-emitting diodes (OLEDs): Diodes based on organic materials with an electron-injecting cathode instead of the n-layer and a hole-injecting anode. These devices have considerable promise for energy-efficient lighting and are also used in flat-screen displays.

Output characteristics: The plot of the drain current (I_D) as a function of the source to drain voltage (V_{DS}) for a given value of gate bias. These are different from the transfer characteristics.

p-channel: The conduction from the source to the drain in the channel (which is n-type) occurs by holes; the channel is known as p-channel or PMOS.

Photocurrent (I_L): A current directed from the n- to the p-side of a p-n junction resulting from the process of photogeneration. Also known as the short-circuit current, it flows in an external circuit with no resistance.

Photogeneration: See **Optical generation**.

Photovoltaic effect: The appearance of forward voltage across an illuminated p-n junction.

Photovoltaics: A field of research and development involving the conversion of light energy into electricity.

PMOS: A MOSFET in which the current from source to drain is carried by holes.

pnp transistor: A BJT with an n-type base sandwiched between a p-type emitter and source.

Polymer organic light-emitting diodes (PLEDs): See **Organic light-emitting diodes**.

Polysilicon (poly-Si or polysil): A polycrystalline form of silicon. A relatively high-conductivity form of this material is used as a gate electrode for MOSFET and is referred to as a metal in this context.

Richardson constant (R^*): See **Effective Richardson constant**.

Saturation mode: A mode for operating an npn transistor in which both the emitter–base and base–collector junctions are forward-biased.

Schottky barrier (ϕ_B): The barrier that prevents injection of electrons from a metal into a semiconductor. Often, this barrier height is independent of the metal because the Fermi energy is pinned at the interface.

Schottky contact: The rectifying contact between a metal and a semiconductor, seen if $\phi_M < \phi_S$ for a p-type semiconductor or if $\phi_M > \phi_S$ for an n-type semiconductor.

Schottky diode: A diode based on a metal–semiconductor junction, requiring $\phi_M < \phi_S$ for a p-type semiconductor or $\phi_M > \phi_S$ for an n-type semiconductor.

Short-circuit current (I_{sc}): The same as a photocurrent (I_L), this refers to the maximum current flowing when a solar cell is connected to an external circuit with zero external resistance.

Silicides: Intermetallic compounds formed by reactions of elements with silicon.

Solar cell: A p-n junction–based device that generates electric voltage or current upon optical illumination.

Solar cell conversion efficiency (η_{conv})**:** The ratio of maximum power delivered to the power that is incident (P_{in}):

$$\eta_{conv} = \frac{I_m \times V_m}{P_{in}}$$

where I_m and V_m are the current and voltage, respectively, leading to maximum power.

Spontaneous emission: Radiation of photons by electron–hole recombination such that the emitted photons have no particular phase relationship with one another and are thus incoherent.

Stimulated emission: Radiation of photons by electron–hole recombination such that the emitted photons are coherent; that is, they have the same energy and wave vector as the photons that cause the emission to occur.

Surface pinning: A constant energy level that does not depend upon doping. Defects at the surface or interface of a semiconductor cause the Fermi energy of the semiconductor to be pinned at a band gap level. Surface pinning of the Fermi energy level causes the same Schottky barrier heights for different metals deposited on a semiconductor.

Thermionic emission: The process in which electrons on the semiconductor side that have a high enough energy overcome the built-in potential (V_0) and flow onto the metal side in a Schottky contact. This creates a current known as the thermionic current.

Threshold voltage (V_{TH}): The gate voltage that is required to either just form the channel or just deplete it.

Transfer characteristics: The plot of drain current (I_D) as a function of the gate voltage for an NMOS or a PMOS. These are different from the output characteristics that refer to the change in drain current (I_D) as a function of V_{DS} for a fixed value of gate bias.

Transistor: An abbreviation for a transfer resistor, a microelectronic device based on p-n junctions that is used as a tiny switch, as an amplifier, or for other functions.

Transistor action: In a transistor, the current at one terminal is controlled by the voltage at the other two terminals.

Unipolar device: A device in which the current is carried mainly by one type of carrier, electrons or holes. An FET is a unipolar device. This is one of the main distinctions between an FET and a BJT.

Work function ($q\phi$): The energy required to remove an electron from its Fermi energy level and set it free.

REFERENCES

Baliga, B. J. 2005. *Silicon Carbide Power Devices*. Singapore: World Scientific.

Bergh, A., G. Craford, A. Duggal, and R. Haitz. 2001. The promise and challenge of solid-state lighting. *Physics Today* 54:12.

Dimitrijev, S. 2006. *Principles of Semiconductor Devices*. Oxford, UK: Oxford University Press.

Edwards-Shea, L. 1996. *The Essence of Solid State Electronics*. Englewood Cliffs, NJ: Prentice Hall.

Floyd, T. L. 1998. *Electronics Fundamentals: Circuits, Devices, and Applications*, 4th ed. Boca Raton, FL: Chapman & Hall.

Floyd, T. L. 2006. *Electric Circuit Fundamentals*, 7th ed. Boca Raton, FL: Chapman & Hall.

Green, M. A. 2003. Crystalline and thin-film silicon solar cells: State of the art and future potential. *Solar Energy* 74:181–92.

Kasap, S. O. 2002. *Principles of Electronic Materials and Devices*. New York: McGraw Hill.

Mahajan, S., and K. S. Sree Harsha. 1998. *Principles of Growth and Processing of Semiconductors*. New York: McGraw Hill.

Nakamura, S., M. Senoh, N. Iwasa, S. Nagahama, T. Yamaka, and T. Mukai. 1995. Superbright Green InGaN Single-Quantum-Well-Structure Light-Emitting Diodes. *Jpn J Appl Phys* 34:L1332.

Neamen, D. 2006. *An Introduction to Semiconductor Devices*. New York: McGraw Hill.

Schubert, E. F. 2006. *Light-Emitting Diodes*. Cambridge, UK: Cambridge University Press.

Singh, J. 1996. *Optoelectronics: An Introduction to Materials and Devices*. New York: McGraw Hill.
Singh, J. 2001. *Semiconductor Devices, Basic Principles*. New York: Wiley.
Solymar, L. and D. Walsh. 1998. *Electrical Properties of Materials*, 6th ed. Oxford, UK: Oxford University Press.
Sparkes, J. J. 1994. *Semiconductor Devices*. London: Chapman and Hall.
Streetman, B. G., and S. Banerjee. 2000. *Solid State Electronic Devices*. Englewood Cliffs, NJ: Prentice-Hall.
Sze, S. M. 2002. *Semiconductor Devices, Physics and Technology*, 2nd ed. New York: Wiley.

7 Linear Dielectric Materials

KEY TOPICS

- Dielectric materials
- Induction, free, and bound charges
- Dielectric constant and capacitor
- Polarization mechanisms in dielectrics
- Dielectric constant and refractive index
- Ideal versus real dielectrics
- Complex dielectric constant
- Dielectric losses
- Frequency and temperature dependence of dielectric properties
- Linear and nonlinear dielectrics

7.1 DIELECTRIC MATERIALS

A *dielectric material* typically is a large-bandgap semiconductor ($E_g \sim >4$ eV) that exhibits high resistivity (ρ). The prefix *dia* means *through* in the Greek language. The word *dielectric* refers to a material that normally does not allow electricity (electrons, ions, and so on) to pass through it. There are special situations (for example, exposure to very high electric fields or changes in the composition or microstructure) that may lead to a *dielectric material* exhibiting semiconducting or metallic behavior. However, when the term dielectric material is used, it generally is understood that the material essentially is a nonconductor of electricity. An electrical *insulator* is a dielectric material that exhibits a high breakdown field.

7.1.1 ELECTROSTATIC INDUCTION

To better understand the behavior of nonconducting materials, let us first examine the concept of *electrostatic induction* and what is meant by the terms *free charge* and *bound charge*. First, consider a dielectric such as a typical ceramic or a plastic that has a net positive charge on its surface. Now, assume that we bring a Conductor B near this charged Insulator A (Figure 7.1a). The electric field associated with the positively charged Insulator A *pulls* the electrons toward it from Conductor B. This is also described as the atoms in Conductor B being *polarized* or affected by the presence of an electric field.

The process of the development of a negative charge on Conductor B is known as electrostatic induction. In this case, the negative charge developed on Conductor B is the bound charge because it is bound by the electric field caused by the presence of the charged Insulator A next to it.

The creation of a bound negative charge on Conductor B, in turn, creates a net positive charge on the other side of the conductor because the conductor itself cannot have any net electric field within it. If we connect this Conductor B with a grounded wire, then electrons will flow from the ground to this conductor and make up for the positive charge (Figure 7.1b). The flow of electrons from the ground to the conductor can also be described as the flow of the positive charge from Conductor B to the ground. Thus, the positive charge on Conductor B is considered the free charge. The word *free* in this context means that these charges are mobile, that is, they are free to move.

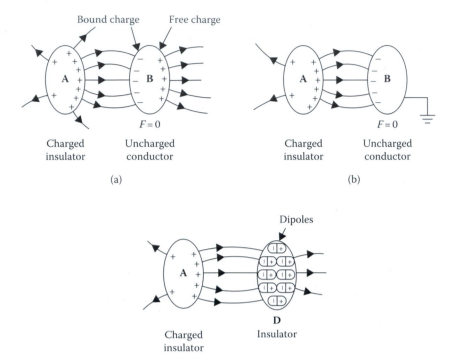

FIGURE 7.1 Illustration of electrostatic induction and bound and free charges, where A is the charged insulator, B is the conductor, and D is another dielectric. (a) Isolated induction, (b) grounded induction, and (c) induction with isolated Insulator D. (From Kao, K.C., *Dielectric Phenomena in Solids*, Elsevier-Academic Press, London, 2004. With permission.)

If, instead of bringing in Conductor B near the charged Insulator A, we bring in another dielectric material, D, then no induced charge is created on this dielectric material because no free carriers are available (Figure 7.1c). If we could look into this dielectric material, D, at an atomic scale, we would see that the electronic clouds are not perfectly symmetrical around the nuclei of atoms. Instead, the electronic clouds are tilted toward the positive charges in Insulator A. The presence of an electric field creates dipoles within the atoms of Material D, which is polarized by the electric field emanating from Dielectric A. (Readers can find further details of the polarization mechanisms and quantitative description of the electric displacement in Sections 7.6 and 8.1.) If we now move Dielectric D away from Dielectric A, we find that, after some time, the dipole moments induced in this material fade away because of fluctuations in thermal energy.

7.2 CAPACITANCE AND DIELECTRIC CONSTANT

7.2.1 Parallel-Plate Capacitor Filled with a Vacuum

A capacitor is a device that stores electrical charge. Consider two parallel conductive plates of area (A), separated by a distance d, and carrying charges of $+Q$ and $-Q$; assume that there is a vacuum between these plates. This is the basic structure of a parallel-plate capacitor. A schematic on the charge storage and the free/bound charges is also found in Figure 8.6.

The charge (Q) on the plates creates a potential difference V. We define the proportionality constant as the capacitance (C). Thus,

$$Q = C \times V \tag{7.1}$$

The SI unit of capacitance is a Farad (or coulombs per volt [C/V]). One Farad is a very large capacitance. Some *supercapacitors* have capacitances in this range. Most capacitors in microelectronics have a capacitance that is expressed in microfarads (10^{-6} F), nanofarads (10^{-9} F), picofarads (10^{-12} F), or femtofarads (10^{-15} F).

One of the laws of electrostatics is *Gauss's law*, which states that the area integral of the electric field (E) over any closed surface is equal to the net charge (Q) enclosed in the surface divided by the permittivity (ε).

$$\oint \vec{E} \times d\vec{A} = \frac{Q}{\varepsilon_0} \tag{7.2}$$

Gauss's law and the fundamentals of many electrical and magnetic properties and phenomena are derived from Maxwell's equations.

We define the surface charge density (σ or σ_s) as the charge per unit area.

$$\sigma_s = \frac{Q}{A} \tag{7.3}$$

The SI unit of charge density is coulombs per square meter (C/m^2). Note that the letter "C" represents the capacitance of a capacitor or charge in coulombs. The charge density is usually expressed as, for example, microcoulombs per square centimeter (μC/cm^2) or picofarads per square nanometer (pF/nm^2).

The generation of a voltage (V) between two plates separated by a distance d is also represented using the electric field (E) as follows:

$$E = \frac{V}{d} \tag{7.4}$$

We can rewrite Equation 7.1 for the capacitance (C) using Equations 7.3 and 7.4 as follows:

$$C = \frac{Q}{V} = \frac{\sigma_s \times A}{E \times d} \tag{7.5}$$

This equation tells us that the capacitance of a capacitor—that is, its ability to hold charge—depends on geometric factors, namely, the areas (A) of the plates and the distance (d) between them.

We define the *dielectric permittivity* (ε) of the material between the plates as

$$\varepsilon = \frac{\sigma_s}{E} \tag{7.6}$$

Therefore, from Equations 7.5 and 7.6,

$$C = \varepsilon \frac{A}{d} \tag{7.7}$$

We use the special symbol ε_0 for permittivity of the vacuum or free space, which is equal to 8.85×10^{-12} F/m.

We can write an expression for the capacitance of a capacitor filled with a vacuum (C_0) between the conductive plates as

$$C_0 = \frac{Q}{V_0} = \frac{\sigma_{s,0} A}{E_0 d} = \varepsilon_0 \frac{A}{d} \tag{7.8}$$

The subscript "0" has been added to indicate that this equation refers to a capacitor filled with a vacuum. Thus,

$$\varepsilon_0 = \frac{\sigma_{s,0}}{E_0} \tag{7.9}$$

We define the *dielectric flux density* (*D*) or *dielectric displacement* (*D*) as the total surface charge density.

$$D = \sigma_s \tag{7.10}$$

We will see in Section 7.3 that the dielectric flux density (*D*) can also be written as a sum of the free charge density and the *bound charge density* (σ_b). The SI unit for dielectric displacement (*D*) is the same as that for charge density, that is, C/m^2.

7.2.2 PARALLEL-PLATE CAPACITORS WITH AN IDEAL DIELECTRIC MATERIAL

Consider a capacitor in which the space between the two plates is filled with an *ideal dielectric material* (Figure 7.3). The term ideal dielectric means that, when charge is stored in a capacitor, no energy is lost in the processes that lead to storage of the electrical charge. In reality, this is not possible. In Section 7.13, we will define the term *dielectric loss*, which represents the electrical energy that is wasted or used when charge-storage processes occur in a dielectric material. These processes are known as polarization mechanisms.

In one form of polarization, a tiny dipole is induced within an atom. As shown in Figure 7.2, an atom without an external electric field has a symmetric electronic cloud around the nucleus. The centers of the positive and negative charges coincide, and there is no *dipole moment* (μ). However, when the same atom is exposed to an electric field, the electronic cloud becomes asymmetrical, that is, the electronic cloud of a polarized atom moves toward the positive end of the electric field. The nucleus essentially remains at the same location. Therefore, the centers of the positive and negative charges are now separated by a smaller distance (*x*). This process of creating or inducing a dipole in an atom is known as *electronic polarization* (Figure 7.2).

For a dipole with charges $+q$ and $-q$ separated by a distance *x*, the dipole moment is $\mu = q \times x$. The international system (SI) unit of dipole moment is $C \cdot m$. We define one *Debye* (D) as being equal to 3.3356×10^{-30} $C \cdot m$. Please note that the symbol D is also commonly used for dielectric displacement.

In a dielectric material exposed to an electric field, these induced atomic dipoles align in such a way that all the negative ends of the dipole line up near the positively charged plate (Figure 7.3b). This process means that some of the surface charges on the plates, which were originally free when there was a vacuum between the plates, will now become *bound* to the charges inside the dielectric material. When the capacitor is filled with a vacuum, all of the charges on the plates are free. The original free surface charge density (σ_s) with a vacuum-filled capacitor is now reduced to ($\sigma_s - \sigma_b$), where σ_b is the bound charge density (see Section 7.3). This reduced charge density means that the

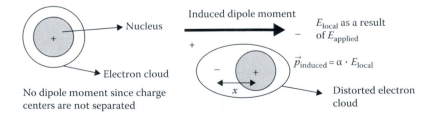

FIGURE 7.2 Illustration of electronic polarization of an atom that is exposed to an electric field.

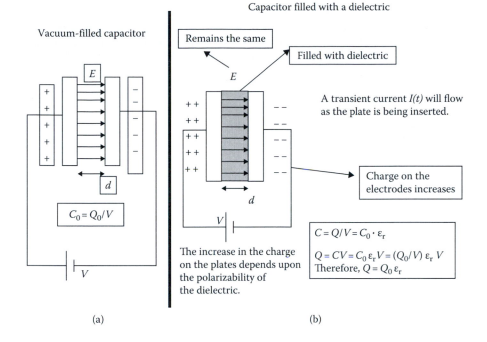

FIGURE 7.3 Structure of a parallel-plate capacitor filled with (a) a vacuum and (b) a real dielectric material; C_0 = capacitance without the dielectric material, C = capacitance with the dielectric material.

voltage (V) across the plates is also reduced. Therefore, the new electric field (E) for a capacitor filled with a dielectric is smaller compared to E_0. A lower electric field is needed to maintain the same charge because part of the surface charge is now held or bound by the dielectric material (also see Figure 8.6).

The ratio E_0/E—the electric fields existing in a capacitor filled with a vacuum and a dielectric material, respectively—is defined as the dielectric constant (k) or relative dielectric permittivity (ε_r).

$$\varepsilon_r = k = \frac{E_0}{E} \tag{7.11}$$

Now because the total charge on the plates is the same,

$$Q = C \times V = C_0 \times V_0$$

The distance between the parallel plates (d) is the same.
Therefore, $Q = C \times E = C_0 \times E_0$ or $(E_0/E) = (C/C_0)$
We can write Equation 7.11 as

$$\varepsilon_r = k = \frac{C}{C_0} = \frac{\varepsilon}{\varepsilon_0} \tag{7.12}$$

Thus, the dielectric constant (k) is also defined as the ratio of the capacitance of a capacitor filled with a dielectric to that of an identical capacitor filled with a vacuum. One advantage of defining a dielectric constant (k) or relative dielectric permittivity (ε_r) as a dimensionless number is that it becomes easy to compare the abilities of different materials to store charges. For example, the dielectric constant of the vacuum becomes 1. The dielectric constant of silicon (Si), alumina (Al_2O_3), and polyethylene are approximately 11, 9.9, and 2.2, respectively (Table 7.1).

Thus, the capacitance of a parallel capacitor containing a dielectric material with a dielectric constant (k) is given by modifying Equations 7.8 and 7.12 as follows:

$$C = \varepsilon_0 \times k \times \frac{A}{d} \tag{7.13}$$

If the dielectric material consists of atoms, ions, or molecules that are more able to be polarized, that is, if they are easily influenced by the applied electric field, then the dielectric constant (k) will be higher. In Section 7.12, we will examine how the dielectric constant (k) changes with electrical frequency (f), temperature, composition, and the microstructure of a material. Materials with a high dielectric constant are useful for making capacitors. Unlike transistors, diodes, solar cells, and so on, capacitors are considered *passive* components. One of the goals of capacitor manufacturers is to minimize the overall size of the capacitor while enhancing the total capacitance. This usually is achieved by arranging multiple, thin layers of dielectrics in parallel (Figure 7.4). This device is known as a *multilayer capacitor* (MLC).

The *volumetric efficiency* of a single-layer capacitor with area A and thickness d is given by:

$$\left(\frac{\varepsilon_0 \times k \times \dfrac{A}{d}}{A \times d} \right)$$

or

$$\text{volumetric efficiency of a single layer} = \left(\frac{\varepsilon_0 \times k}{2} \right) \tag{7.14}$$

TABLE 7.1

Approximate Room-Temperature Dielectric Constants (or Ranges) for Some Dielectric Materials or Classes of Materials (Frequency ~1 kHz–1 MHz)

Material	Dielectric Constant (k) (Approximate Range)
Vacuum	1 (by definition)
Oxygen (O_2) gas	1.000494
Argon (Ar) gas	1.000517
Mineral oil	2.25
Water (H_2O)	78
Polymers	~2–9
Teflon	2.1
Polyethylene	2.2
Silicon (Si)	11
Linear dielectric ceramics (SiO_2, TiO_2, Al_2O_3, and so on)	~4–200
Silica (SiO_2)	3.8–4.0
Alumina (Al_2O_3)	9.9
96% Al_2O_3 thick film	9.5
Magnesium oxide (MgO)	20
Nonlinear dielectrics or ferroelectrics	10–10,000
Polyvinylidene fluoride (a piezoelectric and ferroelectric polymer)	12–13
Barium titanate ($BaTiO_3$; a ferroelectric ceramic)	~2000–5000

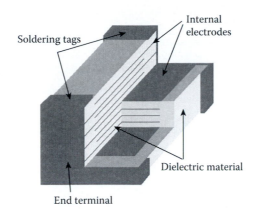

Soldering tags

Internal electrodes

Dielectric material

End terminal

FIGURE 7.4 Multilayer capacitors made using BaTiO$_3$-based formulations. (From Kishi, H., et al., *Jap J Appl Phys.*, 42, 1–15, 2003. With permission.)

Examples 7.1 and 7.2 illustrate how the volumetric efficiency is enhanced using multiple layers of a dielectric connected in parallel instead of using a single, thick layer. An analogy to this is that we stay warmer during winter by wearing multiple layers of clothing rather than by wearing one very thick jacket.

Example 7.1: Capacitors in Parallel and Series

1. Show that the capacitance of two capacitors with capacitances C_1 and C_2 connected in parallel is $C_1 + C_2$.
2. Show that the capacitance of the same capacitors connected in series is

$$\frac{1}{C} = \frac{1}{C_1} + \frac{1}{C_2}$$

(7.15)

or

$$C = \left(\frac{C_1 \times C_2}{C_1 + C_2} \right)$$

(7.16)

Solution

1. When capacitors are connected in parallel (Figure 7.5), the voltage across each capacitor is the same (V). The charge stored on capacitor 1 is Q_1 and that on capacitor 2 is Q_2. These are given by $C_1 = Q_1/V$ and $C_2 = Q_2/V$. The total charge stored on these capacitors connected in parallel is $Q = Q_1 + Q_2$. Thus, $Q = C_1V + C_2V$. If the total capacitance is C, then $C = Q / V$. Therefore,

$$C = \frac{(C_1V + C_2V)}{(V)}$$

or

$$C = C_1 + C_2 \text{ (for capacitors connected in parallel)}$$

(7.17)

Thus, when capacitors are connected in parallel, the capacitances of the individual capacitors add up. This is why the layers in MLCs are arranged and printed with electrodes so that the layers are connected in parallel (Figure 7.4). Note that this is the

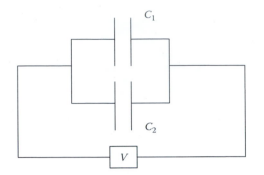

FIGURE 7.5 Capacitors connected in parallel.

opposite of what happens in resistors. Resistances add up when they are in a series. Capacitances add up when connected in parallel.
2. When connected in a series, the total charge on the plates of each capacitor is the same (Q). However, the voltage across each capacitor is different (V_1 and V_2). Thus, we can write $C_1 = Q / V_1$ and $C_2 = Q / V_2$. The total capacitance is given by $C = Q / V$.

$$C = \frac{Q}{(V_1 + V_2)}$$

or

$$C = \frac{Q}{\left(\dfrac{Q}{C_1}\right) + \left(\dfrac{Q}{C_2}\right)} = \frac{1}{\left(\dfrac{1}{C_1} + \dfrac{1}{C_2}\right)}$$

or

$$\frac{1}{C} = \frac{1}{C_1} + \frac{1}{C_2} \ \left(\text{for capacitors in series}\right)$$

or

$$C = \left(\frac{C_1 \times C_2}{C_1 + C_2}\right)$$

Note that when capacitors are connected in a series, the total capacitance (C) decreases (Figure 7.6).

Example 7.2: Multilayer Capacitor Dielectrics

Billions of MLCs are made every year using barium titanate ($BaTiO_3$)-based dielectric materials formulated into temperature-stable dielectrics with high dielectric constants. Dielectric layers of $BaTiO_3$ formulations are connected in parallel in an MLC because this provides the maximum volumetric efficiency, that is, the maximum capacitance per unit volume.
1. Calculate the capacitance of a parallel-plate *single-layer* capacitor made using a $BaTiO_3$ formulation of $k = 2000$. Assume $d = 3$ mm and $A = 5$ mm^2. What is the volumetric efficiency of this capacitor?
2. Calculate the capacitance of an MLC comprised of 60 dielectric layers connected in parallel using $N = 61$ electrodes. The thickness of each layer is 50 μm. The cross-sectional

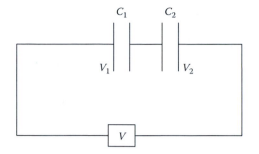

FIGURE 7.6 Capacitors connected in a series.

area of this capacitor is also 5 mm², and $k = 2000$. What is the volumetric efficiency of this capacitor?
3. What is the ratio of the volumetric efficiencies for multilayer and single-layer capacitors?

Solution
1. From Equation 7.13,

$$C = \varepsilon_0 \times k \times \frac{A}{d}$$

$$C = 8.85 \times 10^{-12} \text{ F/m} \times 2000 \times \frac{5 \times 10^{-6} \text{ m}^2}{3 \times 10^{-3} \text{ m}}$$

Thus, the capacitance of a single-layer capacitor with a thickness of 3 mm is 2.95×10^{-11} F or 0.0295 nF.
The volumetric efficiency of this single-layer capacitor = (0.0295 nF)/(5 mm² × 3 mm)

$$= 1.96 \times 10^{-3} \text{ nF/mm}^3$$

2. In the MLC, there are 61 metal electrodes (i.e., $N = 61$) that connect $(N - 1) = 60$ dielectric layers.
The capacitance of one layer of the dielectric in the MLC is given by

$$C_{layer} = 8.85 \times 10^{-12} \text{ F/m} \times 2000 \times \frac{5 \times 10^{-6} \text{ m}^2}{50 \times 10^{-6} \text{ m}}$$

$$= 1.77 \times 10^{-9} \text{ F or } 1.77 \text{ nF}$$

Note that the overall dimensions of the MLC and the single-layer capacitor are about the same (overall thickness is 3 mm, and the cross-sectional area is 5 mm²).
The total capacitance = $(N - 1) \times C_{layer} = (60) \times 1.77$ nF = 106.2 nF.
The volumetric efficiency of the MLC = $(106.2)/(60 \times 50 \times 10^{-3}$ mm × 5 mm²) = 7.080 nF/mm³.
3. The ratio of the volumetric efficiencies (in nF/mm³) is

$$= \frac{7.080}{1.96 \times 10^{-3}}$$

$$= 3600$$

If we have two capacitors that occupy the same volume, the MLC provides significantly more capacitance. This is very important because the goal is to minimize the size of the capacitors in integrated circuits (ICs) or printed circuit boards.

Example 7.3 illustrates some real-world situations in which there is a decrease in the capacitance because the device structure can transform into capacitors in a series.

Example 7.3: High-*K* Gate Dielectrics: HFO$_2$ on Si

IC fabrication technology uses silica (SiO$_2$/SiO$_x$) (k~4) as a gate dielectric. The formula SiO$_x$ indicates that the stoichiometry of the compound is not exactly known. The thickness of this SiO$_2$ gate dielectric in state-of-the-art transistors is ~2 nm. (a) What is the capacitance per unit area of a 2-nm film made using SiO$_2$? Express your answer in fF/μm^2 (1 femtofarad [fF] = 10^{-15} F).

At such small thicknesses, the dielectric layer is prone to charge leakage. The use of materials with higher dielectric constants allows a relatively thicker gate dielectric (for the same capacitance), which will be less prone to electrical breakdown. Hafnium oxide (HfO$_2$; k~25) is one example of such a material. A 10-nm thin film of HfO$_2$ was deposited on Si using a low-temperature chemical vapor deposition (CVD) process. (b) What is the capacitance of this HfO$_2$ film per unit area? (c) When an engineer tested an HfO$_2$ capacitor structure, the thickness of the film formed on the Si was found to be 10 nm. However, the capacitance per unit area was smaller than expected. Explain why this is possible.

Solution

1. The capacitance per unit area of a 2-nm film of SiO$_2$ is given by

$$\frac{C}{A} = \frac{\varepsilon_0 \times k}{d}$$

$$\frac{C}{A} = \frac{8.85 \times 10^{-12} \text{ F/m} \times 4}{2 \times 10^{-9} \text{ m}} = 1.77 \times 10^{-2} \text{ F/m}^2$$

(7.18)

Because 1 F = 10^{15} fF and 1 m^2 = 10^{12} μm^2, 1 F/m^2 = 10^3 fF/μm^2.
Thus, the capacitance per unit area of a 4-nm SiO$_2$ film is 17.7 fF/μm^2.

2. For a 10-nm HfO$_2$ film on Si, the capacitance per unit area is

$$\frac{C}{A} = \frac{8.85 \times 10^{-12} \text{ F/m} \times 25}{10 \times 10^{-9} \text{ m}} = 2.212 \times 10^{-2} \text{ F/m}^2 = 22.12 \text{ fF/μm}^2$$

This is the capacitance per unit area if the film formed is from HfO$_2$ only. Thus, using a high-k material allows us to apply a greater thickness, which means better protection against dielectric breakdown and charge leakage, and still achieve comparable capacitance per unit area.

3. Because the measured capacitance per unit area for the HfO$_2$-on-Si structure is smaller, we expect that something must have caused the *effective* dielectric constant of the *device structure* to become smaller. One possibility is that the HfO$_2$ reacted with the Si and formed a material or phase that had an overall lower dielectric constant, that is, the film was not pure HfO$_2$. However, because a low-temperature CVD process was used, the formation of a different phase, although possible, is unlikely. A careful analysis of the microstructure using transmission electron microscopy (TEM) along with simultaneous nanoscale chemical analysis can resolve this dilemma. When this analysis was performed, it showed that Si had reacted with oxygen (O$_2$) during the CVD process and formed a thin layer (~2 nm) of SiO$_2$ at the interface between the Si and the HfO$_2$. We can assume the thickness of the HfO$_2$ to be (10 − 1.5) = 8.5 nm. We now have a structure that has two dielectric layers forming two capacitors connected in a series (Figure 7.7).

Similar to the previous calculation in part (a) of this example, the capacitance per unit area for a 1.5-nm SiO$_2$ film is 23.6 fF/μm^2. The capacitance of the 8.5-nm HfO$_2$ film

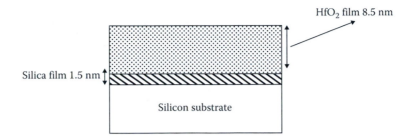

FIGURE 7.7 HfO_2 film on silicon, with a thin layer of SiO_2/SiO_x formed at the interface, thus creating two capacitors in series and reducing the total capacitance per unit area.

is 26.03 fF/μm^2. Because these capacitors are in a series (Figure 7.7), the total capacitance per unit area for this structure is

$$\frac{1}{C_{total}} = \frac{1}{C_{SiO_2}} + \frac{1}{C_{HfO_2}}$$ (7.19)

Note that we are dealing with capacitances per unit area, and thus, each side of this equation is divided by area (A). Therefore, we get

$$\frac{1}{\left(\dfrac{C_{total}}{A}\right)} = \frac{1}{\left(\dfrac{C_{SiO_2}}{A}\right)} + \frac{1}{\left(\dfrac{C_{HfO_2}}{A}\right)} = \frac{1}{23.6} + \frac{1}{26.03}$$

Thus, the new total capacitance per unit area (C_{total}/A) = 12.37 fF/μm^2.

We can see that this value is lower than the value 22.12 fF/μm^2, obtained when there is no interfacial layer of SiO_2 present. Note that in principle, it may seem that thinner layers of SiO_2/SiO_x enhance the capacitance per unit area; however, such layers are prone to increased leakage of currents and electrical breakdown, and are therefore not reliable. We can use strategies to prevent the lowering of capacitance by processing alternative gate dielectric oxide films so that a SiO_2 layer does not form. The Intel Corporation has developed such dielectrics using HfO_2 and zirconium oxides. Another possibility not considered in this example is that the actual dielectric constant of HfO_2 may be lower than 25. The exact value of this apparent dielectric constant of HfO_2 can depend on the thin-film microstructure. Research has shown that HfO_2 films can have an apparent dielectric constant between 18 and 25.

7.3   DIELECTRIC POLARIZATION

We define *dielectric polarization* (*P*) as the magnitude of the bound charge density (σ_b). The application or presence of an electric field (the *cause*) leads to dielectric polarization (the *effect*). The situation is very similar to that encountered while discussing the mechanical properties of materials. The application of a stress (the *cause*) leads to development of a strain (the *effect*). The stress and strain are related by Young's modulus. In this case, the electric field (*E*) applied and the polarization (*P*) created are related by the dielectric constant (*k*).

Assume that the dielectric polarization is caused by *N* number of small (atomic scale) dipoles, each comprising two charges (+q_d and −q_d) separated by a distance *x* (Figure 7.2). Here ($q_d \times x$) is called the *dipole moment*, which represents the polarization of an individual dipole. The dielectric polarization (*P*) is equal to the total dipole moment per unit volume of the material. This is another definition of dielectric polarization. The mechanisms by which such dipoles are created in a material are discussed in Section 7.5.

Assume that the concentration of atoms or molecules in a given dielectric material is N; if each atom or molecule is polarized, then the value of dielectric polarization is

$$P = \sigma_b = N \times <q_d \times x> \tag{7.20}$$

As explained in here and in Figure 7.2, x is the distance between the positive charge center and the negative charge center. If σ is the total charge density for a capacitor, then the other portion of the charge density, that is, $(\sigma - \sigma_b)$, remains the free charge. This creates a dielectric flux density (D_0), such as in the case of a capacitor filled with a vacuum. From Gauss's law,

$$D_0 = \varepsilon_0 \times E \tag{7.21}$$

Thus, the total dielectric flux density (D) for a capacitor filled with a dielectric material originates from two sources: the first source is the bound charge density (σ_b) associated with the polarization (P) in the dielectric material and the other is the free charge density $(\sigma - \sigma_b)$.

Therefore,

$$D = P + (\varepsilon_0 \times E) \tag{7.22}$$

We can also rewrite the dielectric displacement as $D = \varepsilon \times E$. Therefore, we get

$$P = \sigma_b = (\varepsilon - \varepsilon_0) \times E \tag{7.23}$$

If μ is the average dipole moment of the atomic dipoles created in a dielectric and N is the number of such dipoles per unit volume (i.e., the concentration), then we can also write the polarization as follows:

$$P = (N \times \mu) \tag{7.24}$$

where μ is the dipole moment. *Polarizability* (α) describes the ability of an atom, ion, or molecule to create an induced dipole moment in response to the applied electric field. The average dipole moment (μ) of an atom can be written as the product of the polarizability of an atom (α) and the local electric field (E) that an atom within the material experiences.

$$\mu = (\alpha \times E) \tag{7.25}$$

The SI units for dipole moment and electric field are $C \cdot m$ and V/m, respectively. Therefore, the unit of polarizability (α) is $C \cdot V^{-1} \cdot m^2$ or $F \cdot m^2$ (from Equation 7.1). The polarizability is often expressed as *volume polarizability* (α_{volume}) in units of cm^3 or $\mathring{A}^3$:

$$\alpha_{volume} = \text{volume polarizability } (cm^3) = \frac{10^6}{4\pi\varepsilon_0} \times \alpha \left(C \cdot V^{-1} \cdot m^2 \text{ or } F \cdot m^2 \right) \tag{7.26}$$

If the volume polarizability is expressed in $\mathring{A}^3$ (as is often done in the case of atoms or ions), we use the following equation:

$$\alpha_{volume} \text{ in } \mathring{A}^3 = \frac{10^{30}}{4\pi\varepsilon_0} \times \alpha \left(C \cdot V^{-1} \cdot m^2 \text{ or } F \cdot m^2 \right) \tag{7.27}$$

The factors 10^6 and 10^{30} are used in these equations because $1 \, m^3 = 10^6 \, cm^3$ and $1 \, m^3 = 10^{30} \, \mathring{A}^3$.

For now, let us assume that, the electric field, an atom within a material experiences is the same as the applied field (E). From Equations 7.24 and 7.25, we get

$$P = (N \times \alpha \times E) \tag{7.28}$$

From Equations 7.28 and 7.23, we get

$$\alpha = \frac{(\varepsilon - \varepsilon_0)}{N} \tag{7.29}$$

We can rewrite this equation as shown here:

$$\varepsilon_r = \frac{\varepsilon}{\varepsilon_0} = \left[1 + \left(\frac{N\alpha}{\varepsilon_0} \right) \right] \tag{7.30}$$

Equation 7.30 is important because it links the dielectric constant (k or ε_r) of a material to the polarizability of the atoms (α) from which it is made and also to the concentration of dipoles (N).

We also use another parameter called the *dielectric susceptibility* (χ_e) to describe the relationship between polarization (the effect) and the electric field (the cause):

$$P = \chi_e \varepsilon_0 E \tag{7.31}$$

The subscript e in χ_e distinguishes the dielectric susceptibility from the magnetic susceptibility (χ_m), which is defined in Chapter 9. The dielectric susceptibility (χ_e) describes how susceptible or polarizable a material is—that is, how easily the atoms or molecules in the material are polarized by the presence of an electric field. Comparing Equations 7.23 and 7.31,

$$\chi_e = (k - 1) = (\varepsilon_r - 1) \tag{7.32}$$

Also, from Equations 7.31 and 7.32,

$$P = (\varepsilon_r - 1)\varepsilon_0 E \tag{7.33}$$

Another way to express the dielectric susceptibility is as follows:

$$\chi_e = (\varepsilon_r - 1) = \frac{P}{D_0} = \frac{\text{Bound surface charge density}}{\text{Free surface charge density}} \tag{7.34}$$

The dielectric susceptibility (χ_e), which is another way to express the dielectric constant (k), depends on the composition of the material. We can show from Equations 7.29 and 7.31 that

$$\chi_e = \frac{N\alpha}{\varepsilon_0} \tag{7.35}$$

From Equation 7.35, we can see that the dielectric susceptibility (χ_e) of the vacuum is zero or the dielectric constant (k) is 1. This is expected because there are no atoms or molecules in a vacuum, so $N = 0$. The more polarizable the atoms, ions, or molecules in the material are, the higher the bound charge density is, and the higher the dielectric susceptibility (χ_e) or dielectric constant (k) is. In Section 7.5, we will see that several different polarization mechanisms exist for a material (Figure 7.8).

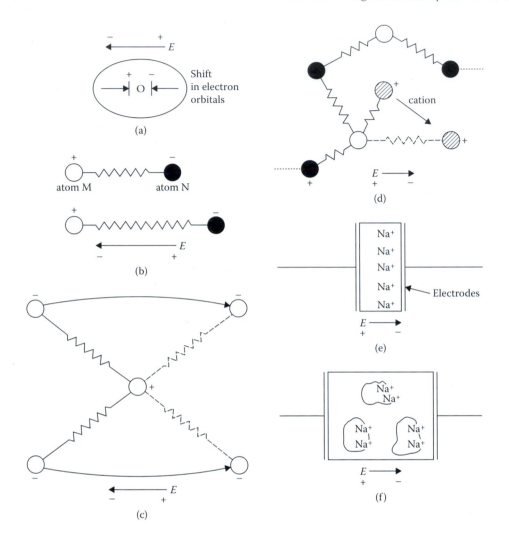

FIGURE 7.8 Schematic representation of polarization mechanisms: (a) electronic, (b) atomic or ionic, (c) high-frequency oscillatory dipoles, (d) low-frequency cation dipole, (e) interfacial space-charge polarization, and (f) interfacial polarization. (From Hench, L.L. and West J.K., *Principles of Electronic Ceramics*, Wiley, New York, 1990. With permission.)

The dielectric susceptibility values (χ_e) of silicon, Al_2O_3, and polyethylene are approximately 10, 8.9, and 1.2, respectively, because the dielectric constants are approximately 11, 9.9, and 2.2, respectively (Table 7.1).

7.4 LOCAL ELECTRIC FIELD (E_{LOCAL})

When we examine the effect of an externally applied electric field (E) on a dielectric material, we need to account for the electric field that exists inside a polarized material. For a solid or liquid exposed to an electric field (E), the actual field experienced by the atoms, molecules, or ions inside the material is different from the external electric field (E) and is referred to as the *internal electric field* or *local electric field* (E_{local}). In general, the greater the polar nature of the material (or the larger the dipole moment), the higher the strength of the dipoles induced, from the external electric field (E) and the larger the local electric field. Furthermore, the magnitude of the electric field experienced by the atoms, ions, or molecules inside a material depends on the arrangement of the dielectric dipoles.

For a cubic-structured isotropic material, a liquid, or an amorphous material, the local electric field is given by

$$E_{\text{local}} = E + \frac{1}{3\varepsilon_0} P \tag{7.36}$$

Equation 7.36 is known as the *local field approximation* and shows that the internal field causing the charge storage is determined by the applied electric field and polarization. Thus, we can rewrite Equation 7.28 by substituting E_{local} for E as

$$P = (N \times \alpha \times E_{\text{local}})$$

or

$$N \times \alpha = \frac{P}{E_{\text{local}}} = \frac{P}{\left(E + \dfrac{P}{3\varepsilon_0}\right)} \tag{7.37}$$

Note that E is the applied electric field.
Therefore,

$$N \times \alpha = \frac{E(\varepsilon - \varepsilon_0)}{E + \dfrac{E(\varepsilon - \varepsilon_0)}{3\varepsilon_0}}$$

Eliminating E, dividing by ε_0, and rewriting $\varepsilon/\varepsilon_0 = \varepsilon_r$, we get

$$\frac{\varepsilon_r - 1}{\varepsilon_r + 2} = \frac{1}{3\varepsilon_0}(N \times \alpha) \tag{7.38}$$

Equation 7.38 is also known as the Clausius–Mossotti equation. It describes the relationship between the dielectric constant (k or ε_r), a *macroscopic* property, and the concentration of polarizable species (N) and their polarizability (α), which are *microscopic properties*. We have used the local field approximation (Equation 7.36), which is valid for either amorphous materials or cubic-structured materials. Strictly speaking, the Clausius–Mossotti equation should be used only for those types of materials where the dielectric dipoles are generated under the electric field. In some materials, molecules have a permanent dipole moment (such as water, ferroelectrics) due to their unique non-cubic crystal structure. In these materials, the role of the electric field is to align the permanent dipoles. In the case of ferroelectrics, spontaneous polarization is developed because of the rearrangement of ions (Section 7.11). For such polar materials with permanent dipoles, the internal electric field within the ferroelectrics is *not* given by Equation 7.37. As a result, the Clausius–Mossotti equation (Equation 7.38) *cannot* be used for polar materials with permanent dipoles.

In Equation 7.38, N is the concentration of dipoles per unit volume (the number of molecules/m^3). If we assume that each molecule or atom becomes a dipole, then N is related to the N_{Avogadro} (6.023×10^{23} molecules/mol), density (ρ in kg/m^3), and molecular weight (M in kg/mol) as follows:

$$N = N_{\text{Avogadro}} \times \frac{\rho}{M} \tag{7.39}$$

Substituting for N in Equation 7.38, we get another form of the Clausius–Mossotti equation:

$$\left(\frac{\varepsilon_r - 1}{\varepsilon_r + 2}\right)\frac{M}{\rho} = \frac{N_{\text{Avogadro}} \times \alpha}{3\varepsilon_0} \tag{7.40}$$

We can verify that this equation is dimensionally balanced. The units are α in $F \cdot m^2$, ε_0 in F/m, and Avogadro's number in number per mole. Thus, the unit on the right-hand side of this expression is m^3/mol. The unit on the left-hand side is also m^3/mol.

We can rewrite Equation 7.40 to get the polarization per mole or *molar polarization* (P_m), defined as follows:

$$\left(\frac{\varepsilon_r - 1}{\varepsilon_r + 2}\right) = \frac{P_m \times \rho}{M} \tag{7.41}$$

The units for molar polarization (P_m) are m^3/mol and cm^3/mol.

As we will see in Section 7.5 and Figure 7.8, there are five polarization mechanisms—electronic, ionic, dipolar, interfacial, and spontaneous (ferroelectric). The total polarizability due to ionic and electronic polarization is additive because these polarizations occur throughout the volume of a material. Thus, we can write the total polarizability as

$$\alpha = \alpha_e + \alpha_{ionic} \tag{7.42}$$

where, α_e and α_{ionic} are the electronic and ionic polarizabilities of the atoms, respectively.

Electronic and ionic polarization are defined in Sections 7.6 and 7.7, respectively.

We can rewrite the Clausius–Mossotti equation (Equation 7.38) to separate out the ionic and electronic polarization effects as follows:

$$\frac{\varepsilon_r - 1}{\varepsilon_r + 2} = \frac{1}{3\varepsilon_0}\left[(N_e \times \alpha_e) + (N_i \times \alpha_i)\right] \tag{7.43}$$

In Equation 7.43, we *cannot* incorporate the effects of dipolar (Section 7.9), interfacial (Section 7.10), and spontaneous or ferroelectric polarization (Section 7.11) because the effects that these polarizations have on the local electric field (E_{local}) are complex. They cannot be described by Equation 7.36; thus, the Clausius–Mossotti equation generally *cannot* be used for polar materials such as water or ferroelectric compositions of materials (e.g., the tetragonal form of $BaTiO_3$). An exception to this is a situation where the Clausius–Mossotti equation may be used for polar materials if the electrical frequency (f) of the applied field is too high for these polarization mechanisms to exist.

7.5 POLARIZATION MECHANISMS—OVERVIEW

We will now examine the different ways in which atoms, ions, and molecules in a material can be polarized by an electric field (Table 7.2 and Figure 7.8).

These polarization mechanisms are shown schematically in Figure 7.8 and discussed in the following sections.

7.6 ELECTRONIC OR OPTICAL POLARIZATION

7.6.1 ELECTRONIC POLARIZATION OF ATOMS

All materials contain atoms (in the form of neutral atoms, ions, or molecules). When subjected to an electric field, each atom is polarized, in that the center of the electronic charge shows a slight shift toward the positively charged electrode.

This very slight elastic displacement of the electronic cloud (a few parts per million of the atomic radius), shown as δ or x, occurs very rapidly with reference to the nucleus ($\sim 10^{-14}$ seconds; Figure 7.9). This means that, even if the electric field changes its polarity $\sim 10^{14}$ times a second, the electronic polarization process can still follow this rapid change in the direction of the electric field. The electric field associated with visible light (which is an electromagnetic wave)

TABLE 7.2

Summary of Polarization Mechanisms in Dielectrics

Polarization Mechanism	Causes of Net Dipole Moment	Approximate Frequency Range (Hz)	Temperature Dependence	Examples
Electronic or optical polarization (Section 7.6)	Displacement of electronic cloud with reference to nucleus	Up to 10^{14} Hz	Not strong	All materials
Ionic, atomic, or vibrational polarization (Section 7.7)	Displacement of ions with reference to each other	Up to ~10^{13} Hz	Not strong	Ionic solids such as oxide ceramics (e.g., Al_2O_3 and TiO_2)
Dipolar or orientational (Section 7.7)	Reorientation of permanent dipoles	Up to ~10^{12} Hz	Yes; decreases with increasing temperature	H_2O
Interfacial, space charge, Maxwell–Wagner, or Maxwell–Wagner–Sillars polarization (Section 7.10)	Movement of ions at an interface such as grain boundaries	Up to several MHz	Yes; involves the short-range movement of atoms or ions	Lithium tantalate ($LiTaO_3$), lithium niobate ($LiNbO_3$), inorganic glasses containing Li^+, Na^+, and so on
Spontaneous, ferroelectric polarization (Section 7.11)	Creation of spontaneous dipoles in the unit cells of a material via small displacements of ions; typically involves a phase transformation between polymorphs	Up to ~10^9–10^{10} Hz	Yes; decreases above a certain temperature, and eventually disappears at very high temperatures	$BaTiO_3$, PZT, and PVDF

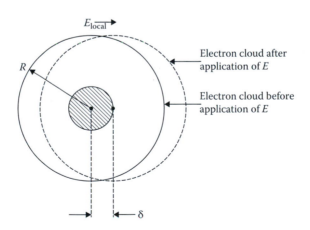

FIGURE 7.9 Displacement of the electronic cloud in an argon atom. (From Kao, K.C., *Dielectric Phenomena in Solids*, Elsevier-Academic Press, London, 2004. With permission.)

oscillates with a frequency of ~10^{14} Hz. This field interacts with dielectric materials and causes electronic polarization. Thus, electronic polarization is related to the optical properties of materials, such as the refractive index. This is why electronic polarization is also known as *optical polarization*. Readers will study how the electronic polarization controls the optical properties of materials in Section 7.12 and Chapter 8.

Because the electrons of the outermost shell are the ones most susceptible to the electric field applied, the *electronic polarizability* (α_e) of an atom or an ion depends primarily on its size and the number of electrons in its outermost shell. The larger the atom or ion, the farther the electrons are from the nucleus. Therefore, the electron clouds surrounding larger atoms or ions are more susceptible to electric fields and are more polarizable. Thus, larger atoms or ions have a higher electronic polarizability. The extent to which an atom or an ion is polarized via this mechanism is measured by the electronic polarizability (α_e).

Consider the nucleus of a monoatomic element (such as argon [Ar] with radius R). When an electric field is applied, the electronic cloud is displaced by a distance δ with respect to the nucleus (Figure 7.9).

A dipole is created between the charge q_1 of the nucleus (Zq; where Z is the number of electrons surrounding the nucleus) and the charge q_2, which is the part of the charge in the electron cloud that no longer surrounds the nucleus because it is displaced. This charge q_2 is contained in a sphere of radius δ and is given by

$$q_2 = -\frac{Zq\left[(4/3)\pi\delta^3\right]}{\left[(4/3)\pi R^3\right]} = -Zq\frac{\delta^3}{R^3} \tag{7.44}$$

Thus, the Coulombic force of attraction between the nuclear charge q_1 and the negative charge q_2, not shielded by the nucleus and separated by distance δ, is given by Coulomb's law:

$$F = \frac{q_1 \times q_2}{4\pi\varepsilon_0\delta^2} \tag{7.45}$$

We can rewrite this as

$$F = \frac{(Zq) \times \left(-Zq\dfrac{\delta^3}{R^3}\right)}{4\pi\varepsilon_0\delta^2} = -\frac{(Zq)^2\,\delta}{4\pi\varepsilon_0 R^3} \tag{7.46}$$

The magnitude of this attractive Coulombic force is balanced by the force (F_d) that causes the displacement.

$$F_d = (Zq) \times E \tag{7.47}$$

Thus, equating the magnitude of the force that causes the displacement of the electronic cloud and the Coulombic restoring force, we get

$$\frac{(Zq)^2\,\delta}{4\pi\varepsilon_0 R^3} = (Zq) \times E \tag{7.48}$$

Solving for displacement (δ), we get

$$\delta = \frac{4\pi\varepsilon_0 R^3 E}{Zq} \tag{7.49}$$

The dipole moment (μ) caused by the electronic polarization is given by

$$\mu = (Zq) \times \delta \tag{7.50}$$

Substituting for δ from Equation 7.49,

$$\mu = 4\pi\varepsilon_0 R^3 E$$

Recalling Equation 7.25, we can write the dipole moment (μ) as

$$\mu = \alpha_e \times E \tag{7.51}$$

where α_e is the electronic polarizability of an atom or an ion. Note that the term electronic polarizability does *not* describe the polarizability of an electron. It describes the polarizability of an atom or an ion.

Comparing Equations 7.51 and 7.54, we get

$$\alpha_e = 4\pi\varepsilon_0 R^3 = 3\varepsilon_0 V_a \tag{7.52}$$

In Equation 7.52, V_a is the volume of the atom or ion being polarized. Because the unit of permittivity is F/m, and the unit of volume is m³, the unit of electronic polarizability (α_e) is F · m².

The electronic polarizability values for atoms of different elements are shown in Figure 7.10. These values are the volume polarizabilities. To convert them into SI units, they should be multiplied by $4\pi\varepsilon_0 \times 10^{-30}$ (Equation 7.27).

$$\alpha\left(\text{volume polarizability}\right) \text{in Å}^3 = \frac{10^{30}}{4\pi\varepsilon_0} \times \alpha\left(\text{in } C \cdot V^{-1} \cdot m^2 \text{ or } F \cdot m^2\right) \tag{7.53}$$

Example 7.4 will give us an idea of the magnitudes of the electronic polarizability of atoms and the electron-cloud displacement distances.

Example 7.4: Electronic Polarizability of the Ar Atom

The atomic radius of an Ar atom is 1.15 Å.
1. What is the electronic polarizability (α_e) of an Ar atom in units of F · m²?
2. What is the volume polarizability of an Ar atom in Å³?
3. What is the volume polarizability of an Ar atom in cm³?
4. If an Ar atom experiences a local electric field (E_{local}) of 10^6 V/m, what is the displacement (δ) when the atom experiences electronic polarization?
5. Calculate the ratio of the radius of atom (R) to the displacement (δ) caused by electronic polarization.

Solution
1. We calculate the electronic polarizability (α_e) of Ar atoms from Equation 7.52 as follows:

$$\alpha_e = 4\pi\varepsilon_0 R^3 = (4 \times \pi \times 8.85 \times 10^{-12} \text{ F/m})(1.15 \times 10^{-10} \text{ m})^3 = 1.69 \times 10^{-40} \text{ F} \cdot \text{m}^2$$

2. We use Equation 7.27 to calculate the polarizability volume in Å³

$$\alpha_{volume} \text{ in Å}^3 = \frac{10^{30}}{4\pi\left(8.85 \times 10^{-12} \text{ F/m}\right)} \times 1.69 \times 10^{-40} \text{ F} \cdot \text{m}^2 = 1.52 \text{ Å}^3$$

3. For calculating the volume polarizability, we use Equation 7.26

$$\alpha_{volume} \text{ in cm}^3 = \frac{10^6}{4\pi\varepsilon_0} \times \alpha \left(C \cdot V^{-1} \cdot m^2 \text{ or } F \cdot m^2\right)$$

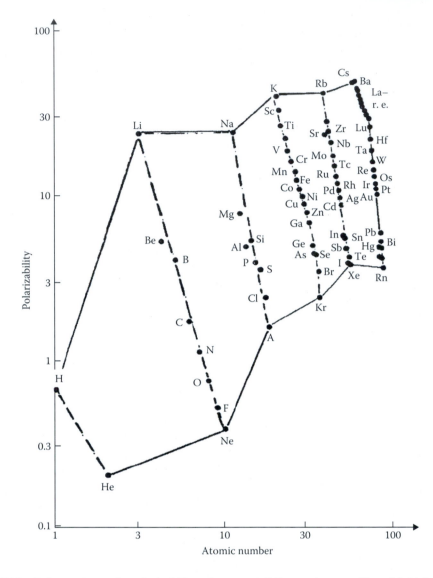

FIGURE 7.10 Volume electronic polarizability of atoms of different elements. Note: Multiply the *y*-axis value by $4\pi\varepsilon_0 \times 10^{-30}$ to get the polarizability in $F \cdot m^2$. (From Raju, G.G., *Dielectrics in Electric Fields*, CRC Press, Boca Raton, FL, 2003. With permission.)

$$\alpha_{volume} \text{ in cm}^3 = \frac{10^6}{4\pi\varepsilon_0} \times \left(1.69 \times 10^{-40} \text{ F} \cdot \text{m}^2\right) = 1.52 \times 10^{-24} \text{ cm}^3$$

This value is similar to the value of 1.642×10^{-24} cm³, reported in literature (Vidal et al. 1984). We use Equation 7.49 to calculate the displacement (δ):

$$\delta = \frac{4\pi\varepsilon_0 R^3 E}{Zq} = \frac{4\pi\varepsilon_0 \left(1.15 \times 10^{-10} \text{ m}\right)^3 \left(10^6 \text{ V/m}\right)}{(8)\left(1.6 \times 10^{-19} \text{ C}\right)}$$

$$= 1.32 \times 10^{-16} \text{ m} = 1.32 \times 10^{-6} \text{ Å}$$

4. The displacement (δ) is 1.32×10^{-6} Å. The radius $R = 1.15$ Å; and hence, the ratio of R/δ is ~870284. Thus, the displacement (δ) (Figure 7.9) is very small compared to the radius of the atoms.

We can use Bohr's model to calculate the electronic polarizability (α_e) of an atom. Using this approach, the electronic polarizability under a static electric field is given by the equation

$$\alpha_{e,static} = \frac{(Z \times q)^2}{m\omega_0^2} \tag{7.54}$$

where Z is the number of electrons orbiting the nucleus, q is the electronic charge, m is the electron mass, and ω_0 is the natural oscillation frequency of the center of the mass of the electron cloud around the nucleus. The *static electronic polarizability* represents the value of electronic polarizability when the electric field causing the polarization is not time-dependent.

The electronic polarizability (α_e) depends on the frequency of the electric field (ω) as follows:

$$\alpha_e = \frac{(Zq)^2}{m(\omega_0^2 - \omega^2) + j\beta\omega} \tag{7.55}$$

where j is the imaginary number and β is a constant that is related to the damping force attempting to pull the electron cloud back toward the nucleus. This equation is derived using a classical mechanics approach. A quantum mechanics–based approach yields a different equation, but the trend in the change in electronic polarizability as a function of frequency is similar. Example 7.5 illustrates a calculation of the value of static electronic polarizability.

Example 7.5: Static Electronic Polarizability of Hydrogen Atoms

If the natural frequency of oscillation (ω_0) of the electron mass around the nucleus in a hydrogen (H) atom is 4.5×10^{16} rad/s, what is the static electronic polarizability (α_e) of the H atom ($R = 1.2$ Å) calculated by Bohr's model and the classical approach? How do these values compare with the polarizability of the Ar atom calculated previously in Example 7.4?

Solution
For Bohr's model, using Equation 7.54,

$$\alpha_e = \frac{(1 \times 1.6 \times 10^{-19}\,\text{C})^2}{(9.11 \times 10^{-31}\,\text{kg})(4.5 \times 10^{16}\,\text{rad/s})^2} = 1.387 \times 10^{-41}\,\text{F} \cdot \text{m}^2$$

The value can also be calculated using the classical approach (Equation 7.52), as shown here:

$$\alpha_e = 4\pi\varepsilon_0 R^3 = (4 \times \pi \times 8.85 \times 10^{-12}\,\text{F/m})(1.20 \times 10^{-10}\,\text{m})^3 = 1.92\ 10^{-40}\,\text{F} \cdot \text{m}^2$$

The value using the classical approach is larger than that predicted from Bohr's model.

7.6.2 ELECTRONIC POLARIZABILITY OF IONS AND MOLECULES

Ions and molecules have electronic polarizability similar to neutral atoms. The electronic polarizability of an ion is nearly equal to that of an atom, with the same number of electrons. Thus, the electronic polarizability of sodium ions (Na^+) (0.2×10^{-40} F $\cdot$ m^2) is comparable to the electronic polarizability of neon (Ne) atoms. From Equation 7.52, we can expect the larger atoms or ions to have a larger electronic polarizability. Because cations typically are smaller than anions, the electronic polarizability of anions generally is larger than that of cations. Similarly, larger ions such as lead (Pb^{2+}) have a larger electronic polarizability. Many real-world technologies make use of these effects, such as in the development of lead crystal and optical fibers.

Electronic polarization is also linked to optical properties such as the index refractive (see Section 7.12). Molecules are composed of several atoms and show electronic polarizability. In general, the polarizability of molecules is larger because they contain more electrons.

7.7 IONIC, ATOMIC, OR VIBRATIONAL POLARIZATION

Many ceramic dielectrics exhibit mixed ionic and covalent bonding. In ceramics with ionic bonds, each ion undergoes *electronic* polarization. In addition to this, ionic solids exhibit *ionic polarization*, also known as *atomic polarization* or *vibrational polarization*. In this mechanism of polarization, the ions themselves are displaced in response to the electric field experienced by the solid, creating a net dipole moment per ion (p_{av}). Consider pairs of ions in an ionic solid such as sodium chloride (NaCl; Figure 7.11a). Assume that these ions have an equilibrium separation distance of a. This is the average separation distance between an anion and a cation. Ions in any material are not stationary. They vibrate around their mean equilibrium positions; these vibrations of ions or atoms are known as phonons.

When an electric field (E) is applied (Figure 7.11b), the electronic polarization of both cations and anions is established almost instantaneously (in ~10^{-14} seconds). This effect (the distortion of the electronic clouds for both anions and cations) is *not* shown in Figure 7.11. Since anions are bigger than cations because of the extra electrons present, anions typically show higher electronic polarizability than cations.

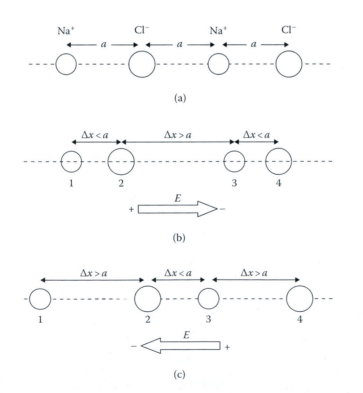

FIGURE 7.11 Illustration of ionic polarization: (a) no electric field; (b) electric field as shown; and (c) electric field direction reversed.

In addition to this electronic polarization, a positively charged cation moves toward the negative end of the electric field. Similarly, anions move closer to the positive end of the electric field. Figure 7.11b shows the displacements of ions; cations 1 and 3 move to the right, that is, toward the negative end of the electric field. Anions 2 and 4 are displaced toward the positive end of the electric field. This means that the separation distance (Δx) between cation 1 and anion 2 is now reduced, compared to their separation (a) without the electric field. The separation Δx between cation 3 and anion 2 is now larger than their equilibrium separation distance (a). The extent to which these ions are displaced also depends on the magnitude of the restoring forces imposed by other neighboring ions. For example, as cation 3 moves toward anion 4 (Figure 7.11b), anion 2 tries to pull cation 3 back, and the cation to the right of anion 4 (not shown) repels it.

Such asymmetric displacements of ions in response to the presence of an electric field create a dipole moment; this effect is known as ionic polarization. The magnitude of ionic displacements encountered in ionic polarization is a fraction of an angstrom (Å). Because ions have a larger inertia than electrons, these movements are a bit sluggish, occurring in about 10^{-13} seconds. When the polarity of the electric field is reversed (Figure 7.11c), the directions of displacement are also reversed. If we have an alternating current (AC) electric field, then the ions move back and forth as long as the electric field does not switch too rapidly, as shown in Figures 7.11b and c. If the frequency (f) of the switching field is greater than ~10^{13} Hz (that is, if the field switches in less than 10^{-13} seconds), the ions cannot follow the changes in the electric field direction. In other words, the ionic polarization mechanism is seen in ionic solids for frequencies up to ~10^{13} Hz (Table 7.2). If the frequency is greater than ~10^{13} Hz, this polarization mechanism *drops out* and does *not* contribute to the total dielectric polarization (P) induced in the dielectric.

Under static electric fields, the magnitude of *ionic polarizability* (α_i) is given by

$$\alpha_i = \frac{(Z \times q)^2}{M_r \times \omega_0^2} \left(\text{for static electric fields} \right) \tag{7.56}$$

where Z is the valence of the ion (not the number of electrons surrounding the nucleus, which Z represents in electronic polarization equations), q is the electronic charge, M_r is the reduced mass of the ion, and ω_0 is the natural frequency of oscillation for a given ion.

For AC fields with a frequency of $\omega = 2\pi f$, where f is the electrical frequency in Hz, the magnitude of the ionic polarizability (α_i) is frequency-dependent and is given by

$$\alpha_i = \frac{(Z \times q)^2}{\left[M_r \times \left(\omega_0^2 - \omega^2 + j\beta\omega \right) \right]} \tag{7.57}$$

where ω is the frequency of the electric field (in rad/s), j is the imaginary number, and β is a coefficient related to the damping force that tries to bring the displaced ion back to its original position.

The details of the treatments for deriving these equations are beyond the scope of this book. However, it is important to recognize that the dielectric polarization mechanism is effective only up to a certain frequency (~10^{13} Hz).

In general, the magnitude of ionic polarizability (α_i) is about ten or more times *larger* than the electronic polarizability (α_e). This is why most solids with considerable ionic bonding character exhibit much higher dielectric constants (Table 7.1). Also, note that in addition to ionic polarization, each ion undergoes *electronic* polarization. Thus, the dielectric constant (k) of ionic solids results from both electronic and ionic polarizations. However, the contributions from ionic polarization tend to be dominant, especially at lower frequencies.

If μ_{av} is the average dipole moment induced by the ionic polarization per ion, then we can write this dipole moment as follows:

$$\mu_i = \alpha_i \times E_{local}$$

where E_{local} is the electric field experienced by the ion. As mentioned in Section 7.4, this is the local electric field (E_{local}), and it is different from the applied electric field (E). The polarization (P) induced in an ionic solid with N_i ions per unit volume is given by the following equation:

$$P = N_i \times \mu_{av} = N_i \times \alpha_i \times E_{local} \tag{7.58}$$

Recall from the Clausius–Mossotti equation (Equation 7.43) that the dielectric constant (k or ε_r) is linked to the different polarizabilities of ions. For an ionic solid with no permanent dipoles, we have contributions from both ionic and electronic polarizations. Note that this equation applies only to amorphous or cubic structures and to nonpolar materials.

In Section 7.8, we will discuss an approach that is useful in predicting the dielectric constants of nonpolar materials from the values of the total (i.e., electronic and ionic) polarizability of ions.

7.8 SHANNON'S POLARIZABILITY APPROACH FOR PREDICTING DIELECTRIC CONSTANTS

7.8.1 OUTLINE OF THE APPROACH

Shannon measured the dielectric constants of several materials and back-calculated their ionic polarizabilities (Shannon 1993). The frequency range for the dielectric-constant measurements was 1 kHz–10 MHz. Thus, both the electronic and ionic polarization mechanisms contributed to the dielectric constant measured.

Shannon used experimentally determined values of dielectric constants to first estimate the polarizability of ions or simple compounds. Then, he used these values to estimate the *total dielectric polarizability* $\left(\alpha_D^T\right)$ of other compounds through the additive nature of ionic and electronic polarizabilities. This calculation requires knowledge of the molecular weight and unit cell volume (i.e., the theoretical densities) of the compound whose dielectric constant is to be estimated. For example, one can measure the dielectric constants of fully dense samples of magnesium oxide (MgO) and Al_2O_3 and estimate the polarizabilities of Al^{3+}, Mg^{2+}, and O^{2-} ions. We can use these polarizability values to calculate the total dielectric polarizability $\left(\alpha_D^T\right)$ and, hence, the dielectric constant of another compound such as magnesium aluminate ($MgAl_2O_4$).

The total dielectric polarizability of $MgAl_2O_4$ can be expressed as follows:

$$\alpha_{MgAl_2O_4} = \alpha_{Mg^{(2+)}} + 2\alpha_{Al^{(3+)}} + 4\alpha_{O^{(2-)}} \tag{7.59}$$

or

$$\alpha_{D,MgAl_2O_4}^T = \alpha_{D,Mgo}^T + \alpha_{D,Al_2O_3}^T \tag{7.60}$$

Following this, we can utilize the value of total polarizability for $MgAl_2O_4$ to estimate its dielectric constant by using the following form of the Clausius–Mossotti equation:

$$\varepsilon_{r,cal} = \frac{3V_m + 8\pi\alpha_D^T}{3V_m - 4\pi\alpha_D^T} \tag{7.61}$$

where V_m is the molar volume of the compound whose dielectric constant, or α_D^T, is being estimated.

Shannon also used the following modified forms of the Clausius–Mossotti equation:

$$\alpha_D^T = \frac{3V_m}{4\pi}\left(\frac{\varepsilon_r - 1}{\varepsilon_r + 2}\right) \tag{7.62}$$

In Equations 7.60 through 7.62, α_D^T is the total dielectric polarizability of a material and is commonly expressed as Å^3.

7.8.2 Limitations of Shannon's Approach

Shannon's approach for predicting the dielectric constant is useful, but it has some limitations. For many compounds, the calculated values of dielectric constants using Shannon's approach are very different from the measured values. The polarizability of the oxygen ion (O^{2-}), as estimated by Shannon (2.01 Å^3; Figure 7.12), is lower than that used by other researchers (2.37 Å^3). This leads to the prediction of lower dielectric constants for some oxides, for example, Al_2O_3. In many other materials exhibiting ferroelectric and piezoelectric behavior or for materials containing compressed or rattling ions, mobile ions, and impurities, the calculated and measured values of the dielectric constants do not match well. This can be due to the ionic or electronic conductivity of the material, the presence of interfacial polarization, the presence of polar molecules such as water (H_2O) or carbon dioxide (CO_2), and the presence of other dipolar impurities. Thus, Shannon's approach cannot be used to calculate the dielectric constants of ferroelectric materials (Section 7.11).

Despite these limitations, Shannon's approach serves as a powerful guide for the experimental development of new formulations of materials with high dielectric constants. The use of Shannon's approach is illustrated in Examples 7.6 and 7.7.

Example 7.6: Dielectric Constant of Li₂SIO₃ from Ion Polarizabilities

Use the ion polarizabilities in Figure 7.12 to estimate the dielectric constant of lithium silicate (Li_2SiO_3). The molar volume (V_m) of Li_2SiO_3 is 59.01 Å^3. The experimental value of the dielectric constant ($\varepsilon_{r,exp}$) for Li_2SiO_3 between 1 kHz and 10 MHz is 6.7. How does the calculated value of the dielectric constant compare with the value estimated using Shannon's approach?

Li 1.20	Be 0.19											B 0.05	C	N	O 2.01	F 1.62	Ne
Na 1.80	Mg 1.32											Al 0.79	Si 0.87	P V 1.22	S	Cl	
K 3.83	Ca 3.16	Sc 2.81	Ti IV 2.93	V V 2.92	Cr III 1.45	Mn II 2.64	Fe II 2.23 III 2.29	Co II 1.65	Ni II 1.23	Cu II 2.11	Zn II 2.04	Ga 1.50	Ge 1.63	As V 1.72	Se	Br	
Rb 5.29	Sr 4.24	Y 3.81	Zr 3.25	Nb 3.97	Mo	Tc				Ag	Cd 3.40	In 2.62	Sn 2.83	Sb III 4.27	Te IV 5.23	I	
Cs 7.43	Ba 6.40	La 6.07	Hf	Ta 4.73	W	Re				Au	Hg	Tl I 7.28	Pb II 6.58	Bi 6.12			

Ce III6.15 IV3.94	Pr 5.32	Nd 5.01	Pm	Sm 4.74	Eu II 4.83 III 4.53	Gd 4.37	Tb 4.25	Dy 4.07	Ho 3.97	Er 3.81	Tm 3.82	Yb 3.58	Lu 3.64
Tn 4.92	Pa	U IV 4.45											

FIGURE 7.12 Polarizabilities of ions expressed in Å^3. Note: The frequency range of 1 kHz to 10 MHz means that the polarizability values include both electronic and ionic components. (From Shannon, R.D., *J. Appl. Phys.*, 73, 348–366, 1993. With permission.)

Solution

From Figure 7.12, the ionic polarizability of O^{2-} ions is 2.01 $\mathring{A}^3$. The ionic polarizabilities of lithium (Li^+) and silicon (Si^{4+}) ions are 1.20 $\mathring{A}^3$ and 0.87 $\mathring{A}^3$, respectively. Thus, the dielectric polarizability $\left(\alpha_D^T\right)$ of Li_2SiO_3 is

$$\alpha_D^T \text{ Lithium silicate} = 2\alpha_{Li^{(+1)}} + \alpha_{Si^{(+4)}} + 3\alpha_{O^{(-2)}}$$

$$= 2(1.2) + 0.87 + 3(2.01) = 9.3\mathring{A}^3$$

The molar volume of Li_2SiO_3 is V_m = 59.01 $\mathring{A}^3$. Therefore, using Equation 7.61, the calculated dielectric constant $(\varepsilon_{r,cal})$ of Li_2SiO_3 is

$$\varepsilon_{r,cal} = \frac{3V_m + 8\pi\alpha_D^T}{3V_m - 4\pi\alpha_D^T}$$

$$\varepsilon_{r,cal} = \frac{(3 \times 59.01) + 8\pi(9.3)}{(3 \times 59.01) - 4\pi(9.3)} = 6.82$$

This compares well with the value $\varepsilon_{r,exp}$ = 6.70. Note that, for many compounds, the calculated and measured values do *not* match very well, for the reasons mentioned in Section 7.8.2. Furthermore, if the temperature is high and the frequency is low, interfacial polarization effects (see Section 7.10) may also develop in Li_2SiO_3. Under such conditions, the experimentally observed and calculated dielectric constants do not match well.

Example 7.7: Dielectric Constant of $MgAl_2O_4$ Using Shannon's Approach

The dielectric constant of MgO is 9.83, and its molar volume (V_m) is 18.69 $\mathring{A}^3$. The dielectric constant of Al_2O_3 is estimated to be 10.126, and its molar volume (V_m) is 42.45 $\mathring{A}^3$. What is the dielectric constant of $MgAl_2O_4$? The molar volume (V_m) of $MgAl_2O_4$ is 66.00 $\mathring{A}^3$.

Solution

We first calculate the total dielectric polarizability of MgO from its dielectric constant and volume using the modified form of the Clausius–Mossotti equation (Equation 7.62) that Shannon used.

$$\alpha_{D,MgO}^T = \frac{3(18.69)}{4\pi}\left(\frac{9.830 - 1}{9.830 + 2}\right) = 3.330$$

Similarly, we calculate the polarizability of Al_2O_3 from its dielectric constant and volume using Equation 7.62:

$$\alpha_{D,Al_2O_3}^T = \frac{3(42.45)}{4\pi}\left(\frac{10.126 - 1}{10.126 + 2}\right) = 7.663$$

From these values of polarizabilities, we estimate the polarizability of $MgAl_2O_4$ using the additivity rule developed by Shannon.

$$\alpha_{D,Al_2O_3}^T = 7.663 + 3.330 = 10.993$$

Now, we have estimated the total dielectric polarizability of $MgAl_2O_4$, and we know its molar volume $(V_m$ = 66 $\mathring{A}^3)$; from these, we calculate the dielectric constant using Equation 7.61.

$$\varepsilon_{r,cal} = \frac{(3 \times 66) + 8\pi(10.993)}{(3 \times 66) - 4\pi(10.993)}$$

$$\varepsilon_r = 7.923$$

Shannon carefully measured the dielectric constant of this material and reported the value to be 8.176. Thus, the value we estimated does not exactly match the experimentally determined value; however, it is relatively close (within a few percent).

7.9 DIPOLAR OR ORIENTATIONAL POLARIZATION

Molecules known as polar molecules have a permanent dipole moment. An example of a well-known polar material is H_2O (water). The polar molecules begin to experience torque when exposed to an external electric field and orient themselves along the electric field. This, in turn, causes an increase in the bound charge density (σ_b) and leads to an increase in polarization (P). The resulting polarization is known as *dipolar polarization* or *orientational polarization*. Orientational or dipolar polarization is the mechanism responsible for the relatively high dielectric constant of H_2O ($k\sim78$; Figure 7.13).

The main feature that distinguishes the orientational polarization mechanism from other mechanisms is the presence of permanent dipoles. This mechanism of dipolar polarization is seen only in materials that have molecules with a permanent dipole moment. Materials in which molecules develop a net polarization or have a permanent or built-in dipole moment are known as polar materials or *polar dielectrics*. Polar materials in which polarization appears spontaneously, even without an electric field, are known as *ferroelectrics* (Chapter 8). They do not contain molecules with a permanent dipole moment.

If each permanent dipole had a dipole moment of μ, and if the concentration of such dipoles was N, then the maximum polarization that can be caused by this mechanism alone is $N \times \mu$. Not all dipoles can remain aligned with the applied electric field because thermal energy tries to randomize their orientations. Thus, this polarization mechanism begins to fade away with increasing temperature. At substantially high temperatures, the orientations of dipoles with respect to the applied electric field become completely randomized. The net polarization begins to decrease, and this polarization mechanism stops. The *dipolar polarizability* (α_d) associated with this mechanism is given by

$$\alpha_d = \frac{1}{3} \frac{\mu^2}{k_B T} \tag{7.63}$$

where μ is the permanent dipole moment of the molecules, k_B is the Boltzmann's constant, and T is the temperature.

Dipole moments associated with polar molecules typically are very large compared to those induced by the polarization of atoms or the displacements of ions. Dipolar or orientational polarizability (α_d) values and the resultant dielectric constants are therefore large for polar materials.

Dipolar polarization typically is seen in polar liquids, gases, or vapors (e.g., H_2O, alcohol, and hydrochloric acid [HCl]). Unlike in their vapor form, molecules with permanent dipoles are not free to rotate

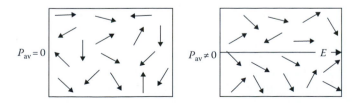

FIGURE 7.13 Illustration of dipolar or orientational polarization. (From Kasap, S.O., *Principles of Electronic Materials and Devices*, McGraw Hill, New York, 2006. With permission.)

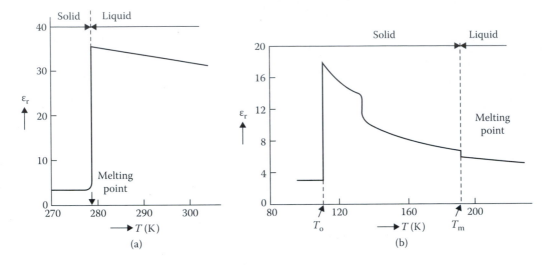

FIGURE 7.14 Dielectric constant of (a) nitrobenzene and (b) hydrogen sulfide. (From Kao, K.C., *Dielectric Phenomena in Solids*, London, Elsevier-Academic Press, 2004. With permission.)

in polar solids (e.g., ice, instead of water vapor), even if they are present. This means that the effect of orientational polarization in polar solids is smaller compared to that in liquids and gases or vapors.

For example, nitrobenzene ($C_6H_5CO_3$) is a polar liquid with $k \sim 35$ (Figure 7.14). The dielectric constant of liquid $C_6H_5CO_3$ is expected to decrease with increasing temperatures (Equation 7.63). When liquid $C_6H_5CO_3$ freezes into a solid, the dipoles are present but are frozen and unable to rotate. This is why the dielectric constant of $C_6H_5CO_3$ decreases to about 3 (Figure 7.14a). This lower value for the dielectric constant reflects the smaller extent of electronic and ionic polarization compared to the extent of orientational polarization and ignoring the effects of the differences in the molar volumes of the solid and liquid phases.

In some materials (such as hydrogen sulfide [H_2S]), the dielectric constant continues to increase with decreasing temperatures, even below the freezing or melting temperature (T_m). This continues up to the critical temperature T_0, below which the dipoles cannot rotate and the orientational polarization mechanism ceases (Figure 7.14b).

7.10 INTERFACIAL, SPACE CHARGE, OR MAXWELL–WAGNER POLARIZATION

Some dielectrics contain relatively mobile ions (e.g., H^+, Li^+, and K^+). At high temperatures, these ions can drift under the influence of an electric field. The movement of such charge carriers is eventually impeded by the existence of interfaces (such as grain boundaries) in a material or a device. This can create a buildup of double-layer-like capacitors at interfaces, such as grain boundaries, in a polycrystalline material or material–electrode interfaces. The increase in polarization due to such movements of mobile ions in a material and the creation of polarization at the interfaces is known as *space-charge polarization*, *interfacial polarization*, or Maxwell–Wagner (M–W) or Maxwell–Wagner–Sillars (M–W–S) polarization. Trapping electrically charged ions, electrons, and holes at the interfaces is the essential process behind interfacial polarization (Figure 7.15).

This polarization mechanism differs from the other mechanisms we have discussed (Figure 7.8). First, the polarization mechanism is usually more prominent at higher temperatures. This is because the rate of the diffusion process, which often occurs through atoms or ions jumping or hopping from one location to another, increases exponentially with increasing temperature (Chapter 2). Second, the diffusion of atoms or ions is relatively slow compared to that of electrons and holes. Thus, this polarization mechanism is usually operated under static (DC) or AC fields, in which the electrical frequency (f) is small (a few mHz to several Hz). If the electrical frequency (f) is too

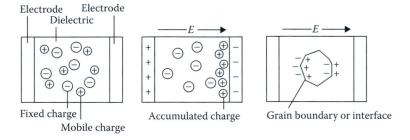

FIGURE 7.15 Illustration of interfacial polarization. (From Kasap, S.O., *Principles of Electronic Materials and Devices*, McGraw Hill, New York, 2006. With permission.)

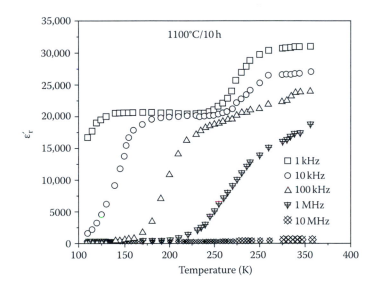

FIGURE 7.16 Apparent dielectric constant of $CaCu_3Ti_4O_{12}$ ceramics processed at 1100°C for 10 hours. The dielectric constant is measured at different temperatures up to 350 K. (From Prakash, B. and Varma K.B.R., *J. Phys. Chem. Solids.*, 68, 490–502, 2007. With permission.)

high, the otherwise mobile ions cannot rapidly follow this frequency. The interfacial polarization mechanism then ceases to exist. Many silicate glasses and crystalline ceramic materials containing mobile ions (e.g., lithium niobate [$LiNbO_3$], lithium tantalate [$LiTaO_3$], and lithium cobalt oxide [$LiCoO_2$]) exhibit this polarization mechanism. This mechanism is also quite common at the liquid electrolyte–electrode interfaces because diffusion in liquids occurs rather readily. It is commonly used in many electrochemical reactions encountered during the operation of *supercapacitors*, batteries, fuel cells, and similar devices.

The existence of interfacial polarization is often considered a strong possibility whenever unusually high dielectric constants are seen, especially at high temperatures and low frequencies. For example, the data for the dielectric constant for a calcium–copper–titanium oxide (CCTO) ceramic are shown in Figure 7.16.

7.11   SPONTANEOUS OR FERROELECTRIC POLARIZATION

A ferroelectric material is defined as a material that exhibits spontaneous and reversible polarization (Chapter 8). This polarization typically is very large in magnitude compared to other polarization mechanisms, such as electronic and ionic polarizations.

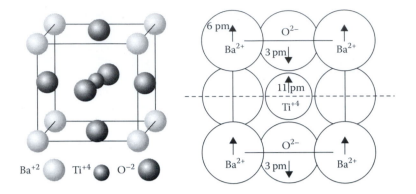

FIGURE 7.17 Small-scale displacements in a cubic structure lead to the tetragonal form of ferroelectric $BaTiO_3$. (Adapted from Moulson, A.J. and Herbert, J.M., *Electroceramics: Materials, Properties, Applications*, Wiley, New York, 2003; Buchanan, R.C., *Ceramic Materials for Electronics*, Marcel Dekker, New York, 2004. With permission.)

A prototypical example of a ferroelectric material is the *tetragonal* polymorph of $BaTiO_3$. The term *polymorph* means the particular crystal structure of a material. Consider the two polymorphs of $BaTiO_3$—one is cubic, and the other is tetragonal. The tetragonal form of $BaTiO_3$ is also known as the *pseudocubic form*. The cation/anion ratio for the tetragonal polymorph of $BaTiO_3$ at room temperature (300 K) is ~1.01. For the cubic polymorph, which is stable at higher temperatures, the cation/anion ratio is 1.00. The lattice constant is ~4 Å. Thus, the physical difference in the dimensions of the unit cells actually is very small. However, the differences in the electrical properties between the cubic and tetragonal forms of $BaTiO_3$ are significant.

For the typical single-crystal or polycrystalline $BaTiO_3$, the centrosymmetric cubic phase is stable at temperatures above ~130°C. The temperature at which a ferroelectric material transforms into a centrosymmetric paraelectric form is known as the *Curie temperature*. In cubic $BaTiO_3$, the titanium (Ti) ion *rattles* very rapidly around several equivalent but off-center positions present around the cube center. For each of these off-center positions of the titanium ion, the unit cell structure has a dipole moment. Thus, at any given time, the time-averaged position of the titanium ion *appears* to be exactly at the cube center. As a result, the cubic phase of $BaTiO_3$ has no net dipole moment from the viewpoint of electrical properties. From a structural viewpoint (e.g., while using X-ray diffraction), the crystal structure appears cubic and is centrosymmetric. In cubic $BaTiO_3$, all the dipole moments that are associated with the barium (Ba^{2+}), titanium (Ti^{4+}), and oxygen (O^{2-}) ions cancel one another out. This high-temperature phase, derived from an originally ferroelectric parent phase that now has no dipole moment per unit cell, is known as the *paraelectric phase*.

When the temperature approaches the Curie temperature (T_c ~ 130°C for $BaTiO_3$), the titanium ions begin to undergo other very small displacements. At temperatures below T_c, barium ions are displaced by a distance of ~6 pm (1 pm = 10^{-12} m). Titanium ions are displaced in the same direction by ~11 pm. Oxygen ions (O^{2-}) are displaced by ~3 pm. After these displacements occur, the unit cell becomes tetragonal. The tetragonal structure is *not* centrosymmetric. It is a polar structure; that is, the tetragonal unit cell of $BaTiO_3$ has a net polarization (Figure 7.17).

7.12   DEPENDENCE OF THE DIELECTRIC CONSTANT ON FREQUENCY

The polarization mechanisms require displacements of ions, electronic clouds, dipoles, and so on (Table 7.2). These displacements are small; nevertheless, they require a small but finite amount of time. Consider a covalently bonded material such as silicon, where the electronic polarization is established quickly (~10^{-14} seconds). Now, consider changing the polarity of the applied electric field. The electronic clouds shift, and the induced dipoles realign with the new field direction within another $\sim 10^{-14}$ seconds. The induced dipoles can align rapidly back and forth in an alternating electric field even if the electrical field switches at a frequency of, for example, 1 MHz. A frequency (f) of 1 MHz means that the field switches back and forth 10^6 times per second, or in one microsecond (μs; 1 μs = 10^{-6} seconds). Under a static (i.e., $f = 0$) field or an electric field oscillating with a frequency of 10^6 Hz, we expect the electronic polarization process to contribute to the dielectric constant (k) of silicon. This continues to very high frequencies, ranging up to 10^{14} Hz because electronic polarization is the only mechanism of polarization that survives up to such high frequencies. Thus, the dielectric constants of silicon and other covalently bonded solids (such as germanium [Ge] and diamond) are expected to remain constant with frequencies up to the range of $\sim 10^{14}$ Hz.

Now, consider a material in which the bonding has some ionic character, such as SiO_2. In this material, we expect both the ionic and electronic polarization mechanisms to contribute to its dielectric constant (k). This will be true as long as the ionic and electronic polarization mechanisms can follow the alternating electric fields. The ionic polarization mechanism is slower compared to that of electronic polarization because it involves the displacement of ions. Thus, up to a frequency of $\sim 10^{12}$–10^{13} Hz, both electronic and ionic polarization mechanisms contribute to the dielectric constant of SiO_2. At higher frequencies, the ions cannot follow the back-and-forth switching of the polarity of the electric field. As a result, the ionic polarization mechanism stops, lowering the dielectric constant. The electronic polarization mechanism survives up to $\sim 10^{14}$ Hz and continues to contribute to the dielectric constant. Thus, we expect the dielectric constant of SiO_2 to become smaller (or for it to relax) as we proceed from low to high frequencies. We can expect this trend for any material with more than one polarization mechanism (Figure 7.18).

The lowering of the dielectric constant with increasing frequency is known as *dielectric relaxation* or *frequency dispersion*. This *low-frequency dielectric constant* is also known as the *static dielectric constant* (k_s), although the value is not necessarily measured under DC fields. The *high-frequency dielectric constant* is designated as k_∞. This is also known as the *optical frequency dielectric constant*, and it is the value of the dielectric constant when only the electronic (optical) polarization mechanism remains.

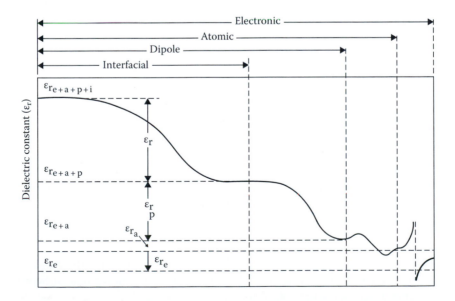

FIGURE 7.18 Variation of dielectric constant with frequency for a hypothetical dielectric with different polarization mechanisms. (From Buchanan, R.C., *Ceramic Materials for Electronics*, New York, 2004. With permission.)

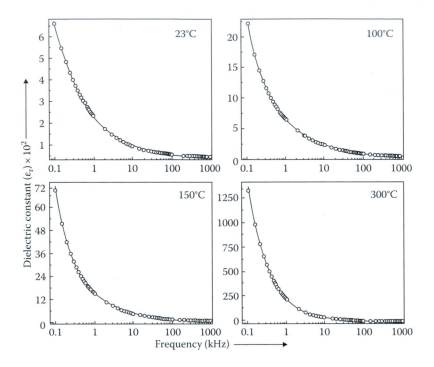

FIGURE 7.19 Relaxation in the dielectric constant of $LiFe_{1/2}Ni_{1/2}VO_4$ ceramics at different temperatures. (From Ram, M., and S. Chakravarty. *J Phys Chem Solids* 69(4):905–912, 2008. With permission.)

In practice, dielectric materials with multiple polarization mechanisms show a decrease in the dielectric constant (k). However, the decrease is rather steady and not as abrupt as that shown in Figure 7.18 because with multiple types of ions or atoms, the polarization mechanisms do not stop at a particular frequency but rather over a range of frequencies, even for the same type of polarization.

As an example, Figure 7.19 shows the lowering of the dielectric constant for lithium–iron–nickel–vanadium oxide ceramics. The data are shown for temperatures ranging from 23°C to 300°C.

At any given temperature, the dielectric constant decreases with increasing frequency. The lower-frequency dielectric constants increase with increasing temperature. This is very much an indication of interfacial polarization, or Maxwell–Wagner polarization. Its presence is not surprising because this material contains relatively mobile lithium ions.

7.12.1 CONNECTION TO THE OPTICAL PROPERTIES: LORENTZ–LORENZ EQUATION

The only polarization mechanism that does not cease to exist at high frequencies is electronic polarization. The high frequencies at which only the electronic polarization mechanism survives correspond to a wavelength (λ) of light (Figure 7.20).

Therefore, electronic polarization is also known as optical polarization (see Sections 7.6 and 8.2). We can show that the *high-frequency dielectric constant* (k_∞) of a material is equal to the square of its refractive index (n):

$$k_\infty = n^2 \tag{7.64}$$

Recall the Clausius–Mossotti equation (Equation 7.43) that correlates the dielectric constant with polarization. Applying this equation for high-frequency conditions, that is, by replacing the term $\varepsilon_r = \varepsilon_\infty$ and removing the ionic polarization term, we get

$$\frac{\varepsilon_\infty - 1}{\varepsilon_\infty + 2} = \frac{1}{3\varepsilon_0}\left[(N_e \times \alpha_e)\right] \tag{7.65}$$

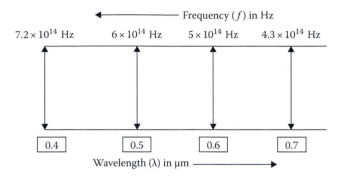

FIGURE 7.20 Relationship between frequency and wavelengths of light.

FIGURE 7.21 Photograph of a lead crystal object.

Combining Equations 7.64 and 7.65, we get the so-called Lorentz–Lorenz equation:

$$\frac{n^2-1}{n^2+2} = \frac{1}{3\varepsilon_0}\left[(N_e \times \alpha_e)\right] \tag{7.66}$$

This can be rewritten by replacing the concentration of atoms (N_e) as follows:

$$\left(\frac{n^2-1}{n^2+2}\right)\frac{M}{\rho} = \frac{N_{Avogadro} \times \alpha_e}{3\varepsilon_0} \tag{7.67}$$

A note of caution is in order regarding the use of the Lorentz–Lorenz equation. Recall that the Clausius–Mossotti equation was derived using the local internal electric field (E_{local}) calculation (Equation 7.36). Because the Lorentz–Lorenz equation is derived using the Clausius–Mossotti equation, it applies to *nonpolar* materials with a cubic symmetry or to amorphous materials. It *cannot* be applied to ferroelectric or other polar dielectric materials.

Materials that contain ions or atoms with a large electronic polarizability (α_e) have a higher refractive index (n). A common example of such a material is lead crystal (Figure 7.21). This actually is an *amorphous* silicate glass that contains substantial (up to 30–40 wt% PbO) concentrations of lead ions (Pb^{2+}). Because lead ions have a large electronic polarizability, the refractive index of

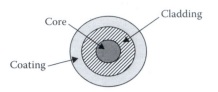

FIGURE 7.22 Core-clad structure of optical fibers.

lead crystal is much higher (n up to 1.7) than that of a common soda-lime glass ($n \sim 1.5$). Note that the polarizability values shown in Figure 7.10 include both electronic and ionic polarizabilities.

Another important example of the use of higher electronic polarizability (α_e) to achieve a higher refractive index (n) is its application in optical fibers (Figure 7.22). In optical fibers, a small yet significant mismatch (~1%) is created between the refractive indices of the core and the cladding. By doping the core of the fibers with dopants such as germanium, the refractive index of the core region is maintained higher than that of the cladding. Optical fibers usually are made from ultra-high-purity SiO_2. The core is doped with germanium oxide (GeO_2), which enhances the refractive index of the core region. This increases the total internal reflection at the core–cladding interface, thereby restricting the light waves (i.e., information) to within the optical-fiber core. Doping the fibers with fluorine (F) causes the refractive index of SiO_2 to decrease.

The following example illustrates the extent of contribution of the electronic and ionic polarization mechanisms to the dielectric constant of ionic materials.

Example 7.8: Dielectric Constant of Silicate Glass

The refractive index (n) of a silicate glass is 1.5. What is the high-frequency dielectric constant of this glass? The dielectric constant of this glass at 1 kHz is $k = 7.6$. Based on the relaxation of the dielectric constant, approximately what fraction of the low-frequency dielectric constant (k_s) can be attributed to ionic polarization?

Solution

The refractive index (n) is 1.5. This means that the high-frequency dielectric constant (k_∞) is $n^2 = 2.25$. The low-frequency dielectric constant is $k_s = 7.6$. Thus, the contribution of ionic polarization to the dielectric constant is $7.6 - 2.25 = 5.35$. The fraction of contribution due to ionic polarization is $= ([7.6 - 5.35]/7.6) = (5.35/7.6) \sim 0.7$. Thus, nearly 70% of the low-frequency dielectric constant (k_s) for this glass is due to ionic polarization, with the rest due to electronic polarization.

7.13 COMPLEX DIELECTRIC CONSTANT AND DIELECTRIC LOSSES

7.13.1 COMPLEX DIELECTRIC CONSTANT

Another feature associated with polarization mechanisms is the notion of dielectric loss. When ions, electron clouds, dipoles, and so on are displaced in response to the electric field (Figure 7.8), these displacements do not occur without resistance. This resistance is similar to the effect of friction on mechanical movement. The electrical energy lost during the displacements of ions, the electronic cloud, or any other entity that causes dielectric polarization, is known as the dielectric loss. One way to represent dielectric losses is to consider the dielectric constant as a complex number. Thus, we define the *complex dielectric constant* $\left(\varepsilon_r^* \right)$ as

$$\varepsilon_r^* = \varepsilon_r' - j\varepsilon_r'' \tag{7.68}$$

In Equation 7.68, j is the imaginary number $\sqrt{-1}$.

Known as the *real part* of the dielectric constant, ε'_r represents the charge-storage process, which is the same quantity that we have referred to so far as ε_r (or k). The *imaginary part* of the complex dielectric constant (ε''_r) is a measure of the dielectric losses that occur during the charge-storage process.

7.13.2 Real Dielectrics and Ideal Dielectrics

An *ideal dielectric* is a hypothetical material with zero dielectric losses (i.e., $\varepsilon'_r = 0$). This means that all the applied electrical energy is used to cause the polarization that leads to charge storage only. A *real dielectric* is a material that does have some dielectric losses. All dielectric materials have some level of dielectric loss because the displacements of ions, electron clouds, and so on, cannot occur without resistance from neighboring atoms or ions. The dielectric losses increase if the applied field switches in such a way that the polarization mechanisms can follow these changes in the applied electric field. It usually is desirable to minimize or lower the dielectric losses for microelectronic devices. However, dielectric losses can be useful for applications in which heat must be generated. A common example is that of a microwave oven. The water molecules in food, which are permanent dipoles, tumble around during the polarization caused by the microwave's electric field. The resultant dielectric losses cause the generation of heat (Vollmer 2004).

While developing materials for capacitors to store charge (Section 7.2.2), we prefer to use a *low-loss dielectric;* in other words, we want a small ε''_r. The need for increasing the dielectric constant (k, or what we now refer to as ε'_r) while maintaining the dielectric losses at small levels poses a problem because polarization processes are required to achieve higher dielectric constants. When these polarization processes occur, they cause dielectric losses. Polarization and dielectric losses originate from the same basic processes (some type of displacement of ions and electron clouds, in addition to the reorientation or rotation of dipoles, etc.; see Figure 7.8).

7.13.3 Frequency Dependence of Dielectric Losses

Because the dielectric constant depends on frequency (f or ω), it is reasonable to assume that dielectric losses are also frequency-dependent.

Consider a hypothetical dielectric material with only the dipolar polarization mechanism. If the electrical frequency (f or ω) is too high, then the dipole cannot rotate or flip back and forth in response to the oscillating electric field. We may also change the magnitude of the electric field from some initial value E_0 to another value E without changing its direction. If this happens, the induced dipole moment (ionic or electronic polarization) adjusts from some starting value of μ_0 (at E_0) to another final value μ (at E). Such changes in the dipole moment require a small but finite amount of time. The *relaxation time* (τ) is the time required for a polarization mechanism to revert and realign a dipole or change its value when the electrical field switches or changes in magnitude.

In a hypothetical dielectric material with a single polarization mechanism and a relaxation time τ, the dielectric losses are very small if the frequency of the electrical field switches too rapidly. When $\omega \gg 1/\tau$, the polarization mechanism is unable to follow the change in E. On the contrary, if $\omega \ll 1/\tau$, then the polarization process can follow the changes in the electric field. However, the induced dipole moment does not switch or readjust that often, so the dielectric losses are again small. When $\omega = 1/\tau$, similar to a resonance condition, the dielectric losses are maximized because the polarization switches or adjusts itself in a synchronized fashion as the electric field switches again. Figure 7.23 shows this variation of dielectric losses with frequency for a hypothetical dielectric with a single polarization mechanism with relaxation time τ.

In any dielectric material, there are different relaxation times for different mechanisms of polarization. For example, the displacement of the electronic cloud around a nucleus occurs very rapidly, and the relaxation time τ is $\sim 10^{-14}$ seconds. On the contrary, for interfacial polarization, the relaxation time (τ) can be several seconds because this mechanism involves a relatively longer-range

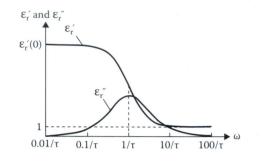

FIGURE 7.23 Dependence of ε_r' and ε_r'' on frequency for a material with a single polarization mechanism and relaxation time τ. (From Buchanan, R.C., *Ceramic Materials for Electronics*, Marcel Dekker, New York, 2004. With permission.)

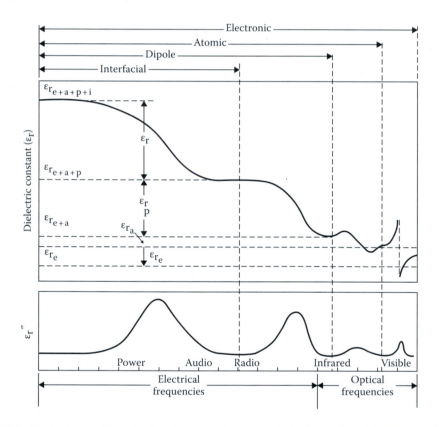

FIGURE 7.24 Dependence of ε_r' and ε_r'' on frequency for a hypothetical material with different polarization mechanisms and relaxation times. (From Buchanan, R.C., *Ceramic Materials for Electronics*, Marcel Dekker, New York, 2004. With permission.)

motion of ions. Thus, for materials with multiple polarization mechanisms, the real part of the dielectric constant (ε_r') shows changes similar to a set of cascades. The dielectric loss component (ε_r'') shows a series of maxima that follow the different polarization mechanisms (Figure 7.24).

For a dielectric material with a single relaxation time (τ), the frequency dependence of ε_r' and ε_r'' can be written quantitatively as follows:

$$\varepsilon_r' = k' = k_\infty + \frac{k_s - k_\infty}{1 + \omega^2 \tau^2} \tag{7.69}$$

$$\varepsilon_r'' = k'' = (k_s - k_\infty)\left(\frac{\omega\tau}{1+\omega^2\tau^2}\right) \tag{7.70}$$

Equations 7.69 and 7.70 describe changes in the real and imaginary parts of the dielectric constant and are also known as *Debye equations*.

We can see from Equation 7.70 that the maximum in k'', known as the *Debye loss peak*, occurs when $\omega = 1/\tau$ (see the problems at the end of this chapter and Example 7.12).

In many materials, there can be different relaxation times for the same polarization mechanism. This is because different atoms and ions may be involved. There is then a distribution of relaxation times due to the different polarization mechanisms, and the distribution of relaxation times applicable for each mechanism for a given dielectric.

The following example illustrates an application of the Debye equations.

Example 7.9: Relaxation of the Dielectric Constant for H_2O

The static dielectric constant (ε_s) of H_2O is 78.4 at 298.15 K (Fernandez et al. 1995). It decreases to about $\varepsilon_r(\omega) = 20$ at a frequency of $f = 40$ GHz. If the high-frequency (optical) dielectric constant (ε_∞) is 5, what is the relaxation time τ for H_2O molecule dipoles? Why is it that ε_∞ is not equal to the square of the refractive index (n) of water?

Solution

We first convert the frequency (f) into angular frequency (ω) by using

$$\omega = 2\pi f \tag{7.71}$$

Therefore, $\omega = 2\pi(40 \times 10^9 \text{ Hz}) = 2.51327 \times 10^{11}$ rad/s.

We assume that relaxation in the real part of the dielectric constant (k') of water follows the Debye equations, Equations 7.69 and 7.70.

For this problem, we substitute the following values in Equation 7.69: $\varepsilon_s = k_s = 78.4$, and $\varepsilon_\infty = k_\infty = 5$. We also know that $k' = 20$, when $f = 40$ GHz, that is, $\omega = 2\pi f = 2.51327 \times 10^{11}$ rad/s.

We rewrite the Debye equation as:

$$\left(1+\omega^2\tau^2\right) = \frac{k_s - k_\infty}{k' - k_\infty}$$

or

$$\omega^2\tau^2 = \left(\frac{k_s - k_\infty}{k' - k_\infty}\right) - 1 \tag{7.72}$$

Substituting these values in Equation 7.72, we get

$$\omega^2\tau^2 = \left[\frac{78.4-5}{20-5}\right] - 1 = 3.893$$

Therefore,

$$\omega\tau = 1.973$$

or

$$\tau = \frac{1.973}{2.51327\times 10^{11}\,\text{rad/s}} = 7.85\times 10^{-12}\,\text{s}$$

One picosecond is equal to 10^{-12} seconds. Thus, the relaxation time (τ) for an H_2O molecule, which is a permanent dipole, is about 7.85 picoseconds, which is the average time that an H_2O molecule needs to "flip" and realign as the direction of the electrical field changes.

Note that because H_2O molecules are polar, neither the Clausius–Mossotti nor the Lorentz–Lorenz equations apply because ε_∞ is not equal to n^2.

7.13.4 Giant Dielectric Constant Materials

The changes in the dielectric properties of dielectrics encountered in applications of microelectronics are more complex. For example, $CaCu_3Ti_4O_{12}$ (CCTO) was reported as a giant dielectric constant material. The changes in the real and imaginary parts of the dielectric constant for CCTO are shown in Figure 7.25.

Note several details from the data in Figure 7.25:

1. First, the dielectric constant is very large. We should therefore immediately suspect that this material may be a ferroelectric or that a space-charge polarization (see Section 7.10) may be present. Crystal structure analysis and other measurements have shown that this material is not a ferroelectric. The measured dielectric constant must be an *apparent dielectric* constant.

   These experimental data show that the apparent or measured dielectric constant is a *microstructure-sensitive* property. If we were to predict the dielectric constant using the Clausius–Mossotti equation, then the dielectric constant, although expected to be a function of frequency, will *not* be expected to show microstructure-sensitive characteristics. This is because the Clausius–Mossotti equation does not account for interfacial or ferroelectric (spontaneous) polarization.

2. Overall, as predicted by the Debye equations, the dielectric constant decreases with increasing frequency (Figure 7.25).

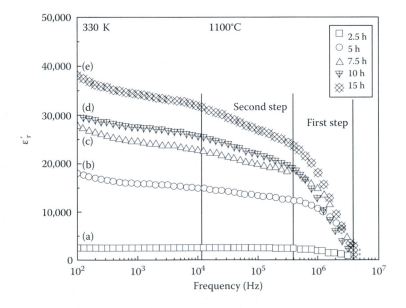

FIGURE 7.25 Changes in the dielectric constant of $CaCu_3Ti_4O_{12}$ ceramics as a function of the sintering time at 1100°C. Measurements of the dielectric constant (k') were carried out at 330 K. (From Prakash, B. and Varma K.B.R., *J. Phys. Chem. Solids.*, 68, 490–502, 2007. With permission.)

3. In this case, the increase in the dielectric constant with the sintering times may be because when the sintering times are low (e.g., 2.5 hours); the samples may not be dense. These materials basically are a composite of a dielectric and air. Because air has a low dielectric constant ($k' \sim 1$), the overall measured values of the dielectric constant are lower.

4. As the measured density increases with increasing sintering times, the overall dielectric constant also increases. The data in Figure 7.25 (marked as the first step) also show that the increase in the dielectric constant with sintering times is higher for the higher-frequency region ($\sim 5 \times 10^5$ to 3×10^6 Hz). In these materials, the interfacial or Maxwell–Wagner polarization at the grain boundaries plays an important role.

5. The relatively moderate increase in the dielectric constant in Step 2 (frequency range $\sim 10^4$ to 5×10^5 Hz; Figure 7.25) is related to the formation of another phase, which involves a process known as liquid-phase sintering. In this process, a liquid phase is formed that can assist densification. However, it can also lead to the formation of grain boundaries or surface phases that have a chemical composition different from that of the original ceramic material.

6. The relaxation of the low-frequency dielectric constant (in the range of 10^2–10^4 Hz—the region to the left of that marked as the second step) may be due to interfacial polarization at the ceramic–electrode interface via a Schottky barrier effect. If this is the case, it may be possible to change the electrode materials to see if the apparent dielectric constant changes.

In summary, the interpretation of changes in the apparent or measured dielectric constant or losses with frequency requires a considerably detailed analysis. We can attempt to correlate these changes with the different polarization mechanisms and microstructures. To determine the controlling polarization mechanisms, the dielectric properties are measured as functions of the temperature and frequency while changing the processing conditions systematically. The change in the dielectric constant of CCTO ceramics as a function of the temperature is shown in Figure 7.16. The technique of measuring k' and k'' as a function of the frequency (f) is known as *impedance spectroscopy*. The measurements of k' and k'' can be accomplished using an instrument called an *impedance analyzer*. The variations of k' (x-axis) and k'' (y-axis) plotted as a function of the frequency appear as sets of arcs or semicircles and are known as Cole–Cole plots.

7.14 EQUIVALENT CIRCUIT OF A REAL DIELECTRIC

For a parallel capacitor, the capacitance is given by

$$C = \varepsilon_0 \times \varepsilon_r'(\omega) \times \frac{A}{d} \tag{7.73}$$

This is similar to Equation 7.13; the difference is that now we explicitly show the dependence of the dielectric constant on electrical frequency by incorporating the variation of the dielectric constant with frequency (Figures 7.18 and 7.23).

We can model a real dielectric material as a pure, lossless capacitor with capacitance C, connected in parallel to a resistor with a resistance (R_p). The subscript "p" tells us that, in this equivalent circuit, the resistor is connected to the capacitor in parallel. This is an equivalent circuit of a real dielectric material (Figure 7.26). A circuit comprising a capacitor and a resistor (in series or parallel) is also known as a resistor–capacitor (RC) circuit. For an ideal dielectric, there is no loss, that is, $R_p = 0$.

The conductance ($G_p = 1/R_p$) of a dielectric is given by

$$G_p = \frac{1}{R_p} = \frac{\omega A \varepsilon_0 \varepsilon_r''(\omega)}{d} \tag{7.74}$$

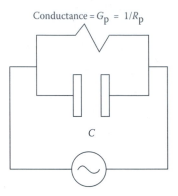

FIGURE 7.26 Equivalent circuit of a real dielectric material. (From Kasap, S.O., *Principles of Electronic Materials and Devices*, McGraw Hill, New York, 2006. With permission.)

Thus, for an ideal capacitor, if $R_p = 0$, then $G_p = \infty$. This means that as the capacitor charges and discharges, the current appears to flow through the capacitor. We have, of course, seen before that a dielectric material is a nonconductor, and very little, if any, current can actually flow through it. Equation 7.74 can be used to calculate the equivalent resistance or conductance of a dielectric material with specific capacitor geometry. The relatively simple appearance of Equations 7.73 and 7.74 is a bit misleading. If we want to calculate these values as a function of the frequency (ω), then we must keep in mind that both ε'_r and ε''_r change with frequency.

7.15 IMPEDANCE (Z) AND ADMITTANCE (Y)

For a real dielectric, we define *impedance* (Z) as a measure of the resistance offered by a circuit under AC fields. The impedance of a real dielectric (Figure 7.26) can be written as follows:

$$Z = R_p + jX_C \tag{7.75}$$

where X_C is the *capacitive reactance*. It is the equivalent of a capacitor's resistance to the flow of current through it. The magnitude of the capacitive reactance (X_C) is given by

$$X_c = \frac{1}{\omega C} = \frac{1}{2\pi f C_p} \tag{7.76}$$

Thus,

$$Z = R_p + j\frac{1}{2\pi f C_p} \tag{7.77}$$

The subscript "p" is used to indicate a parallel arrangement of the resistor and capacitor (Figure 7.26).

The *admittance* (Y) is defined as the inverse of the impedance (Z). Note that the admittance (Y) and impedance (Z) are both complex numbers. The SI unit for admittance is Siemens, also often reported as mho, the inverse of Ohm. The admittance of a capacitor made using a real dielectric material can be written as a combination of a conductor, with conductance G_p, and a capacitor, with capacitance C_p (Figure 7.26). The total admittance is written as a complex number:

$$Y = G_p + j\omega C_p \tag{7.78}$$

or

$$Y = \frac{1}{R_p} + j\omega C_p$$

$$Y = \frac{1}{R_p} + j2\pi f C_p \tag{7.79}$$

The admittance (Y) can be rewritten as follows using Equations 7.73 and 7.74:

$$Y = \frac{\omega A \varepsilon_0 \varepsilon_r''(\omega)}{d} + j\frac{\omega A \varepsilon_0 \varepsilon_r'(\omega)}{d} \tag{7.80}$$

Examples 7.10 and 7.11 illustrate the calculations of capacitive reactance and its frequency dependence.

Example 7.10: Capacitive Reactance (X_c) of a Capacitor

What is the magnitude of the capacitive reactance (X_C) of a 200-pF capacitor at a frequency of 10 MHz?

Solution
Note that one picofarad (pF) is 10^{-12} F. The magnitude of the capacitive reactance (X_C) is given by

$$X_C = \frac{1}{\omega C} = \frac{1}{2\pi f C} = \frac{1}{2\pi \left(10 \times 10^6\right)\left(200 \times 10^{-12} F\right)} = 79.5\ \Omega$$

Thus, this capacitor has a reactance of 79.5 Ω at a frequency of 10 MHz.

Example 7.11: Change of Capacitive Reactance (X_c) with Frequency

A 10-V variable-frequency AC supply is connected to a 200-µF capacitor. What is the value of capacitive reactance for frequencies of (a) 0 Hz, (b) 60 Hz, and (c) 1 kHz? What is the value of the current (I) flowing through the circuit for each frequency?

Solution
1. When the frequency (f) is zero, $X_C = \infty$, and the value of current flowing through the circuit is zero. In other words, when a DC voltage is applied to the capacitor, the dielectric material just blocks this voltage.
2. When $f = 60$ Hz, the value of the capacitive reactance (X_C) is given by

$$X_C = \frac{1}{2\pi f C} = \frac{1}{2 \times \pi \times 60 Hz \times \left(200 \times 10^{-6}\right) F} = 13.26\ \Omega$$

The current flowing through this circuit is given by the modified version of Ohm's law, in which we replace the resistance R with the capacitive reactance X_C. Thus, the current (I) is given by

$$I = \frac{V}{X_C} = \frac{10\ V}{13.26\ \Omega} = 0.754\ A$$

This current is not in phase with the voltage. It leads the voltage by 90° in the circuit of an ideal dielectric capacitor.

3. When $f = 1$ kHz, the value of capacitive reactance is given by

$$X_C = \frac{1}{2\pi f C} = \frac{1}{2 \times \pi \times (1 \times 10^3 \, \text{Hz}) \times (200 \times 10^{-6}) \, \text{F}} = 0.7958 \; \Omega$$

The current (I) is given by the equation

$$I = \frac{V}{X_C} = \frac{10 \, \text{V}}{0.7958 \, \Omega} = 12.56 \, \text{A}$$

Furthermore, the capacitive reactance (X_C) decreases with increasing frequency.

7.16 POWER LOSS IN A REAL DIELECTRIC MATERIAL

We know from Joule's law for a resistor that the power dissipated in a resistor is given as

$$\text{Power dissipated} = V \times I = \frac{V^2}{R} \tag{7.81}$$

The equivalent of this for a capacitor is

$$\text{Power input} = V \times I = V^2 \times Y \tag{7.82}$$

We substitute for Y from Equation 7.79 to get

$$\text{Power input} = j\omega C V^2 + \frac{V^2}{R_p} \tag{7.83}$$

The second term of this expression gives the power dissipated in a capacitor with equivalent resistance R_p (Figure 7.26).

Thus, the power lost in a real dielectric is given by

$$\text{Power lost in a real dielectric} = \frac{V^2}{R_p} \tag{7.84}$$

The dissipated power appears as heat in the dielectric material. It is sometimes important to note how much power is dissipated per unit volume. If we consider a capacitor (Figure 7.3) with volume $= A \times d$, the power loss per unit volume is given by substituting for $1/R_p$ from Equation 7.74 as shown here:

$$\text{Power loss in a real dielectric per unit volume} = \frac{V^2}{R_p \times A \times d} = \frac{V^2}{A \times d} \times \frac{\omega A \varepsilon_0 \varepsilon_r''(\omega)}{d}$$

or

$$\text{Power loss in a real dielectric per unit volume} = \omega \times E^2 \times \varepsilon_0 \varepsilon_r''(\omega) \tag{7.85}$$

7.16.1 Concept of tan δ

Consider a resistor connected to voltage (V). Assure that the voltage changes with time in a sinusoidal fashion as follows:

$$V = V_0 \exp(j\omega t) \tag{7.86}$$

You may recall that the definition of the function $\exp(jx)$ is

$$\exp(jx) = \cos(x) + j\sin(x) \tag{7.87}$$

Thus,

$$\exp(j\omega t) = \cos(\omega t) + j\sin(\omega t) \tag{7.88}$$

In a pure resistor, the current (I) and the voltage are said to be *in phase*. This means that, as the voltage changes over time, the current instantly follows the change in voltage. This is shown in a *phasor diagram* in Figure 7.27. A *phasor* is a rotating vector that shows the phase angle for a particular type of current or voltage in a circuit component. A phasor diagram shows the relative positions of the phasors for currents or voltages (or for both) corresponding to a circuit. The lengths of the phasor arrows generally are in scale with the magnitude of the voltage or current they represent. A phasor diagram is also known as a *vector diagram*.

At $\omega t = 0$, $V = V_0$; at $\omega t = \pi/2$ or 90°, $Vj = V_0$. As shown in Figure 7.27, as the voltage phasor rotates for a pure resistor from $\omega t = 0$ to $\omega t = \pi/2$, the phasor representing the current also rotates from $\omega t = 0$ to $\omega t = \pi/2$. In Figure 7.27, therefore, the voltage phasor and the current phasor point to the same direction.

When an AC voltage V (Equation 7.86) is applied to a pure capacitor, a charge Q appears on this capacitor and is given by

$$Q = C \times V = C \times V_0 \exp(jwt) \tag{7.89}$$

The buildup of charge on the capacitor can be described by defining a *charging current (I_c)*. This current is given by

$$I_c = \frac{dQ}{dt} \tag{7.90}$$

Rewriting the expression for the charging current using Equation 7.89, we get

$$I_c = C\frac{dV}{dt} \tag{7.91}$$

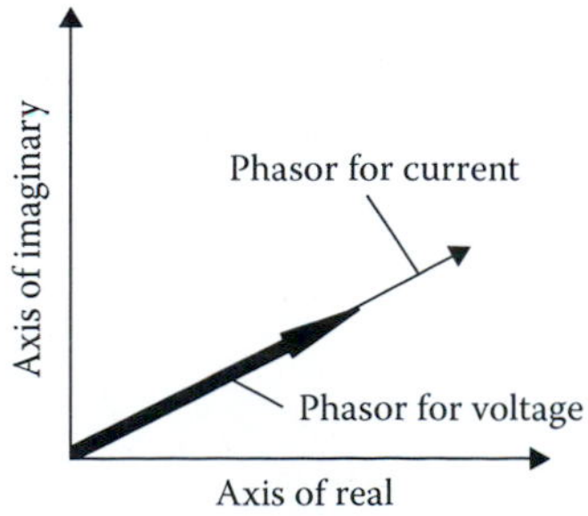

FIGURE 7.27 Phasor diagrams for a pure resistor.

Because $V = V_0 \exp(j\omega t)$, $dV/dt = V_0 j\omega \exp(j\omega t)$. Thus,

$$I_c = jCV_0\omega \exp(j\omega t) = j\omega CV \tag{7.92}$$

Compare Equations 7.86 and 7.92 (for voltage V and I_c, respectively) and check the phase difference between applied voltage (V) and charging current (I_c). You will find that the charging current I_c always leads the applied voltage V by exactly 90°. This is shown in the phasor diagram for an ideal capacitor, Figure 7.28a. The current and voltage waveforms for an ideal capacitor are shown in Figure 7.28b.

Referring to Figure 7.28, if the phase angle $\omega t = 0$, the charging current (I_c) will be along the axis of the imaginary and the voltage (V) will be along the axis of the real (Figure 7.27). As the phasor for I_c rotates through any angle, the phasor for the voltage also rotates by the same amount, thus maintaining a steady phase difference of 90° at all times (Figure 7.28b).

One way to represent a real dielectric is through a parallel combination of a resistor and a lossless capacitor (Figure 7.26). We then describe the total current (I_{total}) as the vector sum of a charging current (I_c) and a *loss current* (I_{loss}). The charging current associated with the ideal capacitor leads the applied voltage by 90°. The loss current originates from two sources. First, the dielectric material has a resistance under DC conditions that is usually very large. However, a very small amount of current still flows through a dielectric material. Second, for an AC voltage, as the polarization processes continue to occur, the back-and-forth displacements of electronic clouds, ions, molecular dipoles, and so on also cause a dissipation of energy and are thus part of the impedance. The loss current (I_{loss}) in a real dielectric due to both of these contributions is always in phase with the voltage because these processes always follow the applied voltage (V).

The loss current (I_{loss}) in a dielectric can be written as follows:

$$I_{loss} = G \times V \tag{7.93}$$

where G is the conductance and V is the voltage. The conductance can be expressed as a sum of the DC and AC components.

The loss current can be written as

$$I_{loss} = (G_{dc} + G_{ac}) \times V \tag{7.94}$$

The total current (I) is given by the *vector sum* of these two currents, namely, the charging current (I_c) and the loss current (I_{loss}). This total current (I_{total}) always leads the voltage but not by 90°. As shown in Figure 7.29, the total current (I_{total}), therefore, leads the voltage at an angle $(90 - \delta)°$.

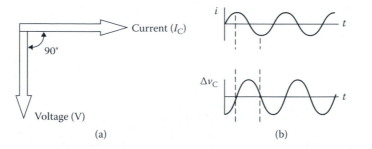

(a)   (b)

FIGURE 7.28 (a) Phasor diagram and (b) waveform relationships for an ideal capacitor.

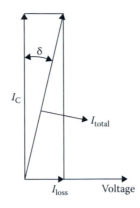

FIGURE 7.29   Corresponding waveforms for a real dielectric.

The angle δ is known as the *loss angle*. Recall that the charging current (I_c) is $j\omega CV$ (from Equation 7.92).The total current (I_{total}) in a capacitor made using a real dielectric can be written as

$$I_{total} = I_c + I_{loss} = (j\omega C + G_{dc} + G_{ac})\,V \tag{7.95}$$

From Figure 7.29, we can also note that

$$\tan \delta = \frac{I_L}{I_c} \tag{7.96}$$

Thus, tan δ (pronounced tan delta) is also a measure of the dielectric losses. There are no dielectric losses in an ideal dielectric. This means that $I_L = 0$, tan $\delta = 0$, and the total current is equal to the charging current.

The concept of tan δ is very important for describing properties of dielectric materials. This parameter can be measured using an *impedance analyzer* similar to measuring the dielectric constant. We can compare how *lossy* different dielectric materials are relative to each other. The real part of the complex dielectric constant is a measure of a material's charge-storing ability. The magnitude of tan δ indicates the inefficiency of the material in terms of the charge-storing ability.

In an ideal dielectric, which does not exist, polarization processes are assumed to occur without any electrical energy waste, with no energy wasted during charge storage, that is, tan $\delta = 0$. In many real dielectrics, the tan δ values range from ~10^{-5} to 10^{-2}. Some special applications, such as ceramic materials used as dielectric resonators for microwave communications, require very low tan δ values. In this case, the values of a parameter defined as the *quality factor* (Q) are reported. The quality factor of a dielectric (Q_d) is defined as

$$Q_d \approx \frac{1}{\tan \delta} \tag{7.97}$$

We can also show that tan δ is the ratio of the imaginary and real parts of the complex dielectric constant (k^*).

The dielectric constant is defined as $\varepsilon_r^* = C / C_0$, and because $Q = CV$, $Q = \varepsilon_r^* C_0 V$.
The current $I_{total} = dQ/dt = C(dV/dt)$, or

$$I_{total} = \varepsilon_r^* C_0 \left(\frac{dV}{dt} \right) \tag{7.98}$$

If $V = V_0 \exp(j\omega t)$, then $dV/dt = j\omega V_0 \exp(j\omega t) = j\omega V$; therefore, I_{total} is

$$I_{total} = \varepsilon_r^* C_0 \left(\frac{dV}{dt} \right) = (\varepsilon_r' - j\varepsilon_r'') C_0 j\omega V \tag{7.99}$$

To derive Equation 7.99, we substituted for the complex dielectric constant in terms of its real and imaginary parts (see Equation 7.68).

$$I_{total} = j\omega\varepsilon_r' C_0 V + \omega\varepsilon_r'' C_0 V \tag{7.100}$$

Note that to derive this equation, we used the value $j^2 = -1$.

Now compare Equations 7.95 and 7.100. The first term of Equation 7.100 represents the charge-storage in the dielectric, that is, the charging current. The second term is the magnitude of the loss current (I_{loss}). Thus, the ratio of the *magnitude* of these two parts is $\tan \delta$:

$$\tan \delta = \frac{\omega\varepsilon_r'' C_0 V}{\omega\varepsilon_r' C_0 V} = \frac{\varepsilon_r''}{\varepsilon_r'}$$

Thus,

$$\tan \delta = \frac{\varepsilon_r''}{\varepsilon_r'} = \frac{k_r''}{k_r'} \tag{7.101}$$

Another interpretation for $\tan \delta$ or loss tangent is that it is the ratio of the imaginary and real parts of the complex dielectric constant or the complex relative dielectric permittivity. One can think of $\tan \delta$ as the ratio of the price we pay for storing the charge in a capacitor (in terms of the energy that is wasted and that appears as heat). Heating due to dielectric losses is useful in some applications. However, in many applications, the generation of heat changes the temperature of the dielectric material, and the dielectric properties may change with changes in temperature. In addition, heating a dielectric material can change its dimensions; the geometrical changes also lead to changes in capacitance. In general, we prefer to use temperature-stable and low-loss dielectric materials.

Because both the real and imaginary parts of the dielectric constant are frequency-dependent (Figure 7.24), $\tan \delta$ also depends on the frequency.

Recall that, for a hypothetical material with a single polarization mechanism with relaxation time (τ), the frequency dependences of the real and imaginary parts of the dielectric constants are given by the Debye equations (Equations 7.69 and 7.70). Therefore, the frequency dependence of $\tan \delta$ for such a hypothetical material is given by

$$\tan \delta = \frac{\varepsilon_r''}{\varepsilon_r'} = \frac{(k_s - k_\infty)\omega\tau}{k_s + k_\infty\omega^2\tau^2} \tag{7.102}$$

where k_s and k_∞ are the values of the low-frequency (static) and high-frequency dielectric constants.

Although the value of ε_r'' reaches a maximum at $\omega t = 1/\tau$ (Figure 7.23), $\tan \delta$ reaches a maximum at a slightly higher frequency, which is given by

$$\omega_{max} \left(\text{for } \tan \delta \right) = \frac{\left(\dfrac{k_s}{k_\infty} \right)^{1/2}}{\tau} \tag{7.103}$$

Recall that the power loss in a real dielectric per unit volume is given by Equation 7.85. We can write this in terms of $\tan \delta$ and ε_r' (dielectric constant) as follows:

$$\text{Power loss in a real dielectric per unit volume} = \omega \times E^2 \times \varepsilon_0 \times \varepsilon_r' \times \tan(\delta) \qquad (7.104)$$

Example 7.12 shows how to calculate the frequency at which the value of $\tan \delta$ reaches a maximum.

Example 7.12: Maximum for tan δ

Show that for a dielectric with only one polarization mechanism and a single relaxation time (τ), the maximum $\tan \delta$ occurs at ω_{max} (for $\tan \delta$) (Equation 7.103).

Solution
We start with Equation 7.102, which describes the dependence of $\tan \delta$ on frequency ω.

To locate the maximum, we take the derivative of $\tan \delta$ with respect to ω and equate it to zero. Because the term $(k_s - k_\infty)$ does not depend on frequency, we can take the derivate of the term $\dfrac{\omega\tau}{\left(k_s + k_\infty\omega^2\tau^2\right)}$ with respect to ω and equate it to zero. Recall the rule for the derivative of a function $h(x)$, which is a ratio of two functions $h(x) = f(x)/g(x)$; then

$$\frac{dh(x)}{dx} = \frac{\left[g(x)\cdot\dfrac{df}{dx} - f(x)\left(\dfrac{dg}{dx}\right)\right]}{g(x)^2} \qquad (7.105)$$

$$\frac{d\left(\dfrac{\omega\tau}{k_s + k_\infty\omega^2\tau^2}\right)}{d\omega} = \frac{\left[\left(k_s + k_\infty\omega^2\tau^2\right)(\tau) - (\omega\tau)\left(\tau^2 k_\infty 2\omega\right)\right]}{\left(k_s + k_\infty\omega^2\tau^2\right)^2}$$

Equating this to zero gives us

$$(k_s + k_\infty\, \omega^2\tau^2)\, (\tau) - (\omega\tau)\, (k_\infty\, \tau^2\, 2\omega) = 0$$

This simplifies to

$$(k_s + k_\infty\, \omega^2\tau^2 - 2k_\infty\, \tau^2\omega^2) = 0$$

or

$$\omega = \frac{\left(\dfrac{k_s}{k_\infty}\right)^{1/2}}{\tau}$$

Tan δ will reach a maximum, that is, ω_{max} (for $\tan \delta$) (Equation 7.103).

This maximum $\tan \delta$ occurs at a frequency higher than $1/\tau$, which is the frequency at which k reaches a maximum.

7.17   EQUIVALENT SERIES RESISTANCE AND EQUIVALENT SERIES CAPACITANCE

A real dielectric can be described as a resistance and a capacitor connected in parallel (Figure 7.26). This description of a real dielectric is useful in understanding both the relationships of loss current with the charging current and the concept of tan δ. However, in many practical applications of capacitors, an important parameter that is often specified is the *equivalent series resistance* (ESR). This essentially entails describing the capacitor as a combination of an equivalent series capacitor (ESC) and an ESR that are connected as shown in Figure 7.30.

Note that in both cases, the dielectric is modeled as either parallel or series connections of a capacitor and resistor, and there is no separate resistor connected to the dielectric for either model. We simply model the real dielectric as an arrangement of a capacitor and a resistor.

We will show that the ESR and ESC are related to their parallel-circuit equivalents (i.e., R_p and C_p) using the following equations:

$$\text{ESR} = \frac{R_p}{\left(2\pi f C_p R_p\right)^2 + 1} \tag{7.106}$$

$$\text{ESC} = C_p\left[1 + \frac{1}{\left(2\pi f C_p R_p\right)^2}\right] \tag{7.107}$$

or

$$\text{ESC} = C_p + \frac{1}{\left(2\pi f R_p\right)^2 C_p} \tag{7.108}$$

In the parallel and series arrangements of the models of a real dielectric shown in Figure 7.30, the total impedance (Z) offered by the dielectric must be the same. The properties of the dielectric material do not depend on how we choose to describe it. Starting with the parallel arrangement, the total admittance of a real dielectric modeled as a parallel arrangement is written as Equation 7.79.

The total impedance of the parallel arrangement is

$$Z = \frac{1}{\dfrac{1}{R_p} + j\omega C_p} \tag{7.109}$$

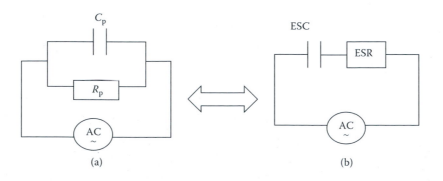

(a)   (b)

FIGURE 7.30   Circuit showing the model of a real dielectric as (a) parallel and (b) equivalent series resistance connected to an equivalent series capacitor.

In the series arrangement (Figure 7.30), we write the impedance as

$$Z = \text{ESR} + jX_{\text{ESC}}$$

or

$$Z = \text{ESR} + j\frac{1}{(\omega \times \text{ESC})} \tag{7.110}$$

Because the total impedance must be the same in both the series and parallel arrangements, the real part of this impedance (Z) in a parallel arrangement must be equal to the ESR. Similarly, the imaginary part of Z is equal to the equivalent series capacitive reactance (X_{ESC}). We start with Equation 7.109, or

$$Z = \frac{R_p}{1 + j\omega C_p R_p} \tag{7.111}$$

To separate the real and imaginary parts of this number, we multiply and divide by $(1 - j\omega C_p R_p)$, which is the complex conjugate of the term in the denominator.

$$Z = \frac{R_p}{1 + j\omega C_p R_p} \times \frac{\left(1 - j\omega C_p R_p\right)}{\left(1 - j\omega C_p R_p\right)}$$

or

$$Z = \frac{R_p}{\left(1 + \omega^2 C_p^2 R_p^2\right)} + j\frac{\left(\omega C_p R_p^2\right)}{\left(1 + \omega^2 C_p^2 R_p^2\right)} \tag{7.112}$$

We can compare Equation 7.112 to Equation 7.110, equate the real part of this derivation to ESR, and substitute $\omega = 2\pi f$ to get Equation 7.106.

We can equate the imaginary part of Z from Equation 7.112 to the equivalent capacitance for the series circuit (X_{ESC}; Equation 7.110) to get:

$$X_{\text{ESC}} = \frac{1}{\omega \times \text{ESC}} = \frac{\left(\omega C_p R_p^2\right)}{\left(1 + \omega^2 C_p^2 R_p^2\right)}$$

$$\omega \times \text{ESC} = \frac{\left(1 + \omega^2 C_p^2 R_p^2\right)}{\left(\omega C_p R_p^2\right)}$$

$$\text{ESC} = \frac{\left(1 + \omega^2 C_p^2 R_p^2\right)}{\left(\omega^2 C_p R_p^2\right)} = C_p\left[1 + \frac{1}{\omega^2 C_p^2 R_p^2}\right]$$

$$\text{ESC} = C_p\left[1 + \frac{1}{\omega^2 C_p^2 R_p^2}\right]$$

$$\text{ESC} = C_p\left[1 + \frac{1}{\left(2\pi f R_p\right)^2 C_p^2}\right]$$

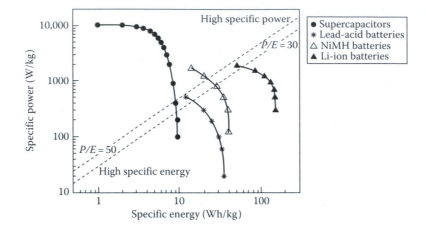

FIGURE 7.31   Ragone plot of electrochemical systems for hybrid electric vehicles. (From Mastragostino, M. and Soavi F., *J. Power Sources*, 174, 89–93, 2007. With permission.)

This equation is also rewritten as Equation 7.108:

$$\text{ESC} = C_p + \frac{1}{\left(2\pi f R_p\right)^2 C_p}$$

In applications involving supercapacitors, the emphasis often is on reducing the ESR, as opposed to increasing the volumetric efficiencies. Supercapacitors, which are based on electrochemical reactions that lead to a double-layer interfacial capacitance, have received considerable attention recently because of their potential for applications in electric hybrid vehicles. In these applications, supercapacitors can provide a high level of energy output in a short time (i.e., high specific power), compared to that provided by lithium ion or nickel hydride batteries and fuel cells, which can provide a high energy density (Mastragostino and Soavi 2007) but a lower energy output. Many supercapacitors have very high *specific capacitance* (that is, capacitance per unit mass), ranging from 700 to 1000 F/g (Yang et al. 2007).

A *Ragone plot* or *Ragone chart* is a diagram that compares the specific power density possible from a specific energy output for different power- or energy-generating devices. This plot is shown for some batteries and compared to supercapacitors in Figure 7.31.

PROBLEMS

7.1   A liquid-level sensor is to be designed for a car. Explain how such a sensor can be designed so that the change in capacitance can be used to measure the liquid level.

7.2   A tantalum oxide (Ta_2O_5) thin film (with a dielectric constant of ~25) was deposited on conductive polysilicon to form a capacitor. Another dielectric material under consideration for this application was SiO_2 (with a dielectric constant of ~3.9). If the thickness of the SiO_2 film required to achieve a certain value of capacitance is d_{silica}, what thickness of Ta_2O_5 (d_{Ta2O5}) will provide the same capacitance per unit area as that for the capacitor made using SiO_2?

7.3   A ferroelectric thin film of a particular composition of lead zirconium titanate (PZT), with an apparent dielectric constant of 400, was manufactured using a sol-gel process. This dielectric constant is lower than that seen for bulk or single crystals of PZT of the same composition. The film was 1 pm thick. What is the capacitance per unit area for this film? Express your answer in μF/cm^2. What is the total capacitance if the electrode area is 10 μm^2?

7.4 Ferroelectric thin films (1000 nm thick) of apparently the same composition were made using a multitarget sputtering process. The dielectric constant of this PZT composition in a thin-film form should be 400. One set of samples was subjected to high temperatures to anneal the films. This caused some lead to evaporate and caused the formation of another layer (~100 nm thick), a phase known as the pyrochlore (assume a dielectric constant of ~50). The total film thickness of the PZT and pyrochlore layers was still 1000 nm. What is the capacitance of this composite film per unit area? Express your answer in μF/cm^2. What possibly can be done to prevent or minimize the formation of the pyrochlore phase having a low dielectric constant?

7.5 The dielectric constant of Si is 11.9. Because Si is covalently bonded, the only polarization mechanism present is electronic polarization. Use the Clausius–Mossotti equation to show that the electronic polarizability of Si atoms is 4.17×10^{-40} F $\cdot$ m^2.

7.6 The density of krypton (Kr) gas at 25°C and one atmosphere is reported to be 3.4322×10^{-3} g/cm^3. The polarizability volume is reported to be 2.479×10^{-24} cm^3 in literature (Vidal et al. 1984). Calculate the dielectric constant of Kr under these conditions, if its molecular weight is 83.8 g.

7.7 For a mixture of Ar (70% by weight) and Kr (30% by weight), show that the average molecular weight (M) is 53.1. Show that the average volume polarizability (using data from Problem 7.5 and Example 7.4) is 1.80×10^{-24} cm^3. What is the dielectric constant of this mixture of gases at 25°C and one atmospheric pressure?

7.8 From the data in Example 7.4, calculate the dielectric constant of Ar gas.

7.9 Assuming that the dipole moment of H_2O molecules is 10^{-29} C $\cdot$ m and the electric field experienced by H_2O molecules is 10^5 V/m, calculate the dipolar polarizability of H_2O molecules in F $\cdot$ m^2. Note:

$$\alpha_d = \frac{1}{3} \frac{\mu^2}{K_B}$$

7.10 Calculate the heat generated from dielectric losses in a polymer cable subjected to an electric field of 80 kV/cm. Assume that the frequency is 60 Hz, the dielectric constant of the polymer is 2.1, and tan δ is 10^{-4}.

7.11 What are the limitations of using the Clausius–Mossotti equation? Explain.

7.12 The low-frequency dielectric constant of a silicate glass is 4.0. The refractive index of this glass is 1.47. What fraction of the dielectric constant can be attributed to ionic polarization?

7.13 The variation in the imaginary part of the dielectric constant (k'') with frequency is given by Equation 7.70. Prove that the maximum k'' occurs when $\omega = 1/\tau$.

7.14 Chalcogenide glasses are amorphous materials based on elements such as selenium (Se), Ge, and arsenic (As). These glasses have promise for many dielectric and optical applications. The dielectric constant of $Se_{55}Ge_{30}As_{15}$ was measured as a function of frequency ranging from 20 kHz to 4 MHz (Figure 7.32).

What is the trend in dielectric constant as the frequency increases at temperatures between 350 and 390 K? At temperatures greater than 400 K, the lower-frequency dielectric constant increases considerably. Suggest a possible reason for this.

7.15 The dependence of the imaginary part of the dielectric constant $\left(\varepsilon_r''(\omega)\right)$ for chalcogenide glasses of composition $Se_{55}Ge_{30}As_{15}$ is shown in Figure 7.33.

From the data shown in Figures 7.32 and 7.33, plot the real and imaginary parts of the complex dielectric constant as a function of ln(ω) for $T = 380$ K. Are the changes in the real and imaginary parts of the dielectric constant sharp, as in Figure 7.24? Explain. Show that,

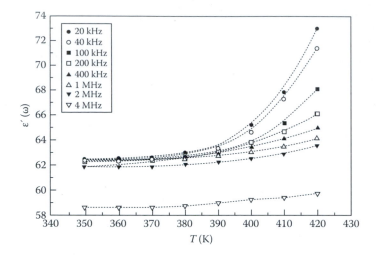

FIGURE 7.32 Dielectric constant of $Se_{55}Ge_{30}As_{15}$ glasses. (From El-Nahass, M.M., et al., *Physica B.*, 388, 26–33, 2007. With permission.)

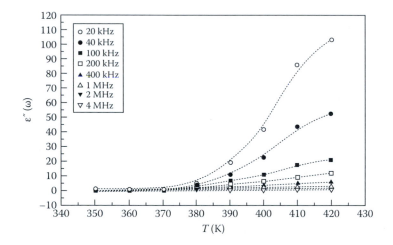

FIGURE 7.33 Dielectric losses ($\varepsilon''_r(\omega)$) of $Se_{55}Ge_{30}As_{15}$ glasses. (From El-Nahass, M.M., et al., *Physica B.*, 388, 26–33, 2007. With permission.)

at 300 K, the maximum $\varepsilon''(\omega)$ occurs at ~381 kHz. What is the value of $\ln(\omega)$ corresponding to this? Show that the most probable relaxation time for polarization in this material is ~4.11×10^{-6} s.

7.16 For the data on chalcogenide glasses shown in Figures 7.32 and 7.33, calculate the values of $\tan \delta$ as a function of the frequency (τ) (ω). Plot these values on a graph. Assume $T = 380$ K.

7.17 Starting with the Debye equations (Equations 7.69 and 7.70) that describe the changes in the real and imaginary parts of the dielectric constant as a function of frequency, show that (ε''/ω) = (τ) × ($\varepsilon' - \varepsilon_\infty$), and that a plot of $\log(\varepsilon''/\omega)$ versus $\log(\varepsilon' - \varepsilon_\infty)$ results in a straight line. The y-axis intercept of this graph gives the value of $\log(\tau)$, from which the value of the relaxation time can be calculated.

7.18 A plot of $\log(\varepsilon''/\omega)$ versus $\log(\varepsilon' - \varepsilon_\infty)$ for a dielectric material of composition $Se_{70}Te_{30}$ is shown in Figure 7.34. The data were collected for temperatures ranging from 298 to 373 K.

What are the $\log(\tau)$ values for $Se_{70}Te_{30}$ for these different temperatures? What happens to the values of the relaxation times as the temperature increases?

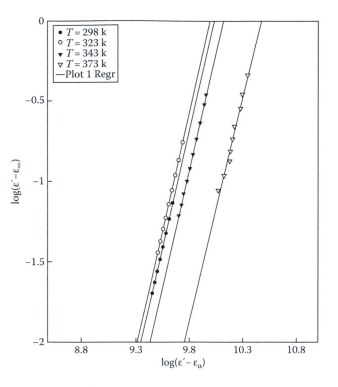

FIGURE 7.34 A plot of $\log(\varepsilon''/\omega)$ versus $\log(\varepsilon' - \varepsilon_\infty)$ for $Se_{70}Te_{30}$. (From Sayed, S.M., *Appl Surf Sci.*, 253, 7089–7093, 2007. With permission.)

7.19 The molar volume of a dielectric known as calcium molybdate ($CaMoO_4$) is 78.064 Å^3. Using Shannon's approach (see Section 7.8), write down the expression for calculating the molar polarizability of $CaMoO_4$. The measured dielectric constant of this material is 10.79 (Choi et al. 2007). What is the polarizability of the Mo^{6+} ion?

7.20 Many microwave ovens work at a frequency of 2.45 GHz. Examine how the real and imaginary parts of the dielectric constant of water change with the frequency for different temperatures (Figure 7.35). Show that the wavelength (λ) of the radiation associated with these waves is ~12.2 cm. The wavelength at which the dielectric losses (i.e., tan δ) are maximized (at room temperature) is 4 cm; what is the corresponding frequency in GHz? The speed of light can be assumed to be 3×10^8 m/s. Why is it that microwave ovens do not use the frequency at which the value of tan δ is maximized? Hint: Microwaves need to penetrate into the food. (See Figure 7.36.)

7.21 Many foods contain H_2O and salt, and the salt content of the food decreases the dielectric constant of H_2O. What is the effect of the salt content on the dielectric losses of H_2O?

7.22 The molar volume (V_m) for beryllium silicate (Be_2SiO_4) is 61.75 Å^3 (Shannon 1993). What is the expected dielectric constant of Be_2SiO_4 calculated using Shannon's approach? The experimental value of the dielectric constant for Be_2SiO_4 is 6.22.

7.23 What are some of the situations in which Shannon's approach for calculating the dielectric constant probably will not work?

7.24 The low-frequency dielectric constant of Ge is 16. What is its high-frequency dielectric constant? Why? What is the refractive index of Ge?

7.25 If the refractive index of NaCl is 1.5, what is the high-frequency dielectric constant of NaCl? If the low-frequency dielectric constant of NaCl is 5.6, approximately what fraction of the low-frequency dielectric constant can be attributed to ionic polarization?

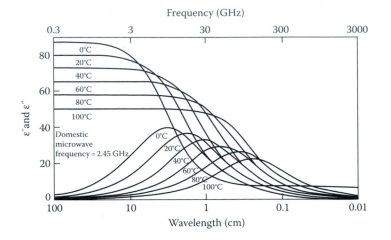

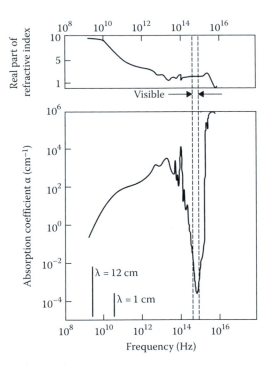

FIGURE 7.35 Dependence of the real and imaginary parts of the dielectric constant of H_2O on frequency for different temperatures. (From Vollmer, M., *Phys. Educ.*, 39(1), 74–81, 2007. With permission.)

FIGURE 7.36 Refractive index (related to the real part of the high-frequency dielectric constant) and absorption coefficient (related to the imaginary part of the complex dielectric constant) of H_2O. (From Vollmer, M., *Phys. Educ.*, 39(1), 74–81, 2004. With permission.)

7.26 Compositions and properties of silicate glasses, as investigated by Wang et al. (2007), are shown in Tables 7.3 and 7.4.
    a. At 10 MHz, what polarization mechanisms are expected to play a role in contributing to the dielectric constant for these glasses?
    b. Why does the glass made using Li^+ (i.e., Sample 11) exhibit the lowest dielectric constant?
    c. Between Samples 12 and 13 (i.e., samples containing sodium and potassium oxides), which sample has a higher total electronic polarizability and refractive index? Why?

TABLE 7.3

Compositions of Glasses (mol%)

Sample Number	SiO_2	B_2O_3	Al_2O_3	CaO	Li_2O	Na_2O	K_2O
11	62	25	1.2	5.4	6.4	–	–
12	62	25	1.2	5.4	–	6.4	–
13	62	25	1.2	5.4	–	–	6.4

Source: Wang, Z., et al., *J. Non Cryst. Solids*, 354(12), 1128–1132, 2008.

TABLE 7.4

Dielectric Properties and Resistivities of Glasses

Sample Number	Dielectric Constant (at 10 MHz)	Dielectric Loss (tan δ) at 10 MHz	Resistivity ($\Omega \cdot$ cm)
11	5.58	4.97×10^{-2}	0.73×10^{13}
12	6.03	4.5×10^{-2}	0.66×10^{13}
13	6.05	4.41×10^{-2}	1.17×10^{13}

Source: Wang, Z., et al., *J. Non Cryst. Solids*, 354(12), 1128–1132, 2008.

d. Because the dielectric constant of Samples 12 and 13 (at 10 MHz) is effectively the same, which sample has a higher ionic polarizability? Does this explain why the dielectric constant of Samples 12 and 13 is approximately the same?

e. The so-called modifier ions, that is, ions that cause disruption of the silicate tetrahedra in the local arrangements within glasses, play a key role in controlling the dielectric losses of these materials. Based on the values of tan δ, which modifier ion seems most mobile?

f. The conductivity of these glasses depends on the type and concentration of modifier ions. For Samples 11, 12, and 13, the concentration of modifier ions (Li^+, Na^+, and K^+) is the same. The conductivity then depends on the mobility of these ions (which, in turn, depends on the size of the ion) and on the strength of the bond between the modifier and the oxygen ions. Explain why sodium oxide–based glass has the lowest resistivity of these compositions.

GLOSSARY

Admittance (Y): The inverse of impedance (Z).

Atomic polarization: See **Ionic polarization**.

Bound charge: Charges on a capacitor plate that are not free to move because they are bound by the dipoles, either induced or present, in the dielectric material.

Capacitive reactance (X_C): A capacitor's effective resistance (in ohms) to the flow of electricity through it, which is given by $X_C = 1/\omega C = 1/2\pi C$. This component of the impedance does not cause any dissipation of electrical energy.

Capacitor: A device or structure filled with a dielectric layer that is separated by electrodes and is used for charge storage.

Charging current (I_c): The current that leads to the storage of charge in a dielectric (or capacitor) when a voltage is applied.

Clausius–Mossotti equation: The equation that relates the dielectric constant (a macroscopic property) to the polarizability of atoms and ions (a microscopic property). This equation *cannot* be used to account for interfacial (Maxwell–Wagner) or ferroelectric polarization.

$$\frac{\varepsilon_r - 1}{\varepsilon_r + 2} = \frac{1}{3\varepsilon_0}\left[(N_e \times \alpha_e) + (N_i \times \alpha_i)\right]$$

Coercive field (E_c): The electric field necessary to cause the domains in a ferroelectric with some remnant polarization to be randomized again in order to achieve a state of zero net polarization.

Cole–Cole plots: The sets of arcs or semicircles that appear when variations of k' (x-axis) and k'' (y-axis) are plotted as a function of the frequency.

Complex dielectric constant $\left(\varepsilon_r^*\right)$: A quantity that describes the ability of a real dielectric material to store an electrical charge ε_r'; also a measure of the energy lost ε_r'' while the polarization processes that lead to charge storage occur. It is given by $\varepsilon_r^* = \varepsilon_r' - j\varepsilon_r''$.

Curie temperature: The temperature at which a ferroelectric material transforms into a centrosymmetric paraelectric material.

Debye: A unit of dipole moment, with one Debye (D) equal to 3.3356×10^{-30} C · m.

Debye equations: Equations describing both the relaxation in a dielectric constant and the dependence of the imaginary part of the dielectric constant on frequency.

Debye loss peak: The frequency $\omega = 1/\tau$, at which the value of k'' reaches a maximum.

Dielectric constant (k or ε_r): A dimensionless parameter that expresses the ability of a dielectric to store charge, which is expressed as a ratio of the capacitance of a capacitor filled with a dielectric material to that filled with a vacuum (C/C_0).

Dielectric displacement (D): The total charge density, which is written as a total of the free and bound charge densities due to the polarization of a dielectric.

Dielectric flux density (D): See **Dielectric displacement**.

Dielectric material: A large-bandgap semiconductor ($E_g \sim$ >4 eV) with high resistivity (ρ).

Dielectric permittivity (ε): A parameter that expresses the ability of a material to store a charge, expressed as $\varepsilon = \sigma_s/E$, where σ_s is the surface charge density and E is the electric field. The unit for permittivity is F/m, with the permittivity of free space being $\varepsilon_0 = 8.85 \times 10^{-12}$ F/m. We often use the dielectric constant $\varepsilon/\varepsilon_0$.

Dielectric polarization (P): The magnitude of the bound charge density in a dielectric material.

Dielectric relaxation: A reduction in the dielectric constant with an increasing frequency of the applied electric field because some of the polarization mechanisms cannot keep up.

Dipolar polarizability (α_d): Polarizability due to the presence of molecules with a dipole moment (μ); given by

$$\alpha_d = \frac{1}{3}\frac{\mu^2}{K_B}$$

Dipolar polarization: Also known as orientational polarization, caused by the realignment of permanent dipoles present in certain materials.

Dipole moment (μ): For a dipole with charges $+q$ and $-q$ separated by a distance d, the dipole moment is $q \times d$, the SI unit of which is C · m.

Electrical insulators: Dielectric materials with high breakdown voltage, also known simply as insulators.

Electronic polarizability (α_e): A measure of the electronic polarization of an atom or an ion, with larger atoms and ions having higher electronic polarizabilities.

Electronic polarization: Creating or inducing a dipole in an atom or an ion by displacing the electronic cloud with respect to the nucleus.

Electrostatic induction: The development of a charge on a material when another material with a net charge is brought near it.

Equivalent series capacitance (ESC): The capacitance of a capacitor when a real dielectric material is modeled as a combination of a resistance and a capacitor connected in a series. It is related to the equivalent elements of a parallel circuit (C_p and R_p), describing the same real dielectric by

$$\text{ESC} = C_p + \frac{1}{\left(2\pi f R_p\right)^2 C_p}$$

Equivalent series resistance (ESR): The resistance of a resistor when a real dielectric material is modeled as a combination of a resistance and a capacitor connected in series. It is related to the equivalent elements of a parallel circuit (C_p and R_p), describing the same real dielectric by

$$\text{ESR} = \frac{R_p}{\left(2\pi f C_p R_p\right)^2 + 1}$$

Ferroelectrics: Materials that show spontaneous and reversible polarization.

Free charge: A charge on a capacitor plate that is free to move and is not bound to the induced or permanent dipole in a dielectric material.

Frequency dispersion: See **Dielectric relaxation**.

Gauss's law: This law states that the area integral of the electric field (E) over any closed surface is equal to the net charge (Q) enclosed within the surface divided by the permittivity of space (ε_0) and is given by the following equation:

$$\oint \vec{E} \times d\vec{A} = \frac{Q}{\varepsilon_0}$$

High-frequency dielectric constant (k_∞): The dielectric constant measured at frequencies at which only the electronic polarization mechanism operates. It is also known as the optical frequency dielectric constant and is equal to the square of its refractive index (n).

Ideal dielectric: A dielectric material with no dielectric losses; such a material does not exist.

Imaginary part of the dielectric constant (k'' or ε_r''): The part of the complex dielectric constant that is associated with dielectric losses and the loss current for a hypothetical dielectric with only one relaxation time (τ). It is given by the following equation:

$$\varepsilon_r'' = k'' = \left(k_s - k_\infty\right)\left(\frac{\omega\tau}{1 + \omega^2\tau^2}\right)$$

Impedance (Z): The equivalent of resistance applicable to a circuit that can have a capacitor and an inductor. The units of impedance are ohms. Its inverse is admittance (Y).

Impedance analyzer: Equipment that can measure the inductance, capacitance, and resistance of a material or device, typically at different frequencies, that in turn helps to calculate the impedance (Z).

Impedance spectroscopy: The technique of measuring k' and k'' as a function of frequency.

Interfacial polarization: Polarization at the heterogeneities in a material or device, typically at the dielectric–electrode interfaces and at the grain boundaries of a polycrystalline material.

Internal electric field: See **Local electric field**.

Ionic polarizability (α_i): A parameter for describing the propensity of ions to undergo ionic polarization.

Ionic polarization: A polarization mechanism in which the ions themselves are displaced in response to the electric field experienced by the solid, creating a net dipole moment per ion (p_{av}). This is also known as atomic or vibrational polarization.

Local electric field (E_{local}): The actual field experienced by the atoms, molecules, or ions within a material, which is different from the applied field and is referred to as the internal or local electric field (E_{local}).

$$E_{local} = E + \frac{1}{3\varepsilon_0} P \left(\text{for cubic structure or amorphous materials} \right)$$

This equation *cannot* be used for materials that either lack a cubic structure or are polar.

Local field approximation: See **Local electric field**.

Lorentz–Lorenz equation: An equation that relates the high-frequency dielectric constant to the refractive index through the electronic polarizability.

$$\frac{n^2 - 1}{n^2 + 2} = \frac{1}{3\varepsilon_0} \left[\left(N_e \times \alpha_e \right) \right]$$

Loss angle (δ): In an ideal dielectric, the charging current leads the voltage by 90°, whereas in a real dielectric, it leads by (90−δ)°. The tangent of this angle represents the dielectric losses.

Loss current (I_{loss}): The dielectric losses that are encountered during the polarization processes in a real dielectric subjected to an electric field.

Loss tangent (tan δ): The ratio of the loss current to the charging current in a real dielectric. It is also the ratio of the imaginary to the real parts of the dielectric constant and is a measure of the quantity of the input electrical energy lost during the polarization processes.

Low-frequency dielectric constant (k_s): See **Static dielectric constant**.

Low-loss dielectric: A material that has a low tan δ (~ <10^{-3}); typically, the value of the quality factor of the dielectric (Q_d) is reported as $Q_d = 1/\tan \delta$.

Maxwell's equations: A set of four equations that describe the basis of many relationships pertaining to the electrical and magnetic properties of materials. These are the basis for many laws, such as Gauss's law.

Maxwell–Wagner polarization: Also known as Maxwell–Wagner–Sillars polarization. See **Interfacial polarization**.

Molar polarization (P_m): Another form of the Clausius–Mossotti equation, expressed such that we obtain the polarization per mole (P_m) with the unit of m^3/mol or cm^3/mol. This is given by the following equation:

$$\left(\frac{\varepsilon_r - 1}{\varepsilon_r + 2} \right) = \frac{P_m \times \rho}{M}$$

Multilayer capacitor (MLC): A capacitor comprising alternating layers of dielectrics and electrodes. The layers are connected in parallel to maximize the volumetric efficiency of the capacitor.

Nonlinear dielectrics: Materials in which the developed polarization is not linearly related to the electric field; thus, the dielectric constant of these materials will be field-dependent. This includes ferroelectrics and other materials, such as water, in which molecules have a permanent dipole moment.

Optical-frequency dielectric constant: See **High-frequency dielectric constant**.

Optical polarization: See **Electronic polarization**.

Orientational polarization: See **Dipolar polarization**.

Paraelectric phase: The high-temperature phase derived from an originally ferroelectric parent phase that now has no dipole moment per unit cell.

Phasor: A rotating vector that shows the phase angle for a particular type of current or voltage in a circuit component.

Phasor diagram: A diagram showing the relative positions of phasors.

Piezoelectric: A material that develops electrical voltage or charge when subjected to stress. The material also develops a relatively large strain when subjected to an electric field.

Polar dielectrics: Materials that have a permanent dipole moment (e.g., water), or ferroelectric materials in which a dipole moment is spontaneously set up.

Polarizability (α): The tendency of an atom or an ion to undergo polarization under the application of an electric field. The dipole moment created (μ) when an electric field (E) is applied is given by $\mu = (\alpha \times E)$.

Polarization: The effect of an electric field on atoms, ions, and molecules in a material, resulting in the creation of induced dipoles or changes in the orientation of permanent dipoles.

Quality factor (Q): A measure of the dielectric losses; the higher the losses, the lower the Q. It is useful for comparing dielectric materials with very low dielectric losses; for these, Q is defined as the inverse of tan δ.

Real dielectric: A dielectric material that exhibits dielectric losses and an ability to store a charge. An ideal dielectric, which does not exist, has no dielectric losses.

Real part of the dielectric constant (k'): The frequency-dependent part of the complex dielectric constant related to the polarization processes, which enables charge storage.

Relaxation time (τ): The time required for a polarization mechanism to revert and realign or to change its value when the electrical field switches or changes in magnitude.

Ragone plot (Ragone chart): A diagram that compares a specific possible power density to a specific energy output for different power- or energy-generating devices.

Specific capacitance: Capacitance per unit mass.

Static dielectric constant (k_s): The value of the dielectric constant under DC fields, sometimes also equated to the low-frequency dielectric constant.

Static electronic polarizability ($\alpha_{e,static}$): The value of electronic polarizability when the electric field causing the polarization is not time-dependent.

Supercapacitors: Capacitors that make use of double-layer, interfacial polarization mechanisms to create a very high (~1 F) level of capacitance. The specific capacitance of these devices can be very high, at ~700–1000 F/g.

Tan delta: See **Tangent delta**.

Tangent delta (tan δ): The ratio of the loss current to the charging current in a dielectric material; also the ratio of the magnitudes of the imaginary and real parts of the complex dielectric constant.

Total dielectric polarizability $\left(\alpha_D^T\right)$: The dielectric polarizability of a compound calculated using Shannon's approach. This can be estimated from the individual polarizabilities (both electronic and ionic) of the ions, the unit cell volume, and the molecular weight; or from the total dielectric polarizabilities of other compounds.

Vector diagram: See **Phasor diagram**.

Vibrational polarization: See **Ionic polarization**.

Volume polarizability (α_{volume}): Polarizability that is often expressed in the unit of cm^3 or Å^3:

$$\alpha_{volume} \text{ in cm}^3 = \frac{10^6}{4\pi\varepsilon_0} \times \alpha\left(C \cdot V^{-1} \cdot m^2\right)$$

$$\alpha_{volume} \text{ in Å}^3 = \frac{10^{30}}{4\pi\varepsilon_0} \times \alpha\left(C \cdot V^{-1} \cdot m^2\right)$$

Volumetric efficiency: The total capacitance per unit volume. This is maximized by using a large number of thinner dielectric layers connected in parallel and by using materials with higher dielectric constants.

REFERENCES

Buchanan, R. C. 2004. *Ceramic Materials for Electronics*. New York: Marcel Dekker.

Choi, G. K., J. R. Kim, S. H. Yoon, and K. S. Hong. 2007. Microwave dielectric properties of scheelite (A = Ca, Sr, Ba) and wolframite (A = Mg, Zn, Mn) $AMoO_4$ compounds. *J Eur Ceram Soc* 27:3063–7.

El-Nahass, M. M., A. F. El-Deeb, H. E. A. El-Sayed, and A. M. Hassanien. 2007. Electrical conductivity and dielectric properties of bulk glass $Se_{55}Ge_{30}As_{15}$ chalcogenide. *Physica B* 388:26–33.

Fernandez, D. P., Y. Mulev, et al. 1995. A database for the static dielectric constant for water and steam. *J Phys Chem Ref Data* 24(1):33–69.

Hench, L. L., and J. K. West. 1990. *Principles of Electronic Ceramics*. New York: Wiley.

Kao, K. C. 2004. *Dielectric Phenomena in Solids*. London: Elsevier-Academic Press.

Kasap, S. O. 2006. *Principles of Electronic Materials and Devices*. New York: McGraw Hill.

Kishi, H., Y. Mizuno, and H. Chazono. 2003. *Jpn J Appl Phys* 42:1–15.

Mastragostino, M., and F. Soavi. 2007. Strategies for high-performance supercapacitors for HEV. *J Power Sources* 174:89–93.

Moulson, A. J., and J. M. Herbert. 2003. *Electroceramics: Materials, Properties, Applications*. New York: Wiley.

Prakash, B., and K. B. R. Varma. 2007. Influence of sintering conditions and doping on the dielectric relaxation originating from the surface layer effects in $CaCu_3Ti_4O_{12}$ ceramics. *J Phys Chem Solids* 68:490–502.

Raju, G. G. 2003. *Dielectrics in Electric Fields*. Boca Raton, FL: CRC Press.

Ram, M., and S. Chakravarty. 2008. Dielectric and modulus behavior of $LiFe_{1/2}Ni_{1/2}VO_4$ ceramics. *J Phys Chem Solids* 69(4):905–12.

Sayed, S. M. 2007. The study of dielectric relaxation and glass forming tendency in Cd-Se-Te glassy system. *Appl Surf Sci* 253:7089–93.

Shannon, R. D. 1993. Dielectric polarizabilities of ions in oxides and fluorides. *J Appl Phys* 73:348–66.

Vidal, D., L. Guengant, and J. Vermesse. 1984. Density and dielectric constant of dense argon and krypton mixtures. *Physica A* 127:574–86.

Vollmer, M. 2004. Physics of the microwave oven. *Phys Educ* 39(1):74–81.

Wang, Z., Y. Hu, H. Lu, and F. Yu. 2008. Dielectric properties and crystalline characteristics of borosilicate glasses. *J Non-Cryst Solids* 354(12):1128–32.

Wang, Z., Y. Hu, H. Lu, and F. Yu. 2008. *J Non-Cryst Solids* 354:1128–32.

Yang, X., Y. Wang, H. Xionga, and Y. Xia. 2007. Interfacial synthesis of porous MnO_2 and its application in electrochemical capacitor. *Electrochim Acta* 53:752–7.

8 Optical Properties of Materials

KEY TOPICS

- Basic properties of light as an electromagnetic wave
- Refractive index and its physical origin
- Drude model and Lorentz model
- Refraction, reflectance, and transmittance
- Extinction: scattering, absorbance, and attenuation of light
- Color and radiative recombination
- Application of light–matter interaction: anti-reflection coating, optical fiber, LASER, LED

We care about optical properties of materials because they determine how materials look to human eyes. Light is an electromagnetic wave. Figure 8.1 shows the wavelengths of electromagnetic waves including visible light in air. The wavelength of visible light that humans can see ranges from 400 nm (violet light) to 700 nm (red light). The propagation of light through materials is controlled by the materials' optical properties. For example, color in nature is a result of interactions between white light and materials. When white light enters matter, the electric and magnetic fields of light induce the electric and magnetic polarizations in materials. Since electromagnetic waves with different wavelengths interact differently with the induced dielectric and magnetic polarizations, only a part of the incident white light is reflected, scattered, refracted, absorbed, or transmitted. All types of these light–matter interactions contribute to the colors that we find in nature.

As we learned and will learn in Chapters 7 and 11, materials possess several dielectric and magnetic polarization mechanisms. For instance, dielectric polarization comes from the appearance or alignment (or from both) of electronic dipoles, ionic dipoles, orientational dipoles, and space-charge dipoles under an electric field. It is noted that each dipole has different operation frequency ranges. Since the frequency of visible light (~10^{15} Hz) is very high (i.e., the wavelength of 400 ~ 700 nm), only the electronic dipole can respond to the electric field of visible light and control the appearance of color. Most importantly, the formation of dielectric and magnetic dipoles by an electromagnetic wave is not constant even in the visible light regime. This indicates that materials respond differently to different parts of the solar spectrum.

For example, when only a red component of light is absorbed through material–light interactions (e.g., other parts of the visible light are transmitted or scattered), a material exhibits blue or green colors that are complementary to the red. In a transparent material such as window glass, visible light experiences negligible absorption, reflection, and scattering. If most of the incoming light is reflected on the surface of a material such as a metal, the material looks very shiny. The color of the sky is determined by the interaction of light with gas molecules (N_2, O_2) in the atmosphere. Even when the size of gas molecules (<1 nm) is much smaller than the visible light wavelength (400–700 nm), gas molecules can scatter light. Lord Rayleigh quantitatively explained light scattering by very fine particles. In the Rayleigh scattering regime, the phase of the electromagnetic wave is assumed to be constant in scattering particles, and the intensity of the light scattered by very small particles is proportional to λ^{-4} (with λ representing the wavelength of light). Therefore, a blue component of the solar spectrum, with its shorter wavelength, is scattered more than the red component (Figure 8.2). Scattered blue light travels in every direction and reaches human eyes on the Earth.

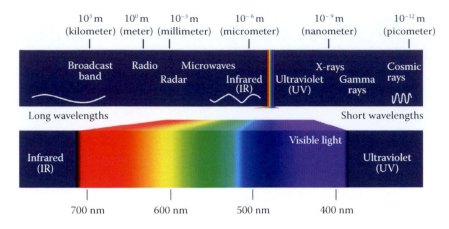

FIGURE 8.1 Wavelengths of electromagnetic waves including visible light. (From Amateur Astronomers Group, US. With permission.)

This is explains why the sky looks blue during the daytime. However, just before sunset, the color of the sky—especially in the region nearest the sun—turns red. This is attributed to the fact that the traveling distance of solar light in the atmosphere is increased before sunset as compared to at noon. Since the traveling path of sunlight in the atmosphere increases, blue light is scattered multiple times and the most incident photons are filtered completely. Thus, only red and yellow light reach human eyes on Earth and the sky near the sun turns red just before the sunset (Figure 8.2).

A material's property-controlling, visible light–matter interaction is expressed in the refractive index, which is a frequency-dependent complex function. The refractive index is correlated to dielectric and magnetic polarizations at a frequency of ~10^{15} Hz/s, which will be discussed in Section 8.1. In addition to the intrinsic properties of materials, extrinsic properties such as size and surface roughness also influence light–matter interactions. The extrinsic effects will be studied in Section 8.5.

An interesting example showing intrinsic and extrinsic effects involves the different colors of ocean and cloud—both of which are composed of water. The ocean's color comes from the intrinsic properties of water. Since a hydrogen bond of water (Section 1.4) absorbs the red part of the solar spectrum, water in the ocean exhibits a blue or green color. Therefore, even when white light enters, only the blue light is reflected back; thus, the ocean looks blue (Figure 8.3). However, water drops in clouds scatter the entire range of the solar spectrum rather than absorbing only red light because the individual drop size (1 ~ 10 μm) in clouds is slightly larger than the wavelength of visible light. Hence, the extrinsic property of water drops (i.e., size effect) makes clouds look white to the human eye (Figure 8.3). This type of scattering by relatively large particles is called Mie scattering. In contrast to Rayleigh scattering, Mie scattering explains the situation when the wavelength of light and the size of scattering particles are comparable. In the case of Mie scattering, the phases of the alternate electric and magnetic fields of the wave vary within the scattering particles. The difference between Rayleigh and Mie scattering mechanisms will be explained in Section 8.5.

Another example of Mie scattering is the stained glass often used for windows in medieval churches. Metal salts added into the raw materials of glass (mainly sands) are converted to metal nanoparticles (size: 10~100 nm) during the melting and cooling of glass. In comparison to water drops, the metal nanoparticles have many more free electrons that can be excited by an electromagnetic wave. Due to the free electrons' contribution, the absorption and scattering of the metal nanoparticles are frequency-dependent, which is different from Mie scattering by water drops. The selective absorption and scattering of visible light by the metal nanoparticles in glass result in

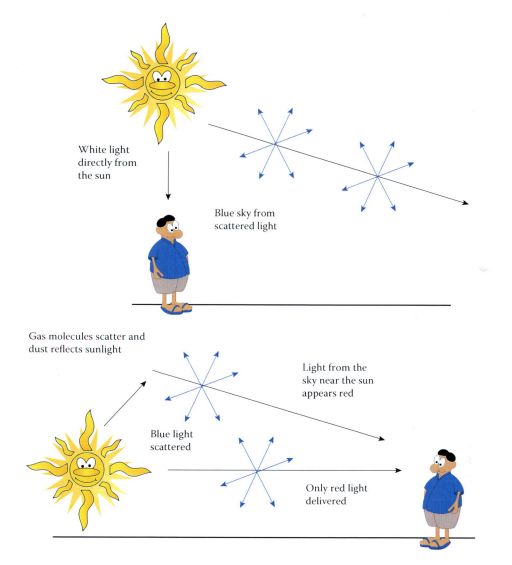

White light
directly from
the sun

Blue sky from
scattered light

Gas molecules scatter and
dust reflects sunlight

Light from the
sky near the sun
appears red

Blue light
scattered

Only red light
delivered

FIGURE 8.2 Schematics explaining the color of the sky at noon and sunset. (From NASA Space Place, US. With permission.)

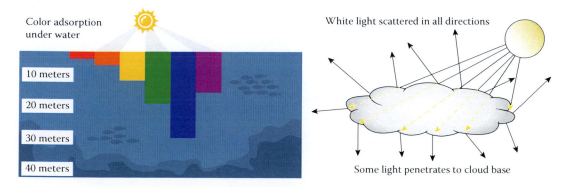

Color adsorption
under water

White light scattered in all directions

10 meters

20 meters

30 meters

40 meters

Some light penetrates to cloud base

FIGURE 8.3 Schematics explaining light absorption in the ocean and light scattering in clouds. (From NASA Space Place, US; US and Hong Kong Observatory, Hong Kong. With permission.)

the unique colors of stained glass, which are not observed in the shiny surface of bulk metals or in the insulating water drops of a cloud.

In this chapter, we will first study the basic properties of light, the electron-induced polarization of various materials, and the results of polarization–light interactions (e.g., absorption and scattering). Then, optically or electrically pumped light emissions and several applications of optical materials will be introduced in the latter part of the chapter.

8.1 DESCRIPTION OF LIGHT AS AN ELECTROMAGNETIC WAVE AND ITS CONNECTION TO THE PHYSICAL PROPERTIES OF OPTICAL MATERIALS

Light is an electromagnetic wave in which an electric field and a magnetic field oscillate in normal directions (Figure 8.4). The difference between electromagnetic waves and mechanical waves is that an electromagnetic wave does not require a medium. This suggests that the electromagnetic wave can travel in a vacuum (e.g., the free space outside the Earth's atmosphere). In contrast, the mechanical wave cannot propagate in a vacuum because there is no medium for mechanical vibration. This unique property of the electromagnetic wave is due to the coupling of the electric field and the magnetic field. That is, an oscillating electric field induces an oscillating magnetic field and vice versa. The coupling between two fields enables a self-propagation of the electromagnetic field without the medium. Therefore, it is necessary to review the electric field and the magnetic field to understand the electromagnetic wave.

Macroscopic features of the electric field and the magnetic field are described by Maxwell's equations, which integrate important laws and empirical observations on electricity and magnetism from the nineteenth century.

$$\nabla \cdot D = \rho \tag{8.1}$$

$$\nabla \cdot B = 0 \tag{8.2}$$

$$\nabla \times E = -\frac{\partial B}{\partial t} \tag{8.3}$$

$$\nabla \times H = J + \frac{\partial D}{\partial t} \tag{8.4}$$

$$\left(\nabla = i\frac{\partial}{\partial x} + j\frac{\partial}{\partial y} + k\frac{\partial}{\partial z} \right)$$

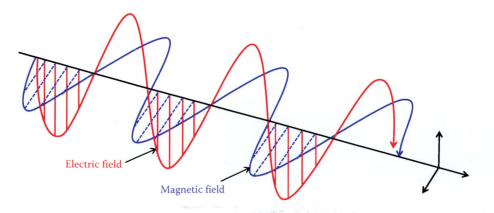

Electric field

Magnetic field

FIGURE 8.4 A schematic of an electromagnetic wave.

In Equations 8.1–8.4, E and H are the electric field vector (unit: volt/meter) and magnetic field vector (unit: ampere/meter), respectively, and show external driving forces that induce a change in materials; D and B are electric displacement (unit: coulomb/m^2) and magnetic induction (weber/m^2), respectively, which represent the responses of a material to either the electric field or magnetic field of the electromagnetic wave; J and ρ are current density (unit: ampere/m^2) and charge density (unit: coulomb/m^3), respectively, which are often considered as sources of electric and magnetic fields in materials.

The dot product of del and electric displacement (e.g., $\nabla \cdot D$) in the first Maxwell's equation (Equation 8.1) is called divergence of electric displacement (D), which measures a change in the magnitude of the electric displacement. As discussed in Chapter 7, D equals the sum of $\varepsilon_0 E$ and P, where P shows the effect of the charge that is bound in materials exposed to the electric field. Note that an electron can be separated from a hole and accumulated to a material's surface, which makes that material's surface negatively charged. This explains why $\nabla \cdot D$ is not zero. Equation 8.1 implies that the electric field can diverge or converge and that the nonuniform distribution of the net electric charge (i.e., the sum of the negative and positive charges) is the origin of electric displacement in the material. In other words, a change in electric displacement leaves the electric charge on the material's surface and produces the net electric field flux out of the close surface. This is known as Gauss's law of electrostatics or simply Gauss's law.

In contrast to Equation 8.1, where $\nabla \cdot D$ is equivalent to ρ, $\nabla \cdot B$ in Equation 8.2 is zero in any case. This is because the source of the magnetic induction is a magnetic dipole rather than a magnetic monopole. The north pole always pairs with the south pole to generate the magnetic field, and neither the north pole nor the south pole exists by itself. Therefore, a change in the magnetic induction does not accompany a change in the density of the sum of the north poles and the south poles. The qualitative meaning of Equation 8.2, which is also known as Gauss's law for magnetostatics, is that magnetic induction does not diverge and that magnetic field lines entering the volume of interest do not disappear or converge before they exit that volume (i.e., the net magnetic flux is zero). In Figures 8.5 and 8.6, different microscopic origins of magnetic polarization and dielectric polarization are compared. As shown in Figure 8.6, an external electric field generates or aligns (or both) the dielectric dipoles that hold electrons and holes on the surface of the conductive plates. The charge held by the dipoles is called the bound charge and the remaining charge is called the free charge.

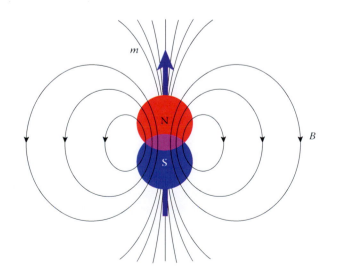

FIGURE 8.5 A schematic of a magnetic dipole with a north pole and south pole pair.

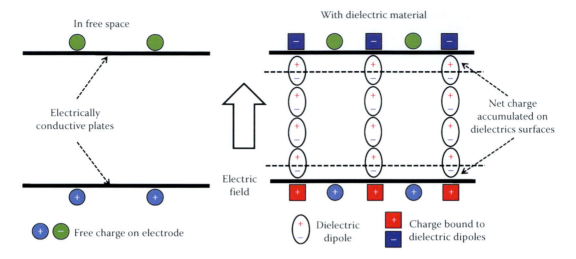

FIGURE 8.6 A schematic of charge accumulation on the surface of dielectric materials (the bound charge on the right plot is related to electric displacement).

Example 8.1: Surface Electric Charge and Electric Potential

1. If the surface charge of a material is not zero, electric potential is built up near the surface of the charged material. Show the relation between the electric potential and the charge density of the surface using Gauss's law.
2. Both sides of an alumina plate are coated with a metal electrode. The surface area of the electrode is 5 mm² and the relative dielectric constant and dielectric strength of the alumina are 9.1 and 17 volt/mm, respectively. What is the maximum amount of electric charge that can be stored on the alumina plate's surface?

Solution

1. In Gauss's law (or the first Maxwell's equation),

$$\nabla \cdot \boldsymbol{D} = \rho \text{ and } \boldsymbol{D} = \varepsilon \boldsymbol{E}$$

Since the electric field equals the gradient of electric potential,

$$\boldsymbol{E} = -\nabla V$$

By combining these two equations,

$$-\nabla \cdot (\varepsilon \nabla V) = \rho$$

$$\nabla^2 V = -\frac{\rho}{\varepsilon}$$

This shows that the gradient of electric potential is proportional to the charge density. If the material's surface is electrically conducting, the surface is equipotential and the potential gradient is developed along the normal direction of the charged surface. Also, when the left and right sides of the aforementioned equation are integrated, we can find a correlation between electric field and electric charge, which is called Poisson's equation

$$\oint_{\text{surface}} E_n \, dA = \frac{Q_{\text{total}}}{\varepsilon_0}.$$

where E_n is the electric field normal to the surface and Q_{total} is the total charge inside the surface.

2. Note that only the electric charge on the surface of the dielectric layer is not compensated. This means that the material is electrically neutralized except at its surface. Therefore, the surface charge density per unit area (σ) is correlated to the charge density (ρ). By combining these two equations,

$$\rho = \int \sigma \, (dA)$$

From $\nabla^2 V = -\dfrac{\rho}{\varepsilon}$, we can get an equation, $\nabla V = E = -\dfrac{\sigma}{\varepsilon}$. Consequently, the maximum surface charge density per unit area (σ) that can be stored on the dielectric material's surface equals the product of the maximum dielectric strength (17 volt/mm) and the permittivity of alumina (=9.62 × ε_0). In this example, the charge stored on the alumina with the electrode area of 5 mm^2 is $\varepsilon E A$:

$$Q = 17,000 \left(\frac{\text{volt}}{\text{m}} \right) \times 9.62 \times 8.85 \times 10^{-12} \left(\frac{A^2}{\text{m}^3 \text{kg}^1 \text{s}^4} \right) \times 5.0 \times 10^{-6} \ (\text{m}^2)$$

$$= 7.23 \times 10^{-12} \ (\text{F})$$

The cross product of del and the electric field ($\nabla \times E$) is called the curl of the electric field, which is the torque (rotational flow). In Equation 8.3, Faraday's law on magnetic field induction and Lenz's law on the direction of the induced magnetic field are combined. Equation 8.3 states that a change in the magnetic induction over time generates an electromotive force in a closed conductor line circuit. The third Maxwell's equation explains how an electric generator converts mechanical energy to electricity in a power plant. While Equation 8.3 is related to generation of the electric field, the fourth Maxwell's equation shows how the magnetic field is produced by circulating electric current. Equation 8.4 shows that integration of a magnetic field induced along a closed loop equals the sum of the electric current (J) and displacement current $\left(\dfrac{\partial D}{\partial t} \right)$ passing through the surface within the loop. To develop Equation 8.4, Maxwell added the term $\dfrac{\partial D}{\partial t}$ to Ampere's law. This indicates that not only electric current but also a change in the electric field leaving the closed surface of the material can generate the magnetic field.

Maxwell's equations in Equations 8.1–8.4 demonstrate how an electric charge and acceleration of an electric charge (i.e., electric current) generate and modify the electric field and the magnetic field. From these equations, Maxwell first postulated the electromagnetic wave. Simple manipulation of Maxwell's equations results in a general relation that quantitatively describes how the electric field and magnetic field are coupled to produce a propagating wave. This wave is called an electromagnetic wave—where the electric field and magnetic field oscillate in normal directions.

As shown in discussions of linear dielectric materials and magnetic materials (Chapters 7 and 11), the following constitutional relations exist between D, E, B, and H.

$$D = \varepsilon_0 E + P = \varepsilon E \tag{8.5}$$

$$B = \mu_0 H + M = \mu H \tag{8.6}$$

where ε_0 and ε are the permittivity of free space and materials, and μ_0 and μ are the permeability of free space and materials. When constitutive relations in Equations 8.5 and 8.6 are combined with Maxwell's equations, D and B in Equation 8.3 and Equation 8.4 are eliminated and the following equations are obtained:

$$\nabla \times E = -\mu \frac{\partial H}{\partial t} \tag{8.7}$$

$$\nabla \times H = J + \varepsilon \frac{\partial E}{\partial t} \tag{8.8}$$

We then take a curl operator on both sides of Equation 8.7. It is noted that J becomes zero and $\nabla \times H$ is the same as $\varepsilon \dfrac{\partial E}{\partial t}$ in free space (Equation 8.8). This gives

$$\nabla \times (\nabla \times E) = -\mu \frac{\partial}{\partial t}(\nabla \times H)$$

$$= -\mu\varepsilon \frac{\partial^2 E}{\partial t^2} \tag{8.9}$$

where $\nabla \times (\nabla \times E)$ in the left-hand side of the equation can be rewritten as $\nabla(\nabla \cdot E) - \nabla^2 E$. Since the surface charge density is zero in free space, the dot product of del and the electric field becomes zero $\left(\nabla \cdot E = \nabla \cdot \left(\dfrac{1}{\varepsilon} D \right) = \dfrac{\rho}{\varepsilon} = 0 \right)$. Then, Equation 8.9 becomes

$$\nabla \times (\nabla \times E) = \nabla(\nabla \cdot E) - \nabla^2 E$$

$$= -\nabla^2 E$$

$$= -\mu\varepsilon \frac{\partial^2 E}{\partial t^2} \tag{8.10}$$

In Equation 8.10, we find that $\nabla^2 E = \mu\varepsilon \dfrac{\partial^2 E}{\partial t^2}$, which yields a general form for the electric field of an electromagnetic wave. If the wave is assumed to propagate along the x-direction, a solution of $\nabla^2 E = \mu\varepsilon \dfrac{\partial^2 E}{\partial t^2}$ is:

$$E = E_0 e^{i(kx - wt)} \tag{8.11}$$

where w is the angular frequency and is equal to $2\pi\nu$ (ν: ordinary frequency with a unit of Hz). In the equation for a propagating electric field, $\mathbf{k}$ is called the wave vector and its magnitude is

$$|\mathbf{k}| = w\sqrt{\mu\varepsilon} = \frac{2\pi}{\lambda} \tag{8.12}$$

where λ is the wavelength of the electromagnetic wave. Similarly, a curl operator is applied to Equation 8.8. This gives us

$$\nabla^2 H = \mu\varepsilon \frac{\partial^2 H}{\partial t^2} \tag{8.13}$$

$$H = H_0 e^{i(kx - wt)} \tag{8.14}$$

It must be emphasized that we started from Maxwell's equations describing the relations among electric charge, magnetic induction, current generation, and the magnetic field. This leads to an equation that describes a general form of a traveling electromagnetic wave.

8.2 REFRACTIVE INDEX: A FACTOR DETERMINING THE SPEED OF AN ELECTROMAGNETIC WAVE

Given that the speed (s) of a wave is a product of wavelength (λ) and frequency (ν), Equation 8.12 is rewritten as:

$$s = \lambda \frac{w}{2\pi} = \frac{1}{\sqrt{\mu\varepsilon}} = \frac{1}{\sqrt{\mu_r \mu_0 \varepsilon_r \varepsilon_0}} \tag{8.15}$$

This shows the velocity of an electromagnetic wave in free space as well as in a medium. In free space, both ε_r and μ_r are 1 and the velocity of the electromagnetic wave is $\left(\sqrt{\mu_0\varepsilon_0}\right)^{-1}$. This gives the speed of light in a vacuum (e.g., $c = 2.9979 \times 10^8$ m/s).

Expressions on the waveform of an electric field vector and a magnetic field vector also show how the magnitude of the electric field and the magnetic field in free space are correlated. When Equations 8.11 and 8.14 are combined, we find:

$$E_0 = \sqrt{\frac{\mu}{\varepsilon}}H_0 \tag{8.16}$$

A correlation of $\sqrt{\dfrac{\mu}{\varepsilon}}$ showing a relation between the electric field and the magnetic field is called the intrinsic impedance (Z) of a material. In free space, Z is 120 $\pi\Omega$ (~377 Ω). More importantly, Equation 8.15 indicates that light's speed changes when light enters the medium from free space. The relative permittivity and permeability determine light speed (s) in the medium as follows:

$$s = \frac{1}{\sqrt{\mu_r\varepsilon_r}}c \tag{8.17}$$

Equation 8.17 (change in light speed) introduces an important parameter exhibiting the optical property of materials. The ratio of the electromagnetic wave speed in media to that in vacuum is called the refractive index (n) of the medium. Most media that allow for propagation of an electromagnetic wave are not magnetic materials and μ_r is close to 1. In this case, the refractive index of a medium is expressed as:

$$n = \frac{c}{s} = \sqrt{\varepsilon_r} = \sqrt{\frac{\varepsilon}{\varepsilon_0}} \tag{8.18}$$

Equation 8.18 shows that the speed of an electromagnetic wave is inversely proportional to $\sqrt{\varepsilon_r}$ and the refractive index is related to ε_r. To understand the n better, it is important to recall what has been learned in physics about linear dielectric materials. The permittivity of dielectric materials is a function of frequency, and only electron oscillation contributes to dielectric polarization in the frequency of visible light. This implies that the origin of the speed change in the dielectric medium is due to the electron oscillation that is synchronized to the alternate electric field of the electromagnetic wave.

Example 8.2: Change in Surface Electric Charge and Electric Potential

Two electromagnetic waves in a visible light regime enter from air (refractive index, n~1) and water (refractive index, n~1.33). Their wavelengths are 470 nm and 600 nm, respectively. Do the speed and frequency of these two electromagnetic waves remain the same when the medium of propagation changes? If not, calculate the change quantitatively.

Solution
The energy of the electromagnetic wave is determined by

$$E = h\nu$$

Since the energy of the electromagnetic wave is conserved, the frequency of the wave is maintained in different propagating mediums. This means that the wavelength and speed of the wave are different in air and water. As shown in Equation 8.17, the speed (s) of the electromagnetic wave is $s = \dfrac{1}{\sqrt{\mu\varepsilon}} = \dfrac{1}{\sqrt{\mu_r\mu_0\varepsilon_r\varepsilon_0}}$, which indicates that the speed does not depend on the electromagnetic wave's wavelength.

Because air and water do not exhibit strong magnetic field responses and the relative magnetic permeability (μ_r) is approximately 1 for both air and water, the speed of electromagnetic waves is

$$s(\text{air}) = \frac{1}{\sqrt{\varepsilon_r \mu_0 \varepsilon_0}} = \frac{1}{\sqrt{1.0 \times 1.27 \times 10^{-6} \times 8.85 \times 10^{-12}}} = 3.00 \times 10^8 \left(\frac{m}{s}\right)$$

$$s(\text{water}) = \frac{1}{\sqrt{1.33 \times 1.27 \times 10^{-6} \times 8.85 \times 10^{-12}}} = 2.25 \times 10^8 \left(\frac{m}{s}\right)$$

At this point, some readers may wonder whether or not the frequency and the wavelength of light change when light travels from one medium to another one. The answer to this question is that the frequency of light is constant and only the wavelength of light changes. A change in the light speed at the interface of different media is due only to the change in the wavelength of light. The frequency of the electromagnetic wave is constant regardless of its propagating medium.

One more important and unique feature related to the frequency of light is particle–light duality. This means that light behaves like a particle as well as a wave. To represent the particle-like property of an electromagnetic wave, physicists introduced the concept of a photon, which is an elementary unit of the electromagnetic wave. The photon can be understood as a discrete wave packet that behaves like a particle. As schematically illustrated in the left of Figure 8.7, when multiple waves with slightly different wavelengths are superimposed, the wave is localized instead of uniformly spreading out in space. Hence, the superimposition of waves creates a discrete wave pocket that looks like a particle. As more waves are added, the wave packet is more localized (i.e., Δx in Figure 8.7 decreases). Smaller Δx means that the wave packet gets closer to the particle, although uncertainty of the wave vector of the wave packet is increased at the same time. This is because the product of Δx and Δk is a constant or larger than a constant, which is predicted in quantum mechanics. As seen in the right of Figure 8.7, the particle-like photon is visualized as a wave pocket and the frequency of the wave pocket is considered the frequency of the photon.

Since the photon is an elementary unit of light and the energy of light is quantized, the minimum energy unit of light is the energy of a single photon. The energy of a photon is proportional to its frequency:

$$\text{Energy} = h\nu \tag{8.19}$$

where h is the Planck's constant ($6.62606957 \times 10^{-34}$ m^2 kg/s) and ν is the frequency of the light. When the photon travels in different media or when they come across the boundary between two media, the energy of a photon (i.e., frequency of the electromagnetic wave) is maintained. Therefore, a change in the refractive index of the medium only leads to a change in the wavelength of light; the light speed is altered without modifying the frequency of the light.

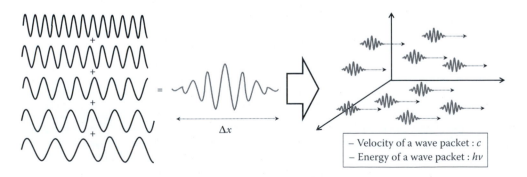

FIGURE 8.7 A schematic explanation of a photon that can be viewed as a wave pocket rather than a simple particle.

8.3 ORIGIN OF THE REFRACTIVE INDEX: INDUCED POLARIZATION OF A MEDIUM

In Sections 8.1 and 8.2, we learned how the refractive index of a medium is defined (i.e., the ratio of light speed in materials to free space, $s = c/n$) and how dielectric permittivity relates to the refractive index ($n^2 = \varepsilon/\varepsilon_0$). In this section, we will study the microscopic origin of the refractive index. For this purpose, we need to revisit Chapter 7 on linear dielectric materials and find out what factors determine dielectric permittivity. In materials with a negligible magnetic property, the response of dielectric dipoles to the electric field component of the electromagnetic wave is the microscopic origin of the refractive index.

When the dielectric material is exposed to an electric field, dielectric dipoles are aligned inside the material and an electric charge is stored on the material's surface. As readers learned in Chapter 7, there are four different polarization mechanisms responsible for dipole alignment: interfacial polarization (or space-charge polarization), dipole polarization (or orientation polarization), atomic polarization (ionic polarization), and electronic polarization (Figure 8.8). Each polarization mechanism has its own unique response time, which is called the relaxation time (or resonance time). The response time of different polarization mechanisms shows how quickly the dipoles are aligned when a material is exposed to an alternating electric field. As illustrated in the right plot of Figure 8.8, the upper limit of an operation frequency window varies. A wavelength of visible light ($\lambda = 400$ nm ~ 700 nm) corresponds to a frequency range of 4×10^{14} Hz $- 7 \times 10^{14}$ Hz. At such a high frequency regime, only 1/(response time) of electronic polarization is shorter than the frequency of visible light; so only the electronic polarization works well in the visible light regime. This means that the other three polarization mechanisms do not contribute to the refractive index in the visible light regime. The origin of electronic polarization is the shift of electron clouds surrounding

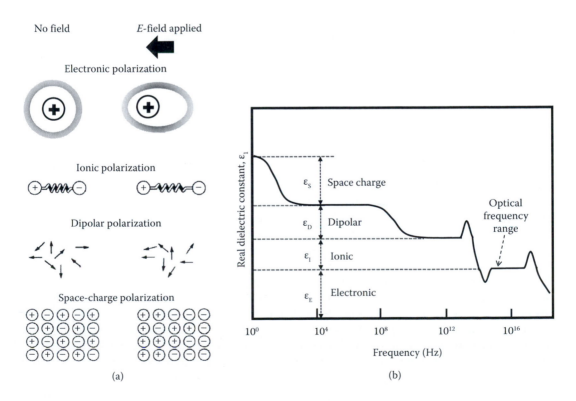

FIGURE 8.8 Schematics of (a) four polarization mechanisms and (b) their response time and operating frequency. (From *Electrochemistry Encyclopedia of the Electrochemical Society*. With permission.)

a nucleus under the external electric field. Due to the shift induced by the electric field, the center of the electron clouds (e.g., the center of the negative charge) is displaced from the position of the nucleus (the center of the positive charge) and polarization is generated. Given that the nuclear mass is much larger than the electron mass, we can assume that the charge displacement in electronic polarization is caused by the motion of the electrons. In Section 8.3, we will study how an alternating electric field moves electrons in solids.

Example 8.3: Dielectric Constant Versus Frequency of Electromagnetic Waves

(a) Although the dielectric constant of a diamond with strong covalent bonding characteristics does not exhibit strong frequency dependence at room temperature, that of sodium chloride (NaCl) with strong ionic bonding characteristics changes significantly as a function of frequency. Explain this different frequency dependence. (b) Relative permittivity of water at 10^8 Hz and 10^{11} Hz is ~90 and ~6, respectively. Estimate the relative contribution of dipolar polarization, ionic polarization, and electronic polarization to the permittivity of water. The refractive index of water is 1.33 at room temperature.

Solution
1. Since the valence electrons form bonds among atoms, covalent materials do not exhibit ionic polarization and dipolar polarization. Dielectric properties of strong covalent materials are determined by electronic polarization. Until the frequency of the incident wave reaches the natural resonance frequency of electronic polarization (>10^{16} Hz), the dielectric response of the electric field does not change significantly. In contrast, the natural resonance frequency of ionic bonding is near 10^{13} Hz, and the ionic polarization mechanism does not work above the resonance frequency. This leads to a large frequency dependence of the permittivity near the resonance frequency.
2. Dipolar polarization, ionic polarization, and electronic polarization contribute to the relative permittivity (~90) of water at a low frequency (10^8 Hz). As the frequency increases to 10^{11} Hz, the relative permittivity is attributed to the sum of ionic polarization and electronic polarization (~6). The contribution of the electronic polarization to the relative permittivity at a high frequency can be calculated from the square of the refractive index ($\varepsilon_r = n^2 = 1.8$). Therefore, 84:4.2(= 6 − 1.8):1.8 is the relative magnitude ratio of dipolar polarization, ionic polarization, and electronic polarization in water.

8.3.1 RESPONSE OF FREE ELECTRONS TO AN ELECTROMAGNETIC WAVE: A CASE OF METALS

Electrons in solids can travel freely in the conduction band (i.e., free electrons) or can be tightly bound to a nucleus (i.e., bound electrons) in a solid. Responses of free electrons and bound electrons to an electric field are different. We will first study a case of metals (and degenerate semiconductors) where free electron motion is dominant. Free electrons in a conduction band do not form a strong pair with the nucleus. When an electric field is applied to metals, electrons in the conduction band move freely within a solid without electric interaction with the positive charge of the nucleus. In metals or semiconductors with a high free electron density, repulsion among free electrons shields (or cancels) the electrostatic field of positively charged ions or the nucleus (called a screening effect, see Figure 8.9). Hence, in a traditional free electron model (Drude model), it is postulated that free electrons are detached from positively charged ions or the nucleus. From previous chapters, you will recall how we treat the transport of free electrons in metals to calculate electric conductivity. Only when electrons collide with positively charged ions do positively charged ions delay electron transport through scattering. We do not take into account a term representing electron-ion attraction to estimate the response of free electrons to an electric field. Due to the lack of electron–ion interaction (except in a collision), free electrons displaced by an alternating current (AC) electric field do not return to their initial locations even after the AC electric field is removed. This means that

Electrons near the nucleus shields
the nuclear charge for electrons outside

Positive nucleus

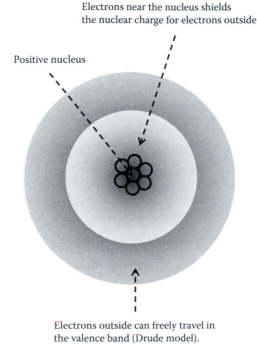

Electrons outside can freely travel in
the valence band (Drude model).

FIGURE 8.9 Schematics of the shielding of free electrons from positive ions.

the restoration of displaced electrons by positive ions is not necessary to quantitatively estimate the effect of an electric field on electron displacement.

Therefore, electron vibration, which is forced by the AC electric field of the electromagnetic wave, is described using the constitutive relation of classical mechanics:

$$\boldsymbol{F} = e\boldsymbol{E} = m_e a + \gamma v_e \tag{8.20}$$

where $\boldsymbol{F}$ is force, m_e is electron mass, a is acceleration vector, γ is a damping factor, and v_e is electron velocity. Damping is added to Equation 8.20 to consider collisions of electrons with ions and the subsequent delay of electron motion. In an ideal system where energy loss is neglected, the damping term is set to zero. Since the atom is much smaller than the wavelength of the electromagnetic wave, it is assumed that electric field intensity is constant through the atom. If that is the case, $E = E_0 e^{i(kx - wt)}$ is reduced to $E = E_0' e^{-iwt}$ and the electron displacement (x) is as follows:

$$m_e \frac{d^2 x}{dt^2} + \gamma \frac{dx}{dt} = -eE = -eE_0 e^{iwt} \tag{8.21}$$

In Equation 8.20, γv_e is a frictional force that results from the collision of electrons with atoms. In classical theory, electrons cannot be accelerated infinitely under an electric field. It is assumed that electrons are accelerated only in the time period between two collision events. This assumption allows for quantitative analysis of a steady state using a damping factor. Electron velocity at a steady state $\left(\text{when } \dfrac{d^2 x}{dt^2} = 0 \right)$ is called drift velocity (v_d), which is related to electric current density (J) as follows:

$$v_d = \frac{eE}{\gamma} \quad \text{and} \quad J = N_f e v_d = \sigma E \tag{8.22}$$

Then, the damping factor can be described using free electron concentration (N_f), electric conductivity (σ), and electron mobility:

$$\gamma = \frac{N_f e^2}{\sigma} = \frac{e}{\mu} \tag{8.23}$$

This indicates that the damping is inversely proportional to the electron mobility. We find a solution of Equation 8.24 by plugging Equations 8.22 and 8.23 into Equation 8.21. Then, charge displacement ($x = x_0 e^{iwt}$) is given by:

$$x = \frac{-eE}{\gamma wi - m_e w^2} = \frac{-eE}{\dfrac{ew}{\mu}i - m_e w^2} \tag{8.24}$$

According to the calculation of polarization in Chapter 7, charge displacement is related to polarization per unit volume as follows:

$$P = N_d qx = N_d \alpha E = \varepsilon_0 (\varepsilon_r - 1)E \tag{8.25}$$

$$\varepsilon_r = 1 + \frac{N_d \alpha}{\varepsilon_0} = 1 + \frac{N_d qx}{\varepsilon_0 E} \tag{8.26}$$

where N_d is dielectric dipole density, α is the polarizability, and q is the charge of each pole ($q = -e$ for electronic polarization). Then, the refractive index in Equation 8.18 can be rewritten by plugging the charge displacement in Equation 8.24 into Equation 8.26. If only free electron motion contributes to dielectric polarization, N_d is the same as free electron density (N_f), and we find the refractive index as:

$$\varepsilon_r = n^2 = 1 - \frac{N_f e^2}{\varepsilon_0} \left(\frac{1}{m_e w^2 - \gamma wi} \right) = 1 - \frac{N_f e^2}{\varepsilon_0} \left(\frac{1}{m_e w^2 - \dfrac{ew}{\mu}i} \right) \tag{8.27}$$

In Equation 8.27, it is noted that n can reach 0 at a specific frequency called plasma frequency (w_p). If we assume that free electrons do not experience any collision with atoms, the carrier mobility is infinitely high ($\mu = \infty$), and w_p is given by:

$$w_p = \left(\frac{N_f e^2}{m_e \varepsilon_0} \right)^{\frac{1}{2}} \tag{8.28}$$

If the frequency of an electromagnetic wave is smaller, then the plasma frequency ($w < w_p$), n^2 in Equation 8.27) becomes smaller than 0 and only an imaginary part is left in n. This implies that free electrons react quickly to an electromagnetic wave at a low frequency and that no incident energy is stored (i.e., materials do not work as a capacitor). In addition, such prompt motion of free electrons near metal surface shields materials from incident light because a vibrating electron functions as a dipole antenna and generates a secondary electromagnetic wave (radiation). More importantly, an electrodynamic calculation using Maxwell's equations shows that there is a 180° phase difference between incident light and forward radiation by the electron. Due to the out-of-phase relation, incident light and forward radiations cancel each other in the deeper region of bulk metals. This means that the propagation of the incident light toward the inside material is blocked. Instead of transmission, at a frequency lower than w_p, strong radiation by free electrons causes reflectance of the electromagnetic wave. Since there is a 180° phase difference between the incident light and the remitted electromagnetic wave, reflected light by metal surface is out-of-phase in comparison to incident light. Figure 8.10 schematically illustrates the oscillation of a free electron and the 180° phase difference between the incident light and the reemitted electromagnetic wave.

For high frequency electromagnetic waves ($w > w_p$), however, the frequency of the wave is too fast, and free electrons do not follow completely the incident electromagnetic wave. In that case, materials start to work as capacitors rather than conductors. Also, n^2 becomes larger than 0. In this frequency regime, n no longer is an imaginary number. A positive sign for n^2 means that the incident wave can propagate in the medium because free electrons have no response. This change in sign for n^2 leads to an abrupt decrease in the reflectance near the plasma frequency, which is schematically shown in Figure 8.11. The plasma frequency is intuitively understood as the upper frequency limit below which free electrons vibrate collectively without time lag in response to the electromagnetic wave. However, as the frequency of the electromagnetic wave becomes larger than the plasma frequency, forced electron vibration is delayed and unique metallic behavior starts disappearing.

Equation 8.28 shows that w_p is proportional to $\sqrt{N_f}$. In metals, w_p lies near the visible light regime due to the high free electron density, which explains why the surface of metals looks very shiny (or very reflective) to our eyes. (Reflection at the surface and its relation to a complex refractive index will be covered in detail in Section 8.4.) As the free electron density decreases, the plasma frequency shifts from the visible light regime to the infrared (IR) regime. If the free electron density is comparable or smaller than the bound electron density, we need to take into account the bound

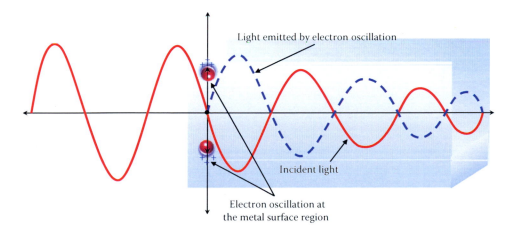

FIGURE 8.10 A schematic on the oscillation of a free electron and an 180° phase difference between the incident light and the reemitted electromagnetic wave.

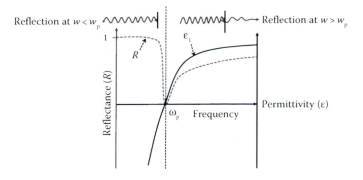

FIGURE 8.11 Schematics on the frequency dependence of the reflectance (left axis) and the permittivity (right axis) in materials where only free electron motion contributes to dielectric polarization (w_p: the plasma frequency).

electron motion to calculate the refractive index. This is a case of dielectric materials with ionic and covalent bond characteristics. In addition, if free electron density is very low in materials, propagation of visible light is not prevented in the medium and reflectance at the materials' surface is reduced. This explains why silica glass made of covalent bonds looks transparent to the human eye.

Example 8.4: Plasma Frequency of Metals

1. Estimate the plasma frequency (v_p) of lithium metal using the atomic density of lithium metal.
2. Show that the reflectance of the metal dramatically changes near v_p.
3. Explain how the defect density of lithium metals influences a slope of frequency versus the reflectance curve near v_p.

Solution

1. As shown in Equation 8.26, the permittivity of the dielectric is given by

$$\varepsilon_r = 1 + \frac{N_d \alpha}{\varepsilon_0} = 1 + \frac{N_d qx}{\varepsilon_0 E}$$

where N_d is dielectric dipole density (= electron density for electronic polarization), α is polarizability, and q is the charge of each pole ($q = -e$ for electronic polarization). Since each lithium ion has one valence electron, N_d is the same as the atomic density (ρ_A), which equals (mass density, ρ) × (molecular weight, M) ÷ (Avogadro's number, N_A). In this case,

$$\varepsilon_r = n^2 = 1 - \frac{\rho_A e^2}{\varepsilon_0} \left(\frac{1}{m_e w^2 - \frac{ew}{\mu} i} \right)$$

If electron collision is ignored (i.e., μ is very large), ε_r (= n^2) changes near the plasma frequency, $w_p = \left(\dfrac{\rho_A e^2}{m_e \varepsilon_0} \right)^{\frac{1}{2}}$, from negative to positive

$$w_p = 2\pi v = \left(\frac{\rho_A e^2}{m_e \varepsilon_0} \right)^{\frac{1}{2}} = 1.22 \times 10^{16} \text{ Hz (for lithium metal)}$$

2. Note that reflectance (R) follows a relation, $= \left(\dfrac{\bar{n} - 1}{\bar{n} + 1} \right)^2 = \dfrac{(n-1)^2 + \kappa^2}{(n+1)^2 + \kappa^2}$. In ideal metals where electron collision is ignored (i.e., μ is very large), the complex refractive index, $\bar{n}$ is given by

$$\bar{n}^2 = 1 - \left(\frac{w_p^2}{w^2} \right)$$

For $w < w_p$, $\bar{n}^2 (= \varepsilon_r)$ is negative, which requires that a real part of $\bar{n}$ is zero. Then,

$$R = \frac{(-1)^2 + \kappa^2}{(+1)^2 + \kappa^2} = 1$$

Reflectivity of 1 shows that the incident radiation energy is totally reflected at the surface of the bulk metal. In other words, electronic dipoles are not generated in the bulk part of metals (no energy storage) and incident light cannot penetrate into the deeper side of the bulk metal. If w is larger than w_p, $\bar{n}^2(=\varepsilon_r)$ becomes positive and n is not zero anymore. In other words, the incident light can travel inside metals and oscillate free electrons inside metals. Therefore, the reflectivity of the metal is smaller than 1. This indicates that a small increase in the frequency of the incident electromagnetic wave near w_p dramatically changes the function of a metal from a total reflector to a mixture of a reflector and a capacitor (e.g., $\varepsilon_r > 1$ and $\mathbf{R} < 1$).

3. In an ideal metal where the damping of electron oscillation is ignored, the refractive index (n) and permittivity (ε_r) dramatically change near $w = w_p$. However, note that the damping cannot be ignored in real metals. Therefore, the frictional force is not zero, and a transition of the function of the metal occurs over a broad range of the frequency domain in real materials. In this case, $\bar{n}^2(=\varepsilon_r)$ is rewritten as:

$$\varepsilon_r = n^2 = 1 - w_p^2 \left(\frac{1}{w^2 - w\dfrac{e}{m_e\mu}i} \right)$$

This equation shows that μ (electron mobility) is the main material property that determines the shape of the frequency versus the reflectance curve near $w = w_p$. Hence, an increase in the defect density makes the reflectance change more gradually in the frequency domain by reducing the electron mobility.

8.3.2 Response of Bound Electrons over an Electromagnetic Wave: A Case of Dielectrics

So far, we have studied free electron motion over a traveling electromagnetic wave. In contrast to free electrons, electrons bound to a nucleus are not completely shielded from the positive charge of ions. Therefore, there is attraction between the positive charge of ions (nucleus) and the negative charge of electrons. This interaction generates a restoring force that prevents electrons from freely leaving an atom. Since free electron concentration is very small in dielectrics, the aforementioned assumption is applicable for explaining the optical properties of dielectric materials. In this case, an electron bound to a positively charged ion can be treated as a harmonic oscillator, as shown in Figure 8.12.

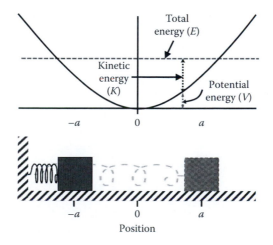

FIGURE 8.12 A schematic on a harmonic oscillator; a mass (electron) is attached to the immovable wall (positively charged ion) through a spring (Coulomb force).

Lorentz developed a model that explains light–matter interaction through bound electrons. An electron connected to a positively charged ion via Coulomb force is similar to a mass attached to an immovable wall through a spring. If there is no damping, the sum of the kinetic energy and potential energy of an electron is maintained. In this scenario, the restoring force of a displaced electron is proportional to electron displacement. Hence, constitutive Equation 8.21 can be rewritten for the damped oscillation of an electron by Coulomb force:

$$m_e \frac{d^2 x}{dt^2} + \gamma \frac{dx}{dt} + \kappa x = -eE = -eE_0 e^{iwt} \tag{8.29}$$

where κ is a spring constant that is a factor quantifying interactions between a bound electron and positively charged ions. In an ideal harmonic oscillator without damping $\left(\frac{d^2 x}{dt^2} + \frac{\kappa}{m_e} x = 0 \right)$, the spring constant ($\kappa$) and the resonance frequency of the harmonic oscillator (w_r) are related as:

$$w_r^2 = \frac{\kappa}{m_e} \tag{8.30}$$

Plugging $w_r = (\kappa/m_e)^{1/2}$ and $x = x_0 \exp(iwt)$ into the constitutive relation, we find the following steady-state solution:

$$x = \frac{-eE}{m_e \left(w_r^2 - w^2 \right) + iw\gamma} \tag{8.31}$$

Since one atom exposed to the electric field forms one dipole, N_d in Equation 8.26 is the same as the atomic density (N_a), and n^2 is rewritten as:

$$n^2 = \varepsilon_r = 1 - \frac{N_a e^2}{\varepsilon_0 \left[m_e \left(w^2 - w_r^2 \right) - i\gamma w \right]} \, 1 - \frac{N_a e^2}{\varepsilon_0 \left[m_e \left(w^2 - w_r^2 \right) - \frac{ew}{\mu} i \right]} \tag{8.32}$$

Equation 8.32 indicates that n and ε_r are complex numbers. Their general forms are:

$$n = n + ik \tag{8.33}$$

$$\varepsilon_r = \varepsilon_1 + i\varepsilon_2 \tag{8.34}$$

This gives the real and imaginary parts of ε_r:

$$\varepsilon_1 = 1 + \frac{N_a e^2 \left(w_r^2 - w^2 \right)}{\varepsilon_0 m_e \left[\left(w_r^2 - w^2 \right)^2 + \left(\frac{\gamma w}{m_e} \right)^2 \right]} \tag{8.35}$$

$$\varepsilon_2 = \frac{N_a e^2 \gamma w}{\varepsilon_0 m_e \left[\left(w_r^2 - w^2 \right)^2 + \left(\frac{\gamma w}{m_e} \right)^2 \right]} \tag{8.36}$$

Equations 8.31 and 8.32, which connect the electron displacement with the motion of the damped oscillator, indicate that the origin of the imaginary part of n and ε_r is the electron collision that results in a decay of the propagating waves. For this reason, **k** is named the extinction coefficient and represents energy dissipation during electron oscillation and the consequent attenuation of the propagating electromagnetic waves. In Equations 8.32–8.34, it is also important to understand that n and k are not independent. They are correlated through ε_1 and ε_2. In the equation $(n + ik)^2 = \varepsilon_1 + \varepsilon_2$, n and k are given by:

$$n = \sqrt{\frac{\varepsilon_1 + \left(\varepsilon_1^2 + \varepsilon_2^2\right)^{1/2}}{2}}, \quad k = \sqrt{\frac{-\varepsilon_1 + \left(\varepsilon_1^2 + \varepsilon_2^2\right)^{1/2}}{2}} \tag{8.37}$$

Comparing Equations 8.35–8.37, we find that an increase in ε_1 from an increase in the in-phase displacement of electrons increases n and decreases k. This, in turn, slows down the electromagnetic wave in the media (see Equation 8.15). Due to the in-phase motion, an oscillating electron bound to a positively charged ion behaves like a dipole antenna and reemits the electromagnetic wave. This light–electron interaction decreases the wavelength of the electromagnetic wave, which reduces the speed of the propagating electromagnetic wave. On the other hand, out-of-phase displacement of electrons, which is the origin of ε_2, increases both n and k. The out-of-phase component of electrons does not contribute to a function of electron motion as the dipole antenna. An increase in ε_2 only causes the energy loss that leads to attenuation of the propagating light (i.e., a decrease in light intensity) and a decrease in light speed in the media.

8.4 CHANGE IN LIGHT TRAVELING DIRECTION AT A MATERIAL INTERFACE: REFRACTION AND REFLECTANCE

In Section 8.3, we learned why the speed of light is different in free space and dielectric materials. The electric field component of light displaces free and atom-bound electrons of media. Electron displacement, in turn, changes the speed and wavelength of light. Inelastic displacement of electrons dissipates the energy of propagating light as well as slows down the light. We also calculated a relation between dielectric permittivity and electron displacement at the regime where the electronic polarization dominates the material's response to the electric field. For a UV-Vis-IR regime where the frequency of the electromagnetic wave is $10^{14} \sim 10^{16}$ Hz, relaxation or resonance time of other polarization mechanisms (ionic, orientational, and space-charge polarizations) is too large to follow the high frequency electromagnetic wave.

In this section, we will quantitatively study refraction, reflectance, scattering, and transmittance of light traveling in materials. We will show how the direction and intensity of propagating light change at the interface of two materials with different refractive indices and how the microscopic view of the refractive index (vibration of free electrons and oscillation of bound electrons) is correlated to the macroscopic property of light.

8.4.1 REFRACTION AND REFLECTION OF AN ELECTROMAGNETIC WAVE AT A FLAT INTERFACE

When light enters from air to water, we know that the traveling direction of light is altered at the water–air interface unless the traveling path of the incident light is normal to the water–air interface. This light bending is called refraction, and it is a consequence of a change in light speed (or wavelength). Refraction is common in our daily life. For example, white light is refracted at the prism–air interface and visible light with color is separated (see the left plot of Figure 8.13). In addition, refraction on the surface of a convex lens helps us to see objects more clearly. Also, silica glass fibers deliver light without energy loss by using refraction.

The right plot of Figure 8.13 intuitively explains the refraction of light. If light travels from air to glass at an incident angle of θ_1, light experiences refraction due to the change in light speed. As shown in Figure 8.13, light traveling a distance along Path 1 $\left(\overline{BD}\right)$ and Path 2 $\left(\overline{AC}\right)$ is not the same because light speed is different in air and glass. Let us represent the traveling distance in air and glass as d_1 and d_2, respectively. If Medium 2 has a higher n than that of Medium 1,

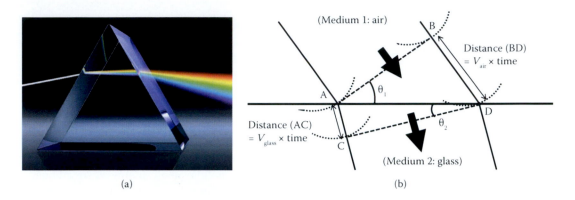

(a)   (b)

FIGURE 8.13 (a) Due to refraction by a glass prism, red, yellow, green, blue, and purple components of white light are split at glass–air interfaces. (From *Encyclopedia Britannica*. With Permission.) (b) A schematic showing the refraction of light at the interface of two media.

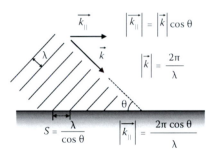

FIGURE 8.14 A schematic showing the wave vector component parallel to the interface on which light impinges.

light speed (or wavelength) in Medium 2 decreases and d_1 is larger than d_2. Because of a difference in d_1 and d_2, the refraction angle of θ_2 is not the same as that of θ_1, and their relations are obtained from geometric consideration:

$$\frac{d_1}{\sin\theta_1} = \frac{d_2}{\sin\theta_2} \tag{8.38}$$

Since $d_1/d_2 = V_1/V_2 = \left(\dfrac{c}{n_1} \Big/ \dfrac{c}{n_2}\right)$, Equation 8.38 is rewritten as:

$$n_1\sin\theta_1 = n_2\sin\theta_2 \tag{8.39}$$

Equation 8.39 is known as Snell's law and gives the angles of incidence and refraction for the electromagnetic wave impinging on the interface of media with different refractive indices. This means that light entering the media with a higher refractive index is slowed down and the refraction angle is smaller than the incidence angle.

We can also derive Snell's law on refraction using wave vectors of light. First, let us consider a change in the wave vector at the interface of two media. Since the tangential components of incident and refracted wave vectors must be continuous at the interface (see Figure 8.14), the wave vector component ($k_{\parallel}$) that is parallel to the interface must satisfy the following relation:

$$|k_{\parallel}| = |k_1| \, \sin\theta_1 = |k_2| \, \sin\theta_2 \tag{8.40}$$

where $\mathbf{k}_1$ and $\mathbf{k}_2$ are the wave vectors of the light traveling in Medium 1 and Medium 2, respectively. Since the absolute value of a wave propagation vector is the same as $\dfrac{2\pi}{\lambda}\left(\text{i.e.,}|\mathbf{k}| = \dfrac{2\pi}{\lambda}\right)$, $|\mathbf{k}|$ can be expressed using the refractive index $\left(n = \dfrac{c}{v}\right)$ and frequency (v) as follows:

$$|\mathbf{k}| = \frac{2\pi}{\lambda} = \frac{2\pi v}{v} = \frac{2\pi v}{c} n \tag{8.41}$$

Here, photon energy does not depend on a traveling medium and refraction does not change the frequency of propagating electromagnetic waves. Therefore, we can derive Snell's law again by plugging Equation 8.41 into Equation 8.40.

In addition to refraction, the electromagnetic wave traveling from one medium to another experiences reflection due to a difference in the refractive indices. If we consider both reflection and refraction, the right plot of Figure 8.13 is redrawn as Figure 8.15. To circumvent the discontinuity of the wave vector, a tangential component of all wave vectors for incident, reflected, and transmitted waves (i.e., $\mathbf{k}_i$, $\mathbf{k}_r$, $\mathbf{k}_t$) in Figure 8.13 must be the same in any interface between two different media. Then, θ_i, θ_r, and θ_t must satisfy the following relation:

$$|\mathbf{k}_i|\sin\theta_i = |\mathbf{k}_r|\sin\theta_r = |\mathbf{k}_t|\sin\theta_t \tag{8.42}$$

As shown in Equation 8.42, $|\mathbf{k}|$ is determined by n, and n_i is the same as n_r. Therefore, Equation 8.42 depicts the important relation that incident angle is equal to reflected angle at the interface ($\theta_i = \theta_r$), in addition to Snell's law ($n_i\sin\theta_i = n_t\sin\theta_t$).

Reflection can be understood by using the concept of the dipole antenna that is a consequence of electron oscillation. In Section 8.3.1, we noted that the electric field of a light swings both free electrons and bound electrons in an illuminated medium. This electron motion by light induces AC electric current into the shined medium. If the electron–light interaction is elastic (e.g., no energy loss in the energy conversion process), AC electric current leads to reemission of the electromagnetic wave (called radiation). Though radiation from an individual electron spreads out circularly, the sum of reemitted waves originating from many electrons is not circular. This is because the interference

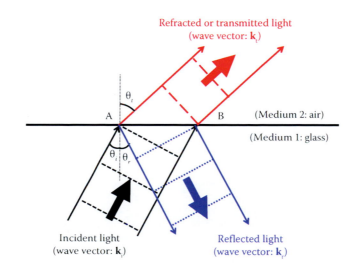

FIGURE 8.15 A schematic showing the relation among incident light, refracted light, and reflected light; the refractive index of an incident medium (n_1) is larger than that of a transmitted medium (n_2).

between reemitted waves reconstructs the light propagation direction. Reconstruction of electromagnetic waves originating from many electrons through interference can be explained using the Huygens–Fresnel principle. Reflected light is a part of the reconstructed electromagnetic wave that travels backward. In other words, backward radiation by oscillating (or vibrating) electrons is the microscopic origin of reflection.

If the size of the material on which light impinges is comparable to the wavelength of the incident light, backward radiation by oscillating electrons is called light scattering rather than light reflectance. Scattering will be studied in Section 8.5. In comparison, reconstruction of the reemitted electromagnetic waves along a forward direction is important to understand refraction, which is the sum of incident light and forward radiation by electrons.

One interesting question arising from Snell's law is what if θ_i is large enough to make θ_t equal to 90° when light travels from a medium with higher refractive index to one with lower refractive index (i.e., $n_i > n_t$)? This incidence angle for $\theta_t = 90°$ is called the critical angle θ_c. At this condition, $\sin\theta_c$ is the same as n_t/n_i and the refracted wave travels along the interface. For a critical angle, Snell's rule can be rewritten:

$$\sin\theta_t = \frac{n_i}{n_t}\sin\theta_i = \frac{\sin\theta_i}{\sin\theta_c} \tag{8.43}$$

where $\sin\theta_c$ is a constant ($= n_t/n_i$). If θ_i is larger than θ_c in Equation 8.43, we reach a strange conclusion that $\sin\theta_t$ must be larger than 1. This requirement cannot be met, which means that such a transmitted light component does not exist. Consequently, all of the incident wave is reflected at the interface. This phenomenon, which occurs in the case of $\theta_i > \theta_c$, is called total internal reflection. Figure 8.16 explains schematically the total internal reflection of the wave.

Example 8.5: Condition for Total Internal Reflection

Green light with the wavelength of 530 nm travels in a waveguide. The refractive index of the core is 1.45 and that of the cladding is 1.43. What is the condition of total internal reflection at the core–cladding interface?

Solution
The critical angle (θ_c) for total internal reflectance meets the condition:

$$\sin\theta_c = n_t/n_i$$

Since n_i and n_t are 1.45 and 1.43, respectively,

$$\theta_c = \sin^{-1}(n_t/n_i) = 80.5°$$

If the incident angle is larger than 80.5°, total power of the incident green light is reflected back to the core at the core–cladding interface.

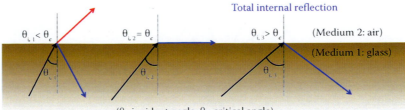

(θ_i: incident angle, θ_c: critical angle)

FIGURE 8.16 Schematics showing light propagation when an incident angle is smaller, same as, and larger than the critical angle (θ_c).

8.4.2 Power Distribution Between Refracted Light and Reflected Light at a Flat Interface

8.4.2.1 A Case of Normal Incidence

From Figure 8.15, we learn that the energy of an incident electromagnetic wave is distributed between reflected light and transmitted light at the interface of the two different media. Also, the light propagating direction of the transmitted light is deflected (called refraction). Both reflection and refraction are due to the oscillation or vibration of bound and fee electrons. We also see that the transmittance is not allowed for a certain range of incident angle ($\theta > \theta_c$ in Figure 8.16). Here, we will quantitatively estimate how the energy of incident light is distributed to that of reflected and transmitted light.

First, we start with the simple case of an electromagnetic wave whose traveling direction is normal to the interface of two different media, which is schematically shown in Figure 8.17. The incident electric field (E_i), reflected electric field (E_r), and transmitted electric field (E_t) are given by:

$$E_i(z,t) = E_{i,0} e^{j(wt - k_1 z)} \tag{8.44}$$

$$E_r(z,t) = E_{r,0} e^{j(wt + k_2 z)} \tag{8.45}$$

$$E_t(z,t) = E_{t,0} e^{j(wt - k_1 z)} \tag{8.46}$$

Since they propagate in opposite directions, the sign of the wave vector is different for E_i and E_r in Equations 8.44–8.46. If there is no energy loss at the interface, the tangential components of the electric field vector (i.e., vector components parallel to the interface of two media) must be continuous at the interface. In addition, the tangential components of the magnetic field intensity must differ at the interface by any surface current that is located on the interface. Since surface electric current is negligible on the surface of dielectrics with high resistivity, we can write the electric field and the magnetic field of the normal incident light at the interface using a condition of continuity (see Figure 8.17). The electric field and magnetic field at the interface are:

$$E_i + E_r = E_t \tag{8.47}$$

$$H_i - H_r = H_t \tag{8.48}$$

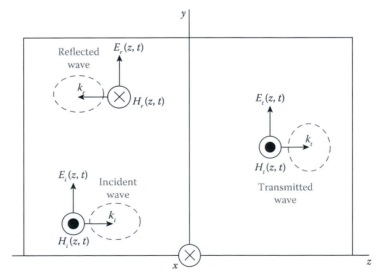

FIGURE 8.17 Schematics on the wave vector, electric field, and magnetic field of incident light, reflected light, and transmitted light. (Note that incident and reflected light have a 180° phase difference.)

The magnitude of the electric field vector is related to that of the magnetic field vector as $|H| = \left(\varepsilon_r/\mu_r\right)^{1/2}|E|$. In addition, $\varepsilon_r^{1/2}$ equals the refractive index ratio (n_2/n_1) of two media. Therefore, if μ_r is assumed to be 1, Equations 8.44 and 8.48 regarding the electric field vectors of incident, reflected, and transmitted light are reduced to:

$$E_i = \frac{1}{2}\left(1 + \frac{n_2}{n_1}\right)E_t \tag{8.49}$$

$$E_r = \frac{1}{2}\left(1 - \frac{n_2}{n_1}\right)E_t \tag{8.50}$$

$$E_{i,0} + E_{r,0} = E_{t,0} \tag{8.51}$$

$$\frac{E_{i,0}}{n_1} - \frac{E_{r,0}}{n_1} = \frac{E_{t,0}}{n_2} \tag{8.52}$$

where n_1 and n_2 are the refractive index of the incident medium and transmitted medium, respectively.

Electric fields obtained from Equations 8.49–8.52 also enable us to calculate the intensity ratio of incident, reflected, and transmitted light. In the late nineteenth century, Poynting defined a vector term (S) to express the magnitude and direction of the energy flow in electromagnetic waves as follows:

$$S = E \times H \tag{8.53}$$

The Poynting vector (S) indicates the rate of energy transfer through the unit surface area. Hence, integration of S over the area is the power that the electromagnetic wave delivers to the surface, which is often called light intensity. Since the magnitude of the electric field and the magnetic field has a relation of $|H| = \left(\varepsilon_r/\mu_r\right)^{1/2}|E|$, S is proportional to the square of the electric field magnitude:

$$\text{Light intensity} = s\varepsilon\frac{|E|^2}{2} \tag{8.54}$$

where s and ε are the light speed in the medium and in the permittivity of medium, respectively. According to Equations 8.49, 8.50, and 8.54, reflectance (R), which is the intensity ratio of incident light and reflected light, is given by:

$$R = \frac{\text{Reflected light intensity}}{\text{Incident light intensity}} = \frac{\left(s\varepsilon|E_r|^2\right)/2}{\left(s\varepsilon|E_i|^2\right)/2} = \frac{|E_r^2|}{|E_i^2|} = \left(\frac{n_1 - n_2}{n_1 + n_2}\right)^2 \tag{8.55}$$

If light absorption by materials is negligible, the sum of reflected and transmitted light intensity should be the same as the incident light intensity. Therefore,

$$T = 1 - \frac{|E_r^2|}{|E_i^2|} = \frac{4n_1 n_2}{(n_1 + n_2)^2} \tag{8.56}$$

Equations 8.55 and 8.56 clearly demonstrate that the reflectance and transmittance of light at the interface of two different media are determined by the difference in the refractive index. As the difference between two media increases, the power of the incident light is distributed more to reflected light, and the intensity of transmitted light intensity is weakened.

These calculations assume that materials do not absorb light (i.e., the imaginary part of the refractive index is neglected). However, if the energy of light is lost while it travels in materials, the imaginary part of the refractive index must be taken into account. For light traveling from free space ($n = 1$) to a material with a complex refractive index, $\bar{n}$, Equation 8.55 for reflectance changes to:

$$R = \left(\frac{\bar{n}-1}{\bar{n}+1}\right)^2 = \frac{(n-1)^2 + \kappa^2}{(n+1)^2 + \kappa^2} \tag{8.57}$$

where the complex refractive index $\bar{n}$ is equal to ($n + i\kappa$). A comparison of Equations 8.55 and 8.57 demonstrates that an increase in the imaginary part of the refractive index (κ) increases the reflectance of materials. This explains why metals have highly reflective surfaces when the frequency of the incident wave is smaller than the plasma frequency ($w < w_p$). As shown in Equation 8.27, the oscillatory response of free electrons in the regime of $w < w_p$ increases κ and makes n much smaller than 1. Therefore, R in Equation 8.57 is rewritten as $[(-1)^2 + \kappa^2]/(1^2 + \kappa^2)$, which equals 1. A reflectance of 1 implies that all incoming light is reflected at the air–metal interface due to the electric field induced motion of electrons.

Example 8.6: Reflectance at Glass Window

1. Sunlight is incident through the glass window. If there is no antireflection layer on the glass surface, how much radiation energy is transmitted through the glass for blue ray ($\lambda = 450$ nm, $n = 1.47$) and infrared ray ($\lambda = 3$ μm, $n = 1.42$)? Assume that absorption is negligible.
2. Can you reduce the transmittance of infrared light, which delivers heat, without changing the transmittance of blue ray?

Solution

1. When light is incident from air to the medium, $R = \left(\dfrac{\bar{n}-1}{\bar{n}+1}\right)^2 = \dfrac{(n-1)^2 + \kappa^2}{(n+1)^2 + \kappa^2}$

   For light of $\lambda = 450$ nm and 3 μm, reflectance (R) is 3.6% and 3.0%, respectively.
2. Yes, we can control the transmittance of infrared (IR) light by coating a very thin oxide layer on the glass surface. If the coated layer exhibits high transmittance for visible light and high reflectance for IR light, the glass has selective reflectance for infrared light. This is a principle of low emissivity glass, which reflects a significant amount of the energy of infrared light and preserves heat inside the building. A well-known coating material for low emissivity glass is SnO_2 film and ceramic (e.g., ZnO, SnO_2, TiO_2, Si_3N_4)/Ag/ceramic (e.g., ZnO, SnO_2, TiO_2, Si_3N_4) multilayers. Due to a difference in the refractive index (n) of the coating layer for visible light and infrared light, transmittance of low emissivity glass drops significantly for infrared light, while the transmittance of visible light stays high.

8.4.2.2 A Case of Oblique Incidence

Previously, we discussed only normal incidence. What would happen if an incidence direction is not normal at a material's interface? Fresnel studied reflection and transmission of an electromagnetic wave at an interface with a more general geometry and found an intensity ratio for reflected waves and transmitted waves.

In a transverse electric (TE) mode, the electric field vector is normal for a traveling direction of the electromagnetic wave and parallel to the interface of the media, while the magnetic field vector is parallel to the plane of incidence (see the left plot of Figure 8.18). In that case, the tangential components of the electric field and magnetic field are continuous and the following relations are valid:

$$E_i + E_r = E_t \tag{8.58}$$

$$H_i \cos\theta_i - H_r \cos\theta_r = H_t \cos\theta_t \tag{8.59}$$

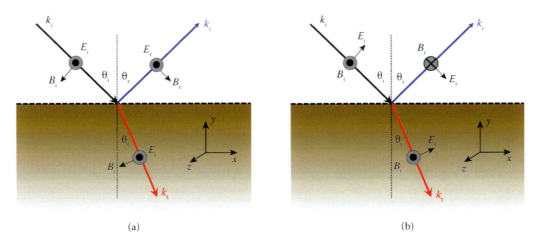

FIGURE 8.18 Schematics on the wave vector, electric field, and magnetic field of light for (a) transverse electric (TE) mode and (b) transverse magnetic (TM) mode; incident angle = θ_i, reflected angle = θ_r, transmitted angle = θ_t.

From relations of $H_0 = \sqrt{\mu_r \varepsilon_r}\, E_0$ and $\sqrt{\varepsilon_r} = n$, the substitution of H with E for the light traveling in nonmagnetic media provides the relation:

$$n_1(E_r - E_i)\cos\theta_i = -n_2(E_r + E_i)\cos\theta_t \tag{8.60}$$

Then, the reflection coefficient (r) and transmission coefficient (t) for the perpendicularly polarized light (i.e., light with the electric field vector normal to an incident plane containing directions of incident and reflected light) are given by

$$r_\perp = \frac{E_r}{E_i} = \frac{n_1\cos\theta_i - n_2\cos\theta_t}{n_1\cos\theta_i + n_2\cos\theta_t} \tag{8.61}$$

$$t_\perp = \frac{E_t}{E_i} = \frac{2n_1\cos\theta_i}{n_1\cos\theta_i + n_2\cos\theta_t} \tag{8.62}$$

These relations are called Fresnel's equations for light with perpendicular polarization. Equations 8.61 and 8.62 show a change in the electric field and magnetic field of the reflected and transmitted light, which are separated at the boundary of two media with different refractive indices. If the boundary of the two media is not flat, incoming light is reflected through diffuse reflectance.

In a transverse magnetic (TM) mode, the magnetic field vector is normal to a traveling direction of the electromagnetic wave and parallel to the incident plane, while the electric field vector is parallel to the incident plane (see the right plot of Figure 8.18). The relations to maintain the continuity of the tangential components of the fields are slightly changed, and Fresnel's equations for the light with parallel polarized light (i.e., light with the electric field vector parallel to the incidence plane) are summarized as follows:

$$r_\parallel = \frac{E_r}{E_i} = \frac{n_1\cos\theta_t - n_2\cos\theta_i}{n_1\cos\theta_t + n_2\cos\theta_i} \tag{8.63}$$

$$t_\parallel = \frac{E_t}{E_i} = \frac{2n_1\cos\theta_i}{n_1\cos\theta_t + n_2\cos\theta_i} \tag{8.64}$$

As Poynting postulated, the energy ratio of reflected light and incident light is given by the square of the reflection coefficients in Equations 8.61–8.64. It is noted that neither $r_\parallel$ or $r_\perp$ can be used to describe the reflectance and transmittance of natural light correctly. Since the natural light has both TE and TM modes, the electric field of the natural light has both perpendicular and parallel components. This is called an unpolarized state. For unpolarized natural light, therefore, the reflection coefficient is determined as the arithmetic mean of $r_\parallel$ and $r_\perp$ $\left(r = \dfrac{r_\perp + r_\parallel}{2} \right)$.

Figure 8.19 shows the effect of the incident light angle on $r_\parallel$ and $r_\perp$ when light travels from air ($n\sim1$) to glass ($n\sim1.5$). It is noted that $r_\parallel$ and $r_\perp$ may have different signs. This indicates that the phase of reflected light is shifted from that of the incident light. The right plot of Figure 8.19 suggests that the phase of perpendicular polarized light is always flipped at the event of reflection when the medium with a higher refractive index reflects the light. For the parallel polarized light, incident light and reflected light have a 180° phase difference only when the incident angle is smaller than the Brewster's angle. If θ_i = Brewster's angle, $r_\parallel$ becomes zero, and reflected light has only one polarization component ($r_\perp$). Therefore, the reflected component of natural light is polarized when the incidence of θ_i equals Brewster's angle. This explains why Brewster's angle is also called the polarization angle. A change in the phase of reflected light is intuitively explained in Figure 8.20, which shows the motion of a rope tied to a boundary. Depending on the nature of rope motion at the boundary, the incident wave and reflected wave along the rope may have different phases. When a transverse wave is fixed at an end, a reflected wave experiences a phase change of 180°. However, if the boundary is flexible, the phase change decreases. For an end where the rope moves freely, there is no phase change between the incident wave and reflected wave. This implies that the level of refractive index difference at the boundary controls the phase shift of reflected light. If $n_1 > n_2$, the reflection coefficient and phase change are very different from the data in Figure 8.19. In this case, the phase of reflected light does not change for an incident angle smaller than the Brewster's angle, which is analogous to the case of free end motion in Figure 8.20.

One last factor to be considered about light reflectance (shown in Figure 8.18) is how E_t (the electric field component of transmitted light) would change if the incident light is completely reflected (for a case of total reflectance). If the incidence angle exceeds the critical angle (see Equation 8.43), the transmitted light intensity must be zero. However, this does not mean that the electric field and magnetic field in Medium 2 are zero, too. If E_t and H_t are zero, there is no analytic solution for Equations 8.47, 8.48, 8.58, and 8.59, which are driven from Maxwell's equations. This contradiction

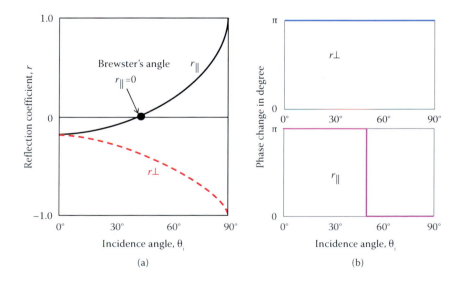

FIGURE 8.19 Dependence of (a) the reflection coefficient and (b) phase change on incident light angle, θ_i; $r_\perp$ = for perpendicular polarization, $r_\parallel$ = for parallel polarization.

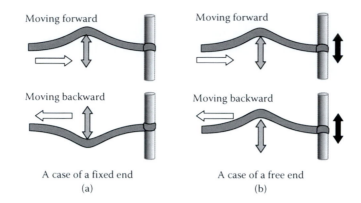

FIGURE 8.20   The motion of a rope tied to a boundary; there is (a) a 180° phase difference for a fixed end and (b) a zero phase difference for a free end.

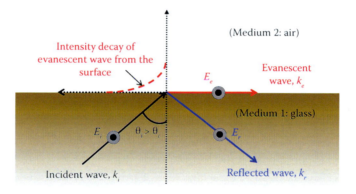

FIGURE 8.21   A schematic of an evanescent wave; when an incident angle of the plane wave is larger than the critical angle ($\theta_i > \theta_c$), a decaying and standing transmitted wave is generated.

between no transmitted light and non-zero electric/magnetic field is resolved by introducing a non-vanishing electromagnetic surface wave that is bound to the surface region of Medium 2. This surface wave is called an evanescent wave. Since the evanescent wave neither propagates in Medium 2 nor delivers energy, its wave vector (**k**) in Medium 2 is purely imaginary. Then, **k** can be expressed as ($i\alpha$) and the evanescent wave is given by:

$$E = E_0 e^{i(kx - wt)} = E_0 e^{i(i\alpha x - wt)} = E_0 e^{(-\alpha x)} e^{(-iwt)} \tag{8.65}$$

Equation 8.65 shows that the evanescent wave does not oscillate as a function of position (it oscillates only over time). The evanescent wave is a *standing* transmitted wave that decays exponentially from the interface (see Figure 8.21). When total internal reflection occurs, the evanescent wave is generated in the transmitted medium side to provide a solution to Maxwell's equation as well as to satisfy the energy conservation law.

8.5   EXTINCTION OF LIGHT: SCATTERING AND ABSORPTION

In Section 8.4, only reflection and transmittance of light were considered and light absorption was neglected. The motion of free and bound electrons in response to incident light was pointed out as an origin of reflection and transmittance. Excitation of electrons from a valence band or lattice vibration by incident light was not discussed in Section 8.4. However, when light travels in real materials, it is attenuated. Extinction is a process by which the radiant intensity of traveling light is decreased. Two different mechanisms are responsible for the extinction process. One is scattering, by which the energy of incident light is radiated in all directions. Scattering occurs because vibration (oscillation) of dipoles by light reemits secondary electromagnetic waves in every direction. The other is absorption, which decreases the number of photons in the light. Absorbed energy of light is converted to heat (i.e., lattice vibration) or light with different wavelengths. In this section, we will study how incident light is scattered or absorbed in materials and what the consequences of scattering and absorption are.

8.5.1 SCATTERING AT ROUGH SURFACE OR FINE OBJECTS

Scattering of light means that incident light is bent from a straight trajectory and travels toward all different directions. The incident light is scattered when it hits the irregular surface of a large object or a very fine object with a size close to or smaller than the wavelength of incident light.

8.5.1.1 Diffuse Reflection

Strictly speaking, the scattering of large objects on a non-flat surface is called diffuse reflection. Since the surface is rough or granular, the incident light is reflected in all directions, as schematically shown in Figure 8.22. The exact form of scattered light from diffuse reflection depends on surface features. An individual reflection occurring on a local surface follows the well-known mirror-like behavior of reflectance, and the total scattered light equals the sum of multiple reflections bouncing at different parts of the objects having irregular surfaces. The intensity of the diffuse reflection on a rough surface is expressed using Lambert's cosine law (see Figure 8.22). In this

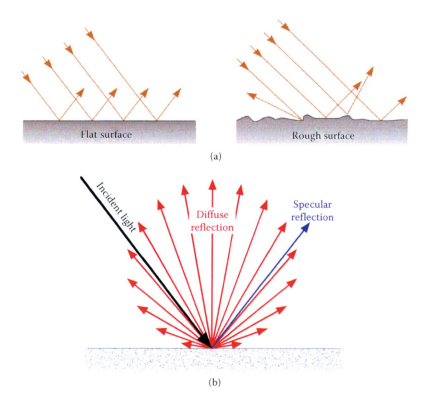

FIGURE 8.22 Schematics on specular reflection on (a) a flat surface and diffuse reflection on a rough surface and (b) reflected light power by diffuse reflection versus reflected light angle—the largest power along the surface–normal direction.

model, the angle dependence of the radiant power by the diffuse scattering is estimated by the cosine of an angle θ between the surface–normal direction and the observer's line of sight. According to Lambert's cosine law, an increase in the angle θ (i.e., a deviation of the reflected light direction from the surface–normal direction) decreases the power of the reflected light. However, the areal power density (i.e., power/area) of reflected light by the diffuse reflection is uniform, regardless of angles. Since an increase in the angle θ decreases the area exposed to the reflected light, a decrease in the power of the reflected light is compensated by a decrease in the area exposed to light. Therefore, the areal power density of the reflected light is the same over all the angles.

8.5.1.2 Rayleigh Scattering

Light is also bent and scattered when it strikes a very small subject. As explained in Section 4.2, the incident light oscillates the electrons of very fine materials, and the electron oscillation emits secondary electromagnetic waves in all directions. Since the material size is very small, emitted electromagnetic waves from fine materials do not exhibit the constructive interference of bulky materials. Instead, the reemitted light by fine objects spreads out in all directions, which is similar to the diffuse reflection by a rough surface. This spread of incident light via light–matter interaction is also called scattering. If the scattering is elastic, the energy of the incident photon is conserved and only the light-traveling direction is changed by the induced polarization of materials. In an inelastic scattering process, however, a part of the energy is consumed by the scattering media and converted to a different kind of energy such as heat.

Since the electron oscillation by the incident waves and the interference of the emitted electromagnetic waves play important roles in the scattering, the intensity of the scattered light depends on the polarizability and size of the scattering media. As explained in Section 8.3, polarizability is a parameter quantifying the electron oscillation and is determined by the dielectric constant of materials ($P = N_d qx = N_d \alpha E = \varepsilon_0 (\varepsilon_r - 1)E$). Hence, the real and imaginary part of the dielectric constant is the first material parameter that influences the scattered light intensity at different angles. The other important parameter of the scattering media is a ratio of the object size to the light wavelength. As the size (more strictly speaking, the surface feature) of the scattering object becomes much larger than the light wavelength, the light propagation at the interface is governed by the reflection that is described in the Huygens–Fresnel principle. Therefore, the following dimensionless size parameter (x) was developed to express the characteristic dimension of the scattering object properly:

$$x = \frac{2\pi r}{\lambda} \tag{8.66}$$

where r is the radius of the spherical scattering object and λ is the light wavelength. Depending on the value of x, the scattering of a fine object is classified into two groups. The first mechanism is Rayleigh scattering, which is applied to the dielectric object of $x \ll 1$. This means that the radius (r) of the object is much smaller than light wavelength (λ). In Rayleigh scattering, the electric field of light is uniform in the scattering object. The ideal system, which is explained by the Rayleigh scattering model, is elastic scattering by a single dipole, which does not absorb the incident light at all. In that case, the intensity (I_s) of the scattered light coming from the induced polarization of the dipole is given by:

$$I_s = I_0 \frac{8\pi^4 N \alpha_p^2}{R^2 \lambda^4} \left(1 + \cos^2 \theta\right) \tag{8.67}$$

where I_0 is the incident light intensity, N is the scattering particle density, α_p is the polarizability, R is the distance from the scattering center, λ is the wavelength of the incident light, and θ is the angle between incident light and scattered light in the plane of the electric field of the incident light. In Equation 8.67, it is noted that the I_s of forward scattering is equal to that of back scattering for the same cos θ. In other words, the intensity of scattered light is symmetric in the scattering plane, which is schematically shown in Figure 8.23.

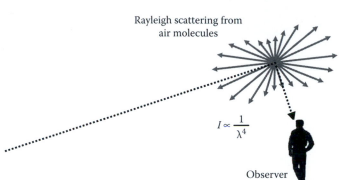

Rayleigh scattering from
air molecules

$I \propto \dfrac{1}{\lambda^4}$

Observer

FIGURE 8.23 A schematic on Rayleigh scattering of light by a very fine object ($\lambda \gg$ r).

In addition, Equation 8.67 indicates that the scattering becomes stronger as the wavelength decreases $\left(I_s \propto \dfrac{1}{\lambda^4} \right)$. This is why blue light is scattered much more than the red light when sunlight passes through the atmosphere where N_2 and O_2 molecules function as light scattering dipoles. Consequently, an observer at sea level sees blue light scattered from all directions of the sky and feels that the color of the sky is blue. Though the scattering power of violet light is stronger than that of blue light, the intensity of violet light in incident sunlight is much weaker than that of blue light; therefore, the sky does not look violet. In contrast to the sky, astronauts feel that the color of space is black because there are no scatterers, such as N_2 and O_2 molecules, in space.

8.5.1.3 Mie Scattering

The Rayleigh scattering model fits well with the scattering behavior of nonabsorbing fine particles that have a single dipole. However, this model is limited in its explanation of scattering when (1) the particle size is comparable to the wavelength of light (e.g., $x \approx 1$) or (2) particles absorb light. In these conditions, the assumption that the electric field is uniform throughout is no longer valid.

Gustav Mie explained scattering beyond the Rayleigh range. He solved Maxwell's equations for general cases of scattering and formulated solutions for scattering and absorption by spherical particles without particular conditions of size and light absorption. Mie scattering can be applied to very fine particles ($x \ll 1$) that are encompassed by the Rayleigh scattering model. However, since it is more difficult to obtain solutions for the Mie scattering model, the use of the Rayleigh scattering model is still preferred for very fine particles ($x \ll 1$). One problem with the Mie scattering model is that it only provides solutions to the problem of scattering by spherical particles. Later, some modification was made to span other shapes such as spheroids, ellipsoids, and rings.

If the particle shape is arbitrary, general solutions are not available. In this theory, the incident plane wave and the scattering field are expanded into radiating spherical vector wave functions, while the field inside the scattering particles is expanded into regular spherical vector wave functions. Then, the expansion coefficients of the scattered field can be calculated by applying the boundary condition of the spherical surface to the radiating spherical vector wave functions. Based on this process, Mie scattering theory provides the exact solution of the light scattering and absorption by spherical particles. The optical efficiencies Q_i, which describe the interaction of radiation with a scattering sphere, can be calculated by the cross section σ_i and geometrical particle cross section πa^2, shown as

$$Q_i = \frac{\sigma_i}{\pi a^2} \tag{8.68}$$

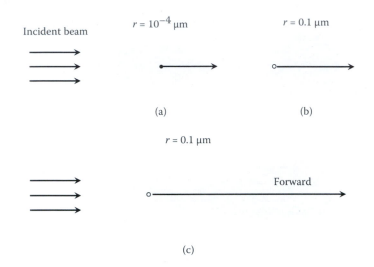

FIGURE 8.24 Schematics showing spatial intensity distribution of scattered visible light: (a) By Rayleigh scattering of fine particles ($r = 10$ nm), (b) by Mie scattering of medium-sized particles ($r = 100$ nm), and (c) by Mie scattering of relatively large particles ($r = 100$ μm).

Since Mie scattering encompasses both scattering and absorption of light, the extinction efficiency (Q_{ext}) of the incident light equals the sum of scattering efficiency (Q_{sca}) and absorption efficiency (Q_{abs}), which are given by:

$$Q_{ext} = Q_{sca} + Q_{abs} \tag{8.69}$$

In Equation 8.69, Q_{sca} is obtained from the integration of the scattered waves in all directions, Q_{ext} is deduced from the forward-radiation theorem of the incident light, and Q_{abs} equals the difference between Q_{ext} and Q_{sca}. Q_{ext} indicates the amount of light that is scattered or absorbed by the particle. Readers can find detailed calculations of radiating vector wave functions in various references.

The relation of Q_{ext} versus x and the spatial distribution of the scatter power are presented in Figure 8.24. Instead of a detailed calculation process, physical meanings of Mie solutions are introduced here. First, the scattering efficiency is less dependent on the wavelength in Mie scattering than in Rayleigh scattering. This means that Q_{ext} of Mie scattering lingers over a wide range of $x \left(= \dfrac{2\pi r}{\lambda} \right)$. Second, Q_{ext} is maximized when the particle size is close to the wavelength of the incident light. Third, a curve of Q_{ext} vs. $x \left(= \dfrac{2\pi r}{\lambda} \right)$ shows an oscillating relation due to interference with the incident light. Near the wavelength of local Q_{ext} maximum (λ_{max}), Q_{ext} approximately depends on $\dfrac{1}{|\lambda - \lambda_{max}|}$. Fourth, Mie solutions show that scattering by relatively large particles redirects the incident light to asymmetric electromagnetic waves in the scattering plane. In comparison to Rayleigh scattering, a ratio of the back scattered power over the forward scattered power is smaller than 1 for Mie scattering. This forward traveling feature of Mie scattering explains why we can see a red sun with a spherical shape in the sky near sunset.

8.5.2 ATTENUATION OF LIGHT BY ABSORPTION

When light travels in materials, it can be absorbed or scattered/reflected. The main difference between absorption and scattering/reflectance is that the absorption process results in dissipation of heat or reemission of light with a different wavelength. In other words, the electromagnetic energy of incoming light is consumed to excite electrons, ions, and atoms from a lower energy state to a higher energy state. When excited electrons, ions, and atoms return to the initial lower energy state, they release heat or light. In a case of light emission, it is noted that the incident and emitted light have a different wavelength. Several phenomena responsible for light absorption are grouped into electronic and lattice contributions. An electronic contribution results from a change in the energy state of electrons.

Light absorption by a lattice involves lattice vibration the frequency of which falls into the regime of infrared light. Though a lattice vibrates in all kinds of solids above 0 K, the absorption of infrared light by lattice vibration heavily depends on the type of atomic bonding. When ionic crystals are exposed to an electromagnetic wave, oppositely charged ions move in opposite directions due to the electric field component of light (see Figure 8.25). If the force between displaced ion pairs is

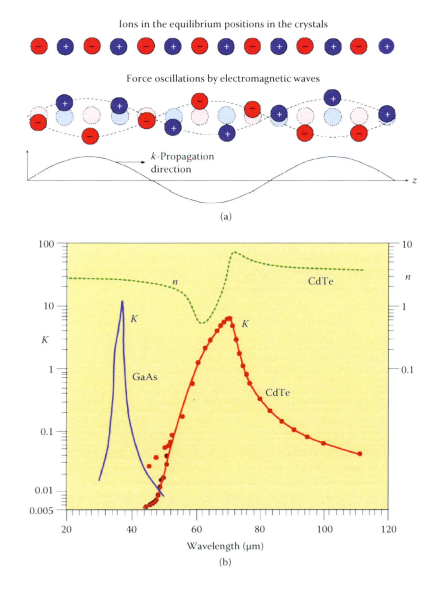

FIGURE 8.25 (a) Lattice vibration of ionic crystals by electromagnetic wave in an IR regime; (b) extinction coefficient of compound semiconductors as a function of wavelength in IR regime—IR light is absorbed through the lattice vibration. (From Kasap, S. O., *Principles of Electronic Materials and Devices*. New York: McGraw Hill, 2006. With permission.)

assumed to be proportional to the displacement (harmonic approximation), the force exerted on the 2nth atom by the (2n+1)th atom approximates $\alpha(u_{n-1} - u_n)$. Then, the following relations between forces and acceleration velocity in the diatomic ionic crystal are found to be:

$$M_1 \frac{d^2 u_{2n+1}}{dt^2} = -\alpha(u_{2n+1} - u_{2n}) + \alpha(u_{2n+2} - u_{2n+1}) + qE \tag{8.70}$$

$$M_2 \frac{d^2 u_{2n}}{dt^2} = -\alpha(u_{2n} - u_{2n-1}) + \alpha(u_{2n+1} - u_{2n}) - qE \tag{8.71}$$

where positive and negative ions have masses (M_1 and M_2) and electric charges (q and $-q$); u_n is the displacement of the nth ion exposed to the electric field, $E = E_0 e^{i(kx - wt)}$; and α is the proportional constant representing an interatomic force. This forced vibration of the ion pairs by the electromagnetic wave is called an optical phonon. The natural frequency range of the optical phonons lies in the infrared region. Interactions of light with optical phonons lead to the absorption and reflection of infrared light in ionic crystals. This indicates that the electric field of IR light oscillates ions and that the electromagnetic energy of light is converted to phonons of ionic crystals. As shown in the bottom plot of Figure 8.25, the extinction coefficient of compound semiconductors exhibits a maximum peak for IR light with a wavelength of 20–100 μm. It is also noted that the response of materials to visible light and IR light takes place through different mechanisms. While visible light activates the electronic polarization mechanism, IR light stimulates the harmonic motion of ions—that is, phonons (see Figure 8.8).

From Equations 8.70 and 8.71, the induced polarization by the displacement of the (2n+1)th ion and the 2nth ion is also given by

$$P_i = n_m q(u_{2n+1} - u_{2n}) = n_m q \left(\frac{1}{M_1} + \frac{1}{M_2} \right) \left(\frac{1}{1 - w^2 / w_t^2} \right) \tag{8.72}$$

where w_t is $2\alpha(1/M_1 + 1/M_2)$. At $w = w_t$, the largest ionic polarization is induced by light and the light absorption by the lattice vibration becomes maximized.

In comparison to the ionic bond, the effect of an electric field of light on atomic displacement is marginal in covalent materials where atoms do not possess a net electric charge. Hence, the absorption by the vibration of the covalent bond is weak and involves higher-order processes. The difference in atomic bonding characteristics explains why compound semiconductors such as GaAs and GaP, which possess ionic bonding characteristics, partially respond to IR light more strongly than pure covalent semiconductors such as Si and Ge.

Although the absorption of IR light by ionic and covalent bonding materials occurs via phonon generation, a major absorption mechanism for visible light is the interband transition of electrons. This mechanism also works for infrared light in a semiconductor with a small band gap ($E_g <$ 1.2 eV). Interband transition can occur in all types of materials if photon energy ($E = h\nu$) is larger than band gap (E_g) of materials (e.g., $h\nu > E_g$). Supra band gap photons (photons with energy larger than the band gap of a material) can excite an electron from a lower energy level to a higher energy level that is unoccupied. From the relations of $E = h\nu$ and $\nu = c/\lambda$ for light, an edge wavelength of the absorption spectrum is calculated as follows:

$$\lambda(\text{nm}) = \frac{1243}{E_g(\text{eV})} (8.57) \tag{8.73}$$

In Equation 8.73, it is noted that the unit of λ and E_g are nm and eV, respectively. Since the band gap of semiconductors ranges from ~0.5 eV to 3.2 eV, Equation 8.73 demonstrates that the absorption spectrum edge lies in the regimes of visible light and IR light, where most solar energy is concentrated. This explains why semiconductors are used as light absorbing materials for solar cells.

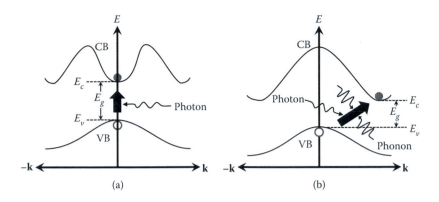

FIGURE 8.26 Photon absorption in (a) direct band gap and (b) indirect band gap materials. VB: valence band, CB: conduction band.

Silicon, which is a dominant material in solar cells, has an E_g of ~1.1 eV and absorbs photons whose wavelength is shorter than ~1100 nm.

The efficacy of light absorption by band-to-band transition is strongly related to the band structures of the semiconductor schematically displayed in Figure 8.26. In direct band gap materials, the maximum in the valence band and the minimum in the conduction band are observed at the same wave vector ($\mathbf{k}$) of electrons. In this case, an electron can be excited from the valence band to the conduction band by absorbing a photon of energy E_g. According to the quantum mechanical approach, an electron has a wave-like property and a traveling electron with the momentum p can be treated as a wave with the wavelength, λ. Relations among the momentum, wavelength, and energy of an electron are given by:

$$p = mv = \frac{h}{\lambda} = \hbar\mathbf{k} \tag{8.74}$$

$$E = \frac{mv^2}{2} = \frac{p^2}{2m} = \frac{(\hbar\mathbf{k})^2}{2m} \tag{8.75}$$

Since p of the electron equals $\hbar \times \mathbf{k}$, an electron moving between two energy states at the same wave vector does not need to change its momentum during the interband transition of direct band gap materials. This means that "the photon energy equivalent or larger than E_g" is a sufficient condition for light absorption in the direct band gap semiconductor.

However, in indirect band gap semiconductors, the minimum point of the conduction band and the maximum point of the valence band have different wave vectors ($\mathbf{k}$). Therefore, the photon energy of E_g is a necessary condition for light absorption. To excite an electron from the valence band edge to the conduction band edge, the difference in electron momentum at two points must be compensated. But, since the photon with energy E_g has negligible momentum, the photon cannot supply or absorb enough momentum when it meets the electron in the valence band. The lattice vibration (e.g., phonon) briefly discussed in this section provides a solution to the problem of the momentum mismatch. It is noted that phonons generally have larger momentum and smaller energy because the mass of atoms and ions is much larger than that of electrons. As shown in Equations 8.74 and 8.75, the energy of the wave is inversely proportional to the mass of the particle if the momentum is fixed. When the electron takes away a photon with an energy $\geq E_g$, the phonon can yield its momentum to the electron or the excess momentum of the electrons can be transferred to the lattice. If that happens, the electron can be excited from the valence band maximum to the conduction band minimum. Therefore, the interband transition in indirect band gap semiconductors requires the absorption of both photon and phonon, which reduces the efficacy of light absorption. For example, if the material thickness is the same, GaAs and CdTe possessing a direct band gap absorb visible light much more than that of Si possessing an indirect band gap by several orders of magnitude.

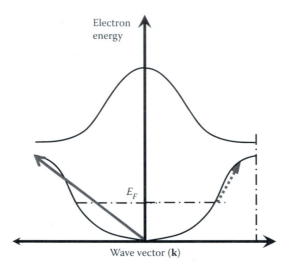

FIGURE 8.27 Schematic explanation of intraband transition of electrons in metals: (i) A dotted line represents a free electron in the Fermi energy level (E_F) that is excited into empty states of the valence band, and (ii) a solid line represents the maximum energy that can be absorbed via interband transition.

In addition to the interband transition, an electron excitation within the same band (known as an intraband transition) also accompanies the absorption of photons. Since a higher empty energy state must be available for the electron excitation within the band, this intraband transition often is found in metals that have a partially filled valence band. Semiconductors can also exhibit an intraband transition in limited cases (for instances, high temperature and/or high impurity doping) that cause the semiconductor to be degenerated (or to have enough free electrons). As schematically shown in Figure 8.27, the intraband transition of the electron results from electron motion within a band consuming photons mainly in IR and visible regimes. The maximum photon energy that can be absorbed via intraband transition is equivalent to the energy width of the electron band.

8.5.3 Quantitative Expression of Extinction: Beer–Lambert law

In previous sections, we learned that the intensity of light traveling in optical materials continuously decreased due to the absorption and scattering phenomena. Since both absorption and scattering reduce the intensity of traveling light, they are grouped as extinction, which is the process of decreasing the intensity of light. However, they have different physical mechanisms. In the absorption process, the energy that the electromagnetic wave loses is transformed to other forms of energy, such as heat or longer wavelength light. In contrast, scattering is a process that redirects the traveling light in other directions through the oscillation of dielectric dipoles with negligible energy conversion. Hence, the scattered light has the same wavelength as the incident light.

A change in light intensity by scattering and absorption can be quantitatively expressed using relations of electric field and electromagnetic energy. Wave vector $\mathbf{k}$ in $E = E_0 e^{i(kx - wt)}$ equals w/v (w: angular frequency, v: velocity of the electromagnetic wave), and v is related to the refractive index (n) of the medium such that $v = c/n$ (c: light velocity in free space). Thus, we find that $\boldsymbol{E} = E_0 e^{i(kx - wt)}$ is rewritten as

$$E = E_0 \exp\left[-iw\left(t - \frac{n}{c}x\right)\right] \tag{8.76}$$

In free space, n has only a real component and a propagating wave is not attenuated. But, the refractive index of optical materials has both real and imaginary components, and n in Equation 8.76

ought to be replaced by $(n - ik)$. Then, the electric field of an electromagnetic wave traveling in the optical material with refractive index n is given by:

$$E = E_0 \exp\left\{-iw\left[t - \frac{x(n-ik)}{c}\right]\right\} = E_0 \exp\left(-\frac{wk}{c}x\right) \cdot \exp\left[-iw\left(t - \frac{xn}{c}\right)\right] \quad (8.77)$$

Equation 8.77 shows an important behavior of the electromagnetic wave. When the wave enters the material (i.e., $k \neq 0$), its electric field decays exponentially from an entering interface, and a decrease in the electric field intensity is described by $E_0 \exp\left(-\frac{wk}{c}x\right)$. This shows again why k is called the extinction coefficient. There are multiple origins of extinctions such as interband transition, lattice vibration, scattering, and intraband transition that are reviewed in Sections 8.4 and 8.5. In Equation 8.77, the undamped oscillating part of the electric field is represented by $\exp\left[-iw\left(t - \frac{xn}{c}\right)\right]$. We can also deduce a change in the intensity of the electromagnetic wave in the material using the change in the electric field intensity. As explained by the Poynting vector in Equation 8.53, the intensity of the light is equal to the square of the electric field, and a combination of Equations 8.53 and 8.77 leads to

$$I = \frac{s\varepsilon}{2}E^2 = \frac{s\varepsilon}{2}E_0^2 \exp[-2iw(t - xn/c)]\exp\left(-\frac{2wk}{c}x\right) \quad (8.78)$$

Equation 8.78 is often rewritten as

$$\frac{I}{I_0} = \exp\left(-\frac{2wk}{c}x\right) \quad (8.79)$$

where I_0 is an intensity of the oscillating electric field. Equation 8.78 explains how the intensity of light is attenuated in optical materials. Extinction of light through the material has a logarithmic dependence on a product of the extinction coefficient and the light traveling distance (x), which is stated in the Beer–Lambert law; $(2wk/c)$ in Equation 8.79 is equivalent to absorbance (α) in the Beer–Lambert law, and Equation 8.79 is rewritten as

$$I = I_0 \exp(-\alpha x) \quad (8.80)$$

The exponential decay of the oscillating wave in Equation 8.80 quantifies the decay of the electromagnetic wave that travels optical materials with a non-zero extinction coefficient.

As described previously, there are several physical origins underlying the extinction process that includes scattering and absorption. In metals, the intraband transition (electric conduction) of free electrons is mainly responsible for extinction. In this case, since the resistivity of metals is not zero, the free electrons that oscillate under the electromagnetic wave experience frictional force (see Section 8.3.1 of this chapter) and a damping term must be introduced to Maxwell equations. A modified relation, taking into account the damping, is given by:

$$c^2 \frac{\partial^2 E_x}{\partial z^2} = \varepsilon \frac{\partial^2 E_x}{\partial t^2} + \frac{\sigma}{\varepsilon_0}\frac{\partial E_x}{\partial t} \quad \text{and} \quad E = E_0 \exp\left[-iw\left(t - \frac{n}{c}x\right)\right] \quad (8.81)$$

where σ is AC conductivity representing free electron motion induced by the electric field of the electromagnetic wave. According to Equations 8.23 and 8.27, displacement (x) obtained from Equation 8.24 is connected to the refractive index and the dielectric constant as follows:

$$\bar{n}^2 = \varepsilon_1 - \frac{\sigma}{\varepsilon_0 w}i = \varepsilon_1 - \frac{\sigma}{2\pi\varepsilon_0 v}i \quad (8.82)$$

Since $\bar{n}^2 = n^2 - k^2 - 2nki$, AC conductivity and extinction coefficient are related by:

$$\sigma = 4\pi\varepsilon_0 nk\nu \tag{8.83}$$

This shows that the extinction coefficient (k) of materials with free electrons is linearly proportional to AC conductivity, which depends on the free electrons' concentration. Equation 8.83 also leads to an important material parameter showing how deeply light can reach from the incident surface. A degree of light penetration into materials is quantified by a characteristic penetration depth (W); W is the depth at which the intensity of the traveling light is equal to $1/e$ of the initial value. From Equations 8.79 and 8.83, the characteristic depth W is given by

$$\frac{I}{I_0} = \frac{1}{e} = \exp\left(-\frac{2wk}{c}W\right) \tag{8.84}$$

$$W = \frac{c}{2wk} = \frac{c}{4\pi\nu k} = \frac{\varepsilon_0 nc}{\sigma} \tag{8.85}$$

Due to the large free electron concentration, AC conductivity of metals is very high and so is the extinction coefficient. Consequently, W of most metals turns out to be very thin. For visible light, W of metals is in the order of 10 nm. Shallow W of metals implies that most of the incident light cannot penetrate into metals. Electron motion excited by the electric field of the light redirects the light instead of transmitting it. In other words, light–free electron interaction in a subsurface region causes light to be reflected (i.e., a change in the light's traveling direction). Also, because of the damping of the electron motion, light–free electron interaction is not a loss-free process, and a small amount of the light is converted to a different kind of energy by the induced oscillation of free electrons. In gold, W for light with the wavelength 600 nm is 15 nm, and 93% of incident light is reflected at the surface. In contrast, CdTe is a direct band gap semiconductor, W is ~ 1 μm for the visible light, and about 70% of light is absorbed in a 1-μm-thick layer. Doped silicon, which is widely used in solar cells, has W of ~ 100 μm due to the combined effect of a low free carrier concentration and an indirect band gap. This penetration depth is an important parameter that determines the thickness of the photoactive component (wafer or thin film) in solar cells and photodetectors.

Example 8.7: Attenuation of Electromagnetic Waves in Metals and Semiconductors

1. If the AC conductivity and the real refractive index of Cu are 3.1×10^2 ($\Omega \cdot$ cm)$^{-1}$ and 0.38 at $\lambda = 1$ μm, respectively, what are the extinction coefficient (k) and absorption coefficient (α)?
2. Experimentally measured α of p-type silicon for light with the wavelength 780 nm is 0.12 μm^{-1}. Can you calculate the AC conductivity of p-type Si from its absorption coefficient?

Solution

1. Light absorption can occur via several mechanisms such as electron band-to-band transition, lattice vibration, and free electron oscillation. In the case of extrinsic silicon (e.g., impurity doped silicon) exposed to light, the role of free electron motion is dominant. Since the frequency of light is much higher than that of the lattice vibration, radiation energy is not absorbed by the lattice vibration. Also, the efficiency of electron excitation from valence band to conduction band is low due to the indirect band gap nature of silicon. Hence, if the free electron concentration is high in doped silicon, the free electron oscillation dominantly dissipates incident radiation energy. In a constituent equation of the free electron oscillation, the friction coefficient (γ) is given by:

$$\gamma = \frac{N_i e^2}{\sigma} = \frac{e}{\mu}$$

If the free electron oscillation is responsible for the dissipation of incident light, an equation of the refractive is rewritten as:

$$\varepsilon_r = \vec{n}^2 = \varepsilon' - i\varepsilon'' = (n - ik)^2 = \varepsilon' - \frac{\sigma}{\varepsilon_0 w}i = \varepsilon' - \frac{\sigma}{2\pi\varepsilon_0 v}i$$

where k is the extinction coefficient (an imaginary part of the complex refractive index). From this equation, ε'' is equal to $2nk$ and k becomes $\dfrac{\sigma}{4\pi\varepsilon_0 nv}$. Note that the intensity of incident electromagnetic wave is expressed as $I = I_0\exp(-\alpha x)$ where absorbance (α) equals to $(2wk/c)$. Therefore, k and α are related to AC conductivity (σ) and the real part of the refractive index (n):

$$k = \frac{\sigma}{4\pi\varepsilon_0 nv}$$

$$\alpha = \frac{4\pi v}{c}\frac{\sigma}{4\pi\varepsilon_0 nv} = \frac{\sigma}{\varepsilon_0 cn}$$

The addition of AC conductivity and frequency into these equations results in an extinction coefficient (k) and absorption coefficient (α);

$$\therefore k = 2.45, \alpha = 3.1 \times 10^7 \text{ (m}^{-1})$$

2. Since the photon energy of light with wavelength 780 nm is larger than the band gap, the attenuation of incident light occurs through electron excitation from the valence band to the conduction band as well as free electron motion. Therefore, the absorption coefficient and AC conductivity do not have a direct relation in p-type Si for light with wavelength 780 nm.

8.6 EFFECTS OF SCATTERING, REFLECTANCE, AND ABSORPTION OF LIGHT

8.6.1 COLOR OF MATERIALS

Absorption, scattering, and transmission depend on the wavelength of light. Therefore, when white light composed of red, green, and blue light is incident upon materials, the relative portion of absorption, scattering, and transmittance is not the same for red, green, and blue light. Since human eyes catch scattered electromagnetic wave that is neither absorbed nor transmitted, the relative ratio of scattering and absorption over electromagnetic waves with different wavelengths controls the color of materials.

For example, a material surface that scatters red light more than blue light looks red. It is noted that both intrinsic and extrinsic properties of material surfaces influence light scattering. As described in the introduction of this chapter, water molecules in the ocean and in clouds result in different colors, though their molecular structures are the same. Since water drops in a cloud cause Mie scattering for the entire visible range, visible light coming from the space experiences forward scattering by the water drops in the cloud, regardless of its wavelength. Hence, humans on land feel that the color of the cloud is white. In contrast, Mie scattering does not occur at the ocean's surface. The color of the ocean is determined by diffuse reflectance, which is complementary to absorption. Since the light absorption efficacy is the highest for the red component of visible light, the diffuse-reflected light contains only green and blue components, which are detected by human eyes.

Scattering of light contributes to the colors of objects, regardless of materials' electric properties. However, the scattering mechanism is different for metals and semiconductors. In addition, the scattering efficacy of metals is higher than that of semiconductors due to the high free electron concentration of metals. The most distinguishable difference between metals and semiconductors is that the plasma frequency, which is related to the free electron density of metals, falls in the UV and visible regimes. When the frequency of incoming photons is smaller than the plasma frequency, photons are absorbed and reemitted through the intraband transition. For photons with higher frequency, metals become transparent. Therefore, the intraband transition and the plasma oscillations are important in the coloration of metals. For example, the plasma frequency of silver with a free electron concentration of 5.9×10^{22} cm^{-3} is ~350 nm, and incoming light collectively oscillates free electrons existing from the surface to a depth of ~350 nm. This oscillation of free electrons in silver reflects all visible light, resulting in the unique color of silver. However, the plasma oscillation itself does not fully explain the color of metals. The free electron concentration is similar for gold and silver; however, their colors are different. The difference in color between gold and silver is due to the interband transition. In addition to plasma oscillation, gold and silver absorb photons through the interband transition from lower energy d-shell (or f-shell) levels to higher energy s-shell levels. The energy gap for this interband transition corresponds to the UV light energy for silver and the blue light energy for gold. Therefore, in gold, the interband transition of gold filters the blue light; the intraband transition of free electrons and light reemission can occur only for green and red light. When both green and red light are emitted from gold, green and red colors are mixed and the gold appears yellow. Examples of gold and silver demonstrate that both interband and intraband transitions are important factors in determining the color of metals, while the color of semiconductors mainly is controlled by the interband transition of electrons.

In contrast to metals, the free electron concentration of semiconductors and insulators is low, and the electron band (valence band and conduction band) is either fully filled or empty. Hence, absorption and reemission of light by free electrons (i.e., intraband transition) are not expected. Instead, the color of semiconductors and metals is determined by the combined effect of diffuse scattering at the surface and the interband transition of electrons from the valence band to the conduction band. Since light with energy higher than the band gap is absorbed, only light with energy smaller than the band gap leaves the material and is recognized as the color of the material.

If the material surface is flat and smooth, the diffuse reflection at the front and back surface is negligible, and the unabsorbed part of the incoming light transmits through the material. This is the case for window glass, which has a very smooth surface. A major component of window glass is SiO_2 with a band gap > 8 eV. Therefore, all visible light is not absorbed by the window glass, and the very smooth surface prevents scattering (diffuse reflection), which makes the window glass transparent. If the material surface is rough, the unabsorbed part of the light is scattered at the surface via the diffuse reflection, and the glass is no longer transparent.

The appearance of rusted steel is another example showing the effect of the band gap and the scattering. The color of a rusted steel surface is due to the diffuse reflection of sub-bandgap light by Fe_2O_3. The main composition of rust on an iron surface is Fe_2O_3 showing semiconducting behaviors. Since the band gap of Fe_2O_3 is 2.2 eV, the iron rust absorbs only blue and green light whose photon energy is larger than 2.2 eV. The unabsorbed part (yellow, orange, and red) of visible light is scattered at the surface of the rusted steel and causes the unique reddish brown color of the iron rust. Inorganic semiconductors such as CdS with band gap 2.42 eV and GaAs with band gap 1.42 eV also exhibit yellow and black colors, respectively, due to the scattering or reflectance of the sub-bandgap light.

Another example showing the role of scattering on the appearance is alumina (Al_2O_3), which is well-known in structural and insulating ceramics. The band gap of Al_2O_3, 7.2 eV, is much larger than the energy of visible light. Therefore, single crystalline Al_2O_3 and polycrystalline Al_2O_3 with a relative density > 99% are transparent as long as the surface is flat and smooth. This is the reason why polycrystalline Al_2O_3 can be used as a window material in the aviation industry. Since dense Al_2O_3 ceramics provide high mechanical strength in addition to transparency, they can replace traditional SiO_2 glass in applications requiring a high safety standard. However, if Al_2O_3 ceramics have micrometer-size pores, they are no longer transparent. As the porosity of polycrystalline Al_2O_3 increases, the micrometer-sized pores diffuse (scatter) the visible light and polycrystalline Al_2O_3 ceramics appear white instead of being transparent (see Figure 8.28). The effect of sintering density and pore scattering on the transparency and color of alumina ceramics indicates that the white color of Al_2O_3 ceramics comes from the same physical origin as that of clouds in the sky.

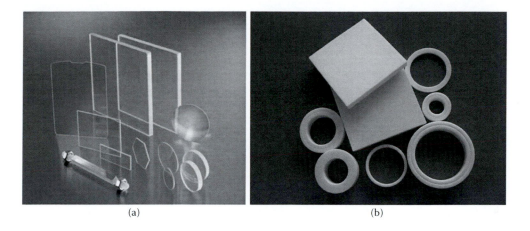

(a)   (b)

FIGURE 8.28   Photos of (a) transparent ceramics and (b) opaque ceramics; though both materials have band gaps larger than the energy of visible light, they exhibit different colors due to the scattering of extrinsic defects. (Courtesy of Ceramtech, Germany.)

8.6.2   Results of Energy Loss of Excited Electrons by Light Absorption

Electrons that are excited to the conduction band via light absorption cannot stay in the conduction band forever because it is not an energetically favorable state. The excited electrons eventually lose energy and return to the lower energy band. This energy released via the de-excitation process generates photons or phonons (or both). The main difference between photon generation and phonon generation is whether the excited electrons directly recombines with the holes. As schematically shown in Figure 8.29, when electrons in the conduction band meet with holes in the valence band at a specific location, they are recombined, and photons with band gap energy E_g are emitted to conserve energy. This process is called radiative recombination or spontaneous light emission, which is exactly the opposite of the interband transition in Figure 8.26. Radiative recombination is widely used in light emitting devices and in material characterization techniques. When electrons are excited via light absorption, the light emission process is called photoluminescence. If the energy source for electron excitation is an electric field, the light emission is referred to as electroluminescence. In the radiative recombination process, it is important to understand that the momentum as well as the energy must be conserved. This means that the momentum of holes in the valence band should be the same as that of electrons in the conduction band. This condition is satisfied only in the direct band gap semiconductor. Therefore, the probability of radiative recombination is very high for the direct semiconductor because an additional process is not required for momentum conservation.

However, the minimum energy of the conduction band and the maximum energy of the valence band do not have the same momentum in the indirect semiconductor. In this case, the energy difference between electrons and holes can be conserved during the recombination process by creating phonons as well as photons. This is the due to the fact that the phonon energy ranges from 10^{-2} eV to 10^{-1} eV near room temperature. Let's look into the case of phonon generation. Since the phonon energy is much smaller than the band gap of most semiconductors (1 eV ~ 3.3 eV), multiple phonons meet with the electron and the hole together to compensate for the energy released during the electron de-excitation. However, when the recombination center serves as a stepping stone between the valence band and the conduction band, the number of phonons involved in the energy conservation decreases. Because fewer phonons are involved during the de-excitation process, the probability of energy conservation by phonon generation increases significantly. The phonon-related recombination is called nonradiative recombination—where the energy difference between holes and electrons is lost to the vibrating lattice.

On the other hand, to make radiative recombination occur in an indirect semiconductor, the momentum difference between the conduction band minimum and the valence band maximum

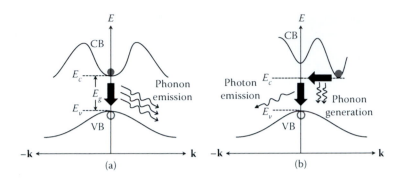

FIGURE 8.29   Schematics of photon and phonon generation in (a) direct and (b) indirect band gap semiconductors; electron–hole recombination releases energy in the form of a photon or a phonon.

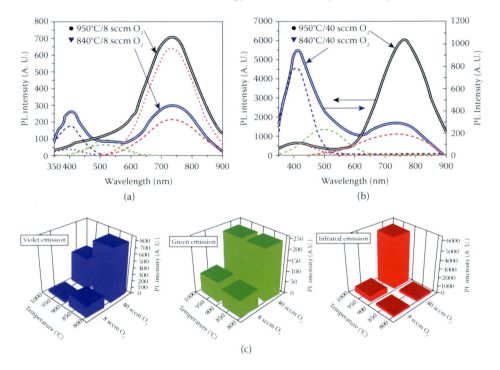

FIGURE 8.30   (a) and (b) Photoluminescence (PL) spectra from annealed porous silicon at different conditions. (c) Relative intensity of PL components. (From Marin, O. et al., *Materials Letters*, 150, 55–58. 2015. With Permission.)

must be compensated for. The momentum conservation for the nonradiative recombination of the indirect semiconductor can happen by exchanging the extra momentum with the lattice vibration (phonons). However, since this requires electrons, holes, and phonons to meet simultaneously, the probability of radiative recombination is low and the lifetime of excited electrons remains high. Low light emission efficiency and slow decay time are features of radiative recombination in indirect semiconductors. One way to facilitate the de-excitation of electrons is to introduce a recombination center between the conduction band and the valence band. This recombination center functions as a trapping center; impurities, point defects (vacancies, interstitials), and extend defects (dislocations) can serve as recombination centers. When electrons approach a recombination center, they are trapped there. Trapped electrons have a higher probability of meeting holes and phonons that satisfy the momentum and energy conservation conditions than free electrons at the conduction band minimum. Therefore, light emission can even be induced from Si crystals, if extended structural defects on the surface of the Si are created by chemical etching (see Figure 8.30). While radiative recombination produces light, phonon generation by nonradiative recombination generates heat inside materials. Nonradiative recombination is undesirable in optical applications of semiconductors (LEDs, phosphors) because it reduces their quantum efficiency (e.g., light generation efficiency).

8.7 APPLICATION OF LIGHT–MATTER INTERACTION

In Sections 8.1–8.6, we have studied the optical properties of materials (e.g., how materials respond to light and how light travels through materials). We also learned which part of the materials controls their optical properties (a major contribution from electrons and a minor contribution from lattice vibrations). The remaining content discusses how the optical properties of materials feature in our daily lives. Optical materials have a wide range of applications—from building glass to renewable energy devices such as solar cells. In this last section, we will review several examples of applications of light–matter interactions.

8.7.1 Antireflection Coating

Scientific knowledge about the optical properties of materials described earlier in this chapter has been applied throughout our daily lives. One well-known example is antireflection coating on material surfaces such as on optical lenses. When light enters the media with a different refractive index, a part of its energy is reflected at the interface. This could be a problem in applications requiring high transmittance of incoming light, such as in lenses of imaging systems (cameras or telescopes). One purpose of antireflection coating is to suppress the reflection and increase the amount of transmitted light by coating the surface layer. This antireflection coating is widely used in glasses for human eyes, lenses for imaging systems, and even in solar cell panels.

The first study on antireflection coating was undertaken by Lord Rayleigh. In contrast to intuitive expectation, he accidentally found out that more light transmitted through a tarnished glass than a brand new glass. His explanation was that the tarnished glass decreased the difference in the refractive media between air and glass since the tarnished layer had a refractive index between glass and air. As shown in Equation 8.55, reflectance (R) of light traveling from Medium 1 to Medium 2 at normal geometry is given by

$$R = \frac{(n_1 - n_2)^2}{(n_1 + n_2)^2} \tag{8.86}$$

(n_1: refractive index of Medium 1, n_2: refractive index of Medium 2).

If a refractive index difference between Mediums 1 and 2 is reduced, the reflectance on the surface is also suppressed. Therefore, deposition of a surface layer with an intermediate refractive index is a simple and effective way to fabricate the antireflective coating on a material's surface. To maximize the effect of the antireflective coating, the refractive index is gradually changed in the coating layer. For this purpose, an inhomogeneous film with a gradual compositional variation along the film thickness direction is often used. As shown in Figure 8.31, light propagating directly is gradually bent in the coating layer with a refractive index gradient and reflection is effectively prevented. The refractive index matching at the film bottom–Medium 2 interface can be achieved by unifying the composition of the film bottom and Medium 2. However, if Medium 1 is air, it is difficult to remove the refractive index difference between Medium 1 and the coated film because the refractive index of solid material is always larger than 1. To avoid the problem of finding materials with $n = 1$, engineers gradually control the porosity of the film. If the porosity of the film increases toward the interface with the air (e.g., the material packing density decreases), the effective refractive index of the film approaches 1 and light incident from the air does not experience a large change in the refractive index at the air–film interface. Though a gradual change in the packing density

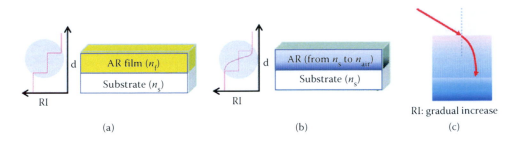

(a)   (b)   (c)

FIGURE 8.31 (a) A sharp drop in the refractive index in a single-layer antireflective film, (b) a smooth change in the refractive index from n_s to n_{air} in antireflective films with a graded refractive index, and (c) the gradual bending of the light propagation direction in antireflective films with a graded refractive index. (From Raut, H.K., et al., *Energy Environ. Sci.*, 4, 3779, 2011. With Permission.)

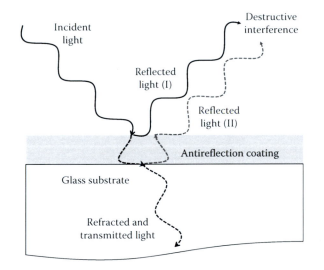

FIGURE 8.32 The principle of antireflection coating of multilayer films: destructive interference of reflected waves.

imparts a gradual refractive index to the film, the low mechanical strength of the antireflective coating layer with high porosity limits its long-term reliability and stability.

A more practical way to enhance transmittance is to utilize the destructive interference of waves. Figure 8.32 schematically shows destructive interference of electromagnetic waves that are reflected at layer interfaces. As illustrated, phases of reflected light at different interfaces are not same. Reflectance at the interface between Medium 2 and the coating layer (R_{c2}) is expressed in a vector form as follows:

$$R_{c2} = \frac{(n_1 - n_c)^2}{(n_1 + n_c)^2} \exp\left(-\frac{2\pi n_c \cos\theta_c d_c}{\lambda}\right) \tag{8.87}$$

where n_c and d_c are the refractive index and thickness of the coating layer, respectively; θ_c is the incident angle; and λ is the wavelength of incident light. For the reflectance at the interface between Medium 1 and the coating layer (R_{1c}), we can use a similar relation. If the phase difference between R_{1c} and R_{c2} is $m\pi/2$ (m: odd number), R_{1c} and R_{c2} interfere destructively, and the total reflectance (a sum of R_{1c} and R_{c2}) is significantly reduced. At normal incidence, this requirement of the destructive interference is satisfied when a film thickness (t) equals $m\lambda/(4n_c)$ (m: odd number, λ: wavelength of light in free space). In addition to the film thickness condition, refractive index matching is also needed for the complete destructive interference of R_{1c} and R_{c2}. If refractive indices meet a matching condition in Equation 8.88, two reflected waves have the same intensity with a 180° phase difference and complete destructive interference takes place (i.e., the total reflectance becomes zero); therefore, the whole incident light transmits to Medium 2.

$$\frac{\left(n_1 - n_c\right)^2}{\left(n_1 + n_c\right)^2} = \frac{\left(n_c - n_2\right)^2}{\left(n_c + n_2\right)^2} \tag{8.88}$$

This multilayer coating approach is commonly applied to UV coating or antireflection coating of lenses for glasses and cameras.

The third way to decrease the reflectance is to exploit surface textures. If the feature size of the surface texture is larger than the light wavelength, incident light is scattered at the surface. As the surface roughness increases, multiple scattering of incident light occurs on the textured surface and specular reflectance is suppressed. Figure 8.33 shows the surface texture on chemically etched ZnO film and its consequence on the haze (i.e., the ratio of the diffused light to the total transmitted light) of the film. As the surface roughness increases (i.e., the surface is etched more), light scattering becomes more pronounced and surface reflectance is suppressed.

If the surface is covered with sub-wavelength surface features, the antireflection of light is still observed, but the antireflection of the nanostructured surface is related to multiple scattering. A good example of antireflection by a nanostructured surface is found in nature. Moths can see objects well even at night using a small amount of light because of their unique eye structure. Figure 8.34 shows a scanning electron microscope (SEM) image of a moth's eye. Very fine protuberances with a size

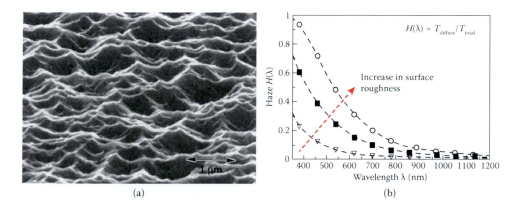

(a) (b)

FIGURE 8.33 (a) An SEM image of the textured surface of a chemically etched ZnO film and (b) the enhancement of light scattering by the texture structure of a ZnO film. (From O. Kluth, et al., *Thin Solid Films*, 351, 247–253, 1999. With Permission.)

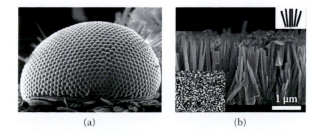

(a) (b)

FIGURE 8.34 (a) An SEM image of a moth's eye covered with nanoscale protuberances with a height of ~200 nm; (b) ZnO nanorods grown on Si; both nanostructures reduce specular and diffuse scattering by creating a gradient of the refractive index on the surface. (From Pignalosa, P., et al., *Optics Letters*, 37, 2808–2810, 2012 and Chao, Y.-C., et al., *Energy Environ. Sci.*, 4, 3436–3441, 2011 With Permission.)

of ~200 nm, which is smaller than the wavelength of visible light, cover the surface of a moth's eye. Since the size is smaller than the wavelength, light cannot detect the individual protuberances and creates only a refractive index gradient profile. The hemispherical shape of the protuberances increases the packing density of materials gradually toward the bottom of the protuberances (or the air content in the surface structure is gradually decreased from the surface). A change in the material content, in turn, gradually modifies the effective refractive index of the protuberance–air mixture layer. As the air content decreases along the depth direction, the effective refractive index in the surface layer gradually increases from the top to the bottom of the mixture layer. According to Equation 8.86, the refractive index gradient at the surface of a moth's eye suppresses light reflectance through the multiple scattering, and more than 99% of incident light transmits from the air to a moth's eye due to the gradual bending of the light propagating direction (see Figure 8.31c). Lessons learned from nature have led to the new area of biomimetic antireflection coating. One example of using fine surface features is the engineered surface of solar cells, which increases light transmittance toward semiconductors. The surface of solar cells is often machined to have texture (e.g., pyramid shape pits) to prevent reflection by the semiconductor surface and to increase power conversion efficiency. In addition, one-dimensional wires or hemispherical domes with sub-micrometer features recently have been investigated to increase light transmittance from air to materials with a high refractive index.

8.7.2 Optical Fibers

Fundamental understanding of light has also been used to develop glass fibers for optical communication. As shown in Figure 8.35, in optical communication, the electric signal in a bit form is first converted to light by a modulator. The modulator encodes a serial bit stream in electrical form and drives a light source such as a laser. Then, pulsed light delivering the encoded information travels through the optical fiber. At the receiver end, the light is fed to a detector, recovered to electrical form, and then is amplified. Optical communication using optical fibers has several advantages over electric communication using copper cables. First, the information capacity of optical communication is enormous. When optical fiber and optical amplification are combined, the network capacity per a single channel of the optical fiber exceeds 10 Tbit/s over a single 160-km line. This means that the world's peak telephone traffic could be carried over a single fiber, which was not imagined in the age of electric communication built on copper cables. Second, optical fibers are cheaper, smaller, and lighter than electric cables, which significantly reduces the capital investment in optical communications. Third, the loss of signals over the length is much smaller in optical fibers than in copper cables. Therefore, the frequency of signal regeneration is much smaller for an optical signal propagating through optical fibers. In addition, signal generation is much easier for an optical signal because a simple optical amplifier is enough to restore the signal.

As described earlier, optical fiber is a key component of optical communication. It is a thin and circular glass fiber composed of core and cladding components; the interface between the core and cladding causes total reflectance of light. The role of the optical fiber is to guide light as it travels so

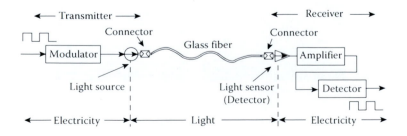

FIGURE 8.35 A schematic of the overall set-up of an optical communication system. (From Dutton, H. J. R., *Understanding Optical Communications*, 1st ed., Prentice Hall PTR, 1998. With Permission.)

that power is delivered from one end of the fiber to the other end without loss. Optical fiber enables modern optical communications where near IR light is exploited to deliver signals. To guide visible and near IR light, the core of the optical fiber is made of silica (SiO_2) glass that is doped with heavy impurities such as GeO_2 to increase the refractive index (typically ~3%). The cladding is pure silica glass that is transparent in visible and IR light. This structure is called a step-index fiber because the refractive index changes in a step at the core–cladding interface. The higher refractive index of the core means that light rays travel slower in the core than in the cladding, and this causes the rays to reflect off the cladding as they travel down the fiber. Therefore, the stepwise change of the refractive index can result in total internal reflectance at the core–cladding interface. In this case, a light ray is confined in the core and guided through the optical fiber. Figure 8.36 schematically shows the structure of an optical fiber. In real optical fibers, a polymer coats the surface of the cladding to provide mechanical protection. Also, this polymer jacket reduces the internal reflection at the outer surface of the cladding so that a light ray is guided only into the core. Since the refractive index difference between the core and the cladding is ~3%, the critical angle (θ_c) for total internal reflectance is ~80° [$\theta_c = \sin^{-1}(n_{cladding}/n_{core})$]. This means that only light at a glancing angle geometry is totally reflected at the core–cladding interface. If the entrance angle is smaller than θ_c, the light ray passes through at the core–cladding interface and is dissipated or refracted at the cladding–polymer jacket interface.

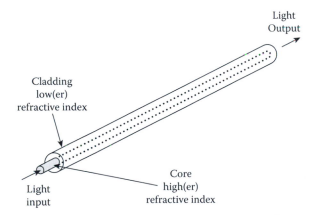

FIGURE 8.36 The structure of an optical fiber and the basic principle of light transmission. (From Dutton, H. J. R., *Understanding Optical Communications*, 1st ed., Prentice Hall PTR, 1998. With Permission.)

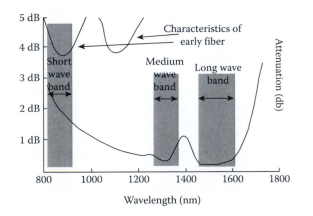

FIGURE 8.37 Attenuation of an electromagnetic wave in silica glass as a function of wavelength. (From Dutton, H. J. R., *Understanding Optical Communications*, 1st ed., Prentice Hall PTR, 1998. With Permission.)

It is noted that, even if the refractive index of the core and the cladding meets the condition of total internal reflectance, light transmission efficiency is lower than 100%. This is ascribed to light absorption by the cladding and the core and to light leakage through the cladding. When light reflects at the internal surface of the cladding, the light ray travels a small distance into the cladding and returns back to the core. This short-distance travel of the light in the cladding can attenuate the signal intensity. More importantly, inherent characteristics of the silica glass cause loss of light intensity, depending on the wavelength of light. Figure 8.37 shows light attenuation in the silica glass as a function of light wavelength. High attenuation loss at mid-IR range ($\lambda > 1.6$ μm) is due to the vibration of Si–O bonds (e.g., phonon). As the frequency gets close to 9 μm (the natural resonance wavelength of Si–O bond in silica), the light attenuation by the phonon absorption increases dramatically. This indicates that the silica is not transparent for mid-IR and far-IR light. On the other hand, as the light wavelength decreases in the visible range, the background of the attenuation curve increases continuously. Attenuation in the visible range cannot be explained by the phonon absorption. Rayleigh scattering is responsible for this background attenuation. As discussed in the earlier section on Mie scattering and Rayleigh scattering, very fine nanoparticles in the light path scatter light and attenuate light intensity. The silica glass is amorphous and its crystal structure does not have long-range symmetry. This means that Si and O are randomly located in the glass. Instead of the long range ordering observed in crystalline silica, the glass only has short-range ordering of Si and O in the scale of several nanometers in size. Consequently, silica glass can be depicted as an assembly of nano-size domains that scatter light. Since Rayleigh scattering becomes more pronounced for light with a shorter wavelength, the background light attenuation in the visible range exhibits the wavelength dependence of Figure 8.37. In addition to the attenuation in the mid-IR range and visible range, two local attenuation peaks between 1.0 μm and 1.5 μm are found in Figure 8.37. These peaks are due to the vibration of the OH group, which inevitably is added to the silica during a manufacturing process (Stone et al. 1982).

When the silica is exposed to water or hydrogen, Si–OH is formed and light with a wavelength of 1.0–1.5 μm excites the first and second overtone vibrations of Si–OH. Therefore, near the IR light of 1.55 μm, the wavelength light exhibits the lowest attenuation in the silica fiber and commonly is used in optical communications. The other important wavelength for optical communications is 1.30 μm, which is between two absorption peaks of Si–OH. To exploit this local minimum of 1.30 μm, the content of the OH group in the silica must be carefully controlled.

In the early stages of optical communications, visible light was used to deliver a signal. However, due to the attenuation of visible light in optical fibers, current optical communications use three different wavelengths in IR range. A short wavelength band at 800–900 nm was first used in the 1970s and early 1980s. This had the benefit of utilizing low-cost optical sources and detectors. In the 1990s, a medium wavelength window was widely used for long-distance communication. Though light sources and detectors are more expensive, the low attenuation (0.4 dB/km) is very advantageous. The attenuation of 0.2 dB/km is very attractive for lossless optical communications, but it was difficult to manufacture light sources and detectors working at this wavelength. Therefore, this long wavelength band was not commonly used until compound semiconductor (indium gallium arsenide, gallium nitride) lasers and photodiodes were developed. Recently, industry and academia have focused on this long wavelength band.

Other important concepts in an optical fiber are the mode and the dispersion. The typical diameter of an optical fiber is about 125 μm without the polymer jacket. The core diameter depends on a number of guided modes. It is 8–10 μm for a single-mode fiber and 50–100 μm for a multimode fiber. As schematically shown in Figure 8.38, the core size and the light entrance angle change the number of possible reflections per unit length of the fiber and allow light ray paths. As the core diameter becomes larger, more reflections can be accommodated in the light path, and more modes of electromagnetic waves can be stabilized in the core (Paschotta 2010). This allowed light path is called the mode.

In multimode fibers, many light paths exist in the core of the optical fiber. For a light of 1.3 μm wavelength, a core 62.5 μm in diameter allows for about 400 modes, depending on the magnitude of the refractive index step at the core–cladding interface. Due to the larger core diameter, the efficiency

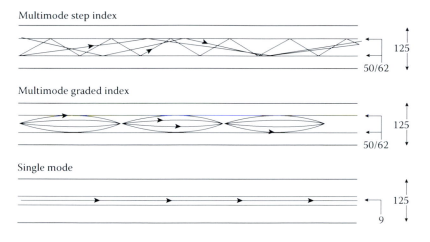

FIGURE 8.38 Different modes of the optical fiber. (From Dutton, H. J. R., *Understanding Optical Communications*, 1st ed., Prentice Hall PTR, 1998. With Permission.)

of capturing the light from the transmitter and giving the light to the receiver is very high. Also, high precision connectors are not needed to join wires together. The problem with the multimode optical fiber is the dispersion. Some of the paths taken by particular modes are longer than other paths and the signal propagation speed is not the same for all modes (Figure 8.38). Therefore, the optical signal—in the form of pulses—spreads out over a period of time and the pulse tends to disperse. A short pulse becomes longer and ultimately joins with the pulse behind, making recovery of a reliable bit stream impossible. The dispersion problem limits the amount of bandwidth in the multimode fibers. However, multimode optical fiber is still an affordable option for short-distance and high-power communications.

In contrast to a multimode fiber, the core diameter of a single-mode fiber is much smaller (~10 μm), and only one mode of light ray can travel through the core. There is no longer any reflection from the core–cladding boundary. Instead, the light path is forced to be parallel to the axis of the fiber. One strength of the single-mode fiber is that the propagation of one mode through the core minimizes the dispersion of the optical signal. Therefore, the single-mode fiber allows for narrow bandwidth and long-distance transmission (50 km or longer) with less attenuation. However, the narrow core generates difficulties in the fiber connection and alignment, which increases the material cost.

8.7.3 LIGHT-EMITTING DIODES

A light-emitting diode (LED) is a device that converts electric energy to light using a radiative recombination of semiconductors. This phenomenon is a kind of electroluminescence, and light color is determined by the band gap of the semiconductor at the junction between p- and n-type semiconductors; III–V compounds such as GaN, InGaN, GaAs, GaP, GaAsP, AlGaAs, and AlGaInP mainly are used as electroluminescence materials. Round observed the electroluminescence from SiC in 1907. However, LEDs did not attract attention until researchers started studying the electroluminescence of GaAs in the 1960s. A turning point for LEDs was when Nakamura successfully fabricated blue GaN-based LEDs in 1993. Since then, a huge amount of effort has been devoted to developing LEDs that generate visible light (blue, green, and red) as well as infrared light. In the 1970s, the motivation of LED research was to develop a highly bright and efficient light source suitable for optical communication because the wavelength of emitted light can be tuned by changing the band gap of semiconductors. In the 2010s, however, the major focus of LED research and development is to formulate a cheaper, brighter, and more efficient lighting that replaces incandescent and florescent bulbs. LED lighting also has been applied to display devices that have been built on liquid crystal displays (LCDs) and plasma display panels (PDPs).

Today, one-fifth of the electrical energy consumed is used to produce light. LEDs could save half of the energy used for lighting. For 800 lumens of light, an incandescent light bulb consumes 60 watts. In contrast, an LED is able to emit the same amount of light by consuming less than 8 watts. Also, the lifetime of an LED is much longer than that of incandescent and florescent light bulbs. The main disadvantages of LED lighting are the relatively high price and cold light quality. Though the manufacturing cost of LED lighting keeps decreasing, it is still higher than that of other bulbs because LEDs require a semiconductor process, sapphire single crystal substrates, and a control circuit. Also, many people feel that LED lights and incandescent bulbs are very different because they have correlated color temperature (CCT). However, these technical barriers are being overcome by ongoing extensive research.

The most practical LED is built on epitaxially grown heterogeneous p-n junctions. This structure, called a double-heterostructure (DH), uses two kinds of semiconductor materials with different band gaps to increase emission. As schematically shown in Figure 8.39, a lightly doped semiconductor with a smaller band gap is sandwiched between cladding layers of a higher band gap semiconductor with a lower refractive index. When an electric field is applied in a forward direction, a potential barrier preventing diffusion of majority carriers decreases by qV (q: charge of electron), and the majority carrier diffusion current increases by a factor of $\exp\left(\dfrac{qV}{k_{\mathrm{B}}T}\right)$. Thus, electrons in an n-type cladding layer and holes in a p-type cladding layer diffuse into a junction layer (i.e., carrier injection into an active layer) where light is produced via radiative recombination. In contrast to a homogeneous junction, injected electrons and holes are confined in the active layer, which has a lower band gap. Therefore, the carrier recombination occurs in a narrow active region and the radiative recombination efficiency is increased. Thinning of the active layer reduces the chances of nonradiative recombination and light reabsorption. In addition, a smaller refractive index difference between the active and cladding layers reduces the reflectance of light incident at a higher angle, and the larger band gap of the cladding layer prevents reabsorption of emitted light at the cladding. Therefore, the emission efficiency of DH-LED is higher than that of LED with a homogeneous structure.

For lighting applications, LED needs to provide a white light, which is close to natural light. There are two ways to generate a white light.

The first is to integrate multiple LED chips that emit light with different colors. Though mixing of either two colors (blue + yellow) or three colors (blue + green + red) can result in white light, a three-color system is preferred. When lights with a peak intensity of 450 nm, 540 nm, and 610 nm

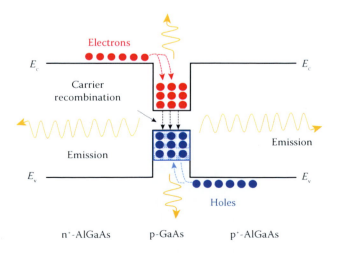

FIGURE 8.39   A schematic on the operation of an LED with a double-heterostructure; injected electrons and holes recombine in a center layer with a low carrier density and small band gap.

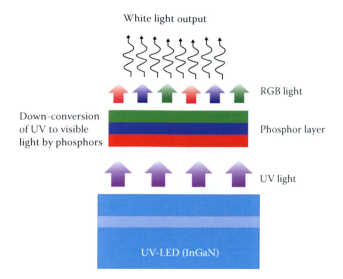

FIGURE 8.40 The principle of white light production in LEDs using phosphors; the mixing of red, green, and blue light results in the emission of white light.

are mixed, white light becomes more luminous to human eyes. One problem of this multichip method is that it requires a sophisticated circuit design to run three chips together and to mix different colors uniformly before light leaves an LED.

The second way to generate white light is to combine high intensity UV or a blue LED with phosphors that absorb UV or blue light and convert it to green and red light. Mixing red, green, and blue light results in an emission of white light. This concept is schematically explained in Figure 8.40. Since only a single chip is used, this method reduces the device cost and provides economically viable solid-state lighting (SSL).

One other benefit of utilizing phosphor is to make up with low efficiency of LEDs emitting green and yellow light. In an LED-based SSL, InGaN semiconductors are used for violet, blue, and green luminescence, and AlInGaP semiconductors are used for red and amber luminescence. One issue is that the blue luminescence of an InGaN LED is much more efficient than other visible light LEDs. Therefore, an LED-based SSL using phosphors could save more energy than an SSL composed of InGaN and AlInGaP multichips. The simple way to produce white light is to blend the blue electroluminescence of an LED and the yellow fluorescence of phosphors. Phosphors commonly exploit the electron decay of rare earth elements or transition metal elements such as Ce^{3+}, Eu^{3+}, Tb^{3+}, and Mn^{2+}. For example, $5d^1 \rightarrow 4f^1$ emission of Ce^{3+} leads to blue or yellow luminescence, depending on the crystal field of the host material where Ce^{3+} is doped as an impurity. In an LED SSL, Ce^{3+} doped yttrium aluminum garnet ($Y_3Al_5O_{12}$, YAG:Ce) has been predominantly used. When excited by UV or blue light, YAG:Ce emits yellow light with a peak wavelength of 560 nm and a broad peak width of more than 100 nm. Blending blue light from an LED and yellow light from YAG:Ce results in white light with a daylight-like color temperature (CCT > 4000 K). Its quantum efficiency reaches 85% even at 200°C. One weakness is that the white light from this combination has a low color rendering index (CRI) due to the lack of red color.

CRI is a quantitative measure of a light source's ability to reveal object colors *naturally* or *faithfully* in comparison to an ideal or natural light source such as sunlight. To improve the CRI of the luminescence from the combination of a blue LED and a yellow phosphor, phosphors emitting red and green have been developed. When Gd^{3+} is added to YAG or Si^{4+}-Mg^{2+} replaces Al^{3+} of YAG:Ce, the phosphor emits a red color at a cost of low quantum efficiency and larger temperature dependence.

To avoid issues of YAG-based phosphors and to improve CRI and CCT, silicon nitride such as Si_3N_4, $Ca_2Si_5N_8$, and $CaAlSiN_3$ and silicon oxynitride such as Ca-SiAlON and $MSi_2O_2N_2$ have been explored as host materials. Though the use of nitride and oxynitride grants significant progress in phosphor performance, researchers still need to circumvent difficulties in the manufacturing process because nitride and oxynitride are synthesized at a high temperature (>1500°C) and under accurately controlled oxygen partial pressure.

In addition to inorganic semiconductors, organic polymers are also utilized to manufacture LED. An organic light emitting diode (OLED) has several important applications, such as flat panel displays and SSL. Current emphasis is on display panels for televisions and mobile phones. Compared with an LCD display, an OLED produces much brighter light and has strength in light and flexible displays. An OLED exploits the electroluminescence of organic materials, where an applied electric field causes the injection of electrons and holes to form emissive states.

A typical OLED consists of an anode, a hole injection layer (HIL), a hole transporting layer (HTL), an emitting layer, an electron transporting layer (ETL), an electron injection layer (EIL), and a cathode. The structure shown in Figure 8.41 has similarities with DHs of inorganic LEDs in that injected electrons and holes are recombined in the organic emitting layer before light is generated.

However, these two devices have several differences due to the unique properties of organic semiconductors. First, organic semiconductors have a low intrinsic carrier concentration and mobility in comparison to inorganic semiconductors. This indicates that (1) delivery of an electron and hole from the electrodes to an emitting layer is critical to the design of high efficiency OLEDs, (2) the dynamic charge equilibrium under the electric field is more important than the electrostatic equilibrium and (3) local charge imbalance can form electric dipoles at the interfaces of an OLED. To promote the charge injection, inorganic and organic layers with a high work function (for hole injection) or a low work function (for electron injection) are inserted between metal electrodes and an organic emitting layer. These are called HIL or EIL, respectively. Polyethylene dioxythiophene polystyrene sulfonate (PEDOT:PSS 4,4', 4'' tris(3-methylphenylphenylamino) triphenylamine (m-MTDATA): 2,3,5,6-tetrafluoro-7,7,8,8,-tetracyanoquinodime thane (F4-TCNQ), MoO_3, CuO_x have been tested as the p-type injection layer, and Li:4,7-diphenyl-1, 10-phenanthroline (BPhen) and $Al(C_9H_6NO)_3$ (abbreviated as Alq3) have been studied as the n-type injection layer.

Second, a pair of an injected electron and a hole forms excitons (e.g., the state of an electron bound to a hole) in the emitting layer, while free electron–hole pairs are produced in an inorganic

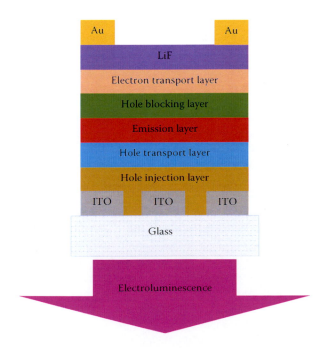

FIGURE 8.41   A schematic illustration of a multilayer structure of small molecule-based OLEDs.

semiconductor such as Si. When neutral molecules are excited from the ground state by taking excitons, excited molecules have either a singlet or triplet state. When the spins of the electron and the hole of the exciton are antiparallel (e.g., the net spin is zero), the state is called a singlet. If the sum of spin momentum is not zero (e.g., parallel spins), the state is called a triplet. Given the spinning direction and phase of the spin momentum, statistical chances of forming a singlet state are 25%. This is important because only singlet exciton can show radiative decay. According to a quantum mechanical selection rule, radiative decay of the triplet excitons is not allowed. Since the electron and hole have the same spin momentum (parallel spins), radiative decay of the triplet excitons breaks Pauli's exclusion principle. Triplet excitons are quenched by molecular vibration, and applied electric energy is converted to heat. Hence, the maximum internal quantum efficiency of a fluorescent OLED is the same as the theoretical concentration of the singlet excitons (25%). Here, fluorescence indicates light emission that results from quantum-mechanically allowed radiative decay of excited states (e.g., singlet exciton). In contrast to an organic OLED, electrons and holes are free in inorganic semiconductors, and spin states of free electrons and holes do not limit the theoretical quantum efficiency of the inorganic LED.

This problem of the low quantum efficiency of an OLED has been solved by promoting phosphorescence. When an organometallic containing heavy metal ions (e.g., Ir, Pt, and Os) are added into the emitting layer, strong spin–orbit coupling around heavy metal ions overcomes spin-prohibition of radiative decay of the triplet excitation. Due to the coupling with the orbital, an exchange between the singlet state and the triplet state is kinetically allowed. Therefore, molecules in the triplet state can return to the ground singlet state, and light is emitted from an active organic layer. This type of light emission from the quantum-mechanically prohibited mechanism is called phosphorescence. A class of OLED using phosphorescence is called a phosphorescent OLED. Since both singlet excitons and triplet excitons contribute to light emission, the internal quantum efficiency of phosphorescent OLEDs is significantly improved.

In addition to quantum efficiency, the quick aging problem of the organic semiconductor, which shortens the lifetime of OLEDs, plagued the introduction of OLEDs to the displays of TV and to the mobile device industry for a decade. The problem is that organic molecules in the active layer of OLEDs easily react with ambient gas and the semiconducting property of molecules disappears. The display industry has solved this inherent problem of OLEDs by developing good sealing technology. Currently, the organic components of OLEDs are sealed with a glass cover in a vacuum ambience or an inert environment (without oxygen). Consequently, OLEDs are now used even for large (> 50 inch) flat panel displays and have quickly replaced LCD display panels for TVs and mobile devices.

8.7.4 LASER

Laser is an acronym of "light amplification by stimulated emission of radiation." As speculated in its name, in laser devices, radiative recombination is stimulated by photons and intensified through multiple reflections. Major differences between a laser and other luminescence is that a laser provides monochromatic, *coherent*, and directional light with strong intensity. *Monochromatic* means that all photons have the same energy (e.g., wavelength and color), and coherent indicates that photons are generated in an organized manner so that electromagnetic waves in a laser have a single phase—though they are not produced at the same time. Due to these unique characteristics, lasers are now widely used in optical communication, data recording and reading, accurate measurement, and in precise machining that require lights with a controlled phase or a highly concentrated and intensified beam (or both).

Stimulated emission is a key process in producing coherent electromagnetic radiation composed of photons that have the same phase, frequency, polarization, and traveling direction. Figure 8.42 schematically explains stimulated emission. When an incident photon is absorbed, an electron is excited from an energy state E_1 to an energy state E_2 and forms an excited state. The radiative decay of this excited state to the low energy state can occur through two different mechanisms.

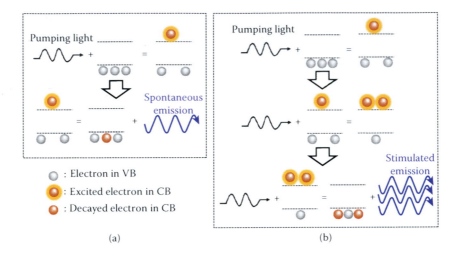

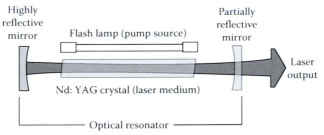

FIGURE 8.42 A schematic illustration of (a) spontaneous emission and (b) stimulated emission; excited electrons decay cohesively for stimulated emission.

FIGURE 8.43 A schematic on the operation of an Nd:YAG solid-state laser; stimulated emission in a laser medium is intensified by repeated reflection and leaves through a partially reflective mirror (Available at https://en.wikipedia.org/wiki/Laser_construction#/media/File:Lasercons.svg.)

First, the electron in the conduction band spontaneously falls to the valence band and emits a photon without an external triggering. This is called spontaneous emission. It is important to know that the phase and polarization of photons coming out of a spontaneous emission are not correlated. In certain kinds of materials, however, a kinetic barrier suppresses the spontaneous recombination of the electron–hole pairs. If that happens, most of the excited electrons stay near the conduction band edge until the second photon stimulates their decay to the valence band edge.

The second radiative recombination mechanism is called stimulated emission because the electromagnetic wave of the second photon triggers the forced oscillation of the electron at E_2 and releases the energy of $(E_2 - E_1)$ in the form of a photon. Due to the force oscillation of the electron, electromagnetic waves of the stimulating photon and emitted photon are correlated and all photons in the stimulated emission have the same phase.

Contrary to the thermodynamically preferred scenario, the electron density at E_2 can be higher than the density at E_1 under specific circumstances (for example, in the application of a high electric field). This reversal of the electron density at two energy states is called population inversion. To intensify light generated by the stimulated emission, the population inversion of electrons between the excited state and the lower energy states is necessary. The other important requirement for a laser is resonance of the emitted light. As shown in Figure 8.43, for lasing, stimulated emission in the laser medium needs to be reflected multiple times by two mirrors of the resonant cavity.

The resonance of the stimulated emission in the cavity produces a cascade effect and a concentration density of photons with the same phase and wavelength. Then, the intensified light (i.e., laser) leaves the cavity through a one-end mirror, which is less reflective (reflectivity of ~98%) than the other end mirror (reflectivity of > 99%), which causes partial transmittance of the stimulated emission at the less reflective end. The resonance process is also responsible for the other important feature of a laser (i.e., directionality). In comparison to a spontaneous emission, a laser is a very tight beam and does not diverge when it travels outside the resonance cavity.

Commercially available lasers are classified into three groups depending on the active medium of the stimulated emission. The first group of laser exploits gases as the laser medium. One example of the gas laser is a helium–neon (He–Ne) laser that generates light with a wavelength of 633 nm. As the name implies, a mixture of He and Ne (10:1 ratio) is used as the active medium. If a high electric field is applied between the cathode and anode of the tube in Figure 8.44, highly energetic electrons are discharged from the cathode. Since most gas atoms filling the tube are inert He, the electrons excite the He atoms first. He–Ne laser exploits inelastic scattering of excited He atoms. Since the energy level of an excited He state ($1s^1 2s^1$) is close to the energy level of Ne ($2p^5 5s^1$), the collision of excited He and ground state Ne transfer the energy from He to Ne via inelastic scattering. Then, excited Ne atoms in ($2p^5 5s^1$) state return to a lower ($2p^5 3p^1$) state by the stimulated emission, with photons of wavelength 633 nm being produced. Subsequently, photons produced by the stimulated emission resonate between two mirrors of the cavity, leading to a coherent, strong, and directional laser beam being generated.

The second group of lasers includes the solid-state laser, which uses solids with wide band gaps as the active laser medium. Glass or crystalline material doped with laser-active ions such as rare earth ions (Nd^{3+}, Yb^{3+}, Er^{3+}) and transition metal ions (Ti^{3+}, Cr^{3+}) commonly are selected as the laser medium. One of the most popular high-power solid-state lasers uses Nd^{3+}-doped ytterbium aluminum garnet (Nd:YAG), which provides light with a wavelength of 1,024 nm. Also, Ti:sapphire and Er:glass are widely used in a solid-state laser. In contrast to the gas laser, the population inversion in the medium occurs through optical pumping by flash lamps, arc lamps, or laser diodes. When incident photons are absorbed, the ions in the gain medium are excited to a higher energy state, and the population densities of lower and higher energy states are inversed.

The benefit of rare earth ions and transition metal ions is that they have several generated energy levels that elongate the lifetime of the excited energy state and prevent the reabsorption of a stimulated emission. If the optically active ions have only two energy levels, it is difficult to enhance both the photon absorption probability and the lifetime of the excited energy state. According

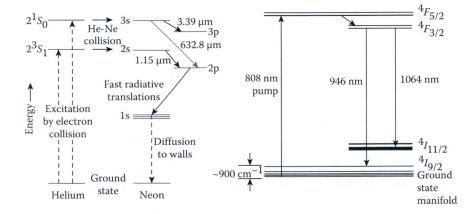

FIGURE 8.44 An energy level diagram of (a) He-Ne gas mixture and (b) Nd:YAG solid; they can produce a laser with a wavelength of 633 nm emission (Ne-Ne) or 1064 nm emission (Nd:YAG). (From Byer, R. L., *Diode Laser-Pumped Solid-State Lasers, Science*, 239: 742–747, 1988. With Permission.)

to Heisenberg's uncertainty principle, the larger width of a higher energy state promotes optical pumping efficiency, but it also increases the chances of spontaneous emission. If the optically active ion has several energy states, a higher energy state with a larger band width is used to improve the optical pumping efficiency. Excited electrons then move quickly to a lower energy state with a narrower band width where electrons can stay longer without spontaneous emission. The right plot of Figure 8.44 shows an energy level diagram of an Nd:YAG system. Optical pumping excites electrons from the ground state to a wider energy level ($^4F_{5/2}$). Then, electrons decay to the $^4F_{3/2}$ level and via nonradiative recombination. Forced oscillation of the electrons at a narrower energy level ($^4F_{3/2}$) by photons causes electron decay from $^4F_{3/2}$ to $^4I_{11/2}$ and the stimulated emission of light with a 1,064 nm wavelength.

In addition to the energy level of the dopant, the dopant concentration is also important for optimizing laser output. If the dopant concentration is too small, the efficacy of optical pumping and lasing is low, and the amplification effect of the optical medium is weak. However, if the doping concentration exceeds the critical condition, dopant ions exchange energy between each other and laser-active ions spend less time in the excited state. This is called concentration quenching, which suppresses the population inversion and causes nonradiative decay instead of the stimulated emission.

The last component to be considered is the host material. High transparency over the pumping and radiation wavelengths and good thermal conductivity are requirements of the host material for low energy loss and stable operation. Though both glass and crystalline materials can be the optical medium, the crystal field of the dopants is more uniform and well-defined when the crystalline matrix is used. Therefore, a crystal medium produces more monochromatic light with a bandwidth of a few nanometers or less, whereas selection of a glass medium broadens the energy levels of the dopants and the bandwidth of the emitted light.

The third kind of laser is a semiconductor laser, which is built on semiconductor materials including GaAs and InP. The semiconductor diode has the capacity of emitting a continuous and high-power (up to W) laser. In addition, a relatively narrow bandwidth of the emitted light is suitable for the optical communication at high bit rates (~10 Gb/s) since the dispersion of the optical fiber is less significant. As do gas and solid-state lasers, a semiconductor laser also exploits optical gain by population inversion, stimulated emission, and the resonance cavity.

Compared with other lasers, however, a semiconductor laser has two unique features. First, the optical gain of a semiconductor laser is attributed to electron–hole recombination in the active layer rather than to a decay of the excited energy states of the optically active ions. In a semiconductor laser, electrons and holes are injected from the heavily doped p-type region and n-type region to the junction layer where the radiative recombination takes place. This means that it is important to increase the probability of the radiative recombination. Therefore, the semiconductor laser diode also utilizes the DH or the multiple quantum well (MQW) structure that is found in LEDs. For laser application, the thickness of the emitting layer and the cladding layer must be thin (in the order of 100 nm) but still thick (in the order of 1 μm) enough so that the electron–hole pairs are confined and the emitted light is not reabsorbed. Second, a semiconductor laser does not install reflective mirrors for lasing. Instead of additional metal mirrors, the cleaved ends of semiconductor materials along their crystal planes reflect the emitted light and create the resonance effect. Due to the sharp change in the refractive index between a semiconductor and the air, the cleaved ends of the semiconductor function as the mirror. Since the resonance structure shown in Figure (8.43) is similar to a Fabry–Perot resonator, the semiconductor laser diode is called a Fabry–Perot laser. In this structure, only waves that can resonate between two mirrors are reinforced. If the distance between two cleaved ends is multiple times the half-wavelength $\left(d = \dfrac{n\lambda}{2} \right)$, the constructive interference of electromagnetic waves takes place and the emission intensity increases. Other waves experience destructive interference between two cleaved ends and are filtered through rather than leaving the resonator. The cavity length of the semiconductor laser diode emitting the visible and near IR light typically is ~100 μm.

As schematically explained in Figure 8.45, the resonance by the cleaved ends also distinguishes a semiconductor laser diode from an LED because spontaneously emitted light of an LED mainly goes through the cladding layer. In the laser diode, however, the spontaneous emission triggers the stimulated emission in the population-inverted region, and only light satisfying the Fabry–Perot cavity mode survives and gets intensified. It is noted that the stimulated lifetime (~0.1–1 ns) is much shorter than the spontaneous lifetime (~2–5 ns). Therefore, once the stimulated emission is triggered, the spontaneous emission is suppressed and the population inversion is maintained for the stimulated emission.

The other important difference between a laser diode (LD) and an LED is the input electric power. Lasing of a coherent emission requires a high electric input power for the population inversion. This electric input power dependence is shown schematically in Figure 8.46. If the electric pumping power is lower than the threshold, the spontaneous emission dominates and the semiconductor device functions as an LED emitting incoherent light. The slope of the curve (light output versus electric current) in Figure 8.46 shows how efficiently the excited electron–hole pairs are extracted as photons. A steeper slope in the LD regime indicates that the extraction efficiency is much higher for the LD than for an LED. This is attributed to the fact that the stimulated emission

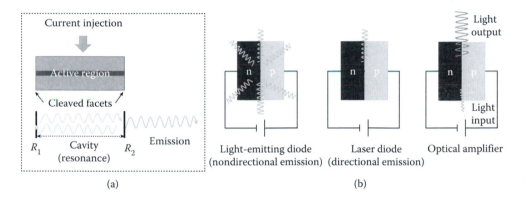

FIGURE 8.45 A schematic illustration of (a) semiconductor lasers and (b) a comparison of a semiconductor laser with semiconductor LEDs, optical amplifier, and LDs.

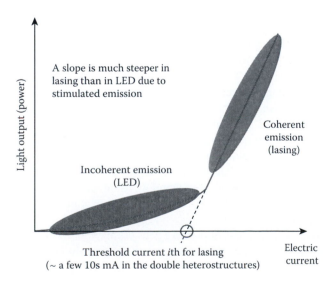

FIGURE 8.46 A schematic illustration of input electric current versus light output; below the threshold current, injected electrons and holes recombine spontaneously.

reduces the probability of nonradiative recombination. Thus, the nonradiative recombination (relaxation via phonons) becomes less critical when more and more electrons and holes are inversely populated. The extraction efficiency of the LD is more than 10%, whereas that of an LED is below 1%.

As summarized here, a semiconductor laser has several features that cannot be found in other lasers. However, a semiconductor laser has also its own weaknesses that are not found in other laser systems. One of them is temperature dependence. As temperature increases, spontaneous emission occurs more easily, and it is difficult to maintain the high degree of the population inversion. This leads to a significant power loss for LEDs at high temperatures. Though DH and MQW structures mitigate this temperature dependence, an increase in the temperature reduces the extraction efficiency of the LD, and the elevated temperature also increases the threshold current; this problem commonly can be found in LEDs and LDs. Consequently, when an LD is heated by lasing, it is necessary to raise the electric input power to maintain the light output of the device.

PROBLEMS

8.1 Based on a band structure, explain why metals are most often highly reflective while insulators are often transparent.

8.2 Spectroscopic ellipsometry measurements on a germanium crystal at a photon energy of 1.5 eV show that the real and imaginary parts of the complex relative permittivity are 20.3 and 2.8, respectively. Find the complex refractive index. What is the reflectance and absorption coefficient at this wavelength? How do your calculations compare with the experimental values of $n = 4.63$ and $K = 0.30$, $R = 0.42$, and $\alpha = 4.53 \times 10^6$ m^{-1}?

8.3 What is a major energy loss mechanism of light traveling inside materials? Please explain this and include a material property parameter representing the dissipation behavior of radiation energy in optical media.

8.4 What is the characteristic penetration depth (depth at which intensity is 37% of original value) of a 600-nm light wave traveling in copper ($n = 0.14$, $k = 3.35$)?

8.5 A plane wave of red light ($\lambda = 600$ nm) enters a nonmagnetic metal with a conductivity of 10^6 S/m. Calculate (i) the reflectance at the surface and (ii) the intensity of the impinging wave that will remain at a depth of 10 nm from the surface. Assume that n and k are 0.15 and 3.5, respectively, at $\lambda = 600$ nm.

8.6 The intensity of a light wave having a wavelength of 750 nm traveling through a material decreases to 37% of its original value after a distance of 9.5 nm. (a) Calculate the absorption coefficient. (b) What is the value of k for this material? (c) Is this material more likely to be a metal or an insulator? Rationalize your answer by calculating AC conductivity.

8.7 (i) Calculate the plasma frequency of Na metal and (ii) explain the optical properties (reflectance and absorbance) of Na metal when the frequency of incoming light is smaller and larger than the plasma frequency. Assume that the atomic mass of Na is 23.

8.8 In a semiconductor-based LED, a heterojunction structure (e.g., p-type GaN/p-type AlGaN/InGaN/n-type AlGaN/n-type GaN) is more popular than a homojunction structure (p-type InGaN/n-type InGaN). Describe the benefits of a heterojunction structure.

8.9 a. Calculate the plasma frequency and damping frequency of Ag (density = 10.49 g/cm^3, electric conductivity = 1.59 μΩ · cm, and electron configuration in the outermost shell = 4s^2 4p^6 4d^{10} 5s^1).

b. If the manufacturing process of bulk Ag goes wrong and the concentration of extended defects is increased, which frequency (v_1 or v_2) mainly is influenced? Justify your choice.

8.10 In alkali metals, we can calculate the plasma frequency of free electrons on the assumption that one atom supplies one free electron. Can this approach also be applied to Mg metal?

8.11 Consider a CsBr crystal that has a CsCl unit cell crystal structure (one Cs^+-Br^- pair per unit cell) with a lattice parameter (a) of 0.430 nm. The electronic polarizability of Cs^+ and Br^- ions are 3.35×10^{-40} F m^2 and 4.5×10^{-40} F m^2, respectively, and the mean ionic polarizability per ion pair is 5.8×10^{-40} F m^2. What is the low frequency dielectric constant and what is it at optical frequencies?

8.12 Transparent conducting oxide (TCO) exhibits the electrical property of a metal and the optical property of a wide band gap insulator. Explain the unique characteristics of TCO using a band structure.

8.13 The following equation is known as the Clausius–Mosotti equation that connects permittivity and polarizability. Please explain (i) the basic assumptions that are used to derive the Clausius–Mosotti equation and (ii) the physical meaning of this equation.

$$\frac{\varepsilon_r - 1}{\varepsilon_r + 2} = \frac{N(\alpha_e + \alpha_i)}{3\varepsilon_0}$$

8.14 You add Sr into the Ba site of $BaTiO_3$ to increase permittivity (ε) at room temperature and to smooth a peak of the ε–T curve at the Curie temperature. Do you expect that pure $BaTiO_3$ and $(Ba,Sr)TiO_3$ each will show a different refractive index (n)?

8.15 The energy loss of an electromagnetic wave traveling in silica (SiO_2)-based fibers is heavily dependent on the wavelength. The energy loss is minimum at $\lambda = 1.55$ μm. Also, the local minimum is found at $\lambda = 1.31$ μm. Please explain the dependence of the energy loss on the wavelength.

8.16 (a) Light traveling the core of optical SiO_2 fibers experiences scattering. What do you think is the physical origin of the light scattering? (b) Can you explain how an increase in the wavelength changes the transmittance of visible light in SiO_2 fibers?

8.17 Metals and insulating materials exhibit the frequency dependence of reflectivity differently. Although metals show a step-like frequency–reflectivity curve, insulators are more likely to have a peak in the reflectivity at a certain wavelength in the infrared range. Why is a peak observed only in the frequency versus reflectivity curve of insulating materials?

8.18 GaAs has a direct band gap structure, and its band gap is 1.42 eV. (a) Calculate the ratio of the absorption coefficient difference for blue light ($\lambda = 450$ nm) and green light ($\lambda = 530$ nm) in GaAs. (b) If the refractive index and the absorption coefficient at $\lambda = 530$ nm are 4.1 and 8.0 cm^{-1}, respectively, how much radiative energy of green light can transmit a 1-μm-thick GaAs film?

REFERENCES

Paschotta, R. *Encyclopedia of Laser Physics and Technology*. Fibers. RP Photonics. Last Update: July, 2010.
Stone, J., Walrafen, G. E. 1982. Overtone vibrations of OH groups in fused silica optical fibers, *J. Chem. Phys.* 76: 1712–1721.

BIBLIOGRAPHY

Chiang, Y.-M., Kingery, D. *Physical Ceramics: Principles for Ceramic Science and Engineering*. Wiley.
Dutton, H. J. R. 1998. *Understanding Optical Communications*, 1st ed. Prentice Hall PTR.
Feynman, R. P., Leighton R. B., Sands, L., 1965. *The Feynman Lectures on Physics: Mainly Electromagnetism and Matter*. Addison-Wesley.
Fleisch, D. 2008. *A Student's Guide to Maxwell's Equations*. Cambridge University Press.
Fox, M. 2001. *Optical Properties of Materials*. Oxford University Press.
Hummel, R. E. 2011. *Electronic Properties of Materials*, 4th ed. Springer.
Kasap, S. O. 2006. *Principles of Electronic Materials and Devices*, 3rd ed. New York: McGraw Hill.

Khanna, V. K. 2014. *Fundamentals of Solid-State Lighting: LEDs, OLEDs, and Their Applications in Illumination and Display.* CRC Press.

Kittel, C., 2004. *Introduction to Solid State Physics*, 8th ed. Wiley.

Livingston, J. D. 1999. *Electronic Properties of Engineering Materials*, 1st ed. Wiley.

Mayer, J. M., Lau, S. S., 1990. *Electronic Materials Science: For Integrated Circuits in Si and GaAs.* Macmillan.

Svelto, O. *Principles of Lasers*, 5th ed. Springer.

White, M. A. 2011. *Physical Properties of Materials*, 2nd ed. CRC Press.

Yeh, P. 1998. *Optical Waves in Layered Media.* Wiley.

9 Electrical and Optical Properties of Solar Cells

KEY TOPICS

- Key aspects of solar cells
- Carrier separation through p-n junction
- Current–voltage characteristics of solar cells
- Fermi energy level change by illumination
- Carrier generation, recombination, and diffusion
- Ideal power conversion efficiency
- Factors that limit and influence solar power to electricity conversion
- Design of high-performance solar cells
- Emerging solar cells

A *solar cell* is a device that generates electric power (electric voltage and current) upon optical illumination. There has been a focus on solar cell technology for the last several decades due to the need for a renewable and sustainable energy source that has a negligible impact on the environment. Problems with traditional fossil fuel energy (global warming and high gas prices) have accelerated efforts to decrease dependence on this source. There are several renewable and alternative energy sources (e.g., solar, wind, and hydropower) that are *greener* than fossil fuel energy. Of them, solar energy is the most abundant and accessible energy source. The field of research and development for converting solar energy into electricity is known as *photovoltaics*.

Most of photons that the sun emits have a wavelength below 1 μm, which corresponds to visible light and near infrared (IR) light. Solar cells are made using semiconductors that can absorb energy of UV, visible and near IR light. A necessary condition for light absorption by the semiconductor is $h\nu > E_g$, where ν is the frequency of light and E_g is the band gap of the semiconductor. *Absorbance* of a semiconductor, therefore, is an important ability of solar cells. At sea level, the peak energy that the sun delivers to the Earth is 100 mJ per second per cm^2. This means that a solar cell with a 15% power conversion efficiency and a 1 m^2 surface area produces 150 W at peak times. Since the solar spectrum is available only during daytime and its intensity varies, the average electric power that a solar cell produces throughout the day is about one-sixth of the peak power output.

A schematic of the structure of a junction-type solar cell built on inorganic semiconductors (e.g., silicon) is shown in Figure 9.1. Usually, the top layer of a crystalline silicon wafer is doped with n-type impurities, and the bottom layer is doped with p-type impurities. In addition to a light absorbing semiconductor, the solar cell has other elements to control light reflectance and to facilitate carrier transport/extraction. An antireflective coating on top of the semiconductor helps capture as much of the light energy incident on the solar cell as possible by minimizing reflection loss (see Figure 9.1). Similarly, in addition to the p-n junction, electrical ohmic contacts are required for operating an electrical circuit. The back contact of the silicon wafer is made using metals such as aluminum or molybdenum. The front contact of the silicon wafer is in the form of metal grids or transparent conductive oxides such as indium tin oxide (ITO) so that light can still get through to the semiconductor below the electrode. A key component for solar energy to electricity conversion

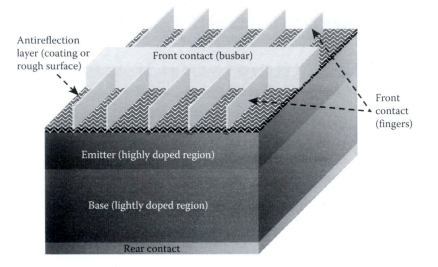

FIGURE 9.1   Basic schematic of a silicon solar cell. The top layer is referred to as the emitter (normally n-type Si) and the bulk material is referred to as the base (normally p-type Si).

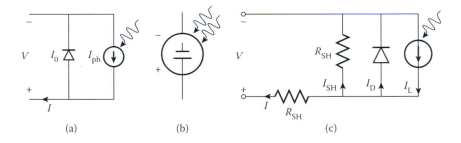

FIGURE 9.2   (a) Equivalent circuit of an ideal solar cell, (b) symbol of an ideal solar cell, and (c) an equivalent circuit of a solar cell with shunt and series resistances.

is a p-n junction that separates electrons and holes that are produced by incident solar energy. The built-in potential at the interface of the p-n junction works as a slide that delivers electrons and holes in different directions. Under illumination, solar light is converted to electricity in both n-type and p-type semiconductors, and carrier transport direction is controlled by the built-in potential of the junction. Therefore, an ideal silicon solar cell is considered a two-terminal device in which an electric current source and a diode are connected in parallel. In dark conditions, the semiconductor does not generate current and the solar cell turns into a p-n semiconductor diode. An equivalent circuit and a symbol of the ideal solar cell are shown in Figure 9.2. In reality, since resistance of electrodes and semiconductors is not zero and manufacturing defects provide alternating current paths (i.e., formation of shunt), series and shunt resistance of the two-terminal device need to be taken into account to model the solar cell and the equivalent circuit of real solar cell changes as shown in Figure 9.2(c).

9.1   WHAT IS A SOLAR CELL?

Light absorbing materials used in solar cells include inorganic semiconductors, organic semiconductors, metal organic compounds, and organic–inorganic hybrid semiconductors. Solar cells are named after what kind of light absorber is used: Si (monocrystalline/polycrystalline/amorphous) solar cells, GaAs solar cells, CdTe solar cells, organic solar cells, dye-sensitized solar cells (DSSCs),

CuInGaSe (CIGS), and organic–inorganic solar cells. Each light absorbing material has a different light absorption spectrum, carrier mobility, long-term stability, and importantly, economic benefit. Carrier transport and mobility is just as important as light absorption in solar cells. In addition to the material type of the energy absorber, the carrier extraction behavior also is used as a criterion to group solar cells. If electrons and holes are strongly bound and separated by a built-in potential of a p-n type semiconductor junction, solar cells are called junction-type solar cells. If electron–hole bound pairs (i.e., excitons) are produced and dissociated at the donor–acceptor interface by a change in the energy level, solar cells are known as excitonic solar cells.

The last way to classify solar cells is based on their development time and the ratio of power conversion efficiency to price. A first generation solar cell is the traditional silicon solar cell made of either monocrystalline silicon or polycrystalline silicon. The power conversion efficiency of silicon solar cells generally is higher than that of other types of solar cells. The highest efficiency of a monocrystalline solar cell using a lab scale is about 25% (as a single cell) and 22% (as a module). Polycrystalline silicon solar cells demonstrate a slightly lower efficiency (~20% as a single cell). The second generation solar cells are thin film solar cells that utilize a few micrometer thick compound semiconductor films (e.g., CdTe and CIGS) as light absorbers. These materials have very high light absorption coefficients and a thin layer is thick enough to fully absorb incident light. Therefore, the advantage of thin film solar cells is that a small amount of material is used, and material cost potentially is lower than that for a crystalline Si solar cell. The highest efficiency of a thin film solar cell in the laboratory is slightly larger than 19% for CdTe and CIGS solar cells as a single cell. The third generation solar cells are built on emerging materials, such as quantum dots, which have the promise of high power conversion efficiency and low material cost. This type of solar cell is not commercialized yet and extensive research is ongoing in the laboratory.

Among the several kinds of solar cells introduced here, wafer-based crystalline silicon cells dominate other solar cells in the commercial market. Global annual production of PV modules from 2000 to 2013 and market shares of monocrystalline Si, polycrystalline Si, and thin film solar cells are shown in Figure 9.3. Cumulative PV module shipment exceeded the 100 GWp landmark in 2012, and cumulative worldwide installed PV module power was about 177 GWp in 2014. Global annual production of solar cells was 39 GWp in 2013 and 45–55 GWp in 2014. Crystalline silicon solar cells account for about 90% of the solar cell market. Polycrystalline silicon is the most widely used because it costs less than monocrystalline silicon with the same power conversion efficiency. In the last 10 years, the efficiency of average commercial silicon solar cell modules has increased

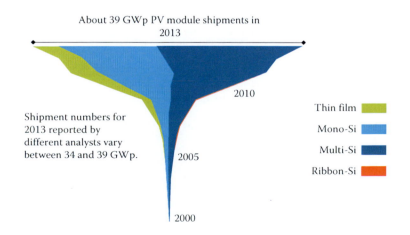

FIGURE 9.3 Global annual production of PV modules from 2000 to 2013; market shares of monocrystalline Si, polycrystalline Si, and thin film solar cells are presented in different colors. (From *Fraunhofer Institue for Solar Energy System (ISE) Annual Report 2014/2015.* With Permission.)

from about 12% to 16%. The commercial market share of thin film solar cells is about 10% using CdTe, CIGS, and amorphous silicon as light absorbers. The recent average module efficiency of a CdTe thin film solar cell is ~13%, which is lower than that of Si. However, the relatively low efficiency of a thin film solar cell is compensated by the lower material cost of thin films. This explains why thin film solar cells can compete with silicon solar cells, even though the raw material cost of silicon wafers has dramatically decreased since 2010.

One of the most important factors affecting widespread use of solar cells is cost. Solar cells need to produce electricity at a cost that is comparable to other energy generation technology (e.g., natural gas/coal power plant and nuclear power plant). The economic viability of solar cells has improved dramatically over the last decade. The learning curve of PV module price shows that the average selling price of these modules decreases 21% for every doubling of cumulative PV module shipments. The module price was 0.72 US\$/Wp in 2013 and 0.62 US\$/Wp in 2014, which is close to the target set up by the U.S. Department of Energy (~0.5 US\$/Wp by 2020). The 2014 PV module price corresponds to the levelized cost of electricity (LCOE) of \$0.05–0.1/kWh. The learning curve suggests that PV module price and LCOE will decrease to 0.33 US\$/Wp and 0.03–0.07 US\$/kWh by 2025. LCOE not only includes PV module cost but also balance of system (BOS) costs that span the wiring, inverter, mounting, and ground. BOS depends on the location of the system installation and the size of the system, and LCOE is not fixed.

As noted, BOS includes the inverter. This means that the inverter is a necessary component for using a solar cell. Since solar cells produce direct current (DC) electric power, an auxiliary part that converts DC to AC electric power is required to connect the solar cells to the electric grid. If the solar cell stands alone, the inverter is not needed. Due to the rapid decrease in PV module price and the need for carbon dioxide (CO_2) reduction, the portion of electricity produced by solar cells continues to increase globally. In Germany, a country that actively seeks renewable energy sources and shuts down nuclear power plants, PV modules supplied 29.7 TWh of electricity (about 5.3 % of the national demand) in 2013. This amount of PV-generated electrical power saved about 20 million tons of CO_2 emissions.

In this chapter, we will study (1) operation of solar cells with an emphasis on semiconductor theory, (2) voltage–current characteristics of the solar cells, (3) factors affecting the power conversion efficiency of solar cells, (4) key components for PV module installation, and (5) different types of emerging solar cells for high energy conversion efficiency.

Example 9.1: Elecricity Generation Capacity and Cost of Solar Cells

Let us say that you want to install solar panels for your house in City A where the yearly solar radiation is 1,600 kWh/m². Each panel costs \$550, and the efficiency and active area of each panel is 16% and 2.5 m².

1. In 2013, the average annual electricity consumption for a U.S. residential utility customer was 10,908 kWh. How many solar panels do you need to cover the electricity demand of your house?
2. The price of electricity in City A is 15 cents per kWh. What is your payback period (in other words, how many years do you need to use the solar panels to get back your investment)? Assume that costs of installation and other auxiliary parts are the same as the panel price.

Solution
1. Total active area of solar cells for yearly production of 10,908 kWh is

$$10,908 \text{ (kWh)} \div [0.16 \times 1600 \text{ (kWh/m}^2)] = 42.9 \text{ (m}^2)$$

Therefore, you need ~17 panels with an active area of 2.5 m² to be self-sufficient in electricity.

2. Each panel produces 640 kWh/year of electricity, which corresponds to $96/year. Since the total cost of the solar panel installation (panel price + installation cost + auxiliary parts) is $1,100/panel, the payback period is

$$\therefore \$1100 \div \$96/year = 11.5 \text{ years.}$$

9.2 OPERATION PRINCIPLE OF p-n JUNCTION-TYPE SEMICONDUCTOR SOLAR CELLS

Basically, a silicon solar cell is a *p-n* junction diode that absorbs energy from light radiation when illuminated (Figure 9.2). Figure 9.4 shows the operating principle of typical p-n junction-type semiconductor solar cells. Absorption of light causes the creation of electron–hole pairs (EHPs). This effect is known as *photogeneration* or *optical generation* of carriers. It is noted that EHPs in silicon do not form an exciton that is a strongly bound electron–hole pair. In contrast, some solar cells (such as organic solar cells) generate excitons that must be broken and converted to the current before electrons and holes recombine. These types of solar cells are sometimes called excitonic solar cells.

To produce electric current, photogenerated electrons and holes must move in the opposite direction. If electrons and holes move together, their movements cancel each other and there is no net flow of electric charge. In Sections 9.2.1 and 9.2.2, general equations for current–voltage characteristics (called an *I–V* curve) and conversion efficiency of solar power to electric power are introduced using the example of p-n junction solar cells. Fundamental physics underlying the operation of the p-n junction-type solar cells is found in Section 9.3.

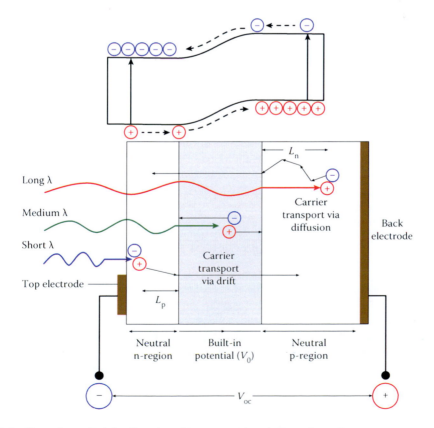

FIGURE 9.4 Operating principle of p-n junction-type semiconductor solar cells.

9.2.1 CURRENT–VOLTAGE CHARACTERISTICS OF p-n JUNCTION-TYPE SEMICONDUCTOR SOLAR CELLS

As schematically shown in Figure 9.1, a silicon solar cell is made mostly from a 200–500 μm base (mainly p-type silicon wafer) that is lightly doped with boron or other acceptor impurities (e.g., ~10^{16} B/cm^3). To form a p-n junction, the surface of a p-type wafer is heavily doped with donor impurities through diffusion (e.g., ~10^{19} P/cm^3). The bulky, lightly doped region is called the base and the top heavily doped region is called the emitter. A key function of the junction is to filter the carriers depending on the sign of their electric charge.

In p-n junction-type solar cells, electrons and holes are separated by the built-in potential that exists in a carrier-depleted interface region (see Figure 9.4 and Section 5.1). This means that photo-generated carriers need to travel to the junction region to provide electric power. This travel to the junction region occurs purely through a diffusion mechanism. Therefore, only the photogenerated carriers within the carrier diffusion distance from the depletion layer (L_p or L_n, for hole or electron, respectively) can travel to the depletion layer and generate electric power. Photogenerated carriers that are recombined during diffusion do not contribute to electricity production. If photogenerated electrons and holes are produced outside the diffusion distance from the depletion region, they mainly are recombined and their energy is dissipated as heat. This is the reason why carrier diffusion (particularly minority carrier diffusion) is very important in silicon solar cells.

If the diffusion distance is short, only a small number of photogenerated carriers are converted to electric power. To increase the carrier diffusion distances, a high quality silicon wafer with a negligible concentration of recombination centers is required. However, this approach increases the production cost of silicon solar cells, and manufacturers are forced to find a balance between the cost and the performance of the solar cell, depending on the application of the solar cells.

Once photogenerated carriers reach the depletion region, the electron and hole are swept up into opposite directions by the internal, built-in electric field. Since the electric field is a driving force, drift is the mechanism responsible for carrier sweeping near the junction. The photogenerated holes (i.e., additionally produced minority carriers on the n-side) drift toward the p-side of the depleted region with the negatively charged space-charge. On the other hand, photogenerated electrons in the p-side drift toward the positively charged space-charge region on the n-side. Both of these motions result in a *photocurrent* (I_L), which is produced through light illumination. As explained earlier, I_L inside the p-n junction semiconductor is directed from the electron-rich side (n-side) to the hole-rich p-side (Figure 9.4), which is the direction of the built-in electric field (E) at the junction.

The photocurrent (I_L) is also known as the *short-circuit current* (I_{sc}), which the solar cell generates when the cell is connected to the resistance of zero ohm. Hence, this is the maximum current that flows through an external circuit. The photocurrent magnitude depends on the rate at which electron–hole pairs are created per unit volume (g_{op}). If A is the area of the p-n junction and L_p and L_n are the diffusion lengths for holes and electrons, respectively, then the photocurrent (I_L) is given by

$$I_L = q \times A \times (L_p + L_n) \times g_{op} \tag{9.1}$$

In Equation 9.1, notice that g_{op} is not constant in the solar cell but dependent on incident photon density; g_{op} is higher near the front surface of the solar cell where light is incident. Since most photons are absorbed, the number of photons reaching the back surface of the solar cell is very small, and g_{op} comes close to zero at the back surface. We also assume in Equation 9.1 that only the carriers that are generated within carrier diffusion distances from the depletion layer will contribute to I_L. The rest of the carriers created outside of the diffusion length from the junction will not result in the generation of electric current because they will recombine before being separated at the junction. In Equation 9.1, we also assume that the number of carriers generated in the depletion layer itself is small compared to those created in the neutral region n- and p-sides of the junction. This assumption is true in most cases except in p-i-n junction solar cells where the depletion region is intentionally expanded to compensate for the weakness of the small diffusion length.

When an external resistance (R) is connected to a solar cell, electric current passing the resistor experiences a voltage drop (($V_{Ideal} - V_{Real}$)~IR). This voltage drop, which is equivalent to applying a forward bias to the diode, reduces the barrier height at the junction. A decrease in the barrier height, in turn, causes an increase in the forward current of the diode (I_F). Since electrons travel toward the p-side of the junction in the diode under the forward bias, the forward current direction inside the diode (I_L from p-side to n-side) is opposite to that of the photogenerated carriers' current by the built-in potential of the junction (I_F from n-side to p-side) (refer to Sections 5.9 and 5.10 on diffusion current and drift current). This indicates that I_L and I_F have opposite signs. The appearance of this forward voltage across an illuminated p-n junction explains how the current and potential of the photogenerated carriers is determined at a practical operating condition. Thus, the net current (I) flowing through the external circuit is

$$I = (I_L - I_F) \tag{9.2}$$

Note that the photocurrent (I_L) that is provided by a current source of Figure 9.2, and the net solar cell current (I) always flow in the opposite direction to that of the forward diode current (I_F).

The maximum value of the solar cell current (I) is I_L; when there is no resistance ($R = 0$), it is also known as the short-circuit current (I_{sc}). The minimum value of the solar cell current (I) is zero when the external resistance is infinite ($R = \infty$), and it is known as an open-circuit condition.

We now derive expressions for the solar cell current (I) and *open-circuit voltage* (V_{oc}), starting with the diode equation

$$I_F = I_S \left[\exp\left(\frac{qV}{kT} \right) - 1 \right] \tag{9.3}$$

Characteristics of the p-n junction diode in Equations 5.56 and 5.63 of Chapter 5 show that the saturation current (I_s) is determined by the diffusion of minority carriers in the neutral region and followed by the drift of the minority carriers through the depletion junction. Therefore, Equation 9.3 can be rewritten as

$$I_S = q \times A \times \left[\frac{D_p p_n}{L_p} + \frac{D_n n_p}{L_n} \right] \tag{9.4}$$

where D_p and D_n are the diffusion coefficients for holes and electrons, and L_p and L_n are the diffusion lengths for holes and electrons. In an equilibrium state, the product of hole and electron concentrations is a constant (n_i^2) for a given semiconductor at fixed temperature T, which means $p_n \times n_n = n_i^2$ and $p_p \times n_p = n_i^2$. Since $n_n = N_d$ and $p_p = N_a$ in the doped semiconductor, $n_p \times N_a = n_i^2$ and $p_n \times N_d = n_i^2$, where N_a and N_d are the dopant concentrations for the p- and n-sides, respectively.

Recall that subscripts for the electron and hole concentrations refer to the side of the p-n junction; for example, p_n is the concentration of holes on the n-side. Thus, Equation 9.4 can be rewritten as

$$I_S = q \times A \times n_i^2 \left[\frac{D_p}{L_p N_d} + \frac{D_n}{L_n N_a} \right] \tag{9.5}$$

Substituting the expression for I_S in Equation 9.4, we can write I_F as

$$I_F = q \times A \times n_i^2 \left[\frac{D_p}{L_p N_d} + \frac{D_n}{L_n N_a} \right] \left[\exp\left(\frac{qV}{k_B T} \right) - 1 \right] \tag{9.6}$$

The diffusion lengths for a carrier are related to the lifetime of the carrier (τ) by the following equations:

$$L_n = \sqrt{D_n \tau_n} \tag{9.7}$$

$$L_p = \sqrt{D_p \tau_p}$$ (9.8)

The longer the lifetime (τ) of a carrier, the higher the diffusion length (L). The longer the carrier can survive without recombining, the higher the probability of it contributing to the photocurrent.

Sometimes, we prefer to express I_F without directly involving diffusion coefficients. For example, using Equations 9.7 and 9.8 and substituting them into Equation 9.6 and rearranging, we get

$$I_F = q \times A \times \left[\frac{L_p}{\tau_p} p_n + \frac{L_n}{\tau_n} n_p \right] \left[\exp\left(\frac{qV}{k_B T} \right) - 1 \right]$$ (9.9)

Now, let us plug I_L in Equation 9.1 and I_F in Equation 9.9 into total current I in Equation 9.2. We get

$$I = -q \times A \times \left[\frac{L_p}{\tau_p} p_n + \frac{L_n}{\tau_n} n_p \right] \left[\exp\left(\frac{qV}{k_B T} \right) - 1 \right] + q \times A \times g_{op} \times (L_p + L_n)$$ (9.10)

In addition to electric current of the solar cells, Equation (9.10) shows the maximum voltage that the solar cell can produce. This upper limit of the voltage can be obtained in a case of $I = 0$ (i.e., open circuit). When the external load connected to the solar cell has a very high resistance, a large voltage drop (V) across the resistor creates the effect of applying a large forward bias to the diode, and the forward current (I_F) increases dramatically. If the magnitude of the forward current (I_F) is equal to the photocurrent (I_L), I_F and I_L cancel each other, and the output voltage of the solar cell reaches the maximum that is called open-circuit voltage (V_{oc}):

$$V_{oc} = (k_B T / q) \ln \left[1 + \left(\frac{I_L}{I_S} \right) \right]$$ (9.11)

where V_{oc} is the maximum voltage that can be generated from the p-n junction in a solar cell. We can see the appearance of this voltage on a band diagram upon the illumination of a p-n junction, called the photovoltaic effect (Figure 9.4).

We can combine Equations 9.1, 9.5, and 9.11 to find the following equation for V_{oc}:

$$V_{oc} = \left(\frac{k_B T}{q} \right) \ln \left\{ 1 + \left[\left(\frac{(L_p + L_n)}{(L_p / \tau_p) p_n + (L_n / \tau_n) n_p} \right) \times g_{op} \right] \right\}$$ (9.12)

In Equations 9.11 and 9.12, it is noted that V_{oc} is not a constant. It increases with the increasing photocurrent (I_L) or with the rate of the optical generation of electron–hole pairs per unit volume (g_{op}). This indicates that an increase in the light intensity increases V_{oc} by increasing g_{op}. However, the open-circuit voltage (V_{oc}) cannot keep increasing indefinitely with the increasing g_{op}. As the rate of the optical carrier generation increases, the minority carrier concentration increases. This causes the lifetime (τ) of the carriers to become shorter. Thus, more carriers recombine, which prevents the voltage from exceeding the built-in potential (V_0). Effects of the increase in the minority carrier concentration and the decrease in the lifetime at large light intensity are described in Equation 9.12. In other words, if V_{oc} becomes the same as V_0, the forward-biased current of the solar cell diode (see the equivalent circuit of Figure 9.2) becomes larger than the photocurrent (I_L). This means that photogenerated carriers are consumed within the solar cells instead of being delivered to the external load when V_{oc} is larger than V_0. Note that V_0 cannot be larger than the band gap (E_g) of the host semiconductor in the p-n junction of the semiconductor (Section 5.1). Now we know (i) that the upper limit of V_{oc} is E_g of the host semiconductor in the p-n junction-type solar cell and (ii) that the electron–hole recombination consumes a part of the photogenerated carriers and makes V_{oc} smaller than E_g. In many cases, the maximum V_{oc} of a single p-n junction semiconductor solar cell is ~85% of E_g of the semiconductor.

In our discussion, we assumed that concentrations of thermally generated electron–hole pairs are negligible. However, if the rate of thermal generation of electron–hole pairs (g_{th}) is comparable to g_{op} (which means operation at a very high temperature) and the p-n junction is symmetric (which means $p_n = n_p$ and $\tau_p = \tau_n$), then V_{oc} can be rewritten as

$$V_{oc} = (k_B T/q)\ln\left(\frac{g_{op}}{g_{th}}\right) \tag{9.13}$$

As the intensity of the illuminating light increases in Equation 9.13, the optical generation rate increases again on both sides of the p-n junction. This leads to an increase in the open-circuit voltage (V_{oc}). We can see this correlation in the $I–V$ curves for a solar cell in Figure 9.5. The maximum voltage from the solar cell is V_{oc} for zero current in the external circuit. The minimum voltage is zero when the current is maximum under a short-circuit condition (I_{sc}). This gives us the diode current (I) as a function of the diode voltage generated as shown in Figure 9.5.

Example 9.2 illustrates the calculation of the open-circuit voltage (V_{oc}) for a solar cell.

Example 9.2: Solar Cell Open-Circuit Voltage

A solar cell is made using a Si p-n junction with $N_a = 10^{18}$ cm^{-3} and $N_d = 10^{16}$ cm^{-3}. The diffusion lengths for electrons and holes are 25 and 10 µm, respectively. Assume that $n_i = 1.5 \times 10^{10}$ cm^{-3} at $T = 300$ K, $D_p = 20$, and $D_n = 10$ cm^2/s, and that the photocurrent density (J_L) is 10 mA/cm^2. (a) What is the open-circuit voltage (V_{oc})? (b) How does this compare with the contact potential (V_0) for this p-n junction?

Solution

1. We rewrite Equation 9.11 by replacing current with current densities:

$$V_{oc} = (k_B T/q)\ln\left[1+\left(\frac{J_L}{J_s}\right)\right]$$

At 300 K:

$$V_{oc} = (0.026 \text{ V})\ln\left[1+\left(\frac{J_L}{J_s}\right)\right]$$

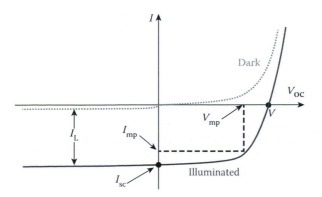

FIGURE 9.5 *I–V* curve of the solar cell in dark and illuminated conditions.

The value of J_L is given as 10 mA/cm^2. To get the value of the saturation current density (J_s) for the solar cell diode (refer to the equivalent circuit of the solar cell in Figure 9.2), we rewrite J_s as

$$J_s = q \times n_i^2 \left[\frac{D_p}{L_p N_d} + \frac{D_n}{L_n N_a} \right]$$

Substituting the values for diffusion coefficients, diffusion lengths, and dopant concentrations:

$$J_s = (1.6 \times 10^{-19}) \times (1.5 \times 10^{10})^2 \left[\frac{20}{(10 \times 10^{-4}\text{cm})(10^{16})} + \frac{10}{(25 \times 10^{-4}\text{cm})(10^{18})} \right]$$

Therefore, for J_s, the saturation current density = 7.21 × 10^{-11} A/cm^2 or 7.21 × 10^{-8} mA/cm^2. We use both current densities in the same units to calculate V_{oc} from Equation 9.13:

$$V_{oc} = (0.026 \text{ V}) \ln \left[1 + \left(\frac{10 \text{ mA/cm}^2}{7.21 \times 10^{-8} \text{ mA/cm}^2} \right) \right]$$

$$\therefore V_{oc} = 0.48 \text{ V}$$

2. From Equation 5.21, the contact potential (V_0) is given by

$$V_0 = (k_B T / q) \ln \left(\frac{N_a N_d}{n_i^2} \right) \tag{9.14}$$

Then,

$$V_0 = (0.026 \text{ V}) \ln \left(\frac{10^{18} \text{atoms/cm}^3 \times 10^{16} \text{atoms/cm}^3}{(1.5 \times 10^{10} \text{electrons/cm}^3)^2} \right)$$

$$\therefore V_0 = 0.813 \text{ V}$$

The contact potential is the maximum forward bias that can appear across the p-n junction. Therefore, the value of the open-circuit voltage (V_{oc}) is always less than V_0.

9.2.2 FILL FACTOR, POWER CONVERSION EFFICIENCY, AND QUANTUM YIELD OF A SOLAR CELL

The power a solar cell delivers to a load is obtained by calculating the value of $I \times V$. Note that the sign of I (output current of the solar cell) is negative, if the forward bias direction of the p-n junction is treated positive. Since the photocurrent (I_L) flows in the reverse-bias direction and the voltage generated in the solar cell has a positive sign, the product of $I \times V$ is negative. This means that the solar cell generates power rather than consuming the power.

From Equations 9.2 and 9.3, the power generated by the solar cell can be written as

$$P = I \times V = (I_L \times V) - \left\{ I_s \left[\exp \left(\frac{qV}{k_B T} \right) - 1 \right] \times V \right\} \tag{9.15}$$

Equation 9.15 shows that the power of the solar cell depends on V (voltage applied to the solar cell diode or output voltage of the solar cell). Given that the external load connected to the solar cell controls V of the solar cell diode, we can state that the power of the solar cell is a function of the external load. From Equation 9.15, we can calculate the current (I_m) and voltage (V_m) that will result in maximum power (P_{max}) of the solar cell. Please note that the subscript "m" in V_m and I_m does not stand for maximum voltage and maximum current. The subscripts instead represent the values of voltage and current that lead to the maximum power.

At the maximum power point, we can equate $dP/dV = 0$. Then,

$$\frac{dP}{dV} = I_L - I_S\left[\exp\left(\frac{qV_m}{k_BT}\right)-1\right]-I_SV_m\left[\left(\frac{q}{k_BT}\right)\right]\exp\left(\frac{qV_m}{k_BT}\right)=0$$

(9.16)

$$1+\left(\frac{I_L}{I_S}\right)=\left[\exp\left(\frac{qV_m}{k_BT}\right)\times\left(1+\frac{qV_m}{k_BT}\right)\right]$$

At $T = 300$ K, $V_t = (k_BT/q) = 0.026$ volt, and Equation 9.16 can be rewritten as follows:

$$1+\left(\frac{I_L}{I_S}\right)=\left[\exp\left(\frac{qV_m}{k_BT}\right)\times\left(1+\frac{V_m}{V_t}\right)\right]$$

(9.17)

Since the right-hand side is known for a given p-n junction, we can solve for V_m by trial and error for a given I_L. The V_m and the corresponding value of I_m are shown in Figure 9.5.

From Figure 9.5, we can see that the product $V_m \times I_m$ is less than $V_{oc} \times I_{sc}$. The ratio of these two products is called the *fill factor* (FF) of a solar cell:

$$FF = \frac{V_m \times I_m}{V_{oc} \times I_{sc}}$$

(9.18)

Good solar cells have an FF of 0.7~0.8. FF is sensitive to processing variables of the solar cell. As shunt resistance and series resistance are added to the ideal solar cell (or carrier recombination increases), FF decreases (see Figure 9.2). In Equation 9.18, it is important to understand that the maximum power of the solar cell is given by $P_{max}= V_m \times I_m = V_{oc} \times I_{sc} \times FF$. Therefore, V_{oc}, I_{sc}, and FF are often used as three factors for evaluating solar cell performance.

Equation 9.18 also helps to quantitatively express power conversion efficiency (η_{conv}); η_{conv} of the solar cell, which shows the fraction of solar energy converted to electric energy, is defined as the ratio of maximum power over the incident power (P_{in}) delivered to the solar cell:

$$\eta_{conv} = \frac{V_m \times I_m}{P_{in}} = \frac{V_{oc} \times I_{sc} \times FF}{P_{in}}$$

(9.19)

The best η_{conv} of commercially available solar cells was close to 0.2 as of 2014. This means that even the best solar cell on the market converts only about 20% of input solar energy to electric energy and 80% of input energy is lost during the conversion process.

In solar cell operation, note that impedance matching between the solar cell and the external load is very important. To extract electric power from the solar cell, the external load should be connected to the solar cell. The amount of the output power from the solar cells depends on the impedance of the load, and there is the resistance maximizing the output power. This impedance is called the characteristic resistance (R_{CH}) of the solar cell. As shown in Figure 9.6, R_{CH} is determined from the slope of V_{mp}/I_{mp}. An impedance smaller than R_{CH} cannot draw out enough output voltage from the solar cell because output voltage of the solar cells becomes smaller (refer to R_2 in Figure 9.6). If the impedance of the external load is larger than R_{CH}, the output voltage of the solar cell could increase, but the output current becomes smaller. This is the case of R_1 in Figure 9.6. Due to a trade-off between current and voltage, there is an R_{CH} corresponding to an optimum operating

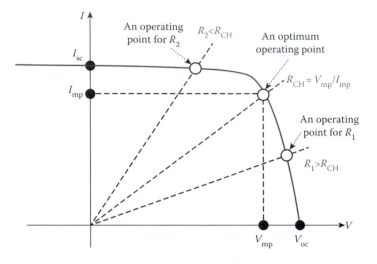

FIGURE 9.6 A schematic illustration of characteristic resistance (R_{CH}) of the solar cell.

point, and the power output from the solar cell is maximized by connecting R_{CH} to the solar cell. Example 9.2 shows how we can find R_{CH} of the solar cell from its $J–V$ curve.

Example 9.3: Performance Parameters of Solar Cells (J_{sc}, V_{oc}, FF)

You are trying to connect an external load to a solar cell. This solar cell exhibits the current density (J) and power (P) versus voltage relations in Figure 9.7. What is the resistance of the external load that allows you to pull out a maximum power from this solar cell? Active area (the area of the p-n junction semiconductor exposed to illuminating light) is 100 cm².

Solution

In Figure 9.7, out power (P) is calculated by multiplying current and voltage at each point of the $J–V$ curve. Then, a relation between P and V shows that maximum power of the solar cell is found at $V_m \sim 0.55$ volt and $J_m \sim 5.5$ mA/cm². Since the active area is 100 cm², total current is 0.55 A (= 0.0055 A/cm² × 100 cm²). This current–voltage relation is obtained when the resistance of 1 Ω ($R = V/I = 0.55/0.55$) is connected to the solar cell with an active area of 100 cm². Therefore, connection of the external resistance 10 kΩ to the solar cell makes the solar cell supply the maximum power to the external resistance.

In addition to η_{conv}, incident photon-to-current efficiency (IPCE) is an important parameter that evaluates the electron conversion process of the solar cell. IPCE is also called external quantum efficiency and represents the ratio of the photogenerated and collected electron number over the incoming photon number at a fixed light wavelength. From the definition, IPCE is given by

$$\text{IPCE} = \frac{\text{Number of photogenerated and collected electrons } (n)}{\text{Number of incident photons } (N)} \tag{9.20}$$

Since incoming power (P_{in}) = $N \times h\nu$ and short-circuit current (I_{sc}) = en/t (e: electron charge, t: time), Equation 9.20 can be rewritten as:

$$\text{IPCE} = \frac{I_{sc} \times t/e}{P_{in}/h\nu} \tag{9.21}$$

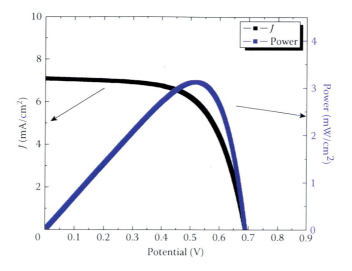

FIGURE 9.7 Exemplary plot showing the dependence of output current and power on the external bias in the solar cell under illumination.

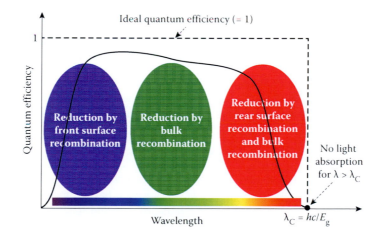

FIGURE 9.8 Incident photon-to-current efficiency (IPCE) of the solar cell.

When all constants are inserted in Equation 9.21, we find that IPCE for light with the wavelength λ is

$$\text{IPCE} = \frac{I_{sc}}{P_{in}} \times \frac{1240}{\lambda} \tag{9.22}$$

where the unit of I_{sc}, P_{in}, and λ are A, watt, and nm, respectively. As the solar cell absorbs more photons (i.e., reflectance and transmittance decreases) and the carrier recombination is suppressed more, IPCE of the solar cell increases. Therefore, high IPCE is indicative of high light absorption and low carrier recombination for a given wavelength. Figure 9.8 shows an exemplary IPCE of a solar cell. When photon energy is smaller than E_g of the semiconductor, IPCE becomes zero. For the longer wavelength light near E_g, IPCE decreases dramatically. This is due to the recombination at the rear surface and in the bulk. Also, IPCE dramatically decreases for light with a very short wavelength because most of photons with the short wavelength are absorbed near the front surface of the solar cell where defects (dangling bonds) of the front surface on which solar light is incident. Facilitate recombination of photogenerated carriers.

Example 9.4: Incident Photon-to-Current Efficiency and Photocurrent

An indium gallium arsenide, (In,Ga)As, photodiode is irradiated by IR light with the wavelength 1 μm and the power density 1.5 mW/cm². Assume that incident photon-to-current efficiency (IPCE or external quantum efficiency) of the photodiode is 90% over IR light, and the light receiving area is 2.5 cm². What is the electric current coming out of the photodiode?

Solution
Since the wavelength λ is 1 μm, the energy of each photon ($E_{ph} = hc/\lambda$) is 1.24 eV. Also, total incident power over the area of 2.5 cm² is 2.8 mW. Therefore, the number of incident photons (N_{ph}) per unit time is

$$N_{ph} = \frac{2.8 \times 10^{-3}\, J/s}{1.24 \times 1.6 \times 10^{-19}\, J} = 1.4 \times 10^{16}\, \text{photons/s}$$

The quantum efficiency is 90% and electric current from the semiconductor is

$$I_{ph} = 1.4 \times 10^{16} \times 1.6 \times 10^{-19} \times 0.9 = 2.0 \text{ mA}$$

9.3 PHYSICAL EVENTS UNDERLYING p-n JUNCTION-TYPE SOLAR CELLS

In Section 9.2, we learned the operating principle of solar cells, the equivalent circuit composed of an electric current source and a diode, and the important factors demonstrating solar cell performance. In this section, we will study details of the physical events occurring in p-n junction-type solar cells.

9.3.1 CHANGES IN FERMI ENERGY LEVEL UNDER ILLUMINATION

According to Section 9.1, photocurrent (I_L) flows from the solar cell to an external resistance (R) that is connected to the solar cell and produces work. However, in Chapter 5, E_F is constant through the p-n junction at equilibrium and there is no net electric current unless external field is biased. How should we modify diode equations in Chapter 5 to explain net current of the solar cell? Let us recall the fact that Fermi energy (E_F) becomes flat at a semiconductor–semiconductor junction and semiconductor–metal junction. Since E_F of different constituents comprising the junction is leveled, E_F must be uniform throughout the junction (see Equation 9.23).

$$V_{bi} = \frac{kT}{q} \ln\left(\frac{p_{p0}}{n_i}\right)_{p-side} + \frac{kT}{q} \ln\left(\frac{n_{n0}}{n_i}\right)_{n-side} = \frac{kT}{q} \ln\left(\frac{N_D N_A}{n_i^2}\right)$$

and

$$qV_{bi} = (E_i - E_F)_{p-side} + (E_i - E_F)_{n-side} \tag{9.23}$$

where E_F is Fermi energy level of *p-n* junction, E_i is intrinsic Fermi energy level, n_i is an intrinsic carrier concentration, N_D is donor impurity concentration in an n-semiconductor, and N_A is donor impurity concentration in a p-semiconductor.

If the p-n junction semiconductor reaches equilibrium even under the illumination, this means that photogeneration of electrons and holes increase n_i and decrease the barrier height at the junction (V_{bi}). Then, E_i is bent less to maintain E_F constant, (see the band diagram of Figure 9.9). Regardless of band bending, if the p-side and n-side of the p-n junction semiconductor have the same E_F

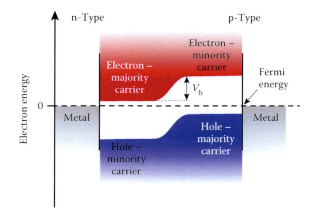

FIGURE 9.9 Band diagram of p-n junction semiconductor at thermal equilibrium.

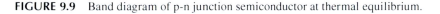

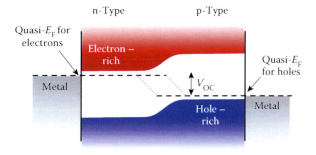

FIGURE 9.10 Band diagram of p-n junction semiconductor at quasi-equilibrium.

(Figure 9.9), a driving force for the net flow of photogenerated carriers cannot be developed. Thus, electrons cannot flow from the n-side to the p-side even though the n-side and the p-side are connected through an external circuit.

To resolve the contradictions of these two facts (the leveling of E_F through the junction region and net electric current under illumination without external bias) in p-n junction-type solar cells, it is important to understand that the solar cell exposed to light is not under thermal equilibrium, and E_F at the quasi-equilibrium state is not uniform so that the charge carrier can flow through an external circuit (Figure 9.10). When the solar cell absorbs photons with energy larger than E_g, photogenerated carriers increase both electron and hole concentrations above the equilibrium concentration. To restore the thermal equilibrium, the semiconductor needs time in the order of the carrier lifetime, which is a millisecond in silicon. However, if the solar cell is continuously exposed to light, the thermal equilibrium cannot be recovered. Instead, electrons in the conduction band (CB) and holes in the valence band (VB) pursue quasi-equilibrium within each band. In this case, free electron concentration in CB and free hole concentration at VB do not change as a function of time as long as a light with constant intensity illuminates. This quasi-equilibrium within the band can be achieved because carriers restore equilibrium through two different relaxation mechanisms. When extra electrons and holes are injected into CB and VB, they rearrange themselves and relax through collision with the lattice (i.e., phonon scattering)—that is, relaxation occurs independently in CB and VB. Of course, free electrons at CB and free holes at VB recombine to regain the thermal equilibrium through the relaxation between CB and VB.

Here, it is important to know that relaxations within the band and between the bands occur at different time scales. Although phonon scattering time is $\sim 10^{-12}$ second or smaller, electron–hole recombination time is at least $\sim 10^{-9}$ second (as large as $\sim 10^{-3}$ second for an indirect semiconductor such as silicon). This means that, if an incident photon flux is constant, the relaxation within the band is achieved much faster than the relaxation between the bands. Consequently, electrons at CB and holes at VB have their own quasi-equilibrium states where quasi-Fermi energy levels are expressed as E_{Fn} and E_{Fp}. Figure 9.10 shows a schematic on quasi-Fermi energy levels. It is noted that electrons and holes in the quasi-equilibrium states do not meet well-known relations of $p_n \times n_n = n_i^2$ and $p_p \times n_p = n_i^2$.

Now, let us figure out how to quantitatively express the quasi-Fermi state that appears as a result of an illumination-induced disturbance. If an extrinsic semiconductor is at thermal equilibrium, electron concentration and hole concentration are given by:

$$n_c = N_c \, \exp\left(-\frac{E_c - E_F}{k_B T}\right) \text{and } n_h = N_v \, \exp\left(-\frac{E_f - E_v}{k_B T}\right) \qquad (9.24)$$

In the quasi-equilibrium state, carrier concentrations still can be formulated by modifying Equation 9.24. This assumption is valid as long as the disturbance of electron and hole profiles is not so great and the time interval between disturbance events (i.e., photogeneration) is not smaller than the carrier scattering time. Strictly speaking, the reasoning of the quasi-Fermi state is valid only when the photogeneration rate of electron and hole is between $\left(\dfrac{1}{\text{scattering time}}\right)$ and $\left(\dfrac{1}{\text{recombination time}}\right)$. If a time interval between electron–hole photogeneration events is shorter than the scattering time, even the relaxation within the band is too slow to achieve a quasi-equilibrium state. On the other hand, if a time interval between electron–hole pair photogeneration events is longer than the recombination time, the relaxation between the bands is fast enough to achieve thermal equilibrium. When the quasi-equilibrium state is established, concentrations of electrons and holes at CB of n-type semiconductor and VB of p-type semiconductor can be expressed using E_{Fn} and E_{Fp}, as follows:

$$n_c = N_c \, \exp\left(-\frac{E_c - E_{Fn}}{k_B T}\right) \text{ and } n_h = N_v \, \exp\left(-\frac{E_{Fp} - E_v}{k_B T}\right) \qquad (9.25)$$

If Equation 9.25 is rewritten for E_{Fn} and E_{Fp},

$$E_{Fn} = E_c - kT \, \ln\frac{N_c}{n_c} \text{ and } E_{Fp} = E_v + kT \, \ln\frac{N_v}{n_h} \qquad (9.26)$$

Equation 9.26 indicates that E_{Fn} and E_{Fp} approaches E_c and E_v as more photogenerated electrons and holes are injected. From Equations 9.25 and 9.26, we also find that the product of n_c and n_h of the quasi-equilibrium state changes from $n_c \times n_h = n_i^2$ to:

$$n_c \times n_h = N_c N_v \, \exp\left(-\frac{(E_c - E_v) + \left(E_{Fp} - E_{Fn}\right)}{k_B T}\right) \qquad (9.27)$$

Given that $N_c N_v \, \exp\left(-\dfrac{E_c - E_v}{k_B T}\right) = n_i^2$, Equation 9.27 can be rewritten as:

$$n_c \times n_h = n_i^2 \, \exp\left(-\frac{E_{Fp} - E_{Fn}}{k_B T}\right) \qquad (9.28)$$

From Equation 9.28, we learn that a new mass–action law is established for electrons and holes at the quasi-equilibrium state. As the gap between E_{Fn} and E_{Fp} increases, a deviation of $n_e \times n_h$ between the quasi-equilibrium state and the thermal equilibrium state increases exponentially.

Note that E_{Fn} of the n-side and E_{Fp} of the p-side are not the same anymore in the quasi-equilibrium state, although the p-side and n-side are in contact. Different quasi-Fermi levels explain why a potential difference is developed between the terminals of the n-side and the p-side. In the open-circuit state, this potential difference becomes V_{oc} and is written as:

$$V_{oc} = E_{Fn} - E_{Fp} \qquad (9.29)$$

Once terminals of the p-side and the n-side are connected with an external circuit, electrons flow from the n-side terminal to the p-side terminal due to V_{oc}.

9.3.2 Generation, Recombination, and Transport of Electrons and Holes

In Equation 9.1, the assumption is that the electric current of the solar cell is due to the flow of electrons and holes that can reach the built-in-potential at the p-n junction. Photogenerated carriers outside the carrier diffusion distances from the depletion layer do not contribute to photocurrent (I_L); instead, they recombine and produce photons or phonons (or both). In this section, we will study the details of recombination and diffusion of photogenerated carriers, which will allow us to calculate I_L, estimated as $I_L = q \times A \times (L_p + L_n) \times g_{op}$ in Equation 9.1.

Before starting our calculations, we need to revisit the concept of carrier recombination. In previous sections, recombination is used to describe a physical event by which electrons and holes meet and combine. The term *recombination* includes several different cases. Electrons in the conduction band and holes in the valence band can meet directly for recombination. Also, electrons can be trapped and stay at localized defect sites between the conduction band edge (E_c) and the valence band edge (E_v) before recombining with holes. The consequences of recombination are not the same for all cases. Energy loss by recombination can be used to produce photons or phonons or to excite electrons from E_c. In order to distinguish the three different types of recombination, the event generating heat is called nonradiative recombination and the other event generating light is called radiative recombination. The energy loss process by which an electron is excited is called Auger recombination. These three different recombination events are schematically described in Figure 9.11.

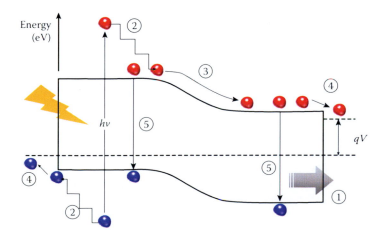

FIGURE 9.11 Loss processes in a standard solar cell: (1) nonabsorption of below band gap photons; (2) lattice thermalization loss; (3) and (4) junction and contact voltage losses, respectively; (5) recombination loss (radiative recombination is unavoidable).

As explained in Section 9.6.2, radiative recombination is dominant in a direct band gap semi-conductor with good crystallinity. When the semiconductor has indirect band gap or possesses high defect density, nonradiative recombination prevails. Although other recombination events are based on combining one electron with one hole, two electrons and one hole need to meet for the occurrence of an Auger recombination. Since more carriers are involved, an Auger recombination becomes pronounced in a semiconductor with high free electron density.

9.3.2.1 Continuity Equation in a Neutral Semiconductor at Quasi-Equilibrium

It is time to quantify electron and hole concentrations of a neutral semiconductor that is exposed to illumination. *Neutral* means that the semiconductor's free electrons and holes are not depleted. In the junction region where the built-in potential is developed, the semiconductor is not neutral. As we learned in Chapter 8, the semiconductor absorbs photons with $E > E_g$ to produce surplus electrons and holes that disturb thermal equilibrium and establish quasi-Fermi energy levels. When a p-type semiconductor is under illumination, the excess electron concentration (Δn_p) can be expressed as the difference between total electron concentration (n_p) and an equilibrium electron concentration (n_{p0}); that is, $\Delta n_p = n_p - n_{p0}$. Note that the minority carrier is selected for the convenience of calculation, because n_p/n_{p0} is much larger than p_p/p_{p0} in p-type semiconductor. If we assume that the rate of electron–hole pair creation per unit volume (g_{op}) is uniform throughout the semiconductor, an excess electron concentration is also uniform throughout the semiconductor. Then, we can connect the time-dependent change of n_p (dn_p/dt) with g_{op} by introducing the recombination rate of photogenerated carriers (U):

$$\frac{dn_p}{dt} = \frac{d(n_{p0}) + d\left(\Delta n_p\right)}{dt} = g_{op} - U \tag{9.30}$$

In Equation 9.30, n_{p0} is a constant at a fixed temperature and dn_{p0}/dt is zero. U is a parameter showing how many free carriers disappear per unit time. Here, let us recall that τ_n is the minority carrier lifetime showing how long it takes for excess electrons to recombine electrons in a p-type semiconductor. If the number of excess minority carriers and their lifetimes are known, you can find that the excess minority carriers disappear at a rate of $\Delta n_p/\tau_n$. Therefore, we can estimate U as $\Delta n_p/\tau_n$ and rewrite Equation 9.30 as:

$$\frac{d\left(\Delta n_p\right)}{dt} = g_{op} - \frac{\Delta n_p}{\tau_n} \tag{9.31}$$

where τ_n is carrier lifetime. Equation 9.31 describes the time dependence of the excess electron concentration with the assumption that the photogeneration rate is constant through the sample (in other words, zero driving force for diffusion). In reality, however, the rate of photogeneration is highest near the front surface of the solar cell and decreases exponentially as a function of the depth. As summarized in the Beer–Lambert law ($I = I_0 \exp(-\alpha x)$) of Section 8.5.3, light intensity continuously decreases though a material and so does the amount of photons absorbed by the semiconductor. Notice that the change in the light intensity as a function of depth implies that the photo-generation rate per volume (g_{op}) also has depth dependence. If the density of photons in the light is N_{ph}, the Beer–Lambert law leads to a relation of $N_{ph} = N_s \exp(-\alpha x)$ (N_s: photon flux density at the surface). Then, we find that g_{op} is related to the depth and the photon density as follows:

$$g_{op} = -\frac{dN_{ph}}{dx} = \alpha N_s \exp(-\alpha x) \tag{9.32}$$

This tells us that the rate of carrier generation by illumination is determined by depth, absorption coefficient, and N_s, with the assumption that there is no surface reflection and all absorbed photons are converted to electron–hole pairs. If there is surface reflection, N_s in Equation 9.32 needs to be changed to $(N_s - R)$, where R is reflectance at the surface.

When the slab in Figure 9.11 is illuminated from one surface, profiles of absorbed photon density and photogenerated carriers exponentially decay from the illuminated surface. This excess electron concentration gradient causes both diffusion current and drift current (that is due to the quasi-Fermi energy level gradient). Therefore, it is necessary to add an effect of carrier transport to Equation 9.30.

Now, let us imagine a slab—one end of which is exposed to light (see Figure 9.4). In this slab, two end surfaces have different excess electron concentrations and electrons diffuse from one surface to the other, which results in an electric current along the x-axis. Then, the excess carrier concentration profile can be obtained from the following time-dependent continuity equation:

$$\frac{d(\Delta n_p)}{dt} = \frac{1}{q}\frac{\partial J_n}{\partial x} + g_{op} - \frac{\Delta n_p}{\tau_n} \tag{9.33}$$

where J_n is electric current density and q is the charge of an electron.

9.3.2.2 Photogenerated Carrier Transport through Diffusion and Drift in a Neutral Semiconductor at Quasi-Equilibrium

In Equation 9.33, $\dfrac{1}{q}\dfrac{\partial J_n}{\partial x}$ is the difference between influx and outflow of electrons and represents the concentration of electrons that are left in the slab due to nonuniformity of carrier transport and photogeneration. In this section, we will study how to quantitatively describe J_n in the neutral semiconductor. For this purpose, we need to recall two previous points:

1. The quasi-Fermi energy level (E_{Fn}) profile is expressed as a function of carrier concentration,
$$E_{Fn} = E_C - kT\ln\frac{N_C}{n_p} = \ln(n_p) + E_C - kT\ln(N_C) \text{ (Equation 9.26).}$$

2. Electric current is driven by a carrier concentration gradient (that is diffusion) and an electric field (that is drift).

According to (1), the concentration gradient of the excess electrons results in the quasi-Fermi energy level gradient along the x-axis of the slab. Note that the quasi-Fermi energy level gradient along the x-axis of the slab, in turn, applies electric field $\left(\sim\dfrac{dE_{Fn}}{dx}\right)$, which induces the movement of free carriers (i.e., drift current). Then, drift current of minority electrons driven by the quasi-Fermi level gradient is $J_{n,drift} = q\mu_n n_p \dfrac{dE_{Fn}}{dx}$, where μ_n is electron mobility. In previous chapters, we learned that the diffusion current ($J_{n,diffusion}$) is a product of electron charge and diffusion flux, $qD_n\dfrac{dn}{dx}$, where D_n is the electron diffusion coefficient. If we combine the diffusion current and drift current, the electric current of minority carriers (J_n) is given by:

$$J_n = qD_n\frac{dn_p}{dx} + q\mu_n n_p\frac{dE_{Fn}}{dx} \tag{9.34}$$

Then, in the steady state (i.e., there is no time dependence), Equation 9.34 can be rewritten by combining Equation 9.26 and Equation 9.33:

$$\left(D_n\frac{d^2 n_p}{dx^2}\right) + \mu_n\frac{d}{dx}\left(n_p\frac{dE_{Fn}}{dx}\right) + g_{op} - \frac{\Delta n_p}{\tau_n} = 0 \tag{9.35}$$

The solution of Equation 9.35 leads to the minority carrier concentration profile ($n_p(x)$) at steady state that takes into account photogeneration, diffusion, drift, and recombination. Once you know $n_p(x)$, you can insert $n_p(x)$ into Equation 9.34 and find the electron current (J_n) in an illuminated p-type semiconductor at the quasi-equilibrium state.

In Equation 9.35, you may question whether D_n and μ_n can be correlated because both of them are a measure of electron movement over different driving forces. D_n shows how easily electrons diffuse when there is a concentration gradient. On the other hand, μ_n represents electron transport over a potential gradient (i.e., electric field). This implies that materials with high D_n may also have high μ_n because both D_n and μ_n represent how electrons transport over external driving forces (electron concentration gradient, electric field). In fact, from basic diffusion and drift equations, a correlation between D_n and μ_n is extracted as:

$$\frac{D_n}{\mu_n} = \frac{kT}{e} \tag{9.36}$$

Equation 9.36 is called the Einstein relation, and its detailed calculation can be found in other textbooks on semiconductor theory. From Equation 9.35, we can calculate excess carrier distribution at the surface and bulk of the illuminated semiconductor.

1. *Excess Carrier Distribution in Bulk* ($g_{op} = 0$)

   Now, let us figure out what would happen if the aforementioned illuminated slab is very thick or the absorption coefficient (α) is very large. Because most photons are absorbed near the illuminated surface and before reaching the opposite side, we can assume that incoming photons do not reach the bulk of the slab (i.e., $g_{op} = 0$) and E_{Fn} is relatively uniform (i.e., no drift current). Then, Equation 9.36 needs to be rewritten as:

$$\left(D_n \frac{d^2 n}{dx^2} \right) - \frac{\Delta n_p}{\tau_n} = 0 \tag{9.37}$$

   Here, we use a well-known diffusion relation $\sqrt{\tau_n D_n} = L_n$ (L_n: electron diffusion length) to couple τ_n and D_n:

$$\frac{d^2 n}{dx^2} - \frac{\Delta n_p}{L_n^2} = 0 \tag{9.38}$$

   A solution of Equation 9.38 is given by:

$$\Delta n_p(x) = A \, \exp\left(\frac{x}{L_n} \right) + B \, \exp\left(-\frac{x}{L_n} \right)$$

or

$$\Delta n_p(x) = \Delta n_p(0) \exp\left(-\frac{x}{L_n} \right) \tag{9.39}$$

   In Equation 9.39, we find that L_n controls the minority carrier profile, and an increase in L_n elongates the minority carrier concentration profile ($\Delta n_p(x)$). Notice that both intrinsic material property (e.g., band gap) and extrinsic material property (e.g., structural defects) change L_n. Indirect semiconductors have longer L_n than direct semiconductors. Materials with high defect density exhibit shorter L_n. Although minority electron diffusion length (L_n) is 100 ~ 300 μm for single crystal p-type silicon, it is >10 μm for single crystal p-type GaAs.

2. Excess Carrier Distribution near Surface ($g_{op} \neq 0$)

An assumption for $\Delta n_p(x)$ in Equation 9.39 is that the slab is very thick and light cannot reach the point of interest (i.e., g_{op} = zero). However, if the slab is thin, the incoming photon flux inside the slab is not zero and g_{op} is not zero. In the case of $g_{op} \neq$ zero, the general form of $\Delta n_p(x)$ is modified as:

$$\Delta n_p(x) = A \, \exp\left(\frac{x}{L_n}\right) + B \, \exp\left(-\frac{x}{L_n}\right) + g_{op}\tau_n \qquad \text{or}$$

$$\Delta n_p(x) = \Delta n_p(0)\left(-\frac{x}{L_n}\right) + g_{op}\tau_n \qquad (9.40)$$

In this section, we have reviewed how photogeneration, diffusion, and recombination determine the electron concentration profile in a neutral p-type semiconductor that is a minority carrier density profile. Once we know Δn_p, we can calculate the electric current by the electrons in a p-type semiconductor (J_n) using Equation 9.34 $\left(J_n = qD_n\dfrac{dn_p}{dx} + q\mu_n n_p\dfrac{dE_{Fn}}{dx}\right)$. Figure 9.12 schematically shows the formation of excess carrier profiles and the diffusion current when its one end is illuminated. From Equations 9.39 and 9.40, we learned the quantitative expressions of the excess carrier concentration profile under illumination and the consequent electric current in the neutral semiconductor.

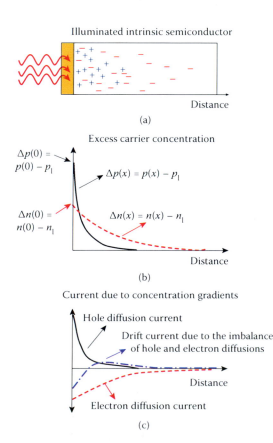

Illuminated intrinsic semiconductor

Distance

(a)

Excess carrier concentration

$\Delta p(0) = p(0) - p_1$

$\Delta p(x) = p(x) - p_1$

$\Delta n(0) = n(0) - n_1$

$\Delta n(x) = n(x) - n_1$

Distance

(b)

Current due to concentration gradients

Hole diffusion current

Drift current due to the imbalance of hole and electron diffusions

Distance

Electron diffusion current

(c)

FIGURE 9.12 (a) Schematic illustration of the carrier distribution in an illuminated slab, (b) Excess carrier concentration vs. distance from the illuminated surface, (c) Electric current due to concentration gradient vs. distance from the illuminated surface. (From Kasap, S. O., *Principles of Electronic Materials and Devices*, 3rd ed., New York: McGraw Hill, 2006. With Permission.)

One remaining question regarding the photogenerated carrier transport is "Is the electron transport influenced by the hole transport?" The answer to this question depends on the difference of diffusion coefficient between electrons and holes. If we use the same reasoning process for Equation 9.40, the excess hole concentration in a p-type semiconductor is given by

$$\Delta p_p(x) = \Delta p_p(0) \exp\left(-\frac{x}{L_p}\right) \text{ for a case of } g_{op} = 0 \text{ in a diffusion path} \tag{9.41}$$

$$\Delta n_p(x) = \Delta n_p(0)\left(-\frac{x}{L_n}\right) + g_{op}\tau_n \text{ for a case of } g_{op} \neq 0 \text{ in a diffusion path} \tag{9.42}$$

A comparison of Equations 9.37 and 9.39 (or Equations 9.38 and 9.40) shows that, if $L_n = L_p$ and $\Delta n_p(0) = \Delta p_p(0)$, $\Delta n_p(x)$ is the same as $\Delta p_p(x)$. Then, electrons and holes have the same concentration profile and quasi-Fermi level gradient, which leads to no interaction between excess electron transport and excess hole transport. In silicon, however, L_n is several times longer than L_p, and excess electrons and excess holes have different concentration profiles. If their concentrations are not the same, different quasi-Fermi level gradients of electrons and holes generate an electric field that causes the drift. In the case of silicon, this electric field arising from different excess carrier profiles suppresses the travel of photogenerated electrons and promotes the transport of the photogenerated holes. This is called the Dember effect. Figure 9.12 schematically illustrates how a difference in excess carrier profiles ($\Delta n_p(x)$ and $\Delta p_p(x)$) develops an additional driving force for the drift. If the mobility of the crystalline semiconductor is large or the semiconductor is in contact with electrode materials, the Dember effect becomes less significant and is omitted when calculating electric current.

In Section 9.3.2.1, we found that the diffusion mechanism is dominant over the drift mechanism in the neutral semiconductor at quasi-equilibrium, based on reasonable assumptions such as uniform doping, small quasi-Fermi level gradient, and high carrier mobility. The last important concept to understand is that the minority carrier diffusion contributes to the electric current much more than the majority carrier diffusion. In the quasi-neutral state, the magnitude of the excess concentration is not much different between the majority carrier and the minority carrier ($\Delta n_p(x) \sim \Delta p_p(x)$). However, the relative change controlling the diffusion is much larger for the minority carrier ($\Delta n_p(x)/n_{p0} \gg \Delta p_p(x)/p_{p0}$). Therefore, it is concluded that the current from the minority carrier diffusion prevails in carrier transport in the neutral semiconductor at quasi-equilibrium.

9.3.2.3 Photogenerated Carrier Transport in a Depletion Region of the p-n Junction

Our understanding from the previous section is that minority carrier diffusion is important to quantify electric current in a neutral semiconductor under illumination. In this section, we will study how photogenerated electrons and holes travel in the depletion region of a p-n junction where space charges build up the built-in-potential. If there is no external electric bias and no illumination (that is, thermal equilibrium), the built-in potential in Equation 9.23 is set up at the interface between the p-semiconductor and the n-semiconductor. As shown in Figure 9.9, this built-in potential at the junction suppresses the diffusion of majority carriers and causes the drift of minority carriers. At equilibrium, electric current by majority carrier diffusion and electric current by minority carrier drift have the same magnitude and opposite flow direction. Consequently, the net current is zero at thermal equilibrium. If that is the case, the hole concentration at the boundary of the depletion region in the n-side (p_{nb}) is given by:

$$p_{nb} = p_{n0} = p_{p0} \exp\left(\frac{qV_B}{kT}\right) = \frac{n_i^2}{N_D} \tag{9.43}$$

where V_B is the equilibrium barrier height at the junction region, n_i is the intrinsic carrier concentration, p_{n0} and n_{n0} are the equilibrium concentration of the hole as a minority carrier and a majority carrier, and N_D is a donor impurity concentration on the n-side.

When the barrier height (V) decreases (or forward bias is applied), the diffusion barrier decreases and there is an exponential increase in the diffusion current. From the viewpoint of the drift current, a decrease in V_B decreases the driving force for the drift current but increases the minority carrier concentration for the drift. Since the drift current is a product of the driving force and the carrier concentration, a decrease in V makes a negligible change in the drift current. Then, the diffusion current becomes larger than the drift current under forward bias, leading to a net current at the junction and an increase in p_{nb}. The effect of the electric field on the electric current density through the junction under no illumination (called dark current density J_{dark}) and the minority carrier concentration at the boundary condition is given by:

$$J_{dark} = \left(q\frac{D_n}{L_n}n_{p0} + q\frac{D_p}{L_p}p_{n0} \right)\left[\exp\left(\frac{qV_A}{kT}\right) - 1 \right] \tag{9.44}$$

$$p_{nb} = p_{n0}\ \exp\left(\frac{qV_A}{kT}\right) = \frac{n_i^2}{N_D}\ \exp\left(\frac{qV_A}{kT}\right)\ \text{and}$$

$$n_{pb} = n_{p0}\ \exp\left(\frac{qV_A}{kT}\right) = \frac{n_i^2}{N_A}\ \exp\left(\frac{qV_A}{kT}\right) \tag{9.45}$$

where V_A is applied voltage ($V_A > 0$ for forward bias and $V_A < 0$ for reverse bias) and n_{pb} is electron concentration at the boundary of the depletion region in the p-side (p_{nb}). A pre-factor $\left(q\frac{D_n}{L_n}n_{p0} + q\frac{D_p}{L_p}p_{n0} \right)$ in Equation 9.44 also implies that the diffusion is responsible for the net current under the bias; p_{nb} in Equation 9.45 also provides a boundary condition for the carrier diffusion in the neutral region in contact with the depletion region. Figure 9.13 shows a schematic on the transport and concentration profiles of an electron and a hole in a p-n junction under forward electric bias. Readers can find that the minority carrier concentrations at the end of the junction boundary (p_{nb}, n_{pb}) are not the same as the bulk minority carrier concentrations (p_{n0}, n_{p0}).

From Equations 9.44 and 9.45, we find that the forward bias to the p-n junction can break the thermal equilibrium. In addition to the bias, there are several ways to break the thermal equilibrium of the p-n junction, and illumination is one of them. When the semiconductor is illuminated, both minority carrier concentration and majority carrier concentration increase in both the n-side and the p-side. Though absolute changes of the carrier concentration are the same for majority and

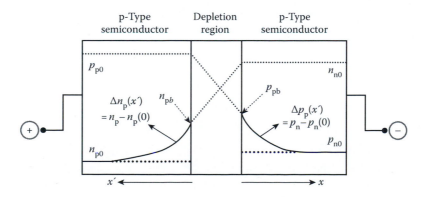

FIGURE 9.13 A schematic on the transport and concentration profiles of an electron and a hole in a p-n junction under forward electric bias.

minority carriers, a relative change is much more significant for the minority carrier in the neutral region. Hence, in contrast to the dark case, the drift current of minority carriers in the illumination case becomes much larger than the diffusion current of the majority carriers and net current flows through the depletion layer. In other words, under illumination, the drift of excess holes from the n-side to the p-side and that of excess electrons from the p-side to the n-side overwhelm the diffusion of majority carriers, leading to the development of quasi-equilibrium. That is schematically explained in Figure 9.10. In an open circuit, there is no external electric current between the p-side and the n-side, and a difference of quasi-Fermi energy levels between E_{Fn} of the n-side and E_{Fp} of the p-side is maintained under continuous photon flux. This explains why the open-circuit voltage (V_{oc}) is developed under illumination. As incident light intensity increases, more majority carriers are accumulated and V_{oc} ($=E_{Fn,n\text{-side}} - E_{Fp,p\text{-side}}$) gets close to the barrier height (V_B) at the junction in a dark condition. If V_{oc} reaches V_B, V_{oc} is saturated. Additionally, photogenerated carriers are consumed through recombination, instead of being used to increase V_{oc}. Therefore, at a steady state, V_{oc} between two terminals of the p-n junction solar cell is constant, and the drift current passing through the depletion region equals the rate of the recombination occurring in both neutral and depletion regions.

If the n-side and the p-side are connected through an external circuit with zero electric impedance (i.e., a short-circuit state), accumulated electrons in the n-side flow toward accumulated holes in the p-side to make the E_{Fn} of the n-side equal to the E_{Fp} of the p-side. The leveling of the E_{Fn} of the n-side with the E_{Fp} of the p-side is the origin of the short-circuit current through an external circuit. In this case, the drift current passing through the depletion region is equivalent to the sum of minority carrier diffusion currents ($J_{total} = J_{n,p\text{-side}} + J_{p,n\text{-side}}$) in the neutral regions. Here, $J_{n,p\text{-side}}$ and $J_{n,p\text{-side}}$ can be calculated from Equations 9.34, 9.40, and 9.41. It is time to recall the photocurrent (I_L) in Equation 9.1 and the equivalent circuit in Figure 9.2. According to Figure 9.2, the electric current of the solar cell consists of the photocurrent (I_L) and the forward bias direction diode current (I_F). However, since there is no voltage drop by the external resistance in the short-circuit state (see Section 9.2.1), we can think that no forward bias is applied, and I_L in Equation 9.1 is equivalent to the short-circuit current ($I_t = A \times J_{total} = A \times J_{n,p\text{-side}} + A \times J_{p,n\text{-side}}$).

Now, let assume that external impedance is connected to the solar cell. Then, electric current flowing through the resistor causes the potential drop across the resistor. Since the solar cell is connected to the impedance in parallel, the potential drop across the impedance also works as the forward bias for the solar cell diode. Then, electric current flowing through the solar cell diode is not zero anymore and is called the forward bias current (I_F). Notice that I_L flows in the drift current direction of the depletion junction, which is opposite to the forward bias current direction of the solar cell diode. Therefore, if we ignore the photogeneration in the depletion region and the drift current in the neutral region, the total current density (J_{total}) of the solar cell is the difference between photocurrent density (J_L) and the forward diode current (J_F). J_{total} is given by:

$$J_{total} = J_L - J_F = qD_n \frac{dn_p}{dx}\bigg|_{p\text{-side}} + qD_p \frac{dp_n}{dx}\bigg|_{n\text{-side}} - \left(q\frac{D_n}{L_n}n_{p0} + q\frac{D_p}{L_p}p_{n0} \right)$$

$$\times \left[\exp\left(\frac{q(I_{total}R)}{kT} \right) - 1 \right]$$

(9.46)

where R is the resistance of the external impedance connected to the solar cell. Equation 9.46 shows that a large diffusion coefficient of the semiconductor is a necessary condition for high current output. If the semiconductor has a small diffusion coefficient (that is, a problem of amorphous silicon), the design of the solar cell can be changed from the p-n junction to the p-i-n junction by inserting an intrinsic layer between the p-side and the n-side. In this case, the built-in potential is extended to the intrinsic layer and the depletion layer thickness is comparable to the diffusion length (L). Then, the photogeneration and the recombination in a wider depletion region become more important and both J_L and J_{dark} in Equation 9.46 need to be modified.

9.4 FACTORS LIMITING POWER CONVERSION OF p-n JUNCTION-TYPE SOLAR CELLS

In Section 9.3, we learned about detailed mechanisms of the power conversion process such as photogeneration, diffusion, recombination, drift, and the competition between diffusion and drift. At this point, we will ask "What is the power conversion efficiency of a solar cell?" As of 2015, the best power conversion efficiency (PCE) is 25.0% for a single crystalline Si solar cell and 20.8% for a multicrystalline Si solar cell. If a solar cell is installed into a module, the best efficiency decreases to ~21%. The PCE of most commercial solar cell modules is still below 20% after 40 years of research. Note that the theoretical upper limit of η_{conv} is ~30% for the silicon solar cell. This means that ~70% of solar energy cannot be converted to electricity even in the best scenario. In Section 9.4, we will study the theoretical limit of power conversion and the energy loss mechanism.

9.4.1 THEORETICAL LIMIT OF PCE

In the solar cell, the theoretical limit of the PCE is determined by the band gap energy (E_g) of the semiconductor because of two factors.

The first limiting factor is that the semiconductor does not absorb all of the incoming photons, and only part of the incoming photons are converted to electrons. If the energy of incident photons ($E = hv$) is less than E_g, these sub-bandgap photons do not produce any electron–hole pairs. Incident photons transmit the semiconductor, and the incident solar energy is not converted to electrical energy. Therefore, if the E_g of the semiconductor comprising the solar cell becomes larger, the ratio of produced electron–hole pairs over incident photons decreases the theoretical conversion efficiency (η_{conv}). Hence, we cannot select a semiconductor with high E_g for a high-performance solar cell.

The second limiting factor is that one supra-bandgap photon (that is, $hv > E_g$) generates only one electron–hole pair no matter how large an amount of energy a photon delivers to the semiconductor. Even though the incident photon energy (hv) is much larger than E_g, only one electron–hole pair is created, and the energy difference ($hv - E_g$) dissipates as heat. This is because free electrons excited by light (called hot electrons) cannot stay in the middle of the conduction band for a long time before producing another free electron at E_c (that is an inverse process of Auger recombination). In nature, hot electrons return to the conduction band edge (E_c) very quickly by donating their surplus kinetic energy to lattice phonons (Figure 9.11). Phonon relaxation of hot electrons by phonons is the second reason why the PCE of the solar cell is lower than 100%. Since surplus energy of photons ($hv - E_g$) is wasted as heat, the energy loss of supra-bandgap photons is called a thermalization loss. As E_g of the semiconductor decreases, the thermalization loss of hot electrons becomes more significant. Hence, a semiconductor with low E_g cannot be chosen for a high-performance solar cell.

Based on the balance of these two physical phenomena (i.e., sub-bandgap photon transmittance without being absorbed and kinetic energy loss of hot electrons as heat), Shockley and Queisser coined the concept of ultimate efficiency (UE). In calculating UE, it is also assumed that (i) all supra-bandgap photons are absorbed, (ii) the mobility of photogenerated carriers is infinitely high (in other words, no recombination, leading to FF = 1), and (iii) the V_{oc} of the solar cell is the same as the E_g of the semiconductor.

$$\text{UE} = \frac{V_{oc} \times I_{sc}}{\text{Incident solar energy}} = \frac{E_g \times Q_S(E_g)}{P_S} \tag{9.47}$$

where $Q_S(E_g)$ is a density of photons with $hv > E_g$ on the Earth's surface (that is I_{sc} for UE calculation) and P_S is the power of the solar irradiation on the Earth's surface. If $S(E)$ is the density of photons with energy E on the Earth's surface, $Q_S(E_g)$ and P_S are given by

$$Q_S\left(E_g\right) = \int_{E_g}^{\infty} S(E)dE \tag{9.48}$$

$$P_S = \int_0^{\infty} E \times S(E)dE = \frac{2\pi h}{c^2} \int_0^{\infty} \frac{v^2 dv}{\exp\left(\dfrac{hv}{kT_s}\right) - 1} \tag{9.49}$$

where v is a frequency of photons and T_S is the sun's temperature. In Equation 9.49, the Sun was treated as a black body and a number of emitted photons obeyed Planck's distribution. If the solar irradiance at a condition of air mass (AM) 1.5 is used, UE in Equation 9.47 can be plotted as follows: The largest UE in Figure 9.14 is ~44 %, when E_g of the semiconductor is ~1.1 eV. One importance of UE calculation is that Planck's law of black body radiation was used to find P_S. Before Shockley and Queisser took a thermodynamic approach, the continuity relation of photogenerated carriers was used to estimate the theoretical efficiency of the solar cell.

Although UE calculation based on the band gap loss provides reasonable theoretical efficiencies, it misses one important fact. Excited electrons and holes can directly recombine to emit photons, and partial recombination of photogenerated carriers is an origin of quasi-Fermi energy levels (refer to Figures 8.29 and 9.10). From a thermodynamic standpoint, recombination is understood as a process following the second law of thermodynamics, which predicts heat loss (i.e., increase in entropy) while work is extracted from the system. Shockley and Queisser considered the radiative recombination effect and calculated the theoretical efficiency of solar cells (Equation 9.50). Since more realistic boundaries were added, the model for Equation 9.50 is called Shockley and Queisser's detailed balance model (SQ-DB model).

$$\eta\left(x_g, x_c, f, t_s\right) = m\left(x_g, x_c, f\right) v\left(x_g, x_c, f\right) u\left(x_g\right) t_s \tag{9.50}$$

where $x_g = \dfrac{E_g}{kT_S}$ (T_S: sun's temperature), $x_c = \dfrac{T_C}{T_S}$ (T_C: Earth's temperature), and f is a factor accounting for the geometric effect of solar irradiance and carrier recombination. In addition, $u\left(x_g\right)$ shows

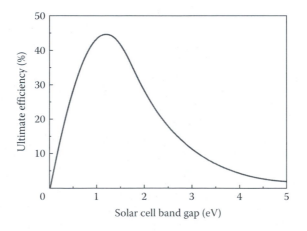

FIGURE 9.14 Ultimate efficiency of the solar cell as a function of semiconductor band gap (E_g).

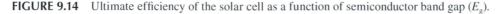

energy gap losses (which are used for UE calculation); $v(x_g, x_c, f)$ is a ratio of the operational voltage to the energy gap (which gives the output voltage of the solar cell); $m(x_g, x_c, f)$ is the impedance matching factor; and t_s is the probability of radiative recombination of an electron–hole pair. Elaborate calculation of SQ-DB is out of the range of this book; however, it is worthwhile to find the meaning of each term in Equation 9.50. In $u(x_g)$, we take into account the energy loss of supra-bandgap photons ($E > h\nu$) and the transmittance of sub-bandgap photons ($E < h\nu$), which results in UE of the solar cell. $m(x_g, x_c, f)$ is controlled by the resistance of the external load and the I–V characteristics of the solar cell (see Example 9.3); $m(x_g, x_c, f)$ can be assumed to be 1 for estimating the theoretical efficiency of the solar cell. New concepts that ask for our attention are f, $v(x_g, x_c, f)$, and t_s.

Let us take a look at these concepts qualitatively. Note that f, $v(x_g, x_c, f)$, and t_s are related to the evolution of quasi-equilibrium. In Equation 9.47, UE is calculated as a product of the band gap and the absorbed photon density. Assumptions rationalizing Equation 9.47 are that (i) the V_{oc} of the solar cell is the same as the band gap and (ii) all of the absorbed photons are converted to photogenerated carriers that are collected by the terminals of the solar cell. Though they are qualitatively reasonable, UE calculations overestimate the efficiency limit. This overestimate is due to the recombination of photogenerated electron–hole pairs. As shown in Section 9.3.1, photogeneration balances with recombination to achieve a quasi-equilibrium state. Consequently, (i) not all of the absorbed photons are converted to electricity, and (ii) the difference between the E_{Fn} of the n-side and the E_{Fp} of the p-side is the maximum output voltage that can be extracted from the system (refer to Section 9.3.2.3). This means that I_{sc} and V_{oc} should be recalculated to obtain a more reliable theoretical efficiency limit.

In the SQ-DB model, entropy in the second law of thermodynamics was used to calculate the minimum recombination rate of photogenerated carriers. Here, the sun and a solar cell on the Earth's surface are considered as black bodies at the temperatures 5800 K and ~300 K. respectively. Based on this scheme of black bodies, we can estimate the minimum entropy increase in the environment (i.e., the minimum electron–hole recombination by radiative recombination) and the maximum work that the system can pull out (i.e., the maximum electricity production by solar cell). In the SQ-DB model, it is assumed that the recombination of electron–hole pairs produces only photons. Once we know the amount of the recombined electron–hole pairs, we can find a new theoretical I_{sc} from the difference between absorbed photon numbers and recombined electron–hole pairs.

Recombination is also important in calculating V_{oc} because the recombination controls E_{Fn} and E_{Fp}. Notice that a higher recombination rate decreases V_{oc} ($= E_{Fn, n\text{-side}} - E_{Fp, p\text{-side}}$) by decreasing the carrier density at the quasi-equilibrium. So far, it has explained why $v(x_g, x_c, f)$ and t_s are required to precisely estimate the theoretical efficiency of a solar cell. The last thing to be mentioned about the SQ-DB model is the geometric factor (f), which allows us to correct solar irradiance. Note that V_{oc} ($E_{Fn, n\text{-side}} - E_{Fp, p\text{-side}}$) is not constant. As the incident light intensity increases, Fermi energy levels get close to E_c and E_v, and V_{oc} ($E_{Fn, n\text{-side}} - E_{Fp, p\text{-side}}$) also increases. This means that light intensity matters in accurately calculating theoretical efficiency. Further details on the light intensity effect appear in Section 9.2. Since the light incident direction is not normal to the surface of the Earth, the incident light spreads out on the solar cell, and the light intensity on the solar cell surface is not the same as the light intensity in extraterrestrial space. As the light incident angle deviates from a normal geometry, light intensity on the solar cell decreases and, thus, the theoretical PCE decreases. This explains why the geometric factor (f) is added in calculating $v(x_g, x_c, f)$, which determines the ratio of the operational voltage to the energy gap. A schematic in Figure 9.15 shows the geometric effect of incident angle on solar light intensity.

Figure 9.16 shows the theoretical efficiency of a semiconductor solar cell as a function of semiconductor band gap (E_g) from the SQ-DB model. A relation between theoretical efficiency and band gap from the SQ-DB model exhibits a parabolic curve that looks similar to the UE curve in Figure 9.14. However, in Figure 9.15, we find that the highest energy conversion efficiency (η_{conv})

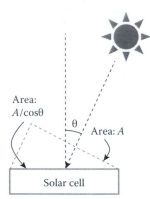

FIGURE 9.15 A schematic explaining the geometric effect on the intensity of the light illuminating a solar cell that is placed on the surface of the Earth.

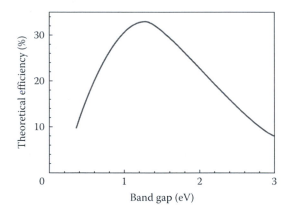

FIGURE 9.16 The theoretical efficiency of a semiconductor solar cell as a function of semiconductor band gap (E_g) calculated from Shockley and Queisser's detailed balance model.

of the solar cell is ~34% and an optimum band gap is ~1.3 V. The theoretical efficiency is smaller than the UE because a part of the absorbed photons are reemitted due to the radiative recombination (i.e., decrease in I_{sc}). An increase in the optimum band gap in the SE-DB model is due to the fact that the open-circuit voltage is the quasi-Fermi energy level difference instead of the semiconductor band gap, which also reduces V_{oc}. If the band gap is the only criterion of material selection for the solar cell, GaAs (E_g~1.4 V) is better than silicon (E_g~1.1 V). In fact, as of 2015, the best PCE of a single crystalline GaAs solar cell was ~27.5%, which is higher than that of a Si solar cell. However, GaAs is more expensive than silicon and thus is used mainly in specialized applications such as extraterrestrial usage.

9.4.2 ADDITIONAL POWER LOSS MECHANISMS IN REAL P-N JUNCTION-TYPE SOLAR CELLS

In Section 9.4.1, we learned the theoretical PCE of a solar cell as a function of the semiconductor band gap. However, as we know, the PCE of real solar cells is smaller than the theoretical efficiency. In this section, we will study factors (reflectance loss, surface/bulk recombination loss, junction/contact voltage loss, shunt/series resistance, impedance mismatch) that are responsible for the difference between ideal and real solar cells.

Reflectance at the surface of the solar cell is the first factor explaining why the power PCE is lowered in the real solar cell. In the ideal solar cell, the surface reflectance is assumed to be zero. However, as shown in Equation 9.55, a change in the refractive index (n) at the interface causes part of the incident light to be reflected; n of silicon ranges from 3.5 to 3.9 in an IR and red light regime, while n of air is ~1. Then, according to Equation 9.55, $\left(R = \left(\dfrac{n_1 - n_2}{n_1 + n_2} \right)^2 \right)$, ~33% of the incident power is reflected at the air–silicon interface, which, in turn, reduces the PCE of the solar cell by ~33%. In addition to the air–semiconductor interface, a metal electrode in the front surface is partially responsible for the reflectance loss. Since light cannot transmit through the metal electrode, photons' incident upon the metal electrode cannot be converted to electron–hole pairs, and the reflectance loss by the metal electrode is proportional to the areal coverage of the top metal electrode. To suppress the reflectance loss, the refractive index of the silicon surface is modified by coating an antireflection film or by creating a texture with a sub-micron feature size or reducing the top electrode area (or a combination thereof). Details of antireflection techniques will be discussed in Section 9.5.

The second loss mechanism is recombination loss (also called collection loss). In addition to the direct recombination that is thermodynamically handled in the SQ-DB model, the nonradiative recombination of photogenerated carriers can occur at the surface (surface recombination), semiconductor–semiconductor interface, semiconductor–metal interface, and at the internal structural defects of the bulk (bulk recombination). In general, surface recombination is more significant than bulk recombination. This nonradiative recombination depends on the crystalline quality, the doping concentration, and on the internal potential profile, which significantly influence minority carrier diffusion length (or lifetime). To prevent the bulk recombination, electron–hole pairs must be produced within one diffusion length from the depletion region. If not, the bulk recombination loss becomes high. In recombination at unpassivated surfaces, the distance between the surface and the junction matters. As the carrier generation location gets closer to the surface, the probability of surface recombination exponentially increases. Note that the front surface and the back surface of the solar cell have a different recombination probability for photons with different energy. Since the absorption coefficient of silicon is larger for blue light than for red light, blue light is absorbed more near the front surface and red light is absorbed more near the back surface. Consequently, electron–hole pairs from blue light mainly recombine at the front surface, while the back surface suppresses the collection of electron–hole pairs produced from red light. Recombination loss results in a decrease in both current and voltage of the solar cell. The effect of the surface recombination on electric current is shown in the IPCE curve of Figure 9.8. Current loss is pronounced in the blue region for the front surface recombination and in the red region for the back surface recombination. In addition, the recombination decreases the quasi-Fermi energy levels, thus there is a loss of output voltage. To reduce the recombination of majority carriers, the defects on the surface are passivated or the semiconductor is heavily doped with impurities to lower the minority carrier concentration. These will be also discussed in Section 9.5.

The third factor related to the decrease in the efficiency is the large series resistance (R_S) of the solar cell, which includes metal electrode resistance, semiconductor resistance, and the contact resistance at semiconductor–semiconductor interface or at the semiconductor–metal interface. The main impact of a high series resistance on the solar cell is a decrease in fill factor (FF) because a voltage drop occurring at the series resistance makes the solar cell deviate from the ideal diode. A change in I–V characteristics of the solar cell by adding R_S is given by

$$I = I_L - I_0 \left\{ \exp\left[\frac{q(V + IR_S)}{kT} \right] - 1 \right\} \tag{9.51}$$

where I is output current and V is voltage applied between two terminals of the solar cell. Equation 9.51 shows that the voltage drop at R_S quickly decreases output current of the solar cell (I) near V_{oc}, leading to reduced FF. We also find that the effect of R_S (i.e., decrease in FF) becomes more pronounced at a high output current (or a large light intensity). If a series resistance is as low as several Ω, V_{oc} and I_{sc} do not change and FF_S (FF of the solar cell having the series resistance) are expressed as

$$FF_S = FF(1 - R_S) \tag{9.52}$$

If the series resistance is too large, I_{sc} can also decrease. An equivalent circuit of R_S and I–V curves of the solar cell, including R_S, is schematically explained in Figure 9.2. Variation of R_S of the solar cell heavily depends on the top metal electrode resistance and the metal–semiconductor contact resistance. Since solar light is incident from the front surface, a reflective metal electrode cannot cover the full surface area. About 10% of the front surface is a covered Ag electrode with a hierarchical structure of bus bars and fingers. As the surface coverage by the metal electrode increases, the series resistance of the solar cell decreases, but the photon absorption also decreases due to surface reflectance. Therefore, there is an optimum design of the top electrode that keeps both series resistance and surface reflectance low. The contact resistance is also an important source of series resistance. Work functions of the metal and the semiconductor are different, and a Schottky barrier is easily formed at the interface. Too much doping of impurities near the surface of the semiconductor decreases the contact resistance; however, this, in turn, reduces the collection efficiency of carriers that are photogenerated in the heavily doped region. This is because the doping concentration approaches a solubility limit of the dopant and precipitates are formed. The precipitates work as a recombination center and significantly decrease the minority carrier's lifetime. Hence, the region with precipitates is a so-called *dead layer* from the viewpoint of minority carriers.

The fourth factor related to a decrease in the efficiency is small shunt resistance (R_{SH}), which is connected to the external load in parallel to an equivalent circuit. To improve the performance of solar cells, R_{SH} in the equivalent circuit of Figure 9.2 must be high so that no charge carriers go through R_{SH}. Low R_{SH} provides a bypath for photogenerated carriers and is a source of leakage current, which reduces the short current. When current passes through an edge or manufacturing defects of the bulk solar cell (or both), R_{SH} decreases. In thin film solar cells, pinholes mostly are a source of low R_{SH}. In bulk solar cells, electric current passing through the edge of the devices easily decreases R_{SH}. In an ideal solar cell, R_{SH} is assumed to be infinitely large and leakage current is zero. I–V curves of the solar cell with reasonably small R_{SH} are given by

$$I = I_L - I_0 \left[\exp\left(\frac{qV}{kT}\right) - 1 \right] - \frac{V}{R_{SH}} \tag{9.53}$$

where I is output current and V is voltage applied between two terminals of the solar cell. Equation 9.53 shows that an obvious effect of R_{SH} is to decrease output current as voltage is applied to the solar cell. We also find in Equation 9.53 that R_{SH} decreases FF because current going through R_{SH} does not contribute to the diode behavior of the solar cell. The effect of R_{SH} becomes important at low current conditions (i.e., small light intensity).

Changes in the I–V curve of solar cells by large R_S and small R_{SH} are schematically illustrated in Figure 9.17.

9.4.3 FACTORS INFLUENCING SOLAR CELL OPERATION

In Sections 9.4.1 and 9.4.2, we learned how the efficiency of the solar cell is determined by intrinsic material properties (e.g., band gap, refractive index) and extrinsic variables (e.g., carrier recombination time, manufacturing defects). In this section, we will study the effect of environment (e.g., external load, temperature, light intensity, light incident angle) on the performance of the solar cell.

The first external factor to consider for solar cell operation is the external impedance that pulls out the power from the solar cell. As we learned in Section 9.1, real output voltage and current of the

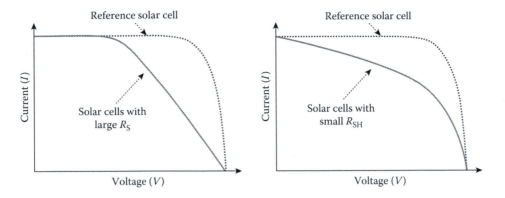

FIGURE 9.17 Schematic illustrations on the effects of large R_S and small R_{SH} on changes in the I–V curve of solar cells.

solar cell connected to the external load are smaller than V_{oc} and I_{sc} and are controlled by the impedance of the external load as well as by incident light intensity. An increase in the external impedance increases output voltage but decreases output current of the solar cell. Notice that there is an optimal voltage (V_m) and current (I_m) at which the output power of the solar cell is maximized (see Example 9.2). If the external impedance is higher than the optimum value, the output power of the solar cell decreases due to low output current. If the external impedance is lower than the optimum value, a decrease in the output voltage decreases the output power of the solar cell. The resistance maximizing the output power is called characteristic resistance (R_{CH}), which is a ratio of $\dfrac{V_m}{I_m}$. Note that R_{CH} is not constant even in the same solar cell because I–V characteristics of the solar cell change as a function of light intensity, temperature, and light incident angle. Therefore, a variable resistor is located amid the solar cell, the invertor, and the battery, and logic and control circuits are utilized to match the resistance of the variable resistor with R_{CH} at different operating conditions. This method is called maximum power point tracking technology (MPTT) and will be discussed further in the section on solar cell modules.

Temperature is the second external factor affecting solar cell performance. The effect of temperature on a solar cell is twofold. One is a change in the forward bias current of the solar cell diode and the other is a change in the band gap (E_g) of the semiconductor. An increase in the forward bias current at a higher temperature is due to an increase in the intrinsic carrier concentration (n_i) of the semiconductor. As shown in an equation of $\left[n_i \propto \exp\left(-\dfrac{E_g}{kT} \right) \right]$, an increase in temperature increases n_i, raises the equilibrium minority carrier concentration $\left(n_i^2 = N_D p_{n0} = N_A n_{p0} \right)$ in the neutral regions, and decreases the built-in potential (V_{bi}) at the depletion region. By increasing temperature, a series of changes enhances the diffusion flow of the carriers at the junction (i.e., a forward bias current of the solar cell diode), leading to a decrease in output voltage, output current, V_{oc}, and an upper limit of V_{oc} (i.e., V_{bi}) of the solar cell (see Equations 9.42 and 9.43). The dependence of V_{oc} on temperature is rewritten in Equation 9.54. In a silicon solar cell, for example, the reduction rate of V_{oc} is ~ 0.4 %/°C.

$$\frac{dV_{oc}}{dT} = -\frac{E_g - qV_{oc} + kT}{qT} \tag{9.54}$$

In contrast to a change in n_i, an increase in the temperature reduces E_g and increases the PCE of the solar cell. As temperature increases, the number of photons that can be absorbed ($h\nu > E_g$) increases and, thus, I_{sc} also increases. In real solar cells, the effect of n_i dominates that of E_g and an increase in temperature deteriorates the PCE by reducing the output voltage of the device.

Example 9.5: Change in Open-Circuit Voltage as a Function of Temperature

As temperature increases, the open-circuit voltage (V_{oc}) of the solar cells decreases. In addition, different semiconductors comprising the solar cell show different dependence of V_{oc} on the temperature. What is the relation between V_{oc} and temperature in Si solar cells and GaAs solar cells? Assume that the V_{oc} of Si solar cells and GaAs solar cells is 0.55 V and 0.9 V, respectively.

Solution

The V_{oc} of the solar cell is shown in Equations 9.12 and 9.13.

Note that $J_s = q \times n_i^2 \left[\dfrac{D_p}{L_p N_d} + \dfrac{D_n}{L_n N_a} \right]$ is related to the band gap (E_g), due to

$n_i^2 \sim \text{constant} \times \exp\left(-\dfrac{E_g}{kT} \right)$.

Then, $\dfrac{dV_{oc}}{dT}$ in Equation 9.55 can be obtained as follows:

$$\frac{dV_{oc}}{dT} \sim \frac{k_B T}{q} \ln\left(\frac{I_L}{I_s} \right) \sim \left[\frac{1}{q} \frac{dE_g}{dT} - \frac{1}{T}\left(\frac{E_g}{q} - V_{oc} \right) \right]$$

$\dfrac{dE_g}{dT}$ is -3×10^{-4} eV/K in Si and -4×10^{-4} eV/K in GaAs. Therefore, $\dfrac{dV_{oc}}{dT}$ is mainly determined

by $\dfrac{1}{T}\left(\dfrac{E_g}{q} - V_{oc} \right)$.

$\therefore \dfrac{dV_{oc}}{dT} = -2$ mV/K (Si), -1.7 mV/K (GaAs)

This indicates that temperature increased by one degree decreases V_{oc} by 2~4%.

The third external factor to consider is a change in light intensity, which varies over time of the day and location of the solar cell. Increasing light intensity increases the photogeneration rate (g_{op}) due to a higher incident photon density, leading to an increase in photocurrent (I_L). Note that higher I_L itself only increases the output power of the solar cell. Since there is a linear correlation between J_{sc} and light intensity (in other words, incident power), increasing I_L does not change the theoretical PCE of the solar cell. In addition to I_L, an increase in g_{op} at a higher light intensity pushes E_{Fn} and E_{Fp} toward E_c and E_v and increases V_{oc} (see Equations 9.9, 9.10, and 9.29). A logarithmic increase in V_{oc} $\left(\propto \ln\left(1 + \dfrac{I_L}{I_s} \right) \right)$ with the light intensity, in turn, improves the theoretical PCE of the solar cell.

Overall performance of the solar cell is improved under higher light intensity. This explains why a concentrator has been used to improve the performance of the solar cell. The role of the concentrator is to focus solar light by an optical element such as a parabolic mirror and increase the output power and efficiency of the solar cell. A concentration factor, X, represents an increase in the photon flux density by the concentrator. A concentration factor of 10 means that the solar cell produces power from concentrated light whose intensity is 10 times larger than that of unconcentrated light. However, if the light intensity is above a critical number, the electron–hole recombination probability increases, which, in turn, reduces V_{oc} and FF. The recombination rate is proportional to the carrier concentration (n) for trapping-assisted recombination, n^2 for radiative recombination (direct electron–hole recombination), and n^3 for Auger recombination. As the photogenerated carrier density is increased, the radiative recombination rate and Auger recombination rate increase dramatically. When all effects noted here are considered, the curve of efficiency versus concentrated light intensity in Figure 9.18 is obtained. As the light intensity becomes larger than the critical value, a decrease in V_{oc} and FF offsets an increase in I_{sc}, and the PCE of the solar cell is lowered.

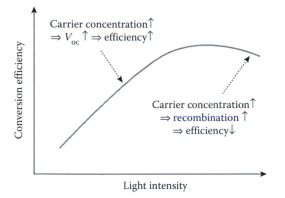

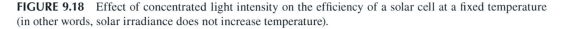

FIGURE 9.18 Effect of concentrated light intensity on the efficiency of a solar cell at a fixed temperature (in other words, solar irradiance does not increase temperature).

In addition to changing I_{sc} and V_{oc}, the light intensity also affects the relative role of parasitic resistances (e.g., shunt resistance and series resistance). Let us assume that a solar cell possesses both series resistance (R_S) and shunt resistance (R_{SH}). R_{SH} is more important at a low light intensity. An increase in the light intensity rebounds to the existence of R_S and decreases FF of the solar cell.

As we learned in SQ-DB for theoretical PCE calculation, the light incident angle is also an important external factor on which the PCE of the solar cell depends. If the surface of the solar cell has a normal to light incident direction, the illuminated area on the solar cell is minimized (i.e., the light intensity is maximized) and the PCE of the solar cell is the highest. When an angle between the solar cell surface and the incident direction deviates from 90°, the solar cell starts decreasing and the best power conversion performance is not achieved. Therefore, a solar cell on the Earth's surface must be tilted to maximize average light intensity and power output. An ideal tilting angle for maximum PCE is related to latitude. As latitude decreases (in other words, as you get closer to the Equator), the angle between incident solar light and the Earth's surface increases and becomes close to 90°. For this reason, a solar cell installed in the region of the Equator does not need to be tilted from the horizontal ground plane. A good rule of thumb for the best performance is that the solar cell panel should be tilted toward the south in the northern hemisphere (and north in the southern hemisphere), and the tilting angle between the solar cell panel and the horizontal ground plane should equal the latitude of the installation location.

9.5 DESIGN OF HIGH-PERFORMANCE p-n JUNCTION-TYPE SEMICONDUCTOR SOLAR CELLS

Difficulties in achieving the PCE of the solar cell close to the theoretical limit are due to poor light absorption and recombination of photogenerated carriers at the surface and in the bulk. There are two reasons for poor low light absorption. One is that light reflectance at the silicon surface is high. Most semiconductors have a high refractive index (usually $n > 3.0$). Given that the reflectance of light entering from air (or free space) is $R = \left(\dfrac{n-1}{n+1}\right)^2$, more than 30% of incident light power is reflected at the surface of Si and GaAs wafers. The other reason for low absorption is that silicon has an indirect band gap, and the light absorption coefficient of silicon is low. This problem becomes more serious for red and infrared light photon energy, which is slightly higher than the band gap energy. The low light absorption coefficient of silicon requires a few 100-nm-thick wafers for complete harvesting of the solar spectrum. An unavoidable side effect of the thick wafer for full solar spectrum absorption is that more carriers recombine during its travel toward the depletion region.

9.5.1 LIGHT MANAGEMENT FOR IMPROVED LIGHT ABSORPTION

The first way to manage light effectively in the solar cell is to put together multiple semiconductors with different band gaps. The Shockley–Queisser (SQ) limit shows that the maximum PCE of single junction Si solar cells is about 31% due to transmittance of sub-bandgap photons and hot electron energy loss of supra-bandgap photons. This problem is universal for all types of solar cells, though detailed balance conditions are different from one to another. The theoretical PCE of the solar cells can be increased if the single junction structure is substituted with the multiple junction structure. To push the SQ limit upward, a tandem structure that uses two different band gap materials has been proposed. In the tandem structure, the lower band gap semiconductor can absorb small energy photons that transmit the higher band gap semiconductor. In addition, the higher band gap semiconductor reduces the energy loss of hot electron because the difference between the excited energy state of hot electrons and the conduction band edge (E_C) decreases. Since the absorption coefficient of the semiconductor mostly is larger for short wavelength light than for longer wavelength light, the semiconductor with the smaller band gap is placed close to the front surface of the solar cell. In other words, incident light hits the smaller band gap semiconductor first. When the band gaps are around 1.5 to 1.7 eV for the top cell and 0.8 to 0.9 eV for the bottom cell in a series-connected double-junction device, the PCE limit can be extended more than 45%. As the number of the semiconductor layer increases, the theoretical PCE increases. As of 2015, the best PCE of the multijunction solar cell obtained with a laboratory scale is 46%, using the four junction structure of GaInP/GaAs/GaInPAs/GaInAs. It is noted that such a high PCE is obtained under concentrated solar light (297×). Since a multijunction solar cell is more expensive than a Si solar cell, the multijunction solar cell normally is operated under concentrated solar light. In the design of the multijunction solar cell, we must be careful to match electric current that is produced from each semiconductor layer. If each layer generates very different electric current under light, carrier recombination becomes serious at the junction between light absorbing semiconductors. Figure 9.19 shows a schematic illustration and an *I–V* curve of the multijunction solar cell.

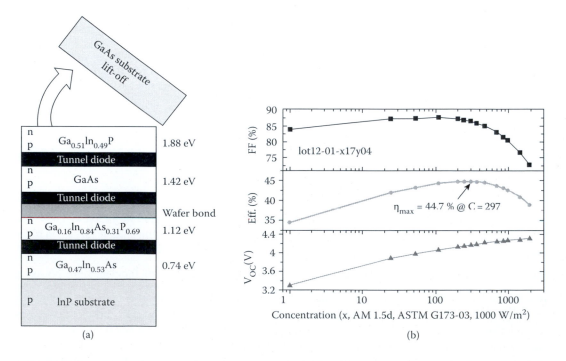

(a)   (b)

FIGURE 9.19 Schematic illustration of the multijunction solar cell (a) and a summary of its *I–V* characteristics (b). (From Dimorth, F., et al., *Prog. Photovolt: Res. Appl.*, 22, 277–282, 2014. With Permission.)

Surface texturing, antireflective coating, and front contact size reduction (i.e., redesign of the metal electrode on the front surface) are the most representative ways to enhance the light absorption of silicon wafers. All of them suppress reflectance at the front surface of silicon, due to the refractive index mismatch, and they increase the density of photons that enter silicon. In addition to lowering reflectance, the surface texturing compensates for the weakness of the low absorption coefficient of silicon by elongating an optical path.

Roughness of the surface is an important factor that controls the diffuse reflectance and scattering of light, which we discussed in the section on optical properties. A smooth surface exhibits very high specular reflectance, which is estimated using $R = \dfrac{\left(n_1 - n_2\right)^2}{\left(n_1 + n_2\right)^2}$. As the surface roughness increases, diffuse reflectance occurs on random surface features and reflectance decreases. When the surface has a textured structure, multiple and ordered reflectance by tilted surface features dramatically reduce the surface reflectance. A change in reflectance is shown in Figure 9.20 with schematics explaining light paths.

A textured structure is formed on a Si surface using chemical etching and photolithography. Anisotropic etching of a single crystal Si leaves (111) planes on the surface that exhibit a pyramid shape pattern or an inverted pyramid shape pattern. Grooved planes with an angle of 54.74° from an original (100) plane surface help incident light to be reflected at least twice before light departs the solar cell. In addition, the oblique incident angle on (111) planes increases the optical path for Si. This effect is equivalent to increasing the effective wafer thickness, leading to increased absorbance and decreased transmittance.

In addition to the surface texturing, the semiconductor surface is coated with a dielectric film with the refractive index (~2) between Si and air. TiO_2, Ta_2O_5, SiO_2, and Si_3N_4 are widely used dielectrics for the antireflection (AR) coating. Since a difference in the refractive index decreases, the reflectance is reduced on the surface with the AR coating. It is noted that the absorption coefficient of the semiconductor and the refractive index of the coating layer vary as a function of light wavelength.

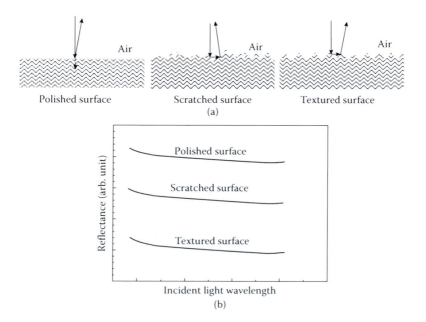

FIGURE 9.20 (a) A schematic on the reflectance of surfaces with different roughness features; (b) reflectance versus wavelength for silicon with different roughness features. (From Angermann H., et al., *Cent. Eur. J. Phys.* 7, 363–370, 2009. With Permission.)

Therefore, it is necessary to consider the spectral variation of the solar cell performance and maximize the AR coating effect for low quantum efficiency wavelength when AR coating is designed. The AR layer usually is coated near the end of the solar cell manufacturing process. Both the optically thin layer (≤ several hundreds of nanometers) and optically thick layers (~several micrometers) are used as coating layers. If the wafer surface is flat and smooth, the thin layer thickness is around one-quarter wavelength of a targeted light to arouse destructive interference between incoming light and reflected light. A thickness of a quarter wavelength means that a phase difference between the incident light and reflected light is 180° at the surface of the coating layer. Hence, two waves cancel out, resulting in destructive interference. A schematic of the solar cell with surface texturing and antireflection coating is shown in Figure 9.21.

Another way to increase the light absorption is to decrease the metal electrode area on the front surface. As discussed in Section 9.4.2, however, there is a trade-off between series resistance (R_S) and reflectance. As the metal electrode area decreases (a decrease in the width of bus bars and fingers or an increase in the distance between bus bars and fingers), reflectance by the metal electrode increases but R_S decreases. Therefore, there is a balanced metal electrode structure optimizing R_S and reflectance. To decrease both R_S and reflectance, the metal electrode design needs to be changed. In some high-performance Si solar cells, the metal electrode is buried inside the Si so that shading by the metal electrode is reduced. The buried metal electrode on the front surface is shown in Figure 9.22. Grooves for metal electrode filling can be manufactured using a laser process or mechanical machining. In this buried contact design, grooves are filled with the contact metals; nickel, copper, and then silver, are deposited using electroless plating. The manufacturing technique for the buried layer is mostly expensive, while the buried electrode can effectively suppress surface reflectance.

The last way introduced in this section as a method to enhance light absorption is to form metal contacts for both holes and electrons only at the back surface of a Si solar cell (see Figure 9.23). This structure is called an integrated backside contact (IBC) solar cell. One advantage of this structure is

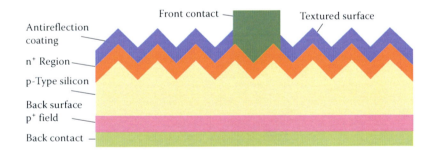

FIGURE 9.21   A simplified cross-section of a commercial single crystalline (monocrystalline) silicon solar cell. (From Saga, T., *NPG Asia Mater.*, 2, 96–102, 2010. With Permission.)

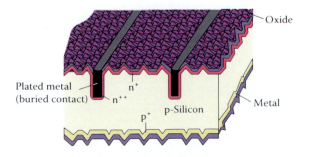

FIGURE 9.22   Buried-contact cell structure of silicon solar cells developed by BP Solar. (From Saga, T., *NPG Asia Mater.*, 2, 96–102, 2010. With Permission.)

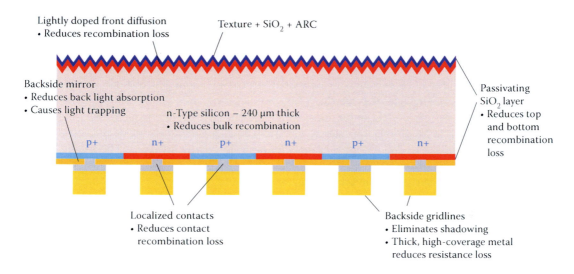

FIGURE 9.23 A schematic illustration of an integrated back contact (IBC) solar cell. (From Eglash, S., *Laser Focus World*, 39–41, 2009. With Permission.)

to maximize the solar absorption area by removing the front electrode and minimize series resistance by reducing the distance between metal contacts. In addition, a lightly doped front surface decreases the dead layer effect, and the carrier recombination and SiO_2 passivation layers are used to reduce losses at both the front and back surfaces. The IBC design needs the modeling of carrier transport and the high quality Si wafer with high carrier mobility to carefully control recombination loss.

9.5.2 ENHANCED COLLECTION OF PHOTOGENERATED CARRIERS

In Section 9.5.1, we learned how to improve the light absorption in the semiconductor. The next important question is how to enhance the collection of electron–hole pairs produced from absorbed photons. Undesired recombination of photogenerated carriers is one of the major factors lowering the PCE of the solar cell. As we learned in the SQ-DB model, thermodynamics determine the radiative recombination probability, which is already taken into account in the theoretical PCE calculation. A type of recombination that we can control is the nonradiative recombination occurring at the wafer surface and at the interface between layers. In particular, dangling bonds on the surface are defects that trap carriers and facilitate the surface recombination. Though the nonradiative recombination also takes place in the bulk of the Si wafer, the bulk recombination is not as significant as the surface recombination in single crystalline Si and polycrystalline Si with high crystallinity. Therefore, suppression of the surface recombination is the major content of this section. However, in Si thin film or in amorphous Si, the minority carrier diffusion length is shorter than the absorption length (~200–500 μm) and bulk recombination cannot be ignored. We will discuss how to handle the problem of short diffusion length at the end of this section.

To prevent the surface recombination, a dielectric film such as a SiO_2 layer is formed on the surface of Si by chemical vapor deposition or oxidation methods. Once the thin dielectric layer is formed, the surface recombination velocity decreases, thereby preventing the loss of photogenerated carriers. This technique is called surface passivation. The dielectric layer decreases the surface recombination velocity via two mechanisms. One is a chemical effect that removes the dangling bonds on the Si surface. Since the oxide molecules are chemically attached to the dangling bonds, the trapping of carriers at the surface defects does not occur in the passivated Si surface. Because the interface between Si and SiO_2 is less defective, the total number of recombination centers of the surface passivated Si is much smaller than that of the unpassivated Si. The other is a physical effect.

Since the surface dielectric layer stores the charge under the output voltage of the solar cell, an electric field is established in the dielectric layer. Then, one type of carrier is repelled from the surface and the electric field application direction determines whether an electron or hole moves away from the dielectric layer. A large disparity between electron concentration and hole concentration decreases the number of charge carriers available for recombination, which suppresses the recombination events occurring near the interface of the Si–dielectric layer.

We discussed a similar effect in the section on the metal oxide semiconductor field-effect transistor (MOSFET). In MOSFET, the external electric field is applied to the oxide dielectric layer to deplete one type of carrier and to form a channel for minority carrier diffusion. These chemical and physical effects of the passivation layer increase the collection efficiency of photogenerated carriers at the front and back surface of the semiconductor solar cell.

Another way to suppress recombination is to exploit conduction (valence) band bending in a heavily doped surface. If the back surface is highly doped to improve the response to long wavelength light, this effect is called back surface field (BSF). The BSF region in lightly doped p-type Si is formed on the back surface of a Si wafer by firing a screen-printed Al layer. Then, part of the Al is incorporated into the Si while most of the aluminum layer turns to the Al back contact. A cross-sectional structure, after firing the screen-printed Al layer, is shown in Figure 9.24. In a p-p⁺ interface of the BSF region, the band bending promotes the extraction of carriers toward the back contact and increases the carrier collection efficiency. The heavily doped region also has high electric conductivity and allows carriers to travel through a tunneling behavior, even though there is a Schottky barrier at the interface between the heavily doped region and the metal back contact.

Although the heavily doped region has positive impacts on the carrier collection, too much doping can create a dead layer effect that originates from dopant precipitation and carrier trapping/collision at the activated (i.e., electrically charged) dopant site. To reduce the adverse effect of heavy doping, the interface area between the metal contact and the heavily doped Si region is increased. This is the case of the buried electrode, shown in Figure 9.22. Since the metal contact–Si interface area is increased, shallower donor impurity doping (i.e., a smaller barrier height at p-p⁺ junction) is enough to collect photogenerated carriers at the metal contact. In addition, moderate impurity doping in the subsurface

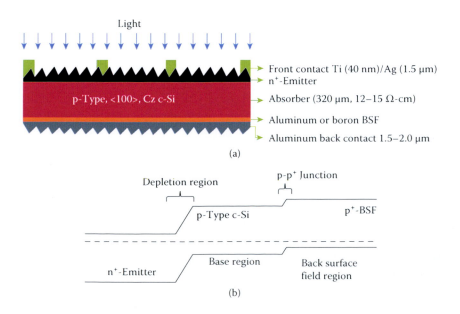

FIGURE 9.24 (a) Cross-section of a single crystalline Si solar cell with a back surface field effect. (b) Energy band diagram of a p⁺-p-n⁺ type junction showing the effect of the surface. (From Singh, G., et al., *RSC Adv.*, 4, 4225–4229, 2014. With Permission.)

FIGURE 9.25 A schematic of an SHJ solar cell on an n-type wafer and its band diagram. (Structure is not drawn to scale.) (From Wolf, S. D., et al., *Green.*, 2, 7–24, 2012. With Permission.)

region reduces the dead layer effect and suppresses the carrier recombination. This, in turn, prevents degradation of open-circuit voltage and improves the short wavelength response of the solar cell.

Similar to the dangling bonds of the Si surface, the highly recombination-active metal contacts with Si work as a recombination center and decrease the carrier collection efficiency. One way to avoid the problem of highly recombinative metal contact is to insert a few nanometer thick layer of the semiconductor (mainly hydrogenated amorphous Si, a-Si:H) and displace highly recombination-active contacts from the Si surface. The solar cell with the structure of Figure 9.25 is called a silicon heterojunction solar cell (SHJ); a-Si:H layer of SHJ passivates the interface between metal contacts and crystalline Si. Also, the function of conducting electrode can be awarded to a-Si:H layer by doping donor impurities or acceptor impurities. From the viewpoint of light absorption, a-Si:H has a wider band gap and a-Si:H does not limit the light absorption ability of crystalline Si. This means that a-Si:H plays the role of a window. Though the fabrication process of SHJ is costly due to additional thin film deposition steps, the SHS structure allows for solar cells with energy conversion efficiencies above 20% at the industrial-production scale.

So far, we have discussed how to decrease the front and back surface recombination in a crystalline Si solar cell where the minority carrier diffusion length is comparable to or larger than the wafer thickness. However, in Si thin films, the minority carrier diffusion length is short, and the bulk recombination cannot be ignored. Though the best solution to the bulk recombination problem is to improve the crystalline quality and elongate the diffusion length, there is a limitation to increasing the crystalline quality of the thin films. In this case, an intrinsic Si layer is inserted between the p-side and the n-side to make p-i-n junction solar cells. An advantage of p-i-n junction over p-n junction is that the depletion region is extended to the intrinsic layer, as shown in Figure 9.26. Note that the inserted intrinsic layer does not change the height of the built-in potential. In the p-i-n junction, the drift in the depletion region compensates for the small minority carrier diffusion length, and carriers that are photogenerated in the elongated depletion region can travel longer with the aid of the drift.

One more way to improve the carrier transport behavior of the solar cell is to replace a p-type wafer with an n-type wafer. This is still in the research and development stage but has attracted considerable amount of interest from industry. P-type Si wafers have been widely used because the minority carrier mobility is higher in p-type Si than in n-type Si. However, boron impurities in p-type Si with a traceable amount of oxygen impurities form boron–oxygen complexes under illumination. Since boron–oxygen complexes are strong recombination centers, the PCE of the solar cell degrades 1 to 3% in the initial period of solar cell operation. This is called light-induced degradation (LID). In the solar cells using an n-type Si wafer, LID is not observed.

In addition, since a p-type dopant (boron) diffuses faster than an n-type dopant (phosphor) at high temperature, the precise control of the doping profile during a thermal annealing process is more difficult for p-type Si. This problem gets worsened due to the base impurities present in Si. Metallic

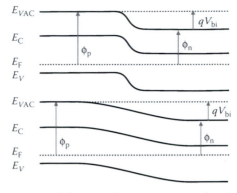

ϕ_p, ϕ_n: work function of p-, n-type semiconductor

FIGURE 9.26 Comparison of p-n junction (top) and p-i-n junction (bottom), showing the effect of the i-region on the extended depletion region.

impurities inevitably added to Si mainly work as n-type donors, which means that higher amounts of p-type impurities are required in p-type Si than in n-type Si for high electric conductivity. Several problems of p-type dopants require stricter quality control for p-type Si wafers than for n-type Si wafers to fabricate high efficiency solar cells (PCE > 20%). This, in turn, increases the production cost of the solar cells. To address the problem of p-type Si wafers, several companies have tested a potential n-type Si wafer for high efficiency solar cells and some of them have shown n-type Si base solar cells with PCE > 20%.

9.6 EMERGING SOLAR CELLS NOT USING p-n JUNCTION OF INORGANIC SEMICONDUCTORS

In addition to Si solar cells, there are several emerging solar cells that are not based on the semiconductor wafer. Among promising solar solutions with great potential for solar energy harvesting are TiO$_2$ nanoparticle-based DSSCs. DSSCs offer significant economic and environmental advantages over conventional photovoltaic devices because they can be manufactured relatively inexpensively and in an energy-efficient and environment-friendly manner. The typical structure of DSSCs using a liquid electrolyte is shown in Figure 9.27. The DSSC is created by coating nanoparticles comprising a wide band gap material (such as TiO$_2$) on transparent conducting oxide (TCO) electrodes. This is followed by dipping the nanoparticles and TCO pairs in a solution containing small organic dye molecules. When the DSSC is exposed to solar light, an incident photon creates a bound electron–hole pair (exciton) in the organic dye. This electron–hole pair dissociates at the organic–inorganic interface and electrons then flow into the TiO$_2$ nanoparticle photoelectrode and the TCO film. The full potential of DSSCs has not yet been realized due to unresolved limitations in absorption of solar spectrum and transport of photogenerated carriers. In fact, the efficiency of current DSSCs has plateaued at 11–12%, which is well below their theoretical limit of 33%. The current yield at a given wavelength for DSSCs can be expressed as:

$$\eta(\lambda) = LHE(\lambda) \cdot \varphi_{inj} \cdot \eta_c \qquad (9.55)$$

where light harvesting efficiency (LHE) is a fraction of the incident photons absorbed by the dye, ϕ_{inj} is the quantum yield for charge injection, and η_c is the efficiency of collecting the injected charge at the back contact.

In conventional designs, one of the limiting factors contributing to low conversion efficiency of DSSCs is long-term stability due to their liquid electrolyte component. If sealing of the device is not

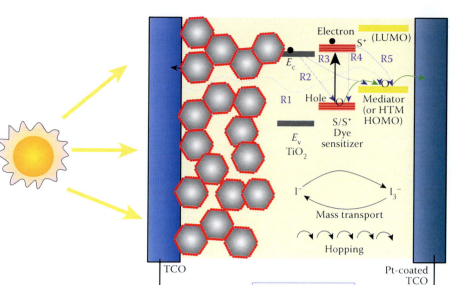

FIGURE 9.27 A schematic on the structure and operating principle of dye-sensitized solar cells (DSSCs). (From Lee et al., *Materials Science and Engineering: B*, 176, 1142–1160, 2011. With Permission.)

perfect, the liquid electrolyte gradually evaporates away and impurities such as water and oxygen molecules permeate into the cell. Therefore, the assembly of DSSCs containing the liquid electrolyte requires a critical sealing technique to reduce solvent leakage/vaporization, which increases the difficulty in the manufacturing process and decreases the durability of the device. In order to improve the stability of DSSCs, different types of electrolytes have been extensively studied to supersede the liquid-type electrolyte. A solid electrolyte is an ideal form for commercialization of DSSCs since this addresses the problem of conventional liquid electrolytes, such as leakage and evaporation. Therefore, the compatibility of p-type inorganic semiconductors and organic hole conductors with DSSCs has been widely investigated. In 1996, Matsumoto et al. introduced the polymer solid electrolyte, oligoethylene glycol methacrylate for SDSSCs. Since then, many different solid electrolytes including pyrrole, epichlormer-16, polyaniline (PANI), 2,2',7,7'-tetranis (N,N-di-p-methoxyphenyl-amine) 9,9'-spirobifluorene (Spiro-OMeTAD), poly(3-hexylthiophene (P3HT), and poly(3,4-ethylenedioxythiophene) (PEDOT) based (PEDOT-PSS, PEDOT/X) were examined to develop SDSSCs. At present, Spiro-OMeTAD is the most commonly used organic hole conductor for SDSSCs. This solid electrolyte has a small molecular size (around 2 nm), high solubility in organic media, and an amorphous structure, which make it suitable for being impregnated into the mesopores of the thick film. In addition, the redox potential of Spiro-OMeTAD is more positive than that of an I^-/I_3^- couple, which is beneficial in increasing open-circuit voltage (V_{oc}) of SDSSCs.

Although the solid state electrolyte increases the stability and reliability of DSSCs, the solid electrolyte has two serious problems. One of them is the low charge carrier mobility. In Spiro-OMeTAD, the hole mobility is only 10^{-4} cm²/Vs, which is much smaller than the charge carrier diffusivity of the liquid electrolyte. This increases the probability of carrier recombination during the transport process and reduces the photocurrent density. Though certain p-type organic semiconductors such as PEDOT/PSS have better electrical conductivity ($10^{-3} \sim 500$ S/cm), the size of PEDOT/PSS in a secondary or tertiary structure is too large to pass through the mesopores of TiO_2 nanoparticle films. The other question regarding the solid state electrolyte is how to fill the pores inside the mesoporous photoanode. While Spiro-OMeTAD dissolved in the organic media has a better pore-filling capability than other solid electrolytes, it is difficult to fully fill the pores of the thick photoelectrode even with Spiro-OMeTAD. When the Spiro-OMeTAD solution is spin-coated on a 2.5-µm-thick photoanode, ~35 % of pores inside the mesoporous film are still empty. This partial filling of the pores is caused by the small pore size of the mesoporous films.

The unfilled portion of the mesoporous photoelectrode and the low carrier mobility of the solid electrolyte are the source for electron–hole recombination and parasitic current in SDSSCs where the charge transport is controlled via a trap-limited diffusion. Since photoexcited electrons and holes in the dye do not diffuse fast, they recombine at the interface of solid electrolyte–TiO_2 or solid electrolyte–dye (or both) before reaching the counter electrode. The poor extraction of carriers in SDSSCs prevents the implementation of the very thick photoelectrode (≥ 10 µm) that is used in the liquid-electrolyte-based DSSCs to make up with weak light absorption. Because of short carrier diffusion length, the optimum photoelectrode thickness of SDSSCs is ~2 µm, and the best efficiency of SDSSCs is still lower than that of the liquid-electrolyte-based DSSCs (~12 %).

As summarized here, the PCE of traditional DSSCs is far off the theoretical value (~20 %), even after extensive research for the past few decades. The breakthrough of solar cells using the device physics of DSSCs came about very recently. A series of studies have proved that organolead trihalide perovskite, $(CH_3NH_3)PbX_3$ (X; halogen ions such as I^- and Br^-) can be an excellent light absorber with a very long carrier diffusion length and therefore is suitable for sensitizing the mesoporous photoanode of the solar cells. Since its potential was shown in 2012, this inorganic–organic hybrid halide material has increased the PCE of DSSCs more than two times (from ~9 to ~20%). A $CH_3NH_3PbI_3$ semiconductor has high light absorption coefficient (approximately 10^5 cm^{-1}), long absorption wavelength edge (> 800 nm), and unique electrical properties. Also, the band gap, absorption coefficient, and electron diffusion length of the halide perovskite can be tuned either by changing the organic cation group, or metal atom, and halide ion.

In the organic–inorganic halide perovskite $[(CH_3NH_3)AX_3]$, the A-site of the perovskite is occupied by an organic cation, which is enclosed by twelve nearest X (halide) anions. The prerequisite for a closed packed perovskite structure is that the organic cation must fit in the hole formed by the eight adjacent octahedra connected through the shared X corners as shown in Figure 9.28. The size of the organic cation and metal ion (M) is an important parameter to modulate the optical and electronic properties of perovskite materials. Any sort of distortion will affect the physical properties of perovskite materials, such as electronic, optical, magnetic, and electric properties. Though there are many kinds of halide perovskites, perovskites containing metal halides in the fourth main group (4A, including Ge^{2+}, Sn^{2+}, Pb^{2+}) of a periodic table have attracted more interest due to their good optoelectronic properties and potential for low-temperature device fabrication.

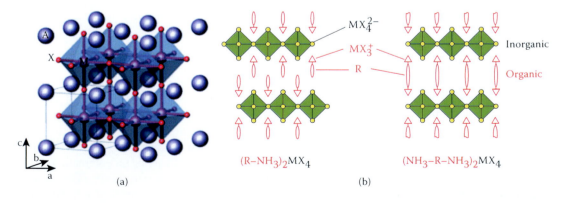

FIGURE 9.28 Crystal structure of halide perovskite (AMX_3: A-organic group, M-divalent metal ion, X-halide ions). (From Gao, P., *Energy Environ. Sci.*, 7, 2448, 2014; Mitzi, D.B., *J. Chem. Soc., Dalton Trans.*, 1–12, 2001. With Permission.)

A typical organic–inorganic perovskite solar cell is composed of a mesoporous oxide semiconductor for the electron transport layer, an organic hole transport layer (HTL), a perovskite-based absorber, a metal electrode, and a TCO. Usually, TiO_2 is chosen as the electron transport material because of its superior optical and photoelectronic properties as well as its chemical stability.

Figure 9.29 shows the schematic structure and energy diagram of perovskite solar cells. The operating principle of a perovskite solar cell is similar to that of SDSSCs. First, organic–inorganic halide perovskite ($CH_3NH_3PbX_3$) absorbs incident photon flux that exit from the highest occupied molecular orbital (HOMO) level to the lowest unoccupied molecular orbital (LUMO) level. Second, the exited electron–hole pairs undergo rapid charge separation at the perovskite–TiO_2 interface or hole transport material (Spiro-OMeTAD). Energy levels in the TiO_2–$CH_3NH_3PbI_3$–Spiro-OMeTAD junction are well matched for charge separation, as shown Figure 9.29. Third, electrons and holes are injected into the electron transport TiO_2 film and hole transport layer, respectively. Fourth, the injected electron in the conduction band of TiO_2 is transported through the surface toward the TCO (FTO) and the injected hole in the HOMO level of Spiro-OMeTAD is transported toward the metal electrode. Consequently, electrons and holes reach metal terminals that are connected to the external load via wiring.

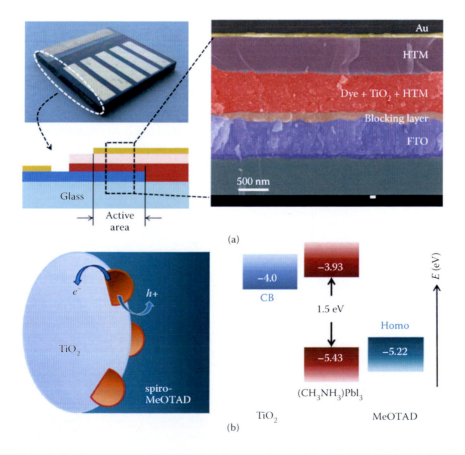

FIGURE 9.29 (a) Device structure of SDSSCs and (b) energy levels of the TiO_2/$CH_3NH_3PbI_3$/spiro-MeOTAD, demonstrating the band alignment for high power conversion efficiency over 15%. (From Park et al., *J. Phys. Chem. Lett.*, 15, 2423–2429, 2013. With Permission.)

PROBLEMS

9.1 There are multiple ways to harness solar energy. Please pick out three different technologies for solar energy harvesting and explain their operating principles.

9.2 The junction between p-type and n-type materials is used to collect photogenerated charge carriers.

   a. Please explain how a built-in potential is developed at the junction of p-type and n-type Si semiconductors.

   b. When external electric voltage is applied to a p-n junction, the effect of the applied voltage on the electric current can be estimated by considering only a diffusion component. Please explain why the effect of applied voltage on a drift component is negligible in either a qualitative or quantitative way.

9.3 A bulk heterojunction (BHJ) structure is crucial for high efficiency organic solar cells. Please summarize the benefit of a BHJ over a normal donor/acceptor bilayer structure organic solar cell.

9.4 One of the methods to evaluate the performance of solar cells is to measure electric current as a function of external electric voltage.

   a. Draw $I-V$ curves of the solar cell under dark and illuminated conditions with underlying principles.

   b. Point out important parameters in the $I-V$ curve of the solar cell that is exposed to solar light. Also, show how these parameters are influenced by the change in the illuminating light intensity.

9.5 Describe what shunt and series resistances in solar cells are and how they change the performance of ideal solar cells.

9.6 A buried-contact cell structure and an integrated back contact structure significantly increase the performance of Si solar cells. Please explain the pros and cons of these two structures over a traditional solar cell design.

9.7 The right plot is the $I-V$ curve of the solar cells measured in a dark condition when forward bias is applied. Though $n = 1$ in an ideal diode, n often deviates from 1 and gets close to 2, as shown in the low voltage regime. Please explain one reason for this non-ideal behavior.

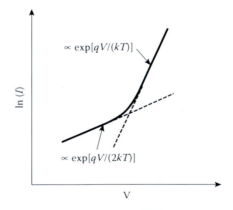

9.8 Since you have used your solar cells in a very harsh environment for a long time, metal electrodes on the front side of your solar cells are partially oxidized. Show how the $I-V$ curve of your solar cell has changed over time due to the degradation of metal electrode. Which property (voltage or current) of your solar cell is mainly influenced?

9.9 Semiconducting Si is widely used as the main component of commercial solar cells.

   a. What is the major reason limiting the theoretical PCE of Si-based solar cells to slightly higher than 30%?

    b. The refractive index of Si is much larger than that of air. Is this a problem when fabricating a high efficiency solar cell? If so, explain why a difference in the refractive index matters and demonstrate how this problem can be circumvented.

9.10 How does an increase in temperature change the energy conversion efficiency of silicon solar cells? Please explain this change qualitatively.

9.11 To improve the energy conversion efficiency of Si solar cells, a rear side of Si is coated with a thin SiO_2 layer. What is the advantage of this heaving doping?

9.12 Traditionally, the solar cell industry has used p-type Si wafers to produce Si solar cells, but now there are ongoing efforts to develop a process using an n-type Si. What are the pros and cons of p- and n-type Si substrates in the solar cell industry?

REFERENCES

Angermann, H., Rappich, J., and Klimm, C. Wet-chemical treatment and electronic interface properties of silicon solar cell substrates. *Cent. Eur. J. Phys.* 2009; 7: 363–370.

Chopra, K. L., Paulson, P. D., and Dutta, V. Thin-film solar cells: An overview. *Prog. Photovolt.* 2004; 12(2–3): 69–92.

Fonash, S. 2010. *Solar Cell Device Physics*, 2nd ed. Academic Press, Cambridge, US.

Green, M. A. 2003. *Third Generation Photovoltaics*. Springer.

Green, M. A., Ho-Baillie, A., and Snaith, H. J. The emergence of perovskite solar cells. *Nat. Photonics.* 2014; 8: 506–514.

Hummel, R. E. 2011. *Electronic Properties of Materials*, 4th ed. Springer.

Jeon, N. J., Noh, J. H., Kim, Y. C., Yang, W. S., Ryu, S. and Seok, S. I. Solvent engineering for high-performance inorganic–organic hybrid perovskite solar cells. *Nat. Mater.* 2014; 13: 897–903.

Kasap, S. O. 2006. *Principles of Electronic Materials and Devices*, 3rd ed. New York: McGraw Hill.

Lee, J. K., and Yang, M. J. Progress in light harvesting and charge injection of dye-sensitized solar cells. *Mater. Sci. Eng. B.* 2011; 176: 1142–1160.

Livingston, J. D. 1999. *Electronic Properties of Engineering Materials*, 1st ed. Wiley.

McEvoy A, Castaner, L., and Markvart, T. 2012. *Solar Cells: Materials, Manufacturing and Operation*, 2nd ed. Elsevier.

Mitzi, D. B. Templating and structural engineering in organic–inorganic perovskites. *J. Chem. Soc. Dalton Trans.* 2001: 1–12.

Nelson, J. 2003. *The Physics of Solar Cells, UK*. Imperial College Press.

Photovoltaics Report from Fraunhofer-Institute for Solar Energy System (ISE), Annual Report 2015, Freiburg, Germany.

Powell, D. M. Winkler, M. T., Choi, H. J., Simmons, C. B., Berney Needleman, D., and Buonassisi, T. Crystalline silicon photovoltaics: A cost analysis framework for determining technology pathways to reach baseload electricity costs. *Energy Environ. Sci.* 2012; 5(3): 5874–5883.

Saga, T. Advances in crystalline silicon solar cell technology for industrial mass production. *NPG Asia Mater.* 2010; 2: 96–102.

Singh, G., Verma, A., and Jeyakumar, R. *RSC Adv.* 2014; 4: 4225–4229.

Travino, M. R. 2012. *Dye-Sensitized Solar Cells and Solar Cell Performance*. Nova Scientific.

Tres, W. 2014. *Organic Solar Cells: Theory, Experiment, and Device Simulation*. Springer.

Wolf, S. D., Descoeudres, A., Holman, Z. C., and Ballif, C. High-efficiency Silicon Heterojunction Solar Cells: A Review. *Green.* 2012; 2: 7–24.

Würfel, P. 2009. *Physics of Solar Cells: From Basic Principles to Advanced Concepts*. Wiley.

10 Ferroelectrics, Piezoelectrics, and Pyroelectrics

KEY TOPICS

- Ferroelectric materials
- Origin of ferroelectricity
- Ferroelectric hysteresis loop
- Properties and applications of ferroelectrics
- Electrostriction
- Piezoelectric materials
- Direct and converse piezoelectric effects and applications
- Properties and applications of soft and hard piezoelectrics
- Strain-tuned ferroelectrics
- Lead-free piezoelectrics
- Pyroelectric materials and applications

10.1 FERROELECTRIC MATERIALS

10.1.1 FERROELECTRICITY IN BARIUM TITANATE

Ferroelectrics are materials that possess a macroscopic spontaneous polarization that can be reoriented through the application of an external electric field (Schlom et al. 2007). Polarization in ferroelectric materials can exist in the absence of an electric field under certain ranges of temperature and pressure. This is the biggest difference between ferroelectric materials and dielectric materials. In dielectric materials, polarization is also generated by applying an electric field. However, the polarization of dielectric materials disappears when the electric field is not applied. Ferroelectric materials have crystal structures that lack inversion symmetry. This broken inversion symmetry is a source of permanent polarization that is observed in ferroelectrics.

We can summarize the origin of ferroelectricity in tetragonal barium titanate ($BaTiO_3$), an archetypal ferroelectric, as follows.

The temperature at which the transformation occurs from a nonpolar, paraelectric phase to a polar, ferroelectric phase is known as the *Curie temperature* (T_c). At a given temperature higher than the Curie temperature, materials may exhibit a cubic structure. For $BaTiO_3$, this temperature is ~120°C. In this cubic phase, also called the *paraelectric phase,* the titanium (Ti^{4+}) ion appears to be exactly at the center of the cube. In reality, of course, ions are not stationary. The titanium ion, for example, vibrates very rapidly around several equivalent off-center positions. Each of these off-center configurations has a net dipole moment, which rotates very rapidly in space. The result is that the time-averaged position of the titanium ion in the cubic, paraelectric phase appears to be at the center, and this higher-temperature polymorph of $BaTiO_3$ has no net dipole moment. The cubic or paraelectric phase of the structure is described as centrosymmetric and nonpolar.

In the case of $BaTiO_3$, the paraelectric phase is cubic, and the ferroelectric phase is tetragonal (Figure 10.1). As the temperature is lowered, a phase transformation occurs at $T = T_c$, in which the

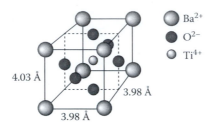

FIGURE 10.1 Schematic representation of the ionic arrangements in ferroelectric tetragonal barium titanate (BaTiO$_3$). (From Askeland, D. and Fulay P., *The Science and Engineering of Materials*, Thomson, Washington, DC, 2006. With permission.)

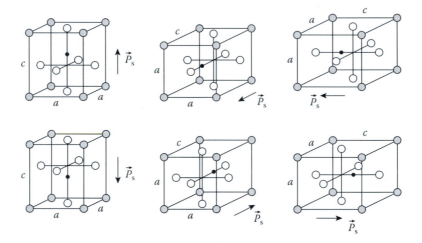

FIGURE 10.2 Six variants of the positions in the tetragonal phase of barium titanate (BaTiO$_3$) leading to six spontaneous polarization states and three spontaneous strain states. (From Mehling, V., et al., *J. Mech. Phys. Solids.*, 55, 2106–2141, 2007. With permission.)

nonpolar, paraelectric phase transforms into a polar, ferroelectric phase at the Curie temperature. The tetragonal structure develops when the titanium ions are locked into any of these six variants (Figure 10.2).

Other ions, such as Ba^{2+} and O^{2-}, are also displaced during the transformation from the cubic to the tetragonal phase in BaTiO$_3$. The actual displacement of ions is very small—only a few picometers (pm), shown for BaTiO$_3$ in Figure 10.3. Recall that 1 pm = 10^{-12} m = 0.01 Å. As the titanium (Ti^{4+}) and barium (Ba^{2+}) ions in one unit cell move in one of the possible off-center directions (e.g., titanium ions moving up by 11 pm and barium ions moving up by 6 pm, as shown in Figure 10.3), the titanium and barium ions in the unit cells above the unit cell shown in Figure 10.4 will also be pushed up.

These unit cells, with the asymmetric positions of titanium ions, create a net dipole moment in the tetragonal structure of BaTiO$_3$, which can be calculated using an equation of $\mu = q_d \times x$ as readers already studied in Chapter 7 (Mehling et al. 2007).

Other materials, such as lead zirconium titanate (PZT), also show a development of ferroelectric polarization. At higher temperatures, the dipole moments created in a ferroelectric structure become randomized, and the overall ferroelectric polarization begins to decrease. The ionic arrangements for the tetragonal ferroelectric and cubic paraelectric forms of PZT are shown in Figure 10.4.

Both BaTiO$_3$ and PZT have crystal structures other than the tetragonal form that exhibit ferroelectric behavior. Note that, for these materials, the ferroelectric-to-paraelectric transition is not always from the tetragonal to the cubic structure.

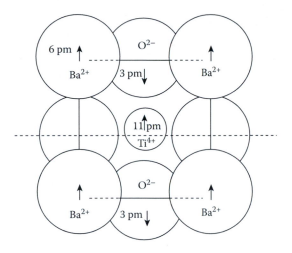

FIGURE 10.3 Actual picometer displacements of ions, leading to ferroelectric polarization in barium titanate (BaTiO$_3$). (From Moulson, A.J. and Herbert J.M., *Electroceramics: Materials, Properties, and Applications*, Wiley, New York, 2003. With permission.)

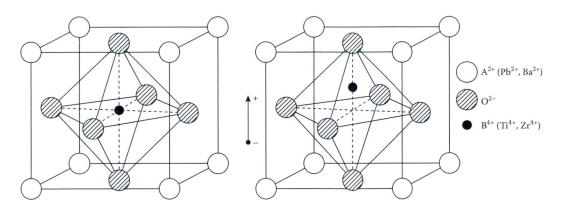

FIGURE 10.4 Ionic arrangements in the paraelectric and ferroelectric forms of lead zirconium titanate (PZT). (Adapted from Morgan Technical Ceramics, *Guide to Piezoelectric and Dielectric Ceramics*. Available at http://www.morganelectroceramics.com/pzbook.html.)

10.1.2 FERROELECTRIC DOMAINS

The cooperative displacement of the ions of several unit cells occurs during the formation of tetragonal BaTiO$_3$ from its cubic phase. This leads to the spontaneous formation of a region consisting of several unit cells, with all the unit cells' dipole moments lined up in the same direction. This region of a ferroelectric material, in which the polarization is in a given direction, is known as the *ferroelectric domain* or a *Weiss domain*.

Ferroelectrics are clearly different from other polar materials that have orientational or dipolar polarizations (e.g., water). In a ferroelectric material, there are no built-in or permanent dipole molecules to start with. Instead, a spontaneous polarization occurs in these materials as the atoms or ions self-assemble in an arrangement that causes the unit cells to develop a net dipole moment.

The process that causes the existence of a polar region or a ferroelectric domain in one part of a material cannot continue indefinitely because the process will gradually increase the free energy of the system from the standpoint of electric and mechanical energy. Alignment of the permanent dipoles increases the electrostatic energy of the system in comparison with the randomly distributed state.

In addition, the aligned dipoles result in a change in the shape of the ferroelectric materials and increases the strain energy (Figure 10.2). Therefore, ferroelectric materials cannot be composed of a single large ferroelectric domain for the purpose of energy minimization. The growth of a ferroelectric domain during the phase transition from the paraelectric state to the ferroelectric state stops when the domain size reaches a certain value. This means that ferroelectric materials normally consist of many fine ferroelectric domains ranging in size from ~100 nm to ~1 mm. If the unit cell of the ferroelectrics has the tetragonal symmetry, the polarizations in the neighboring domains may be at an angle of 90° or 180° in relation to one another to decrease the electrostatic or strain energy. The electric energy of the dielectric polarization in one domain is compensated for by another domain next to it, in which the polarization is in the opposite direction. If the polarization directions of two domains have a 180° difference, the electrostatic energy is decreased. In addition, when the angle between the polarization directions of ferroelectric domains is 90°, a change in the shape and strain energy of the ferroelectric materials can be minimized in comparison with the single domain ferroelectrics. Thus, a ferroelectric material comprises many polar regions (called ferroelectric domains), in which each region has polarization in a certain direction to suppress an increase in the electrostatic and strain energy. This is shown schematically in Figure 10.5. Domain walls (or domain boundaries) in ferroelectrics are very thin boundaries (just a few angstroms) that separate domains polarized in different directions. When the polarization directions of two domains sharing the same domain wall have a 180° difference for the electrostatic energy reduction or a 90° difference for the strain energy reduction, the domains are called 180° or 90° domains, respectively.

A scanning electron microscope (SEM) image of the ferroelectric domains in the tetragonal form of PZT is shown in Figure 10.6. The contrast among the domains has been obtained by carefully etching the sample with chemicals.

Unless we apply a substantial electric field, the polarizations associated with the different regions cancel each other out, and the ferroelectric material has no net polarization. Thus, when the domains are randomized and no substantial electric field has ever been applied to a ferroelectric material, the net dielectric polarization is zero. When an electric field is applied to a ferroelectric material, its domains begin to align with the electric field, and the material develops a net polarization. This

FIGURE 10.5 Schematic representation of the randomly arranged ferroelectric domains in an unpoled ferroelectric. (From Mehling, V., et al., *J. Mech. Phys. Solids.*, 55, 2106–2141, 2007. With permission.)

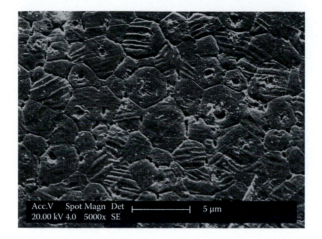

FIGURE 10.6 A scanning electron microscope (SEM) image of the microstructure of a lead zirconium titanate (PZT) ceramic modified using strontium (Sr). Sets of parallel lines within grains show the ferroelectric domain. (Courtesy of Dr. Raj Singh and I. Dutta, University of Cincinnati).

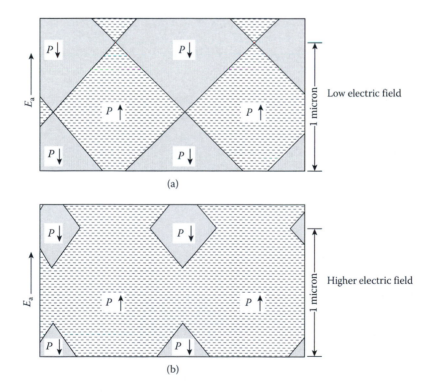

FIGURE 10.7 Schematic image of the domain growth in a ferroelectric material with (a) a low electric field and (b) a higher electric field. (From Hench, L.L. and West J.K., *Principles of Electronic Ceramics*, Wiley, New York, 1990. With permission.)

is similar to the alignment of permanent dipoles in a polar material. It is shown in Figure 10.7, in which the electric field is relatively lower for the top part and higher for the bottom part.

The application of an electric field helps to align the domains or polar regions. This development of polarization in a ferroelectric material is nonlinear. This is a very important distinction between a linear dielectric material, such as silica (SiO_2), alumina (Al_2O_3), titanium oxide (TiO_2), or polyethylene, which shows only nonferroelectric polarization, and a nonlinear dielectric material such as the tetragonal form of $BaTiO_3$. Thus, Equation 10.1, which we have shown to be applicable to linear dielectrics, cannot be used to describe polarization as a function of the electric field in ferroelectric materials.

$$P = \chi_e \varepsilon_0 E \text{ (for linear dielectrics only, not for ferroelectrics)} \qquad (10.1)$$

As will become clear in Section 10.1.3, the relationship polarization develops in a ferroelectric material is not linear. Ferroelectric materials that have never been exposed to an electric field are said to be in a *virgin* or *unpoled state*.

10.1.3   DEPENDENCE OF THE DIELECTRIC CONSTANT OF FERROELECTRICS ON TEMPERATURE AND COMPOSITION

When an electric field is applied, domains begin to undergo alignment. This leads to the development of considerable ferroelectric polarization, which causes ferroelectrics such as $BaTiO_3$ to have a relatively large dielectric constant (Figure 10.8). As we continue to increase the magnitude of the electric field applied to a ferroelectric material, an increasing number of domains become aligned. However, similar to the law of diminishing returns, the additional increase in electric field does not produce much additional dielectric polarization. This means that at higher electric fields, the dielectric constant (k) of ferroelectrics decreases. This is another important distinction between regular dielectric materials, also known as linear dielectrics, and nonlinear ferroelectric materials.

As we saw in Figure 10.8 and in Chapter 7, the measured or apparent dielectric constant of ferroelectrics and other dielectrics is a microstructure-sensitive property. The dielectric constant also strongly depends on the temperature. As the temperature increases beyond T_c, that is, if there is a switch from the ferroelectric phase to the paraelectric phase, the dielectric constant decreases. Above T_c, in the paraelectric state, the dielectric susceptibility ($1/\chi_e = 1/(\varepsilon_r-1)$) changes linearly

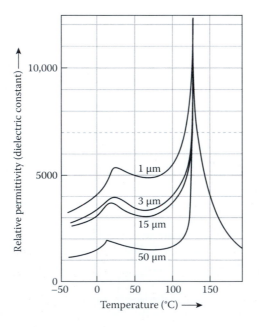

FIGURE 10.8 Dielectric constant of polycrystalline barium titanate ($BaTiO_3$) as a function of the temperature and average grain size. (From Moulson, A.J. and Herbert J.M., *Electroceramics: Materials, Properties, and Applications*, Wiley, New York, 2003. With permission.)

with temperature. This behavior is known as the *Curie–Weiss law*. This variation in the dielectric constant of the paraelectric phase is written as follows:

$$\varepsilon = \varepsilon_0 + \frac{C}{(T - T_0)}\left(\text{applicable for } T > T_0\right)$$

where T_0 is equal to the Curie temperature (T_c) if the phase transformation from the ferroelectric to the paraelectric phase is of the second order. A second-order phase transformation means that the volume, energy, and structure change continuously as we go from the paraelectric phase to the ferroelectric phase. For $BaTiO_3$, the transformation is of the second order. If this phase transition is of the first order, then $T_0 < T_c$. The constant C is known as the Curie–Weiss coefficient or the Curie constant. The first term, ε_0, is the temperature-independent part of ε. It is related to the polarization of ions. Near T_0 or T_c, the second term dominates, and the first term can be ignored (Xu 1991). Thus, the Curie–Weiss law is sometimes also written as

$$\varepsilon = \frac{C}{(T - T_0)}\left(\text{applicable for } T > T_0\right)$$

In Chapter 11, we will learn that essentially the same type of behavior is seen for magnetic materials.

An example of Curie–Weiss behavior in a material that is a solid solution of $BaTiO_3$ and strontium titanate $(SrTiO_3)$ is shown in Figure 10.9. Note that the Curie temperature of this material is $\sim +10°C$. Although the data for dielectric constants are shown for temperatures below T_c, the Curie–Weiss law does *not* apply to the region below $T = T_c$. Note that, even in a paraelectric state, the dielectric constant of materials is very high compared to other linear dielectrics such as Al_2O_3 and SiO_2. This property is very important for applications such as multilayer capacitors.

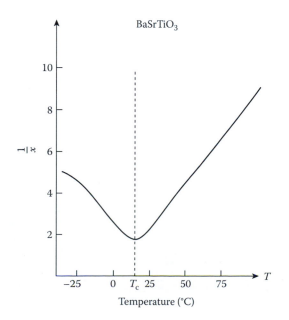

FIGURE 10.9 Curie–Weiss behavior in barium titanate–strontium titanate $(BaTiO_3–SrTiO_3)$ ceramic with $T_0 = T_c \sim +10°C$. (From Hench, L.L. and West J.K., *Principles of Electronic Ceramics*, Wiley, New York, 1990. With permission.)

Lowering the Curie temperature of a ferroelectric material is useful for the applications of such dielectrics in capacitors. It is common to add *Curie-point shifters* to $BaTiO_3$. As we can see in Figure 10.9, the addition of $SrTiO_3$ to $BaTiO_3$ causes an overall lowering of the Curie temperature. Consequently, this dielectric formulation has a relatively high dielectric constant at and around room temperature (~27°C, or 300 K), which offers high volumetric efficiency. At the same time, the material is paraelectric and thus is not piezoelectric (Section 10.11). This means that a capacitor made from such a material will not produce any spurious voltages. We can form solid solutions of $BaTiO_3$ with lead titanate (PT, $PbTiO_3$) and, in this case, the Curie temperature shifts to temperatures higher than T_c ~120°C (for $BaTiO_3$). Similarly, compounds such as barium zirconate ($BaZrO_3$) can form solid solutions with $BaTiO_3$, which can help suppress the Curie temperature. Such *Curie-point suppressors* can also be useful because they help to develop temperature-stable materials with a high dielectric constant. The dielectric constant of such materials will not vary significantly with temperature. This ability to form solid solutions allows the development of different formulations of capacitors.

The Electronic Industries Alliance in the Unites States has developed a scheme for the classification of different dielectrics based on their temperature stability. For example, an X7R dielectric means that, in the temperature range of −55°C to +125°C, the capacitance of the capacitor will be within ±15% relative to its value at 25°C. A Z5U dielectric formulation means that, in the temperature range +10°C to +85°C, the capacitance can change between +22% and −56% relative to its value at 25°C.

In some instances, we can create a composition and a nanostructure so that the resultant material shows a very high apparent dielectric constant, with a broad Curie peak. The dielectric constant of such a material significantly depends on the frequency of measurements. Such materials are known as *relaxor ferroelectrics*. The broadening of the Curie temperature occurring in these materials can be attributed to the disorder of the ions at different crystallographic sites in the crystal structures and to the nanoscale fluctuations in the compositions. These lead to nanoscale polar regions that have, in effect, a Curie temperature of their own. Collectively, the ferroelectric to paraelectric transition becomes diffused (Figures 10.10 and 10.11).

Relaxor ferroelectric materials include ceramics such as lead magnesium niobate (PMN) that can be formulated in combination with other ferroelectrics such as $PbTiO_3$ and are useful in actuator applications because of their high electrostriction coefficients. The so-called coupling coefficient of these materials (see Section 10.16) is also better than PZT. Many novel relaxor ferroelectric compositions in the form of both single-crystal and polycrystalline materials have been made recently. The acoustic impedance of relaxor materials is better matched to human tissue than it is

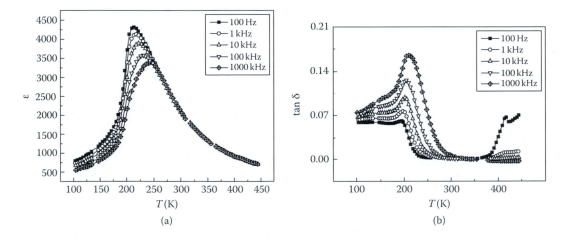

FIGURE 10.10   (a) Dielectric constant and (b) tan δ for strontium-bismuth-barium titanate ($BaTiO_3$) ceramics. (From Chen, W., et al., *Solid State Commun.*, 141, 84–88, 2007. With permission.)

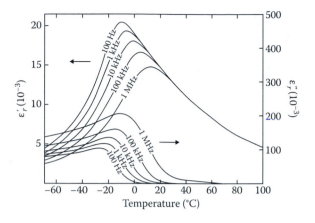

FIGURE 10.11   Dielectric properties as a function of the frequency of lead magnesium niobate ceramics. (From Moulson, A.J. and Herbert J.M., *Electroceramics: Materials, Properties, and Applications*, Wiley, New York, 2003. With permission.)

to PZT (Harvey et al. 2002). In principle, these materials are better suited for applications such as ultrasound imaging (see Section 10.11).

As an example, the dielectric constant (real part) and tan δ, a measure of the dielectric losses, are shown for strontium-bismuth-barium titanate (SBBT) ceramics in Figure 10.10.

Because of compositional fluctuations occurring at the nanoscale, relaxor materials exhibit a diffuse phase transition and show a Curie range rather than a specific Curie temperature. A modified Curie–Weiss law expression is used to describe the variations in dielectric constants of these materials above the Curie temperature range.

The real part (ε_r') for the dielectric constant and the imaginary part (ε_r'') for PMN ceramics is shown in Figure 10.11. Note the multipliers on the *y*-axis. The values of the dielectric constant range up to 20,000.

10.2   RELATIONSHIP OF FERROELECTRICS AND PIEZOELECTRICS TO CRYSTAL SYMMETRY

In a ferroelectric material, dielectric polarization appears spontaneously. This appearance of spontaneous dielectric polarization also causes a strain to develop in the ferroelectric material. For the six polarization direction variations shown in Figure 10.2, there are three spontaneous strain states: one in the *c* direction and two in the *a* direction. A *piezoelectric* material develops a voltage when subjected to stress. A prefix *piezo* derived from Greek word, *piezein*, means *pressure*.

All ferroelectric materials are also piezoelectric. However, not all piezoelectric materials are ferroelectric. For example, zinc oxide (ZnO), gallium arsenide (GaAs), and quartz (SiO_2) are piezoelectrics, but they are not ferroelectrics.

The seven crystal systems are triclinic, monoclinic, orthorhombic, tetragonal, trigonal, hexagonal, and cubic, and lead to 32 crystallographic point groups (Gupta and Ballato 2007). Of these, 21 point groups have no center of symmetry (Figure 10.12). Of the 21 point groups without a center of symmetry, only 20 are piezoelectric. Among the 20 point groups that show piezoelectricity, 10 point groups are polar or *pyroelectric*. A pyroelectric is a material that shows a flow of charge to and from its surface consequent to a temperature change (see Section 10.20).

Thus, a ferroelectric material is also a pyroelectric; however, the converse is not true; that is, a pyroelectric material is not necessarily a ferroelectric (e.g., ZnO and cadmium selenide [CdSe]). Ferroelectrics are pyroelectric, piezoelectric, and lack inversion symmetry.

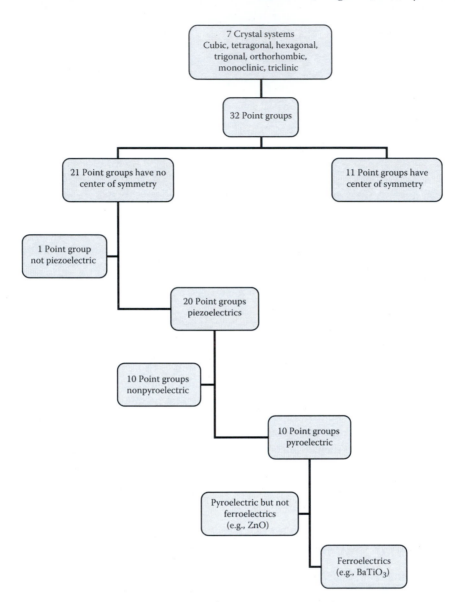

FIGURE 10.12 Crystal classes, point groups, and ferroelectrics.

Another definition of ferroelectrics is that they are pyroelectric materials whose direction of polarization can be reversed by a sufficiently strong electric field (Figure 10.12). Ferroelectric materials are also characterized by the presence of a Curie temperature (or a Curie temperature range for relaxor ferroelectrics), above which they become paraelectric.

We will discuss piezoelectric and pyroelectric materials in detail in Sections 10.12 and 10.20.

10.3 ELECTROSTRICTION

Every material, including liquids and amorphous materials such as glass, undergoes polarization when subjected to an electric field (E). This involves the movement of electron clouds, ions, and so on. Such movements associated with the polarization processes also cause the development of a small strain in the material. *Electrostriction* is a phenomenon that occurs in all materials in which

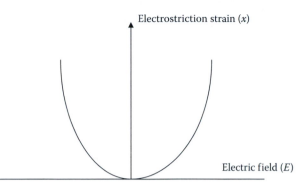

FIGURE 10.13   Electrostriction strain development with the application of an electric field.

the application of an electric field induces an elastic strain. This strain (ε) is proportional to the *square* of the electric field (E; Figure 10.13). The development of strain in piezoelectrics that are also ferroelectric is quite different (Figure 10.26).

Electrostriction produces a strain that is independent of the applied field's sign. The strain that develops due to ferroelectric polarization (Figure 10.2) is different from electrostriction in important ways. First, the strain produced because of ferroelectric polarization is proportional to the electric field (E). Second, typically, it is much larger than that produced by electrostriction. Third, the strain produced by ferroelectric polarization is strongly temperature-dependent, especially as the material transitions from the paraelectric phase to the ferroelectric phase. In comparison, the strain produced by electrostriction does not show significant temperature dependence. Finally, because every material undergoes some sort of polarization, all materials show electrostriction.

In piezoelectric materials, the total strain developed will be the result of both the piezoelectric effects and electrostriction (Section 10.7). Electrostriction-induced strain contributes significantly to the high dielectric constant of ferroelectrics near their Curie temperature.

If Q_{ij} is the electrostriction coefficient for a material with remnant polarization P_r and dielectric constant ε_r, the piezoelectric coefficient d_{ij} is given by the expression

$$d_{ij} \sim 2 \times Q_{ij} \times \varepsilon_r \times \varepsilon_0 \times P_r \tag{10.2}$$

Consequently, relaxor ferroelectrics such as lead magnesium niobate-lead titanate (PMN-PT) exhibit very high strains for a given level of applied electric field and are widely used in electro-mechanical actuators. Many relaxor ferroelectrics, such as PMN-PT, have a high electrostriction coefficient because of their crystal structures and chemical makeup. This is why they have very large piezoelectric coefficients. Even if we do not make use of the electrostriction effect as such, the consequence of this effect is embedded in the piezoelectric coefficients of many materials.

In some compositions of PZT, such as those near the morphotropic phase boundary (MPB; Section 10.14), the total piezoelectric effect strain development plays a major role. Electrostriction brings in a relatively smaller fraction of the total strain. Certain other materials, such as some lead lanthanum zirconium titanate compositions, show very high levels of polarization at high electric fields, and for these, the electrostriction strain is a major fraction of the total strain developed.

10.4   FERROELECTRIC HYSTERESIS LOOP

The dielectric constant (ε_r or k') of ferroelectric materials is dependent on the strength of the electric field (E) applied. When the electric fields applied are relatively high, the entire material can become a single-domain structure. If this stage is reached, the material cannot generate any more ferroelectric polarization; the polarization is said to be saturated, that is, either all domains are aligned or there is one large domain. This state of maximum possible ferroelectric polarization is known as the *saturation polarization* (P_s). The only small increase that is possible beyond P_s is due to the continued ionic and electronic polarization of the atoms or ions that make up a given ferroelectric material.

A ferroelectric material in which a substantial fraction of ferroelectric domains is aligned with the applied electric field due to exposure to a level of electric field is called a *poled ferroelectric*. The process of applying an electric field in order to align a substantial fraction of domains is known as *poling*. A schematic representation of the domain reorientation occurring during poling is shown in Figure 10.14. In Figure 10.14, for the sake of clarity, each grain is shown to have only one ferroelectric domain. In most cases, grains have multiple domains (Figure 10.14). Moreover, the material does not have to be polycrystalline.

The level of electrical field needed for poling a ferroelectric material depends on the field that is needed to move the domain walls. In practice, the poling process often is conducted in an insulating oil bath maintained at a high temperature but below T_c. The higher temperature makes it possible to carry out poling at lower electrical fields because of increased domain wall mobility. The insulating oil bath prevents any arcing that may occur among the electrodes applied on the material being poled.

Higher temperatures help in the alignment of the domains during poling. However, if an already-poled piezoelectric is again exposed to high temperatures, it can also *depole*. Depoling means that the domains undergo randomization, which can cause the net piezoelectric effect to diminish or disappear. A general rule of thumb for utilizing piezoelectrics is that they can be safely used up to a temperature of ½ T_c without significant degradation of the piezoelectric activity. Another factor to consider is whether there are any potential changes in the crystal structure, even though the material will not become paraelectric (Figure 10.35a). Depoling is also caused by stress encountered during applications because applied stress can cause domain switching.

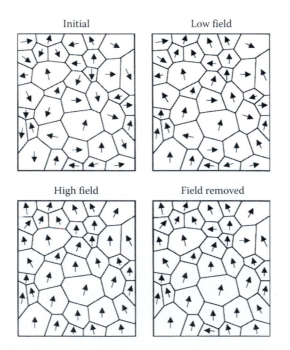

FIGURE 10.14 Schematic figure of the alignment of the domains poling process for a polycrystalline material. (From Buchanan, R.C., *Ceramic Materials for Electronics*, Marcel Dekker, New York, 2004. With permission.)

Even at room temperature, in time, domains in a freshly poled piezoelectric undergo some level of randomization after the poling electric field has been removed. This process is known as *aging*, and it causes a decrease in the piezoelectric properties. The effects of aging on dielectric properties are usually logarithmic with respect to time and are accelerated at higher temperatures.

The dielectric constant (ε_r) of ferroelectrics will be very high at small electric fields and will continue to decrease as the electric field increases. If the applied field is too high, ferroelectric materials will exhibit an electrical breakdown. When the electric field is taken off, not all of the domains will return to their original random states of polarization (Figure 10.14). As a result, the ferroelectric material shows *remnant polarization* (P_r). We need to apply a magnitude equivalent in the opposite direction to what is known as the *coercive field* (E_c) in order to bring the polarization back to zero by again randomizing all the domains. This behavior, involving development of polarization in a ferroelectric material, is captured in what is described as a *ferroelectric hysteresis loop*. This trace of polarization (P) or displacement (D) as a function of the electric field is known as the polarization–electric field (P–E) or dielectric displacement–electric field (D–E) hysteresis loop (Figure 10.15).

There is a difference between the P–E and D–E loops. In a P–E loop, after all the domains are aligned (that is, when $P = P_s$), the trace describing P–E will become flat.

Recall that dielectric displacement (D) and polarization (P) have the following relationship:

$$D = \varepsilon_0 E + P \tag{10.3}$$

The ferroelectric polarization (P) that is typically induced in ferroelectric materials is significantly larger than $\varepsilon_0 E$, and therefore $D \cong P$. Thus, the difference between a D–E and P–E loop may not appear very large. However, it is important to note the fundamental difference between a P–E loop and a D–E loop. A P–E loop will show leveling out of P as the electric field causes nearly complete domain alignment. In a D–E loop, the value of D will continue to increase with increasing E, even after domain alignment is complete.

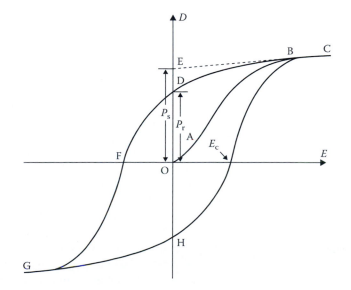

FIGURE 10.15 Typical dielectric displacement (D)–electric field (E) hysteresis loop for an unpoled ferroelectric material. (From Xu, Y., *Ferroelectric Materials and Their Applications*, North Holland, Amsterdam, 1991. With permission.)

10.4.1 TRACE OF THE HYSTERESIS LOOP

We will now follow the trace of the hysteresis loop shown in Figure 10.15. For an unpoled sample, the ferroelectric domains at first are randomly aligned. The starting point is the origin (O), where there is neither an applied electric field nor the development of net polarization or displacement. As an electric field is applied, a very small part of the hysteresis loop initially exhibits a polarization that is linear with the applied electric field. In this region, between Points O and A, the ferroelectric material essentially behaves similarly to a linear dielectric because the applied electric field (E) is too small to cause any changes in the orientation of randomly arranged ferroelectric domains (Figure 10.14).

As we increase the applied field, the ferroelectric domains begin to realign. The realignment of ferroelectric domains between Points A and B results in the development of relatively large polarization. Thus, the dielectric constant of ferroelectrics is relatively large in this region. As we can see from Figure 10.15, the development of this polarization is nonlinear.

As we reach Point B, almost all the domains are aligned (Figure 10.14, high field). Any additional increase in the electric field will cause only a slight increase in polarization because of continued increases in the electronic and ionic polarization mechanisms. The polarization value corresponding to Point B is known as the saturation polarization (P_s). This is the maximum polarization that can be expected from the alignment of domains in a ferroelectric material, ignoring the small levels of electronic and ionic polarizations that can continue to occur at high electric fields. The P_s value depends on the composition of the ferroelectric material; however, in contrast to the coercive field (E_c), P_s is not a microstructure-sensitive property.

Note that the slope of the region between Points B and C is rather small. Thus, the dielectric constant of a ferroelectric at high electric fields is again rather small. A another point that needs to be emphasized is that the dielectric constant (k or ε_r) of ferroelectrics is strongly dependent on the magnitude of the electric field applied. This is not true for so-called linear dielectrics, such as alumina, silica, and polyethylene. If we continue to increase the field indefinitely, the ferroelectric material eventually will experience dielectric breakdown.

Let us assume that we reverse the direction of the applied field after reaching Point C. This state, in which all domains are forced into alignment with the field, is a high-energy state for the material. The natural tendency for the material will be to lower its internal energy by reverting to the random arrangement of domains. As the magnitude of the field decreases while going from Point C toward Point B and on toward Point D, some domains are able to switch their polarization directions and others are not. Thus, as we decrease the magnitude of the applied field, the overall polarization (or dielectric displacement) decreases. However, a fraction of the domains continues to remain aligned (Figure 10.14) with the direction of the field applied (O → A → B → C). This causes the ferroelectric material to develop a remnant polarization (P_r), shown at Point D in Figure 10.15. Please do not confuse this with the symbol used for dielectric displacement.

To remove this remnant polarization (P_r) from the ferroelectric material, we must apply an electric field in a direction opposite to the original direction. The region between Points D and F (Figure 10.15) shows this effect. At Point F, we reach a randomized domain configuration with no net polarization left in the ferroelectric material. The electric field at Point F is known as the coercive field (E_c), which can be defined as the field required to coerce or force all the domains back to a random configuration, causing zero net polarization in a material that was subjected to an electric field.

As the magnitude of the field (applied in a reverse direction) increases from Point F to Point G, the domains realign with the new direction. At Point G, all the domains are aligned with the new (i.e., reversed) field direction (similar to Point C). As the magnitude of the field is decreased, some domains start to become randomized. We reach Point H, where the applied field is reduced to zero. However, similar to Point D, some domains remain aligned, causing a remnant polarization (P_r). The polarizations that remain at Points D and H are essentially of the same magnitude but in opposite directions and are designated as $+P_r$ and $-P_r$, respectively. The polarization that remains at Point H can be reduced to zero by again applying a field in the reverse direction until we reach a point that corresponds to the coercive field E_c.

One of the most widely used piezoelectric polymers is polyvinylidene fluoride (PVDF). Figure 10.16 shows a $D–E$ hysteresis loop for a poled PVDF film with a 25-μm thickness. In practice, such ferroelectric hysteresis loops are recorded using the Sawyer–Tower circuit. Automated instruments are currently available for the measurement of hysteresis loops. Note that because the PVDF sample has been poled, this loop does not have the *initial* region expected for an unpoled or virgin material (Figure 10.15). This means that even at zero electric field, the film already has a built-in dielectric polarization.

Ferroelectric thin films of materials, such as PZT, barium strontium titanate, and PVDF, have been investigated for a number of applications, which include nonvolatile random-access memories, pyroelectric detector and imaging arrays, and microelectromechanical sensors and actuators. An example of a ferroelectric hysteresis loop for a PZT thin film doped with manganese (Mn) and antimony (Sb) is shown in Figure 10.17.

The remnant polarization for this manganese- and antimony-doped PZT film is about 15 μC/cm² (Figure 10.17). The saturation polarization (P_s) is about 34 μC/cm². Note that the x-axis data are shown in voltage and not as an electric field. The coercive voltage is ~3.0 V, and the coercive field based on the thickness of the thin film is 70 kV/cm (see Problem 10.1).

As mentioned before, the saturation polarization (P_s) is dependent on the composition of the ferroelectric material. For example, the polarization–voltage hysteresis loops for different compositions of $Pb(Zr_xTi_{1-x})O_3$ (PZT), with $x = 0.2$, 0.5, and 0.7, are shown in Figure 10.18.

As shown in Figure 10.19, the coercive field associated with 5-μm-thin films of PZT prepared using a sol-gel process changes considerably with changes in the $Zr/(Zr + Ti)$ ratio. From these data, we can see that the coercive field of PZT decreases with increasing zirconium concentrations.

Even for the same composition of a material, the coercive field (E_c) can be made to change significantly with *microstructure* because different microstructural features, such as grain size and point defects, affect the mobility of domain walls. These composition- and microstructure-related effects are used to engineer the so-called *soft* and hard categories of piezoelectrics (Table 10.3).

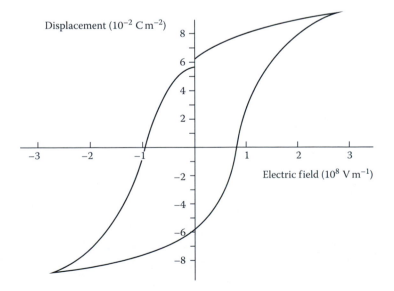

FIGURE 10.16 Hysteresis loop for a polyvinylidene fluoride film. (From Kepler, R.G., In *Ferroelectric Polymers: Chemistry, Physics, and Applications*, Marcel Dekker Inc., New York, 1995. With permission.)

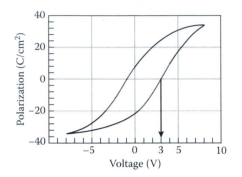

FIGURE 10.17 Hysteresis loop for manganese- and antimony-doped zirconium titanate (PZT) thin films. (From Ignatiev, A., et al., *Mater. Sci. Eng. B.*, 56(2), 191–194, 1998. With permission.)

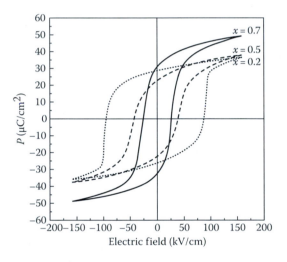

FIGURE 10.18 Composition dependence of ferroelectric properties on ferroelectric hysteresis loops. (From Osone, S., et al., *Thin Solid Films*, 516, 4325–4329, 2008. With permission.)

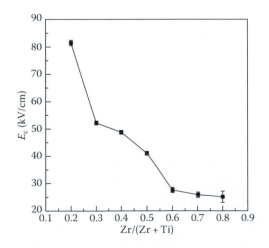

FIGURE 10.19 Coercive field of sol–gel–derived zirconium titanate (PZT) films as a function of the Zr/(Zr⁺ Ti) ratio. (From Osone, S., et al., *Thin Solid Films*, 516, 4325–4329, 2008. With permission.)

In general, the saturation polarization (P_s), which is the maximum polarization that can be obtained from a given composition of a ferroelectric material, is *not* a strong microstructure-dependent quantity. However, the coercive field (E_c) is very much a microstructure-sensitive property. We will see similar trends in ferromagnetic and ferrimagnetic materials (Chapter 11).

The appearance of the ferroelectric loop and the associated values undergo changes with temperature and especially depend on the particular crystal structure of the phases. For example, for $BaTiO_3$, one of the phase transformations is that of a ferroelectric tetragonal structure changing to the paraelectric cubic structure at $\sim T_c = 120°C$ (Figure 10.20).

The corresponding changes in the polymorphic forms of $BaTiO_3$ are shown in Figure 10.21. $BaTiO_3$ also shows a hexagonal polymorph, which is not shown in Figure 10.21. At temperatures above $T_c \sim 120°C$, $BaTiO_3$ is cubic and undergoes a transformation to the tetragonal phase as the temperature decreases to less than 120°C. The crystal structure of $BaTiO_3$ then changes from tetragonal to orthorhombic at $\sim 0°C$. At an even lower temperature ($-90°C$), the crystal structure changes to a rhombohedral form (Figure 10.21).

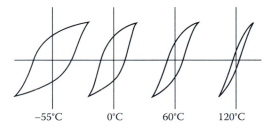

FIGURE 10.20 Qualitative appearance of D–E loops for barium titanate ($BaTiO_3$). (Adapted from Hench, L.L. and West J.K., *Principles of Electronic Ceramics,* Wiley, New York, 1990.)

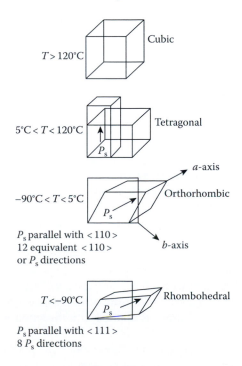

FIGURE 10.21 Polymorphs of barium titanate ($BaTiO_3$) as they change with temperature. (From Hench, L.L. and West J.K., *Principles of Electronic Ceramics,* Wiley, New York, 1990. With permission.)

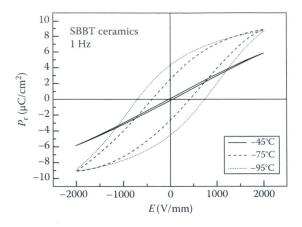

FIGURE 10.22 Changes in the hysteresis loops for strontium-barium-bismuth titanate (SBBT) ceramics. (From Chen, W., et al., *Solid State Commun.*, 141, 84–88, 2007. With permission.)

We will also see similar changes in the ferroelectric hysteresis loop for relaxor ferroelectrics. For example, the changes in the hysteresis loops for SBBT ceramics at −45°C, −75°C, and −95°C are shown in Figure 10.22. As the temperature decreases, the nanopolar domains become stable, and the remnant polarization increases substantially.

10.5 PIEZOELECTRICITY

10.5.1 ORIGIN OF THE PIEZOELECTRIC EFFECT IN FERROELECTRICS

The piezoelectric effect is present in all ferroelectrics; however, it is not limited only to ferroelectrics. Materials such as ZnO and SiO_2 also exhibit piezoelectricity (Figure 10.12). We begin by discussing how piezoelectricity occurs in ferroelectric materials.

A poled piezoelectric ferroelectric material has a net polarization (P_r), which is caused by the alignment of the randomly aligned domains in the ferroelectric material by the poling process. When a poling voltage is applied to a piezoelectric material such as PZT, the positively charged titanium and zirconium ions in the asymmetric crystal structure are attracted toward the negative end of the poling field (Figure 10.23).

When the surfaces of a poled piezoelectric are connected using a wire, no current will flow because there is no free charge on the surfaces. However, when a poled piezoelectric material is subjected to a compressive stress, the dipole moment—and hence, polarization—will change. Polarization is the bound charge density; as it decreases, free charge density increases. If excess free charge is generated on the surface, that charge will flow into an external circuit and create a transient current (Figure 10.24).

If we deliberately do not allow a current to flow by maintaining an *open circuit*, then a voltage with the same polarity as that used in the poling process is developed (Figure 10.24b). As soon as the wires are connected, a current flows as shown in Figure 10.24a.

Note that for PZT and other materials with the same basic crystal structure, essentially the same effect is obtained if a tensile stress is applied in a direction *perpendicular* to the poling direction.

Thus, the application of either a compressive stress along the polarization direction or a tensile stress perpendicular to the polarization direction generates a voltage with the same polarity as the poling voltage.

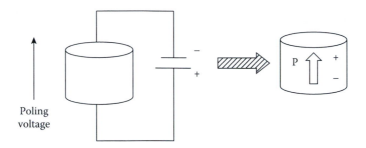

FIGURE 10.23 Poling and development of polarization in a piezoelectric material.

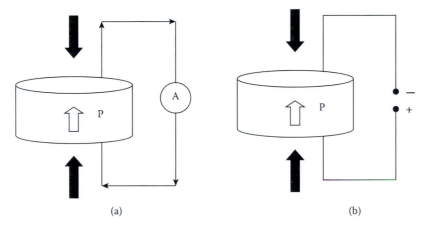

(a)   (b)

FIGURE 10.24 With the application of a compressive stress to a poled piezoelectric along the poling axis or a tensile stress in a direction perpendicular to the polar axis, (a) conventional current flows in a closed circuit as shown. (b) In an open circuit, a voltage with the same polarity as the poling voltage develops.

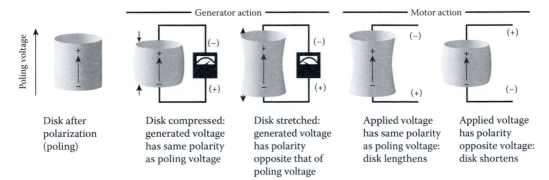

FIGURE 10.25 Schematic representation of the direct and converse piezoelectric effects. (With kind permission from Moheimani, S.O.R. and Fleming A.J., *Piezoelectric Transducers for Vibration Control and Damping*, Springer Science+Business Media, 2006.)

If a tensile stress is applied along the poling direction, these effects are reversed. This means that the current in the closed circuit flows in an opposite direction. If the circuit is open, then a voltage with polarity *opposite* to that of the poling voltage is set up across the piezoelectric. Thus, for materials such as $BaTiO_3$ and PZT, a tensile stress along—or a compressive stress perpendicular to—the poling direction will cause the development of a voltage with polarity opposite to that of the poling voltage. These effects are summarized in Figure 10.25.

If we do not connect the piezoelectric surfaces through an external conductor, the free charges generated on the surfaces of the poled piezoelectric eventually will *leak* inside the high (but finite)-resistivity material. If we bring the wires of the external conductor within 2 to 3 mm of each other but not close to the circuit, we will see an electrical arc! This is because the *voltage* created across the electrode gap is very high (a few thousand volts). This process builds an electric field that causes the electrical breakdown of air, thus creating an arc. This is the principle by which piezoelectric spark igniters work.

10.6   DIRECT AND CONVERSE PIEZOELECTRIC EFFECTS

In the *direct piezoelectric effect*, when a poled piezoelectric material is subjected to a compressive or tensile stress, a net charge develops on the surfaces of the material (Figure 10.25). This creates a voltage across the dielectric material. The direct piezoelectric effect is also known as the *generator effect* because its action generates a voltage or electrical charge.

The *converse piezoelectric effect* refers to the development of a strain when an electric field is applied to a poled piezoelectric (Figure 10.25). The converse piezoelectric effect is also known as the *motor effect* because the application of a voltage creates motion.

When we apply an electric field in the direction of poling (that is, the bottom surface connected to the positive terminal to a cylinder of a poled piezoelectric such as PZT), then the positively charged titanium and zirconium ions will move toward the top surface, which is connected to the negative terminal. This process will cause an extension of the cylinder. If we change the polarity, that is, apply a voltage opposite to the poling voltage and connect the positive electrode to the top surface, then the positively charged titanium and zirconium ions in the unit cells of poled PZT will move away from the top surface. This will cause the cylinder to become shorter in length (Figure 10.25).

10.7   PIEZOELECTRIC BEHAVIOR OF FERROELECTRICS

All ferroelectric materials also show *piezoelectric behavior*. However, not all piezoelectric materials are necessarily ferroelectrics. One common example of this is the quartz (SiO_2) crystal, which is piezoelectric but not ferroelectric.

Ferroelectric materials show a piezoelectric effect. However, for this effect to be measurable and useful, polycrystalline ceramics must be poled. In a polycrystalline ceramic, grains are randomly oriented, and there are ferroelectric domains within each grain. If a ceramic with a randomized domain structure is subjected to stress, no measurable strain—or piezoelectric effect—will be developed. This is because the strains developed in each grain will cancel each other out. Therefore, for a net piezoelectric effect to be observed and to be useful, we must align a majority of the domains in a given direction. This is accomplished using the poling process. As discussed in Section 10.4, poling usually is carried out by heating the piezoelectric material to a higher temperature (not necessarily above the Curie temperature) in an oil bath and then applying an electric field for a few minutes. Poling is necessary in order for a piezoelectric ceramic to be useful for technological applications. Domain switching is a complicated process. It is similar to the manner in which nucleation and growth processes lead to phase transformations in materials.

A poled piezoelectric material that is either a single crystal or polycrystalline develops a voltage when subjected to stress. It also develops a strain when subjected to an electric field. The development of strain in a poled piezoelectric is described in what is known as a *butterfly loop* (Figure 10.26a). The corresponding *P–E* loop is shown in Figure 10.26b.

In Figure 10.18, the ferroelectric hysteresis loops are shown for $Pb(Zr_xTi_{1-x})O_3$ (PZT) with $x = 0.2$, 0.5, and 0.7 (Osone et al. 2008). The corresponding butterfly loops showing the longitudinal strain that develops as a function of the electric field via the piezoelectric effect are shown in Figure 10.27.

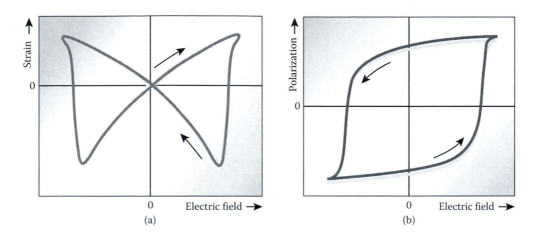

FIGURE 10.26 Relationship between (a) the butterfly loop and (b) the *P-E* loop showing the development of strain as a function of the electric field. (From Cross, E., *Nature*, 432, 24–25, 2004. With permission.)

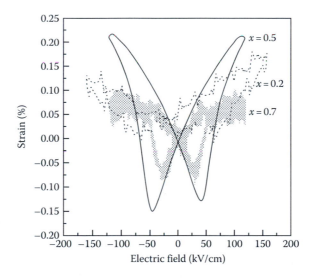

FIGURE 10.27 Longitudinal strain–electric field butterfly loops for $Pb(Zr_xTi_{1-x})O_3$ (PZT), with $x + 0.2$, 0.5, and 0.7. (From Osone, S., et al., *Thin Solid Films*, 516, 4325–4329, 2010. With permission.)

10.8 PIEZOELECTRIC COEFFICIENTS

The direct and indirect effects in piezoelectric materials are commonly described using piezoelectric coefficients. These coefficients are also used to compare different piezoelectric materials with one another.

The converse piezoelectric effect is commonly described by the piezoelectric coefficient (d). Remember, d stands for *displacement*. The piezoelectric coefficient (d) is defined in two ways. One is a ratio of the mechanical strain (x) developed in a piezoelectric material per electric field (E) applied. Mechanical strain (x) means mechanical displacement at constant stress (X) and temperature (T). The other is a ratio of the electric displacement (D) per applied mechanical stress (X) at constant electric field (E) and temperature (T). Electric displacement (D) means the short–circuit charge density stored on the surface of the dielectrics under electric field, which readers learned in

Sections 7.3 and 8.1. If x is the strain developed in a piezoelectric material after the application of an electric field E and D is the electric displacement, then the piezoelectric coefficient (d) is defined by the following equation:

$$d = \left(\frac{\partial x}{\partial E} \right)_{X,T} = \left(\frac{\partial D}{\partial X} \right)_{E,T} \tag{10.4}$$

The SI unit of the coefficient d is m/V if calculated as strain developed per unit electric field. The unit can also be coulombs per newton (C/N), which is the reason why d is sometimes called the *piezoelectric charge constant*.

The generator effect or direct piezoelectric effect is defined by the g coefficient. Remember, g stands for *generator*. The piezoelectric g coefficient is defined as the electric field (E) generated by the application of unit mechanical stress (X), which is given by the following equation:

$$g = -\left(\frac{\partial E}{\partial X} \right)_{D,T} = \left(\frac{\partial x}{\partial D} \right)_{X,T} \tag{10.5}$$

The piezoelectric g coefficient is also known as the *piezoelectric voltage constant* and is equal to the strain (x) induced per unit of dielectric displacement (D) applied. The unit of g coefficient will be $V \cdot m/N$ if expressed as the electric field generated per unit stress or m^2/C if described as the strain induced per dielectric displacement applied.

We have not derived Equations 10.4 and 10.5. However, the rationale for these equations is as follows.

We can induce the development of strain (x) in a dielectric material by using a mechanical stress (X), such as one that is tensile or compressive. The strain that is produced depends on the Young's modulus of the material. The stiffer the material, that is, the higher Young's modulus, the lesser the strain (x) produced for a given level of stress (X). When using a scalar quantity, the inverse of Young's modulus is compliance. Assume that the strain (x) produced in a material will be proportional to its compliance (S) and level of stress (X).

In a piezoelectric material, the strain (x) can also be produced by the application of an electric field (E) via the converse piezoelectric effect. Furthermore, the electrostriction effect also causes strain (Section 10.3). We assume that the strain produced by the converse piezoelectric effect is much greater than that produced by electrostriction.

Thus, we can consider the development of strain (x) as a function of the stress (X) and the electric field (E). If we want to consider only the effect of stress (X) on strain, then we must hold the electric field (E) constant. We show the compliance under a constant electric field as s^E. Similarly, if we want to express only the effect of electric field on the developing strain, then we must hold the stress X constant (this is the other variable that can also produce strain). This is why, in the definition of piezoelectric coefficient d, we specified that X is constant (Equation 10.4).

Therefore, the development of strain x in ferroelectric materials can be written as a function of the two causes: stress (X) and electric field (E):

$$x = s^E X + dE \tag{10.6}$$

Because ferroelectric materials are nonlinear dielectrics, and in some materials, the stress–strain relationship may also be nonlinear, a more appropriate way to express the relationship may be to consider the change in strain (δx) as follows:

$$\delta x = s^E \delta X + d\delta E \tag{10.7}$$

Note that the piezoelectric d coefficient is also defined as the dielectric displacement (D) per unit stress (X; Equation 10.4). Dielectric displacement (D) is also induced when we apply an electric field (E). The extent of the dielectric displacement depends on the dielectric constant or permittivity (ε) of the material. Thus, we can write an equation that relates the dielectric displacement (D) as a function of the applied electric field and stress level (X). We must hold the stress constant when we consider the effect of electric field (E) on the dielectric displacement (D). Therefore, we use the permittivity value at constant stress (ε^X). In most measurements of the dielectric constant, the sample is not clamped or constrained. This allows the sample to expand or contract; thus, the strain varies but the stress (X) is constant.

As seen in the definition of d (Equation 10.4), the value of the electric field (E) is held constant. Thus, the *effect* of the developing dielectric displacement (D) due to two *causes*, (1) electric field and (2) stress, can be written as:

$$D = dX + \varepsilon^X E \qquad (10.8)$$

Considering that ferroelectrics are nonlinear dielectrics, we should write Equation 10.8 as

$$\delta D = d\delta X + \varepsilon^X \, \delta E \qquad (10.9)$$

Using similar arguments, we can write equations for the direct effect, that is, application of stress, causing the generation of an electric field as follows:

$$E = -gX + D/\varepsilon^X \qquad (10.10)$$

Considering that the g coefficient is also expressed as the strain produced per dielectric displacement applied (Equation 10.5), we get

$$x = s^D X + gD \qquad (10.11)$$

We rewrite Equations 10.10 and 10.11, respectively, as follows:

$$\delta E = -g\delta X + \delta D/\varepsilon^X \qquad (10.12)$$

$$\delta x = s^D \delta X + g\delta D \qquad (10.13)$$

In addition to the definitions of the d and g coefficients, we also define another coefficient, e, as

$$e = \left(\frac{\partial D}{\partial x}\right)_{E,T} = -\left(\frac{\partial X}{\partial E}\right)_{X,T} \qquad (10.14)$$

One way to understand the meaning of this e coefficient is that it describes the dielectric displacement caused by the creation of strain.

Similarly, the h coefficient is defined as

$$-h = \left(\frac{\partial E}{\partial x}\right)_{D,T} = \left(\frac{\partial X}{\partial D}\right)_{X,T} \qquad (10.15)$$

Section 10.9 describes the coefficients d and g under *hydrostatic* stress conditions, denoted as d_h and g_h. In these, the subscript h stands for hydrostatic and should not be confused with the coefficient h defined in Equation 10.15.

10.9 TENSOR NATURE OF PIEZOELECTRIC COEFFICIENTS

10.9.1 CONVENTIONS FOR DIRECTIONS

The piezoelectric and pyroelectric coefficients (Section 10.20) are more accurately represented as tensors as are the dielectric constant, compliance, and Young's modulus. However, in this book, we treat them as scalars. It is important to recognize that, in many applications, the effects developed (e.g., development of strain) are in a different direction than the cause (e.g., the stress applied). For example, when we stretch a material along its length, its width also changes. The extent of this effect is given by Poisson's ratio (v). Thus, although the stress is applied along one direction, its effect is felt in the directions perpendicular to it. Similar effects must also be accounted for in piezoelectrics. The notation shown in Figure 10.28 generally is used for directions of piezoelectrics.

According to this convention, the tensile or compressive forces are applied along Directions 1, 2, and 3, also designated as the x-, y-, and z-axes. Planes designated as 4, 5, and 6 are for the shear stresses. Direction 3 is considered the poling direction. This means the electrodes used in the poling process have been applied to the piezoelectric on the faces perpendicular to Direction 3.

10.9.2 GENERAL NOTATION FOR PIEZOELECTRIC COEFFICIENTS

If we apply a stress to the piezoelectric in a direction j, the expression for dielectric displacement (D) induced in the direction i generally would be written as follows:

$$D_i = d_{ij}X_j + \varepsilon_{ii}^x E_i \tag{10.16}$$

This is almost the same equation as Equation 10.8; the only difference is that the tensor nature of the piezoelectric coefficients and dielectric constant is now stated explicitly.

Similarly, we can write an expression for the strain developed in a piezoelectric as

$$x_j = s_{ji}^E X_j + d_{ij}E_i \tag{10.17}$$

Essentially, this is another form of Equation 10.6. The piezoelectric coefficient is d_{ij}. In this symbol, the first subscript (i) shows the applied electrical field that causes a mechanical strain in the direction j. Equation 10.16 also describes the effect of the stress X_j applied in direction j, causing a strain in the direction j. In Equation 10.17, S_{ij}^E is the compliance of the piezoelectric material under zero electric field. This is also known as compliance under a closed-circuit or short-circuit condition.

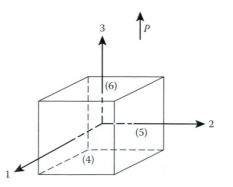

FIGURE 10.28 Notation for designating the directions for piezoelectrics; Direction 3 (z-axis) is taken as the poling direction. (From Buchanan, R.C., *Ceramic Materials for Electronics*, Marcel Dekker, New York, 2004. With permission.)

The value of compliance when the dielectric displacement (D) is constant is shown as S_{ij}^D and is known as compliance under the open circuit condition. Note that the values of compliance under a short circuit and an open circuit are very different. This leads us to the definition of a coefficient that can link the functioning of a piezoelectric as a *transducer*. A transducer is a device that can convert one form of energy into another. For example, a piezoelectric material can convert mechanical energy into electrical energy and vice versa.

Reduced notation is one in which both stress and strain are treated as first-rank tensors, instead of second-rank ones: the first subscript indicates the electrical direction and the second subscript indicates the mechanical direction.

If we apply an electric field to a piezoelectric material in Direction 3 and measure the strain in Direction 1, the corresponding converse piezoelectric coefficient is indicated as d_{31}. Similarly, the coefficient d_{33} will indicate the strain developed in Direction 3 when an electric field is applied in that direction (i.e., along the poling axis; Figure 10.28). If we apply a field (E) in Direction 1, that is, perpendicular to the poling direction, this will cause a shearing strain on the plane shown as 5 (Figure 10.28) and given by the d_{15} coefficient. To use piezoelectrics in this particular shear mode, the original electrodes used for poling in Direction 3 must be removed, and new electrodes need to be applied in a direction perpendicular to Direction 1.

We can subject piezoelectrics to hydrostatic stress. In this case, we use a subscript h with the coefficients. The coefficient d_h is related to the d_{33} and d_{31} coefficients by the following equation:

$$d_h = d_{33} + 2d_{31} \tag{10.18}$$

10.9.3 SIGNS OF PIEZOELECTRIC COEFFICIENTS

The d_{33} values for most commercially used PZT ceramics are in the range of 200 to 550 pm/V. The d_{31} values for PZT are negative and range from about −180 to −220 pm/V. The d_{31} values for PZT ceramics and $BaTiO_3$ and relaxor ferroelectrics, such as lead magnesium niobate-lead titanate (PMN-PT), are also negative. However, the d_{33} values for these materials are positive. This means that, when an external voltage is applied to drive a piezoelectric cylinder, which has been poled previously in Direction 3, in the direction of polarization, the cylinder expands in length, that is, d_{33} is positive (Figure 10.25). When the cylinder expands, the strain in the direction perpendicular to the longitudinal axis is negative, or d_{31} is negative.

In other piezoelectrics, such as PVDF, the mechanisms of piezoelectric polarizations are different. As a result, the d_{31} for PVDF is positive, and the d_{33} coefficients for PVDF are negative. This means that the application of an electric field in the direction of poling (usually the thickness direction) will cause the length of a PVDF sample to increase.

10.10 RELATIONSHIP BETWEEN PIEZOELECTRIC COEFFICIENTS

We will now show that the piezoelectric coefficient d, which defines the strain induced per unit of electric field applied or dielectric displacement per unit stress (Equation 10.4), and the g coefficient, which provides a measure of the electric field produced per unit stress (Equation 10.5), are related such that $d/g = \varepsilon^X$.

From Equation 10.4, the d coefficient is the strain developed by a unit change in the applied electric field. Both the stress (X) and temperature (T), which can also cause strain, are held constant. From Equation 10.4, we get

$$d = \left(\frac{\partial x}{\partial E} \right)_{X,T} \tag{10.19}$$

One definition of the g coefficient is that it describes the generator effect, which is the electric field (E) developed per unit stress (X) or the strain (x) developed per unit of dielectric displacement (D) applied. Again, because strain can be developed in a material by the application of either stress (X) or a change in temperature (expansion or contraction), we hold these constant. Therefore, from Equation 10.5, the following equation is obtained:

$$g = \left(\frac{\partial x}{\partial D}\right)_{X,T} \tag{10.20}$$

Dividing Equation 10.19 by Equation 10.20, we get

$$d/g = \frac{\left(\dfrac{\partial x}{\partial E}\right)_{X,T}}{\left(\dfrac{\partial x}{\partial D}\right)_{X,T}} = \left(\frac{\partial D}{\partial E}\right)_{X,T} \tag{10.21}$$

Recall that the right-hand side of Equation 10.21 is the definition of dielectric permittivity (ε). Dielectric permittivity is the increase in the dielectric displacement (D) with an increase in the electric field (E) applied. Because the application of stress or a change in the temperature can also cause a change in the dipole moment and, consequently, a change in the dielectric displacement, these parameters—stress (X) and temperature (T)—must be held constant. This point was not emphasized previously in this chapter.

Therefore, from Equation 10.21, we get

$$d/g = \varepsilon^x \tag{10.22}$$

We could write a similar equation showing the tensor nature explicitly. For example, the ratio of d_{ij} and g_{ij} will be written as follows:

$$\frac{d_{ij}}{g_{ij}} = \frac{\left(\dfrac{\partial x_j}{\partial E_i}\right)_{X,T}}{\left(\dfrac{\partial x_j}{\partial D_i}\right)_{X,T}} \tag{10.23}$$

This can be rewritten as

$$\frac{d_{ij}}{g_{ij}} = \left(\frac{\partial D_i}{\partial E_i}\right)_{X,T} \tag{10.24}$$

or as

$$\frac{d_{ij}}{g_{ij}} = \varepsilon_{ii}^X \tag{10.25}$$

Applying this for d_{31} and g_{31}, we get

$$\frac{d_{31}}{g_{31}} = \varepsilon_{33}^X \tag{10.26}$$

TABLE 10.1

Relationships between Piezoelectric Coefficients

d Coefficients	*e* Coefficients	*g* Coefficients	*h* Coefficients
$d_{31} = e_{31}\left(s_{11}^E + s_{12}^E\right)e_{33}s_{13}^E$	$e_{31} = \varepsilon_3^x h_{31} = d_{31}\left(C_{11}^E + C_{12}^E\right) + d_{33}C_{13}^E$	$g_{31} = \dfrac{d_{31}}{\varepsilon_3^x}$	$h_{31} = g_{31}\left(C_{11}^D + C_{12}^D\right) + g_{33}C_{13}^E$
$d_{33} = 2e_{31}s_{13}^E + e_{33}s_{33}^E$	$e_{33} = \varepsilon_3^x h_{33} = 2d_{31}C_{13}^E + d_{33}C_{13}^E$	$g_{33} = \dfrac{d_{33}}{\varepsilon_3^x}$	$h_{33} = 2g_{31}C_{13}^D + g_{33}C_{33}^E$
$d_{15} = e_{15}s_{44}^E$	$e_{15} = \varepsilon_1^x h_{15} = d_{15}C_{44}^E$	$g_{15} = \dfrac{d_{15}}{\varepsilon_1^x}$	$h_{15} = g_{15}C_{44}^D$

---

or

$$g_{31} = \frac{d_{31}}{\varepsilon_{33}^X} \tag{10.27}$$

These are essentially the same as Equation 10.22.

Similarly, it can be shown that the following relationships exist among other piezoelectric coefficients.

$$g_{33} = \frac{d_{33}}{\varepsilon_{33}^X} \tag{10.28}$$

$$g_{15} = \frac{d_{15}}{\varepsilon_{11}^X} \tag{10.29}$$

In Table 10.1, a summary of the relationships among different piezoelectric coefficients is listed.

Recall that in the coefficients listed in Table 10.1, *e* relates the dielectric displacement *D* induced per unit of strain or stress generated per unit of electric field. The coefficient *h* represents the electric field generated per unit strain applied or stress generated per unit of dielectric displacement. The term *h* used here should not be confused with the subscript h that is sometimes used to describe piezoelectric coefficients such as *d* or *h* under hydrostatic stress (Equation 10.18). Also note the difference between ε^X, the permittivity under constant stress, and ε^x, the permittivity under constant strain. To maintain constant strain, the sample must be clamped. Compliance is designated by *s* and stiffness by *c*. The terms *E* and *D* represent the electric field and the dielectric displacement, respectively. In addition, the permittivity is shown as a scalar quantity in these equations; that is, we have written ε_3 instead of ε_{33}. Finally, note the difference between the uses of permittivity (ε) and dielectric constant (ε_r). The following examples illustrate the use of piezoelectric coefficients.

Example 10.1: Meaning of the Piezoelectric Coefficients D_{33} and G_{15}

What is the meaning of (a) d_{33} and (b) g_{15}? What are their units?

Solution

1. The coefficient d_{33} is one of the piezoelectric charge constants. It is defined as induced polarization (SI unit: C/m^2) in Direction 3 per unit applied stress (SI unit: N/m^2) also applied in Direction 3. This expresses the direct or the generator piezoelectric effect (Figure 10.25).

$$d_{33} = \frac{\text{Induced polarization in direction 3} \left(\text{C/m}^2\right)}{\text{Applied stress in direction 3} \left(\text{N/m}^2\right)} \qquad (10.30)$$

The unit for the d coefficient is coulomb per newton (typically written as pico-coulomb per newton [pC/N]).

Another way to express the piezoelectric d_{33} coefficient is as the strain induced in Direction 3 (poling direction) by a field applied in Direction 3.

$$d_{33} = \frac{\text{Induced strain in direction 3}}{\text{Applied electric field in direction 3} \left(\text{V/m}\right)} \qquad (10.31)$$

Because strain has no dimensions, another possible unit for the d coefficient is meters per volt, usually written as picometers per volt (pm/V).

2. Recall that the first subscript is the electrical direction, and the second subscript is the mechanical direction. Thus, g_{15} is defined as the induced electric field in Direction 1 per unit shear stress on Plane 5, that is, around Direction 2 (Figure 10.28).

$$d_{15} = \frac{\text{Induced electric field in direction 1} \left(\text{V/m}\right)}{\text{Applied stress on plane 5} \left(\text{N/m}^2\right)} \qquad (10.32)$$

The SI unit for g_{15} would then be Vm/N.

Alternatively, g_{15} is defined as the shear strain induced around Direction 2 (i.e., Plane 5) per unit of dielectric displacement applied in Direction 1 (Figure 10.28). The SI unit would be m²/C.

$$g_{15} = \frac{\text{Induced shear strain around direction 2 on plane 5}}{\text{Applied dielectric displacement in direction 1} \left(\text{C/m}^2\right)}$$

The SI unit will be m²/C.

Example 10.2: Relationship Between Piezoelectrics and Other Coefficients

Show that $g_{33} = d_{33}/\varepsilon_3^x$, that is, prove Equation 10.22.

Solution

To describe the relationship between d_{33} and g_{33}, we start with Equation 10.23.

Therefore,

$$\frac{d_{33}}{g_{33}} = \frac{\left(\frac{\partial X_3}{\partial E_3}\right)_{X,T}}{\left(\frac{\partial X_3}{\partial D_3}\right)_{X,T}} = \left(\frac{\partial D_3}{\partial E_3}\right)_{X,T}$$

The right-hand side of this equation is the dielectric displacement created in Direction 3 per unit of electric field applied in Direction 3. This is ε_3, and we have assumed constant stress (X). Thus, Equation 10.28 is valid.

Note that lower dielectric constants lead to higher values of g. For example, the polymer PVDF has a relatively lower value of the d_{33} coefficient, ~ -30 pC/N (note the negative sign), compared to that of PZT, with $d_{31} \sim 200{-}400$ pC/N. As stated before, the coefficient d describes the converse piezoelectric effect (i.e., the strain developed when a voltage is applied). However, the dielectric constant of PVDF is low ($\varepsilon_r \sim 10$) compared to that of PZT ($\varepsilon_r \sim 1000{-}2800$).

Therefore, the piezoelectric voltage constant g of PVDF is still comparable to that of PZT. The coefficient d is important in applications where the strain developed in piezoelectrics causes a useful action or event.

When a sinusoidal voltage is applied to a piezoelectric disk, for example, the back-and-forth expansion and contraction of this disk (Figure 10.25) can lead to the generation of ultrasonic waves. For this application as an ultrasonic generator, we need materials with a high d value. We would prefer to have a material with a higher g coefficient in applications where we need to have a higher voltage generated when the material undergoes even very small levels of stress. For ultrasound detection, we would prefer to use materials with a higher g coefficient. In many real-life applications that involve piezoelectric materials, we need both ultrasound generation and detection capabilities. In such cases, we use the product $d \times g$ as a figure of merit. In some applications, such as in underwater sound detectors (sonars or hydrophones), piezoelectric materials are exposed to a hydrostatic pressure, that is, they are subjected to stress from all directions. In this case, the *hydrostatic piezoelectric coefficients* are designated as d_h and g_h; thus, the product $d_h \times g_h$ is the important figure of merit.

Piezoelectric composites that make use of materials such as PZT and PVDF have been developed for many such applications (see Section 10.19). In other applications (e.g., pyro-electric detectors; Section 10.20), piezoelectric voltage generation actually is undesirable. For these applications, it is possible to take advantage of the difference between the signs of the d coefficients and create composites in which the piezoelectric effect is substantially reduced. This causes the detector to respond more to changes in temperature and less to vibration or shock.

Example 10.3: Calculations for Evaluating the Direct and Converse Effects in Lead Zirconium Titanate

A poled PZT type I piezoelectric ceramic disk (length $l_0 = 2$ mm and poled in Direction 3) of a certain composition has $d_{33} = 2\ 89$ pC/N, and its dielectric constant $\left(\varepsilon_{r33}^X\right)$ is 1300.

1. What is the value of the piezoelectric voltage constant or the g_{33} coefficient?
2. What is the voltage applied across the thickness of this material if the electric field (E) is 250 kV/m?
3. How much strain (X) will this voltage generate along Direction 3?
4. What will be the increase in length, in micrometers?

Solution

1. For the PZT ceramic sample here, the d_{33} coefficient is 289 pC/N, and the dielectric constant under constant stress $\left(\varepsilon_{r33}^X\right)$ is 1300. Recall that the dielectric constant is the ratio of the permittivity of a material to that of a vacuum, that is,

$$\varepsilon_r = \frac{\varepsilon}{\varepsilon_0} \tag{10.33}$$

Therefore, in this case,

$$\varepsilon_{r33}^X = 1300 = \frac{\varepsilon}{8.85 \times 10^{-12}\ \text{F/m}} \text{ or } \varepsilon_{r33}^X = 1.15 \times 10^{-8}\ \text{F/m}$$

Thus we get (Equation 10.28)

$$g_{33} = \frac{d_{33}}{\varepsilon_{33}^X}$$

Therefore, we get

$$g_{33} = \frac{\left(289 \times 10^{-12} \text{ C/N}\right)}{\left(1.15 \times 10^{-8} \text{ F/m}\right)} = 25.1 \times 10^{-3} \text{ V} \cdot \text{m/N}$$

2. Because the initial length of the material is 2 mm, the voltage applied to the material would be

$$250 \frac{\text{kV}}{\text{m}} \times 10^3 \frac{\text{V}}{\text{kV}} \times 2 \text{ mm} \times 10^{-2} \frac{\text{m}}{\text{mm}} = 5000 \text{ V}$$

3. The strain generated in Direction 3 by applying an electric field of 250 kV/m in the same direction will be given by

$$X = d_{33} \times E_3 = \left(289 \times 10^{-12} \text{C/N}\right)\left(250 \times 10^3 \text{ V/m}\right) = 7.225 \times 10^{-5}$$

$$X = \frac{\Delta l}{l_0} = \frac{\Delta l}{2 \text{ mm}} = 7.255 \times 10^{-5}$$

4. Now, from the definition of strain, that is, $x = \Delta l / l_0$, the change in the dimension will be given by

$$\Delta l = (2 \text{ mm} \times 1000 \text{ } \mu\text{m/mm})(7.225 \times 10^{-5}) = 0.145 \text{ } \mu\text{m}$$

Thus, applying a voltage of 5000 V to this cylinder of height 2 mm will cause an elongation of 0.145 μm.

10.11   APPLICATIONS OF PIEZOELECTRICS

Properties of some piezoelectric materials are summarized in Table 10.2. Various applications of piezoelectrics are summarized in Figure 10.29. We will now discuss some devices based on piezoelectrics.

10.12   DEVICES BASED ON PIEZOELECTRICS

10.12.1   EXPANDER PLATE

The following example helps to illustrate the meaning of some of the equations related to piezoelectric coefficients. Consider a piezoelectric plate of thickness (t) and area (A) made up of electrodes with length l and width w (Figure 10.30).

When a voltage is applied along the thickness (t; also known as Direction 3), a strain develops along the length (described as Direction 1). The corresponding piezoelectric voltage constant (g), which can be described as the strain (x) induced per unit dielectric displacement (D) applied, will be written as

$$g_{31} = \frac{\text{Strain developed in direction 1}}{\text{Dielectric displacement applied in direction 3}} \tag{10.34}$$

We can write Equation 10.34 as follows:

$$g_{31} = \frac{\left(\Delta l / l\right)}{\left(Q / A\right)} \tag{10.35}$$

TABLE 10.2
Approximate Ranges for the Piezoelectric Coefficients and Other Relevant Properties of Some Single-crystal Piezoelectric Materials

	BaTiO$_3$	α-Quartz (SiO$_2$)	LiNbO$_3$	PZN-PT
Piezoelectric				
d_{33} (pC/N or pm/N)	85.6	0	10.69	2140
d_{31} (pC/N or pm/N)	−34.5	0	−1.4	−980
d_{15} (pC/N or pm/N)		0	66.38	130
d_{11} (pC/N or pm/N)	0	2.331		
d_{14} (pC/N or pm/N)	0	−0.7763		
$g33$ (V · m/N)	57.5×10^{-3}	-50×10^{-3}	34.46×10^{-3}	
$g31$ (V · m/N)	-23×10^{-3}	0	-4.110×10^{-3}	
$g14$ (V · m/N)	0	0		
Coupling				
k_{33}	0.56	0.09		0.91
k_{31}	0.315	0		0.50
Dielectric				
"Free" dielectric constant at constant stress $\left(\varepsilon_{r33}^X\right)$	168	4.628	210.5	5242
"Clamped" dielectric constant at constant strain $\left(\varepsilon_{r33}^X\right)$	109	4.628	26.7	869
"Free" dielectric constant at constant stress $\left(\varepsilon_{r11}^X\right)$	2920	4.507	83.3	3099
"Clamped" dielectric constant at constant strain $\left(\varepsilon_{r11}^X\right)$	1970	4.420	44.9	2975
Mechanical				
Compliance $\left(S_{33}^E\right)$ at constant electric field (closed circuit) (m²/N)	15.7×10^{-12}	9.7329×10^{-12}	5.058×10^{-12}	12×10^{-12}
Compliance at constant displacement (open circuit) $\left(S_{33}^D\right)$ (m²/N)	10.8×10^{-12}	9.7329×10^{-12}	4.758×10^{-12}	
Compliance at constant electric field (closed circuit) $\left(S_{11}^E\right)$ m²/N	10.05×10^{-12}	12.7791×10^{-12}	5.854×10^{-12}	10.3×10^{-12}
Compliance at constant displacement (open circuit) $\left(S_{11}^D\right)$ (m²/N)	7.25×10^{-12}	12.6429×10^{-12}	5.291×10^{-12}	

Sources: Buchanan, R.C., *Ceramic Materials for Electronics*, Marcel Dekker, New York, 2004; Shrout, T.S. and Zhang, S.J., *J. Electroceram.* 19, 111–124, 2007.

In Equation 10.35, the numerator is simply the strain in Direction 1—that is, the change in length (Δl) of the plate along Direction 1 per unit length (l). The denominator is the dielectric displacement (D) applied, which is the charge Q divided by the area A. Considering that, for this structure, $Q = V \times C$, and $A = l \times w$, we get

$$g_{31} = \frac{\left(\Delta l / l\right)}{\left(\dfrac{(V \times C)}{(l \times w)}\right)} = \frac{\Delta l \times w}{V \times C} \tag{10.36}$$

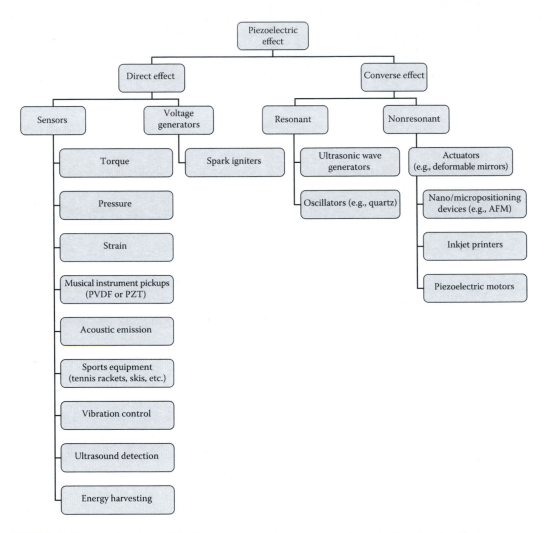

FIGURE 10.29 Applications of the direct (generator) and converse (motor) piezoelectric effects.

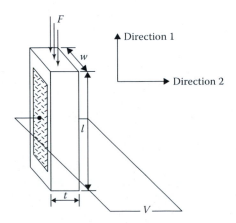

FIGURE 10.30 Piezoelectric plate geometry. The plate is poled in the thickness (*t*) direction. Stress is applied along the length direction. (Adapted from Morgan Technical Ceramics, *Guide to Piezoelectric and Dielectric Ceramics*. Available at http://www.morganelectroceramics.com/pzbook.html.)

Recall that for a parallel-plate capacitor, the capacitance C is given by

$$C = \varepsilon_r \varepsilon_0 \frac{A}{t} = \varepsilon_r \varepsilon_0 \frac{l \times w}{t} \tag{10.37}$$

In this case, we assume that the capacitor structure (that is, the piezoelectrics coated with metal electrodes) is free and not clamped. This means that the stress X to which the capacitor is subjected is constant. Therefore, the dielectric constant (ε_r) or dielectric permittivity (ε) that we use is under constant stress X. Furthermore, the direction of the applied electric field and the poling direction are the same—Direction 3. Thus, the correct value of permittivity to be used will be ε_{33}^X.

Substituting the expression for C from Equation 10.37 into Equation 10.36, we get

$$g_{31} = \frac{\Delta l \times w}{V \times C} = \frac{\Delta l \times w}{V \times \varepsilon_{r33}^X \varepsilon_0 \dfrac{l \times w}{t}}$$

or

$$g_{31} = \frac{\Delta l \times t}{V \times l \times \varepsilon_{r33}^X \times \varepsilon_0} = \Delta l \times \frac{t}{l} \times \frac{1}{V \varepsilon_{r33}^X \varepsilon_0}$$

Rewriting

$$g_{31} = \Delta l \times \frac{t}{l} \times \frac{1}{V \varepsilon_{r33}^X \varepsilon_0} \tag{10.38}$$

Note that in Equations 10.37 and 10.38, we denote the dielectric constant using the symbol ε_r and not k. This is because the symbol k^2 stands for the *coupling efficiency* of a piezoelectric (Section 10.16).

We can write the strain developed in a piezoelectric along its thickness (Direction 3) when a voltage is applied in Direction 1 as follows:

$$\Delta l = g_{31} \times \frac{l}{t} \times V \times \varepsilon_{r33}^X \times \varepsilon_0 \tag{10.39}$$

We can similarly write an expression for the piezoelectric charge coefficient d_{31} as the ratio of strain developed in Direction 1 (length) when a voltage is applied in Direction 3 (thickness).

$$d_{31} = \frac{\text{Strain developed in direction 1}}{\text{Electric field applied in direction 3}} \tag{10.40}$$

$$d_{31} = \frac{\Delta l / l}{V / t} \tag{10.41}$$

or

$$\Delta l = d_{31} \times V \times \frac{l}{t} \tag{10.42}$$

This is the same as Equation 10.39.

A note of caution: These equations, for a change in length, are applicable only under nonresonant conditions. Every physical body or structure, such as the plate considered here, has natural mechanical resonant frequencies. If a piezoelectric material is excited at these resonant frequencies, the changes in dimensions we will get will be larger than those predicted by Equations 10.42 and 10.39. In other words, these equations are applicable to frequencies of excitation well below the resonant frequencies.

Both Equations 10.42 and 10.39 describe the change in the length (Δl) of a dielectric plate. By equating these, we get

$$\Delta l = g_{31} \times \frac{l}{t} \times V \varepsilon_r \varepsilon_0 = d_{31} \times V \times \frac{l}{t}$$

$$\Delta l = d_{31} \times V \times \frac{l}{t} \tag{10.43}$$

Therefore,

$$g_{31} = \frac{d_{31}}{\varepsilon_r \varepsilon_0} = \frac{d_{31}}{\varepsilon_{33}^X}$$

or

$$\frac{d_{31}}{g_{31}} = \varepsilon_{33}^X \tag{10.44}$$

This is the relationship between the g and d coefficients, previously shown in Equations 10.23 and 10.27.

In Example 10.4, we will show how to interpret the meaning of piezoelectric and mechanical coefficients and to calculate the magnitude of strain developed in a poled piezoelectric subjected to voltage.

Example 10.4: Piezoelectric Micropositioners

The small change in dimensions we can obtain by applying a voltage to a piezoelectric can be used to make devices known as micropositioners. A poled PZT ceramic plate, 50 mm long, 5 mm wide, and 2 mm thick, is used as a micropositioner device. Assume that the dielectric constant ε_{r33}^X is 1200 and $g_{31} = 10.5 \times 10^{-3}$ V·m/N.

1. What will be the value of d_{31}?
2. What will be the change in the *length* of this plate if a potential of 100 V is applied across its *thickness*?

Solution

1. We start with the relationship between d_{31} and g_{31} from Equation 10.26.

$$\frac{d_{31}}{g_{31}} = \varepsilon_{33}^X \tag{10.45}$$

Therefore,

$$d_{31} = g_{31} \times \varepsilon_{33}^X = 10.5 \times 10^{-3} \text{ V} \cdot \text{m/N} \times 1200 \times 8.85 \times 10^{-12} \text{ F/m}$$

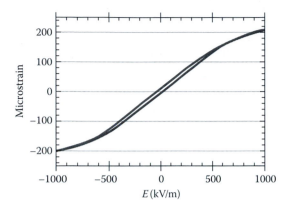

FIGURE 10.31 Typical converse piezoelectric effect in PZT ceramics. (From Pilgrim, S., *Piezoelectric Materials: A Unheralded Component*, 11th ed., FabTech. Available at http://www.fabtech.org/white_papers/_a/piezoelectric_materials_an_unheralded_component. With permission.)

Note the conversion of the dielectric constant to dielectric permittivity using the permittivity of free space ($\varepsilon_0 = 10.85 \times 10^{-12}$ F/m).

Thus, $d_{31} = 139.9 \times 10^{-12}$ m/V.

2. The change in the dimension of this plate along its length can be calculated by the expression

$$\Delta l = d_{31} \times V \times \frac{l}{t} \tag{10.46}$$

$$\Delta l = 139.9 \times 10^{-12} \text{ m/V} \times 10 \text{ V} \times \frac{50 \times 10^{-2} \text{ m}}{2 \times 10^{-2} \text{ m}} = 0.348 \text{ μm}$$

Thus, by applying a voltage of 100 V, we get an increase of about 348 nm in the length of the plate. Such changes in the dimensions of piezoelectrics are used in micropositioning and nanopositioning devices. The relative strain developed in a piezoelectric PZT is shown in Figure 10.31 as a function of the electric field applied.

10.13 TECHNOLOGICALLY IMPORTANT PIEZOELECTRICS

Properties of *single crystals* of some commonly encountered piezoelectric materials are listed in Tables 10.2 and 10.3. Although many piezoelectric materials have been investigated, only a handful of these are useful in commercial applications related to actuators, ultrasonic generators, ultrasound imaging, and vibration control and dampening (Figure 10.29). We will discuss a few of these materials here, such as PZT, the most widely used piezoelectric ceramic.

Another technologically important material is PVDF, which is different because it is a polymer and is therefore quite flexible. Another class of technologically important materials is the group that contains relaxor ferroelectrics, such as lead magnesium niobate-lead titanate (PMN-PT). These materials, in both single-crystal and polycrystalline forms, have become particularly attractive because of their very high piezoelectric coefficients ($d_{33} \sim 2000$ pm/N) and electromechanical coupling coefficients ($k_{33} \sim 0.9$).

One of the limitations of piezoelectrics is that they can depole when they are exposed to temperatures approaching the Curie temperature. This means that the domains aligned during poling will become randomized, and the resulting material will have either a very weak or no piezoelectric response. In general, piezoelectrics cannot be used at temperatures above half the value of their T_c. This is the main reason why $BaTiO_3$, with a Curie temperature of 120°C, is not used widely as

TABLE 10.3

Piezoelectric Properties of Poled Polycrystalline Barium Titanate and Different Grades of PZT

	$BaTiO_3$	PZT-DOD-I/ PZT-4 (Hard PZT)	PZT-DOD-II/ or PZT5A (Soft PZT)	PZT-DOD-III (Hard PZT)	PZT-DOD-VI or PZT5H (Soft PZT)
Piezoelectric					
d_{33} (pC/N or pm/N)	191	289	374	218	593
d_{31} (pC/N or pm/N)	−79	−123	−171	−93	−274
g_{33} (V · m/N)	11.4×10^{-3}	25×10^{-3}	25×10^{-3}	25×10^{-3}	19.7×10^{-3}
g_{31} (V · m/N)	-4.7×10^{-3}				-9.1×10^{-3}
Coupling					
k_p	0.354	0.58	0.60	0.50	0.65
k_{33}	0.493	0.70	0.71	0.64	0.75
k_{31}	0.208	0.33	0.34	0.295	0.39
Dielectric					
Free dielectric constant at constant stress $\left(\varepsilon_{r33}^X\right)$	1900	1300	1700	1000	3400
Clamped dielectric constant at constant strain $\left(\varepsilon_{r33}^X\right)$	1420	635	830	600	1470
Free dielectric constant at constant stress $\left(\varepsilon_{r11}^X\right)$	1620	1475	1730		3130
Clamped dielectric constant at constant strain $\left(\varepsilon_{r11}^X\right)$	1260	730	916		1700
tan δ (dielectric loss)	0.01	0.001	0.02	0.004	0.02
Curie temperature (T_c) (°C)	~120	328	365	300	190
Coercive field (kV/cm)	~1–2	~18	~15	~22	~6–8
Mechanical					
Compliance $\left(S_{33}^E\right)$ (constant electric field, i.e., closed circuit) (m²/N)	10.93×10^{-12}	15.5×10^{-12}	110.8×10^{-12}	13.9×10^{-12}	20.7×10^{-12}
Compliance (constant displacement, i.e., open circuit) $\left(S_{33}^D\right)$ (m²/N)	6.76×10^{-12}	7.9×10^{-12}	9.46×10^{-12}	10.5×10^{-12}	10.99×10^{-12}
Compliance (constant electric field, i.e., closed circuit) $\left(S_{11}^E\right)$ (m²/N)	10.55×10^{-12}	12.3×10^{-12}	16.4×10^{-12}	11.1×10^{-12}	16.5×10^{-12}
Compliance (constant displacement, i.e., open circuit) $\left(S_{33}^D\right)$ (m²/N)	10.18×10^{-12}	10.9×10^{-12}	14.4×10^{-12}	10.1×10^{-12}	14.1

Source: Buchanan, R.C., *Ceramic Materials for Electronics*, Marcel Dekker, New York, 2004; Shrout, T.S. and Zhang, S.J., *J. Electroceram.*, 19, 111–124, 2007.

a piezoelectric material. Other factors that must be considered are changes in the crystal structure and the lowering of dielectric constants. The dielectric constants (ε_r) and the d_{33} coefficients at room temperature for different piezoelectric ceramics are shown in Figures 10.32 and 10.33.

There is a need for piezoelectrics that can function at high temperatures. In Section 10.18, we will briefly discuss the latest developments related to strain-tuned piezoelectrics, which represent a new step in the development of piezoelectric devices that could function at higher temperatures. Another area of concern in the field of piezoelectrics is the presence of lead (Pb) in most commercially useful piezoelectrics. There is considerable interest in the development of lead-free piezoelectrics that are environmentally friendly (Figures 10.32 and 10.33).

10.14 LEAD ZIRCONIUM TITANATE

PZT and PZT-based ceramics are the most widely used piezoelectric materials. These materials are popular because of their relatively large piezoelectric, electromechanical coupling coefficients, and relatively high Curie temperatures (Table 10.3). Table 10.3 also contains data for $BaTiO_3$ for comparison purposes. Note that these are approximate ranges of the values of piezoelectric coefficients and other properties, and have been provided to give a general idea. They should not be relied on for engineering design. The exact values will depend on many factors, including microstructure, processing, exact chemical composition, temperature, and the state and magnitude of any stresses present.

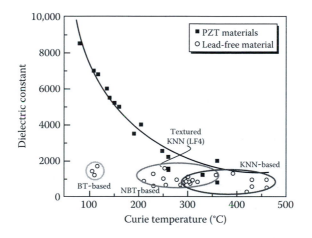

FIGURE 10.32 Room-temperature (300 K) values of dielectric constants as functions of the Curie temperature for zirconium titanate (PZT) materials and novel lead-free piezoelectrics. (From Shrout, T.S. and Zhang S.J., *J. Electroceram.*, 19, 111–124, 2007. With permission.)

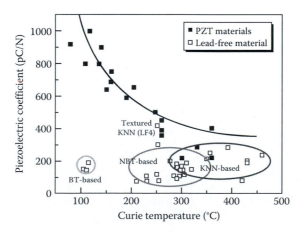

FIGURE 10.33 Room-temperature (300 K) values of the piezoelectric coefficient d_{33} (in pC/N) as a function of the Curie temperature for zirconium titanate (PZT) materials and novel lead-free piezoelectrics. (From Shrout, T.S. and Zhang S.J., *J. Electroceram.*, 19, 111–124, 2007. With permission.)

For PZT ceramics, the maximum strain develops at $x \sim 0.5$, which is a composition near the so-called morphotropic boundary (MPB; Figure 10.34). The MPB defines the change in the crystal structure of a ferroelectric material along with changes in its composition. Important piezoelectric properties of ferroelectric materials, such as the strain developed for a given level of electric field, are maximized when compositions near the MPB are used.

In PZT at room temperature, zirconium-rich compositions have a rhombohedral crystal structure (Figure 10.35a). Titanium-rich compositions exhibit a tetragonal structure. At room temperature and at the mole fraction of zirconium 0.53 (i.e., titanium mole fraction of $x = 0.47$ in $PbZr_{1-x}Ti_xO_3$), the crystal structure changes from rhombohedral to tetragonal. The change in crystal structure traditionally has been used to define and describe the MPB (Figure 10.35).

It was previously believed that such MPB compositions led to maximum piezoelectric properties (e.g., piezoelectric coupling coefficients) and that the polarization vector could switch its orientation in the different variants of the tetragonal and rhombohedral phases. However, it has been suggested recently that the MPB is *not* a boundary but rather a phase with monoclinic symmetry (Ahart et al. 2008), and that there is actually a new monoclinic (and *not* a mixture of nanotwin

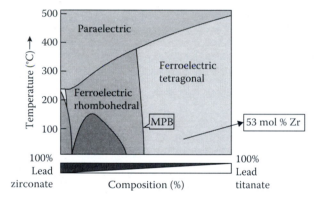

FIGURE 10.34 Phase diagram for the $PbZrO_3$–$PbTiO_3$ system showing the different ferroelectric phases and the morphotropic phase boundary. (From Cross, E., *Nature*, 432, 24–25, 2004. With permission.)

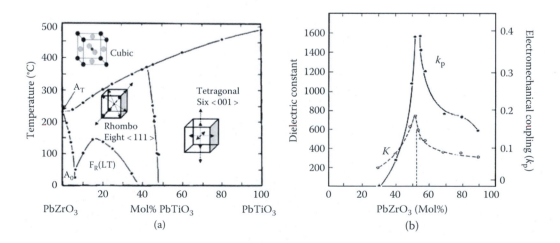

FIGURE 10.35 (a) Crystal structures for zirconium titanate (PZT), their dielectric constants, and (b) the changes in piezoelectric properties with composition. (Adapted from Moulson, A.J. and Herbert J.M., *Electroceramics: Materials, Properties, and Applications*, Wiley, New York, 2003. With permission.)

domains in the tetragonal and rhombohedral phases). This phase is intermediate and between the tetragonal and rhombohedral PZT phases.

Research on essentially pure $PbTiO_3$ has shown changes in its crystal structure caused by increasing pressure (Figure 10.36). Because the original concept of the MPB was linked to changes in its chemical composition, we would have expected that an MPB could *not* exist in a pure compound such as $PbTiO_3$. However, the tetragonal form of $PbTiO_3$ (shown as the space group P4mm) changes to the monoclinic phase (between 11 and 12 GPa; shown as M_c and M_A) with increasing pressure. Above 16 GPa, $PbTiO_3$ changes to a rhombohedral crystal structure (shown as the R3m space group).

Recent research suggests that the classic MPB seen in the PZT system is the result of *chemical pressure* that builds in $PbTiO_3$ as the titanium ions are substituted by zirconium ions. This is similar to the development of strain-tuned ferroelectrics (see Section 10.18.1).

One of the concerns in the use of PZT materials is that they contain lead. A significant amount of recent research has been directed toward development of lead-free piezoelectrics (see Section 10.18.2).

10.14.1 PIEZOELECTRIC POLYMERS

Many polymers are piezoelectric (Table 10.4). The most widely used piezoelectric polymer is PVDF, which exhibits four crystalline structures. The beta (β) phase of PVDF exhibits a polar structure and is ferroelectric, pyroelectric, and piezoelectric. PVDF films are prepared by stretching or rolling the nonpolar alpha (α) phase sheets. During this process, the PVDF films undergo a phase transformation from the nonpolar α to the polar β phase. This phase transformation can also be achieved by annealing or by applying an electric field, also known as poling. Unlike PZT and $BaTiO_3$, the d_{31} coefficient for PVDF is positive, whereas the d_{33} coefficient is negative (Table 10.5). This means that when a voltage is applied in the thickness direction (the poling direction, or Direction 3), the PVDF

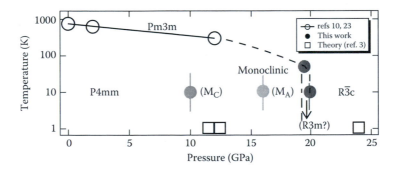

FIGURE 10.36 Changes in the crystal structure with increasing pressure in PT. (From Ahart, M., et al., *Nature*, 451, 545–549, 2010. With permission.)

TABLE 10.4
Monomers or Repeat Units of Some Piezoelectric Polymers

Polymer	Repeat Unit	Polymer	Repeat Unit
Polyvinylidene fluoride	$[CH_2–CF_2]_n$	Nylon-7	$[—NH—(CH_2)_6—CO—]_n$
Polytrifluoro ethylene	$[CHF–CF_2]_n$	Nylon-9	$[—NH—(CH_2)_8—CO—]_n$
Polytetrafluoro ethylene, also known as Teflon	$[CF_2–CF2]_n$	Nylon-11	$[—NH—(CH_2)_{11}—CO—]_n$
		Nylon-5,7	$[—NH—(CH_2)_5—NH—CO—(CH_2)_5—CO—]_n$

TABLE 10.5

Approximate Ranges of Values for the Piezoelectric, Dielectric, and Mechanical Properties of copolymers in comparison with PZT and Quartz

	PVDF	PVDF–55% TrFE	PVDF–75% TrFE	PZT-DOD-VI or PZT5H	Quartz
Piezoelectric					
d_{33} (pC/N or pm/N)	−35			+593	0
d_{31} (pC/N or pm/N)	+20	+25	+10	−274	0
d_h (pC/N or pm/N)	−3			45	
$g31$ (V · m/N)	174×10^{-3}	160×10^{-3}	110×10^{-3}	19.7×10^{-3}	0
$h31$ (V/m)	53×10^{-7}	19×10^{-7}	22×10^{-7}		
Coupling					
				0.75	
k_{31}	0.10	0.07		0.39	0.09
Dielectric					
Dielectric constant	13	18	10	3400	4.5
tan δ (dielectric loss)	0.31		0.15		0.02
Curie temperature (T_c) (°C)				190	
Approximate operating temperature (°C)	80	70 100		110	573
Coercive field (kV/cm)				~6–8	
Mechanical					
Compliance $\left(S_{33}^E \right)$ (constant electric field, i.e., closed circuit) (m²/N)	4.72×10^{-10}			20.7×10^{-12}	
Compliance $\left(S_{11}^E \right)$ (constant electric field, i.e., closed circuit) (m²/N)	3.65×10^{-10}			16.5×10^{-12}	
Density (kg/m³)	1.78×10^{-3}	1.9×10^{-3}	1.88×10^{-3}	7.45×10^{-3}	2.65×10^{-3}
Velocity of sound (m/s)	3×10^3	2×10^3		83.3	

Source: Nalwa, H.S., *Ferroelectric Polymers: Chemistry, Physics, and Applications,* Marcel Dekker Inc., New York, 1995; Buchanan, R.C., *Ceramic Materials for Electronics,* Marcel Dekker, New York, 2004; Moulson, A.J. and Herbert, J.M., *Electroceramics: Materials, Properties, and Applications,* Wiley, New York, 2003.

film will *expand* in the thickness direction and shrink in the length direction. Also, although the piezoelectric coefficients for PVDF and related polymers (~10 pC/N) tend to be lower than those for ceramics (~10^2 pC/N; Table 10.3) by a factor of 10 to 50, the dielectric constants of PVDF and related materials are also lower than PZT and other ferroelectrics. This is why the voltage coefficients (i.e., the *g* values) of PVDF and related materials are similar to those for PZT and other ferroelectric ceramics (Table 10.5).

Blends of PVDF with trifluoroethylene (TrFE) and tetrafluoroethylene (TeFE), known as Teflon™, are also piezoelectric. Similarly, odd-numbered and odd-odd nylons or polyamides (e.g., nylon 9 or nylon 5, 7) and polymers known as cyanopolymers (containing the C–CN group) are also piezoelectric. Ferroelectric nylons exhibit more useful piezoelectric properties (e.g., d_{31} ~15 pC/N) at higher temperatures (up to 200°C; Cheng and Zhang 2008). The monomers or repeat units of some of these are shown in Table 10.4.

TABLE 10.6

Piezoelectric and Other Properties of PVDF Films

Property	Coefficient	Biaxially Oriented Film	Uniaxially Oriented Film
Piezoelectric	d_{31}	4.34	21.4
coefficients (pC/N)	d_{32}	4.36	2.3
	d_{33}	−12.4	−31.5
	d_h	−4.8	−9.6
	d_{33} (calculated from d_h, d_{31}, and d_{32})	−13.5	−33.3
Pyroelectric coefficients (C/m²K)	p_3	-1.25×10^{-5}	-2.74×10^{-5}
Thermal expansion coefficients (K⁻¹)	α_1	1.24×10^{-1}	0.13×10^{-4}
	α_1	1.00×10^{-1}	1.45×10^{-4}
Mechanical properties	Compliance s_{11} (Pa⁻¹)	4×10^{-10}	4×10^{-10}
	Compliance s_{12} (Pa⁻¹)	-1.57×10^{-10}	-1.57×10^{-10}
	Poisson's ratio	0.392	0.392

Source: Nalwa, H. S., *Ferroelectric Polymers: Chemistry, Physics, and Applications.* Marcel Dekker, New York, 1995.

In Table 10.5, the approximate ranges of selected piezoelectric properties of PVDF and some of its blends are shown. Note that these values are for illustration purposes only and should not be used for design. The properties are strongly dependent on the processing methods used, crystal structures, temperature, and on the frequency, and it is therefore important to get more accurate data from the suppliers. For example, the piezoelectric properties of a PVDF film depend strongly on the orientation of the films (Table 10.6). The following example illustrates the use of piezoelectric polymers.

Example 10.5: Polyvinylidene Fluoride Film Sensor

A commercially available poled PVDF film, electroded using silver metallic ink, is 110 μm thick. A stress of 20,000 Pa is applied to this film over a 1-in.² area.

1. Assuming that the film has a rigid backing and deforms only in the thickness direction, what will be the open-circuit voltage generated?
2. If the film is now subjected to the same level of *force* as in part (a) but the film has a compliant backing, with a force acting on the cross section of the film such that the film stretches in the length direction, what will be the voltage generated across the thickness of the film? Assume that $g_{33} = 350 \times 10^{-3}$ and $g_{31} = 220 \times 10^{-3}$ V · m/N.

Solution

1. The stress is acting on a 1-in.² area, and because the film has a rigid backing, it can deform only along the thickness (Direction 3). Thus, we must use the g_{33} coefficient. From Equation 10.66, the electric field

$$(E)\, \text{generated} = -\left(350 \times 10^{-3}\right)\frac{\text{V} \cdot \text{m}}{\text{N}} \times \left(20{,}000\,\text{Pa}\right) = -7000\,\text{V/m}$$

The thickness of the film is 110 μm; therefore, the voltage generated across the film thickness will be

$$= (-7000\,\text{V/m}) \times (110 \times 10^{-6}\,\text{m}) = -0.77\,\text{V}$$

2. For this part, the film has a compliant, not rigid, backing and can change its length. We must use the g_{31} coefficient to calculate the electric field generated.

The example states that the *force* applied is the same as before. A stress of 20,000 Pa was previously applied over a 1-in.^{2} area. Let us first calculate the force that is applied.

The area was = (1 in.^{2}) × (2.54 cm/in.)^2 × (10^{-2} m/cm)^2 = 6.452 × 10^{-4} m^2.

Thus, the force applied = (20,000 N/m^2) × 6.452 × 10^{-4} m^2 = 12.903 N.

The same force is now applied over a cross section of the film that is 2.54 cm in length and 110 μm wide. The cross-sectional area is now 2.794 × 10^{-6} m^2. This is a very small area.

We now have a much higher level of stress for the same force as before:

$$= (12.903 \text{ N})/(2.794 × 10^{-6} \text{ m}^2) = 4.62 × 10^6 \text{ Pa}$$

The electric field generated across the thickness of the film now is

$$= -(220 × 10^{-3} \text{ V·m/N}) × (4.62 × 10^6 \text{ Pa}) = -1.02 × 10^6 \text{ V/m}$$

Now note that this electric field still appears across the thickness of the film of 110 × 10^{-6} m. This is what the coefficient g_{31} represents.

Thus, the voltage generated will be = (-1.02 × 10^6 V/m) × (110 × 10^{-6} m) ~-112 V. This voltage is much higher because of the increased level of stress caused by the smaller cross-sectional area.

10.15 APPLICATIONS AND PROPERTIES OF HARD AND SOFT LEAD ZIRCONIUM TITANATE CERAMICS

Often, PZT ceramics are classified as soft PZT and hard PZT (Table 10.3). This terminology has its origins in the field of magnetism, wherein we refer to soft and hard magnets (Chapter 11). In general, soft PZTs have higher piezoelectric coefficient values. For example, Navy types II and VI are considered soft PZTs (Figure 10.37). These soft PZT compositions are donor-doped (e.g., Nb^{5+} substituting for Ti^{4+}; see Chapter 2) materials that have lower coercive fields (~6–15 kV/cm for PZT). We can move the domains in these materials relatively easily, so they are considered soft ferroelectrics or piezoelectrics. The donor-doping process creates positively charged point defects on which the charge is balanced by cation vacancies or defects with effective negative charge (Chapter 2). Typically, dielectric losses in these soft PZT materials are higher (tan δ ~ 0.02) because domain wall mobility is increased. Soft PZT materials are useful for low-power ultrasonic transducers, force and acoustic pickups, and so on.

Hard PZT compositions are acceptor-doped materials with relatively high coercive fields (~18–22 kV/cm). DOD types I and III are examples of hard PZT (Table 10.3). Their piezoelectric coefficients are smaller. Their coercivity is higher and their dielectric losses are smaller by an order of magnitude (tan δ ~ 0.001–0.004). Hard PZT materials are useful in applications such as resonant mode ultrasonic devices, piezomotors, and so on. Hard PZT materials are doped with acceptor dopants that create point defects with a negative effective charge. This charge is balanced by the presence of oxygen ion vacancies, which are positively charged defects (Chapter 2).

The relationship of electric field versus strain for some soft and hard piezoelectrics is shown in Figure 10.38. This figure also shows data for some relaxor ferroelectrics, such as PMN-PT, that rely on electrostriction as a source for developing strain.

Soft piezoelectrics usually show higher piezoelectric coefficients. Their disadvantage is that their domain-wall motion is easy, so they tend to have higher dielectric losses. This means that when they are used as piezoelectric devices driven by an electric voltage, they will generate a considerable amount of heat. Another problem with soft piezoelectrics is that the electric-field-versus-strain

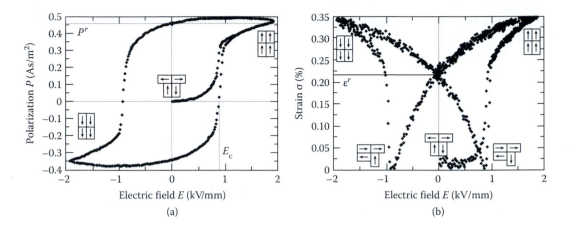

FIGURE 10.37 (a) $P-E$ hysteresis loop and (b) the corresponding strain–electric field butterfly loop for a soft zirconium titanate (PZT) ceramic DOD II or PIC 151. (From Schneider, G.A., *Annu. Rev. Mater. Res.*, 37, 491–538, 2007. With permission.)

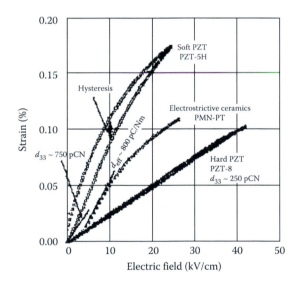
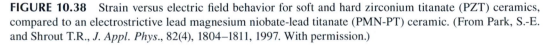

FIGURE 10.38 Strain versus electric field behavior for soft and hard zirconium titanate (PZT) ceramics, compared to an electrostrictive lead magnesium niobate-lead titanate (PMN-PT) ceramic. (From Park, S.-E. and Shrout T.R., *J. Appl. Phys.*, 82(4), 1804–1811, 1997. With permission.)

relationship shows hysteresis. This means that when the material extends from one dimension to another and the field is removed, the material does not go back to its original dimension. This limits their applications as precise micropositioning devices.

Hard piezoelectrics show lower piezoelectric coefficients. Their advantage is that their dielectric losses are very small. These materials can be driven using higher electric fields and will not generate a lot of heat. They also do not exhibit much hysteresis in the electric-field-versus-strain relationship.

10.16 ELECTROMECHANICAL COUPLING COEFFICIENT

Consider a piezoelectric material that is subjected to a stress. An elastic strain will develop and the material will store some elastic energy. When the same piezoelectric material is subjected to an electrical field, it will develop a piezoelectric strain. This causes it to store additional energy. The *electromechanical coupling coefficient* (k) is defined as

$$k^2 = \frac{W_{12}}{W_1 W_{12}}$$

(10.47)

where W_{12}, W_1, and W_2 are the piezoelectric, mechanical, and electrical energy densities, respectively.

The strain development in a piezoelectric material as a function of stress (X) and electric field (E) was given by Equation 10.7.

For a stress δX, the total energy stored can be written as follows:

$$\frac{1}{2}\delta X \delta x = \frac{1}{2}s^E (\delta X)^2 + \frac{1}{2}d\delta X \delta E$$

$$\frac{1}{2}\delta V \delta x = W_1 + W_{12}$$

(10.48)

The first term, that is, ½ s^E $(\delta X)^2$, is the mechanical energy stored (W_1) and the second term, ½ $d\delta X\delta E$, is the piezoelectric energy stored (W_{12}).

The dielectric displacement D can be created or changed in a material by applying an electric field. The cause and effect are related by the dielectric constant. A change in the dielectric displacement can also occur by applying stress, which changes the distance among the entities that cause dipoles. Thus, the change in dielectric displacement is written as shown in Equation 10.9.

Now, if an electric field δE is applied, the total energy stored can be written as

$$\delta E \delta D = d\delta E \delta X + \varepsilon^X(\delta E)^2$$

(10.49)

$$\frac{1}{2}\delta E \delta D = \frac{1}{2}d\delta E \delta X + \frac{1}{2}\varepsilon^X (\delta E)^2$$

$$\delta E \delta D = W_{12} + W_2$$

(10.50)

From the definitions of W_1, W_2, and W_{12} in Equations 10.47, 10.49, and 10.50, we get

$$k^2 = \frac{d^2\delta E^2\delta X^2}{\left[s^E(\delta E)^2\right]\left[\varepsilon^X(\delta E)^2\right]}$$

(10.51)

or

$$k = \frac{d}{\sqrt{s^E \varepsilon^X}}$$

(10.52)

Note that the compliance values for short-circuit (s^E) and open-circuit (s^D) conditions are very different.

We can now develop a relationship between these two as follows.

Now, $\delta x = s^E\delta X + d\delta E$, and also

$$\delta x = s^D\delta X + g\delta D$$

(10.53)

Therefore, we get

$$S^D \delta x + g \delta D = s^E \delta X + d \delta E \tag{10.54}$$

Substituting for δE (as given by Equation 10.12) and from Equations 10.12 and 10.54, we get

$$S^D \delta x + g \delta D = s^E \delta X + d(-g \delta X + \delta D / \varepsilon^X)$$

$$S^D \delta x + g \delta D = s^E \delta X - dg \delta X + (d / \varepsilon^X) \delta D \tag{10.55}$$

Note that $g = d / \varepsilon^X$; therefore, simplifying the right-hand side of the second step in Equation 10.55, we get

$$S^D \delta X + g \delta D = S^E \delta X - \frac{d^2}{\varepsilon^X} \delta X + g \delta D$$

or

$$S^D = S^E - \frac{d^2}{\varepsilon^X} \tag{10.56}$$

This can be rewritten as

$$S^D = S^E \left[1 - \frac{d^2}{S^E \varepsilon^X} \right] \tag{10.57}$$

Recognizing that the last term in Equation 10.57 is k^2 (from Equation 10.52), we get

$$S^D = S^E[1 - k^2] \tag{10.58}$$

Equation 10.58 is very important because it shows that the open- and closed-circuit compliances are related to each other via the electromechanical coupling coefficient. We can show that the stiffness (c) values under open- and closed-circuit conditions are also similarly related.

$$C^E = C^D[1 - k^2] \tag{10.59}$$

Note that usually for stiffness, the left-hand side is the short-circuit value.

Instead of equating δx as we did to write Equation 10.54 and then deriving Equation 10.58, we can equate δE and show that

$$\varepsilon^x \approx \varepsilon^X (1 - k^2) \tag{10.60}$$

This derivation involves an approximation that assumes that $k^4 \ll 1$.

We can rewrite Equation 10.60 as

$$k^2 = \frac{\left[\varepsilon^X - \varepsilon^x \right]}{\varepsilon^X} \tag{10.61}$$

Multiplying both numerator and denominator by $\frac{1}{2}E^2$, we get

$$k^2 = \frac{\left[\dfrac{1}{2}\varepsilon^X E^2 - \dfrac{1}{2}E^2\varepsilon^x\right]}{\dfrac{1}{2}\varepsilon^X E^2} \tag{10.62}$$

Equation 10.62 tells us that the electromechanical coupling coefficient can be written as the ratio of energy densities. The denominator is the total electrical energy that is stored in a piezoelectric body when a piezoelectric that is subjected to an electric field E is free to deform and is not constrained. Under these conditions, the stress X is maintained constant, and the strain x changes. The term $\dfrac{1}{2}E^2\varepsilon^x$ represents the electrical energy stored in a piezoelectric when the material is constrained such that the strain x is constant. Thus, the numerator term in Equation 10.61 represents the difference in electrical energy between a piezoelectric that is free to deform and a piezoelectric that is constrained. This is equal to the mechanical energy that results from the conversion of electrical energy and is stored in the piezoelectric body.

This stored mechanical energy can be recovered from the piezoelectric and used. The electrical energy stored in the piezoelectric also can be used. In this sense, the parameter k^2 should not be thought of as the efficiency in the same sense as we think of the efficiency of an engine. We define k^2 as the ratio of usefully converted energy to the input energy. Thus, for piezoelectric applications where the piezoelectric is used as a transducer, high coupling coefficients are desirable. If, for example, the value of k^2 is 0.7, this does not mean that the transducer is only 70% efficient.

If the piezoelectric material is clamped, it cannot develop any strain; that is, no mechanical deformation is allowed. Thus, no mechanical energy can be stored, and so the only energy stored is electrical energy.

In Equation 10.62, the first term in the numerator is the total input electrical energy and the second term in the numerator is the electrical energy stored at constant strain (i.e., a clamped sample that cannot store any mechanical energy). The difference between these terms is in the conversion of the input electrical energy into mechanical energy. We define the ratio of the electrical energy converted into mechanical energy to the total input electrical energy, as the *effective coupling coefficient* (κ_{eff}), represented as follows:

$$\kappa_{eff}^2 = \frac{\text{Input electrical energy converted into mechanical energy}}{\text{Input electrical energy}} \tag{10.63}$$

Instead of starting with the relationship among compliances under open- and closed-circuit conditions (Equation 10.59), we can start with the stiffness under open- and closed-circuit conditions and show that this coefficient can also be defined as

$$\kappa_{eff}^2 = \frac{\text{Input electrical energy converted into electrical energy}}{\text{Input mechanical energy}} \tag{10.64}$$

10.17 ILLUSTRATION OF AN APPLICATION: PIEZOELECTRIC SPARK IGNITER

Piezoelectrics are used in a variety of applications (Figure 10.28). Most applications involve a relatively sophisticated system that takes advantage of the direct or converse piezoelectric effect.

We have chosen a relatively simple application in the piezoelectric spark igniter to illustrate the applications of some of the equations derived so far. We will show that, for a piezoelectric spark igniter made from a poled piezoelectric cylinder of diameter (d) and length (or height) l, the voltage generated (V) under open-circuit conditions is given by

$$V = g_{33} \frac{F \times l}{\pi d^2} \tag{10.65}$$

When the piezoelectric igniter is subjected to a compressive force F across a cross-sectional area,

$$A = \pi d^2$$

The definition of the piezoelectric voltage constant (g) is

$$g = -\left(\frac{\partial E}{\partial X}\right)_{D,T} = \left(\frac{\partial x}{\partial D}\right)_{X,T}$$

The electric field generated and the stress applied are related by the following expression:

$$E = -g \times X \tag{10.66}$$

Another relationship we have seen in Table 10.1 is

$$g_{33} = \frac{d_{33}}{\varepsilon_{33}^X} \tag{10.67}$$

Note that, in this case, we are dealing with the stress (X) applied in Direction 3, and the electric field is generated across the length of the cylinder; hence, we relate these properties to g_{33}.

Therefore,

$$E = -g_{33} \times \frac{F}{\pi d^2} \tag{10.68}$$

Representing the electric field generated across the length (i.e., height) of the cylinder as $E = V/l$, we get

$$V = -g_{33} \frac{F \times l}{\pi \times d^2} \tag{10.69}$$

This is written as follows:

$$V = \frac{d_{33}}{\varepsilon_{33}^X} \frac{F \times l}{\pi \times d^2} \tag{10.70}$$

When the piezoelectric igniter is first subjected to a compressive stress by using a force F on area A, the strain (x) developed under open-circuit conditions (i.e., constant dielectric displacement D) is given by

$$x = -\frac{\delta l^D}{l} = -S_{33}^D \frac{F}{A} \tag{10.71}$$

The first part of this equation is the definition of strain. The second part relates the strain (x) to the stress (F/A) via compliance S_{33}^D under open-circuit conditions.

The mechanical work (w_m) done by the force F in creating a change in length (δl^D) is given by the equation

$$w_m = \frac{1}{2}F\delta l^D = \frac{1}{2}S_{33}^D \frac{F^2 l}{A} \tag{10.72}$$

From the definition of the electromechanical coupling coefficient (k^2), the result of this mechanical work done and the electrical energy available will be

$$w_{el} = \frac{1}{2}k_{33}^2 S_{33}^D \frac{F^2 l}{A} \tag{10.73}$$

We can also view the compressed piezoelectric cylinder as a capacitor with generated voltage V; for this capacitor with capacitance C and voltage V, the electrical energy stored in this capacitor is

$$w_{el} = \frac{1}{2}CV^2 \tag{10.74}$$

Equating these expressions for w_{el} and substituting for C in terms of dielectric permittivity and geometrical parameters, we get

$$w_{el} = \frac{1}{2}k_{33}^2 S_{33}^D \frac{F^2 l}{A} = \frac{1}{2}CV^2 = \frac{1}{2}\varepsilon_{33}^x \frac{A}{l}V^2$$

Solving for V,

$$k_{33}^2 S_{33}^D F^2 = \varepsilon_{33}^x \frac{A^2}{l^2}V^2$$

or

$$V^2 = \frac{k_{33}^2 S_{33}^D F^2 l^2}{\varepsilon_{33}^x A^2} = \frac{g_{33}^2 S_{33}^D \varepsilon_{33}^x F^2 l^2}{S_{33}^E \varepsilon_{33}^x A^2} \tag{10.75}$$

Substituting $\varepsilon_{33}^x \sim \varepsilon_{33}^x\left(1 - k_{33}^2\right)$ and $S_{33}^D = S_{33}^E\left(1 - k_{33}^2\right)$, we get the following expression:

$$V^2 = \frac{g_{33}^2 S_{33}^E\left(1 - k_{33}^2\right)\varepsilon_{33}^x F^2 l^2}{S_{33}^E \varepsilon_{33}^x\left(1 - k_{33}^2\right)A^2} \tag{10.76}$$

This simplifies to the following equation that we derived previously:

$$V = g_{33}\frac{Fl}{A} = g_{33}\frac{Fl}{\pi d^2} \tag{10.77}$$

If this voltage V generated is high enough to overcome the gap between the two electrodes connected to the piezoelectric, a spark will be created between the ends of the wires connected to the circular electroded faces of the piezoelectric. If this happens, we have a change from an open-circuit condition to a closed-circuit condition. The compliance of the material undergoes a change to its short-circuit value $\left(S_{33}^{D} \right)$ from its open-circuit value $\left(S_{33}^{D} \right)$. Because the short-circuit compliance is smaller, this transition allows for the development of an additional strain. Let us assume that if we were to apply a force F on area A under a closed circuit, then the total change in length would be δl^{E}. What we considered before was the strain x developed under open-circuit conditions, that is, the constant dielectric displacement. Let us refer to this change in length under open-circuit conditions as δl^{D}.

When we apply a force F on an area A under open-circuit conditions, the displacement will be δl^{D}. If a spark is generated such that a closed-circuit condition is created, an additional change in length will occur, with a magnitude of $(\delta l^{E} - \delta l^{D})$.

This additional change in length resulting from the transition from closed- to open-circuit conditions will be given by

$$\delta l^{E} - \delta l^{D} = \left(S_{33}^{E} - S_{33}^{D} \right) \frac{F \times l}{A} = k_{33}^{2} S_{33}^{E} \frac{Fl}{A} \tag{10.78}$$

The corresponding additional mechanical energy stored with this extra deformation, resulting from the formation of spark causing the closed-circuit condition, will be given by the following equation:

$$w_{\text{mechanical, extra}} = \frac{1}{2} F \left(dl^{E} - dl^{D} \right) = \frac{1}{2} F \left(k_{33}^{2} S_{33}^{E} \frac{Fl}{A} \right) \tag{10.79}$$

This corresponds to the extra electrical energy (using the definition of coupling coefficient) as shown here:

$$w_{\text{electrical, extra}} = k_{33}^{2} \frac{1}{2} F \left(k_{33}^{2} S_{33}^{E} \frac{Fl}{A} \right) = \frac{1}{2} k_{33}^{4} S_{33}^{E} \frac{F^{2} l}{A} \tag{10.80}$$

Thus, the total energy stored in a piezoelectric before and after the short-circuit condition will be (from Equations 10.73 and 10.80):

$$w_{\text{total}} = \frac{1}{2} k_{33}^{2} S_{33}^{D} \frac{F^{2} l}{A} + \frac{1}{2} k_{33}^{4} S_{33}^{E} \frac{F^{2} l}{A} \tag{10.81}$$

or

$$w_{\text{total}} = \frac{1}{2} k_{33}^{2} \left(S_{33}^{D} + k_{33}^{2} S_{33}^{E} \right) \frac{F^{2} l}{A} \tag{10.82}$$

In Equation 10.82, if we substitute $S_{33}^{D} = S_{33}^{E} \left[1 - k_{33}^{2} \right]$ from Equation 10.58, we get

$$w_{\text{total}} = \frac{1}{2} k_{33}^{2} \left(S_{33}^{E} - S_{33}^{E} k_{33}^{2} + k_{33}^{2} S_{33}^{E} \right) \frac{F^{2} l}{A}$$

$$= \frac{1}{2} k_{33}^{2} S_{33}^{E} \frac{F^{2} l}{A}$$

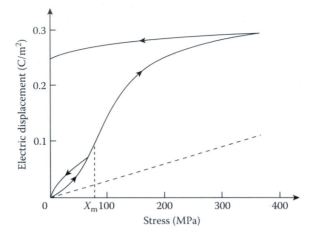

FIGURE 10.39 Increase in the dielectric displacement in certain zirconium titanate (PZT) compositions plotted as a function of the compressive stress. (From Moulson, A.J. and Herbert J.M., *Electroceramics: Materials, Properties and Applications*, Wiley, New York, 2003. With permission.)

Substituting for $k_{33}^2 = \dfrac{d_{33}^2}{S_{33}^E \varepsilon_{33}^X}$, we get

$$w_{\text{total}} = \frac{1}{2}\left(\frac{d_{33}^2}{S_{33}^E \varepsilon_{33}^X}\right) S_{33}^E \frac{F^2 l}{A} = \frac{1}{2}\frac{d_{33}^2}{\varepsilon_{33}^X}\frac{F^2 l^2}{A} = \frac{1}{2}\frac{d_{33}^2}{\varepsilon_{33}^X}\frac{F^2 l}{A} = \frac{1}{2}d_{33}\frac{g_{33}\varepsilon_{33}^X}{\varepsilon_{33}^X}\frac{F^2 l}{A}$$

Therefore,

$$w_{\text{total}} = \frac{1}{2}d_{33}g_{33}\frac{F^2 l}{A} \qquad (10.83)$$

In reality, the electrical energy stored in a piezoelectric igniter is even higher than this because compressive stress causes an additional increase in the dielectric displacement (D). This occurs because some of the so-called 90° ferroelectric domains switch their orientation under the influence of the compressive stress. This increase in dielectric displacement as a function of the compressive stress is shown in Figure 10.39.

Example 10.6: Calculation of the Voltage of a Piezoelectric Igniter

A poled PZT cylinder with a diameter of 2 mm and a thickness of 5 mm is to be used as a spark igniter. What voltage is generated by applying a compressive force of 100 N to the circular cross section of this disk? Assume that the d_{33} for this PZT is 289 pC/N and the dielectric constant in a free (unclamped) state $\left(\varepsilon_{r33}^X\right)$ is 1300.

Solution

The electric field generated and the stress applied are related by Equations 10.66 and 10.67. Therefore, for this ceramic, the g_{33} will be given by

$$g_{33} = \frac{289 \times 10^{-12}\ \text{C/N}}{1300 \times 8.85 \times 10^{-12}\ \text{F/m}} = 0.0255\ \text{V/N} = 25.5\ \text{mV/N} \qquad (10.84)$$

The area of the circular cross section is given by $\pi d^2/4 = 3.1416 \times 10^{-6}$ m^2.

The stress applied is the force divided by the area of the circular cross section and works out to 31,830,981 Pa or 31.83 MPa.

The electric field generated = $-(31,830,981$ Pa$) \times (0.0255$ V/m$) = 811,690$ V/m or 0.812 MV/m.

This field runs across the height of the piezoelectric spark igniter; thus, the voltage this field generates is = $(811,690$ V/m$) \times (5 \times 10^{-3}$ m$) = 4058$ V. If the applied stress is compressive, the voltage generated is in the same direction as the poling voltage (Figure 10.25).

Thus, a relatively small force (~100 N), which can be generated by a human hand, leads to the generation of 4000 V. If the ends of this cylinder are connected by wires and if the two wires are brought within a few millimeters of each other but are not allowed to touch, a spark jumps between the tips of the wire. Although this voltage is high, the amount of charge accumulated because of the direct piezoelectric effect and the resultant current is actually quite small and is not likely to be very dangerous.

10.18 RECENT DEVELOPMENTS

10.18.1 STRAIN-TUNED FERROELECTRICS

One of the limitations of materials developed to date is that most of them are not useful in high-temperature applications due to depoling. *Strain-tuned ferroelectrics* are ferroelectric materials or devices whose dielectric properties have been altered due to a built-in strain (Schlom et al. 2007). This approach could lead to better high-temperature performance of current ferroelectrics.

Strain is used as a method for controlling the electrical properties of materials or devices in several semiconductors (e.g., silicon-germanium) and has recently been applied to ferroelectrics. One of the limitations of many well-known ferroelectrics is that their Curie temperatures (T_c) are low. For example, the T_c in BaTiO$_3$ is only ~120°C. Thus, BaTiO$_3$ loses its ferroelectric and piezoelectric behavior at temperatures greater than 120°C. BaTiO$_3$ epitaxial thin films were recently grown on ReScO$_3$ substrates (Re indicates a rare-earth element). The lattice-constant mismatch (Figure 10.40) leads to the development of a biaxial compressive strain of ~1.7% in these films. This in turn causes the remnant polarization to increase. More importantly, the T_c of such films increases by almost 500°C. Thus, strain-tuned ferroelectrics offer ways to use existing ferroelectric and piezoelectric materials at higher temperatures. In strain-tuned BaTiO$_3$, the increase in T_c also offers the hope of using this environmentally friendly material in applications instead of lead-containing piezoelectrics such as PZT.

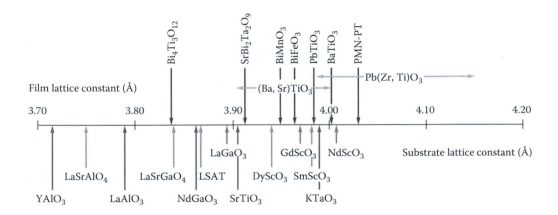

FIGURE 10.40 Lattice constants of several pseudocubic ferroelectrics (above the horizontal line) compared to the lattice constants of several substrate materials. (From Schlom, D.G., et al., *Annu. Rev. Mater. Res.*, 37, 589–626, 2007. With permission.)

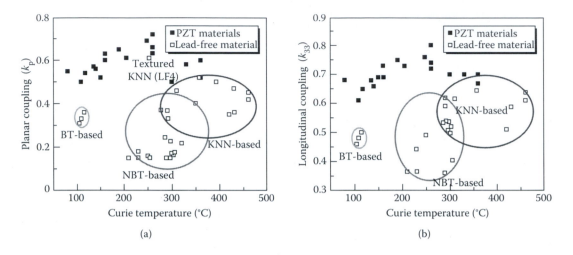

FIGURE 10.41 (a) Room-temperature values of planar (k_p) and (b) longitudinal (k_{33}) coupling coefficients. (From Shrout, T.S. and S.J. Zhang., *J. Electroceram.*, 19, 111–124, 2007. With permission.)

10.18.2 Lead-Free Piezoelectrics

Considerable research has been directed toward the development of lead-free piezoelectrics. Many countries around the world have directives requiring strict control of hazardous materials arising from all sources, including electronics, automotive components, and consumer appliances. This is especially important for materials that contain toxic elements such as lead. The two major classes of materials that have been developed are potassium sodium niobate ($K_{05}Na_{05}NbO_3$, KNN) and sodium bismuth titanate ($Na_{05}Bi_{05}TiO_3$, NBT). The dielectric constants and the d_{33} coefficients for some of the lead-free piezoelectrics were shown in Figures 10.32 and 10.33, respectively. In these diagrams, BT is $BaTiO_3$, and LF is a liquid flux-grown crystal, which refers to a method for growing single crystals.

The planar coupling coefficients (k_p) and k_{33} values for these materials are shown in Figure 10.41.

10.19 PIEZOELECTRIC COMPOSITES

Of all of the piezoelectric materials that have been developed, there is no single *perfect* or *ideal* material. Piezoelectric composites have been developed in order to try to optimize the properties of different piezoelectric and nonpiezoelectric materials.

One major advantage of polymer-ceramic piezoelectric composites is that they are quite flexible. The so-called macro fiber composites (MFCs; Figure 10.42) of ceramic ferroelectrics such as PZT, a brittle material, are arranged in a polymer matrix (the polymer is ferroelectric or linear dielectric; Sodano et al. 2004). Such MFCs have applications in the control of vibrations in defense-related applications such as aircraft and machine guns, and in civil structures such as buildings and cable-stayed bridges (Song et al. 2006). One limitation in the structural applications of MFCs is that piezoelectric actuators require a source of power, which may be disrupted during certain events, such as earthquakes, when control of vibration is needed most.

MFC structures have been used for vibration control in *smart* tennis rackets, smart skis, and other applications. Unlike applications involving vibration control in large structures, the main advantage composites offer in such applications is that they function without the need for an external power source.

Since the dielectric constant of polymers is relatively low, ceramics such as PZT are embedded in the polymer mix. The piezoelectric voltage constant or the *g* coefficient of such composites is

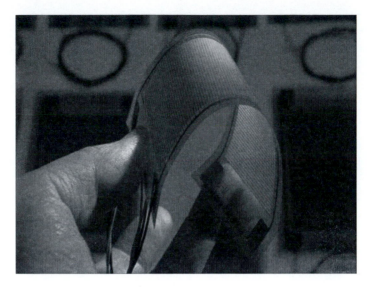

FIGURE 10.42 Photograph of a zirconium titanate (PZT)-based flexible macro fiber composite. (Courtesy of Smart Material Corporation, Sarasota, Florida.)

higher (Equation 10.67). This is similar to a point that has already been made—that PVDF has a lower d coefficient compared to PZT. However, its g coefficient is comparable to that of PZT because PVDF has a lower dielectric constant compared to PZT.

Another advantage of composites is that polymer composites can lower overall acoustic imped-ance. The acoustic impedance of water is ~1.5 MRayl. Many polymers have an acoustic impedance of ~3.5 MRayl. Most muscle, fat, tissue, and so on have an acoustic impedance of ~1.3–1.7 MRayl. Bones have an acoustic impedance of ~3.8–7.4 MRayl (Gururaja 1996). PZT has a high acoustic impedance (~35 Mrayl). If PZT is used by itself for ultrasound detection or imaging, there is a tremendous acoustic impedance mismatch at the PZT–biological tissue or PZT–water interfaces (Harvey et al. 2002). This impedance mismatch causes high reflection or *ringing* of the signal. To better match (i.e., to lower) the acoustic impedance, composites of PZT fibers with epoxy are used. These can be prepared by arranging the fiber bundles and then filling them with epoxy, called *arrange and fill*. This is followed by a dicing step in which cubic blocks are obtained by cutting the cured material perpendicular to the fiber length direction. This provides the so-called 1–3 *connec-tivity* (Figure 10.43). The effect of the arrangement of different phases is known as connectivity, and certain properties in composites of all types depend not only on the relative volume fractions but also on the geometry that is used to connect the different phases.

We can develop composites using different types of piezoelectric materials. As mentioned in Section 10.10, PZT–PVDF composites are formed to decrease the piezoelectric effect in PVDF-based infrared (IR) detectors or imaging elements (Section 10.20). The mechanism by which piezo-electricity occurs in PVDF is different from that in PZT, hence the piezoelectric coefficients of PVDF and PZT are d_{33} ~−30 pC/N (note the negative sign) and ~ +200 to +400 pC/N, respectively. Thus, we can create composites of PZT particles/crystals dispersed in a PVDF matrix (the 0–3 com-posite; Figure 10.43a), in which the overall piezoelectric effect is substantially reduced (Dietze and Es-Souni 2008). This helps make a better pyroelectric detector that responds more to temperature change than to mechanical shock or vibration.

The effective piezoelectric coefficients of such PZT-PVDF-TrFE composites are shown in Figure 10.44. The corresponding changes in the effective pyroelectric coefficients of such compos-ites are shown in Figure 10.45.

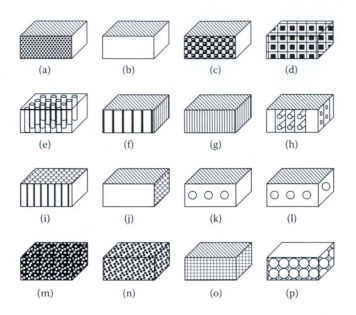

FIGURE 10.43 Illustration of connectivity in composites: (a) particles in a polymer: 0–3; (b) polyvinylidene fluoride (PVDF) composite: 0–3; (c) zirconium titanate (PZT) spheres in a polymer: 1–3; (d) diced composite: 1–3; (e) PZT rods in a polymer: 1–3; (f) sandwich composite: 1–3; (g) glass–ceramic composite: 1–3; (h) transverse reinforced composite: 1–2–0; (i) honeycomb composite: 3–1P; (j) honeycomb composite: 3–1S; (k) perforated composite: 1–3; (l) preformed composite: 3–2; (m) replamine composite: 3–3; (n) burps composite: 3–3; (o) sandwich composite: 3–3; and (p) ladder-structured composite: 3–3. (From Neelkanta, P., *Handbook of Electromagnetic Materials*, CRC Press, Boca Raton, FL, 1995. With permission.)

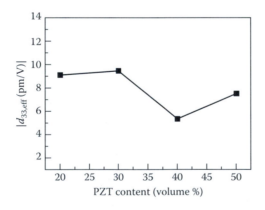

FIGURE 10.44 Effective piezoelectric coefficients of polyvinylidene fluoride-zirconium titanate (PVDF-PZT) composites. Note the minimum in the piezoelectric $d_{33,eff}$ coefficient at about 40% (vol/vol) zirconium titanate (PZT). (From Dietze, M. and Es-Souni M., *Sens. Actuators*, A143, 329–334, 2008. With permission.)

10.20   PYROELECTRIC MATERIALS AND DEVICES

A pyroelectric material is one that shows temperature-dependent spontaneous polarization (Lang 2005). To understand the origin of the pyroelectric effect, consider a disk of poled single-crystal or polycrystalline $BaTiO_3$ such that the polarization axis is perpendicular to the electrodes.

In the poled material or in a single crystal, we have dipoles throughout the volume of the material. These dipoles lead to a certain level of bound charge density on the surface, attracting nearby charges such as ions or electrons toward the faces of the material. This is how the pyroelectric effect

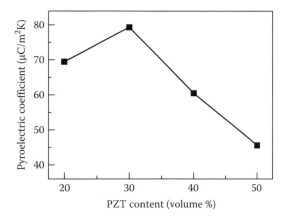

FIGURE 10.45 Effective pyroelectric coefficient (p_{eff}) of polyvinylidene fluoride-zirconium titanate (PVDF-PZT) composites. (From Lang, S.B., *Phys. Today* (August), 31–35, 2005. With permission.)

was discovered in the aluminum borosilicate mineral tourmaline (then known as *lyngourion* in Greek or *lyncurium* in Latin) by the Greek philosopher Theophrastus almost 2400 years ago! These external charges are attracted to the pyroelectric material and neutralize the bound charges on the surfaces of the pyroelectric crystal. If the temperature is constant ($dT/dt = 0$) and the electrodes on this pyroelectric $BaTiO_3$ sample are connected to an ammeter, no current will flow through the sample because no free charges are left behind on the surfaces (Figure 10.46).

If the temperature is increased ($dT/dt > 0$), the dipole moment for the ferroelectric material decreases. This means that the total polarization (*P*) decreases. Consequently, the total bound charge density on the surfaces of the ferroelectric material also decreases (Figure 10.46). This means that there now will be free charge on the surfaces of the pyroelectric material. As shown in Figure 10.46, the negative charges (i.e., electrons) flow from the top surface toward the ammeter, while a conventional current would flow in the opposite direction (Figure 10.46c). The current that flows when the temperature of a pyroelectric material is changing is known as the *pyroelectric current*.

If we further cool the sample (i.e., $dT/dt < 0$), the ferroelectric polarization will increase (Figure 10.46b). This will cause an increase in the bound charge density, and electrons will flow from the wire toward the pyroelectric material surface in order to compensate for the increased bound charge density. Thus, we will again see a pyroelectric current to flow, but its direction will be opposite to that when the material was being heated ($dT/dt > 0$). This is analogous to the development of a transient current due to the development of a piezoelectric voltage (Figure 10.24).

If we do not connect the surfaces of the pyroelectric material to an ammeter or some other resistor using conductive wires, then there will be no electrical current. Instead, we can measure a pyroelectric voltage. Again, the sign of this voltage will change as we heat or cool the material.

If the pyroelectric material is perfectly insulated from its surroundings so that the charges cannot flow in an external circuit, then the charges built on the surface eventually will be neutralized by the intrinsic conductivity of the pyroelectric material. This means that the pyroelectric charge that is developed will leak away inside the material.

The *pyroelectric coefficient* (*p*) of a material is defined as the change in the dielectric displacement (*D*) caused by a unit change in temperature (*T*).

$$dD = pdT \qquad (10.85)$$

The change in dielectric displacement (*D*) or saturation polarization (P_s) as a result of the temperature change is known as the *primary pyroelectric effect*.

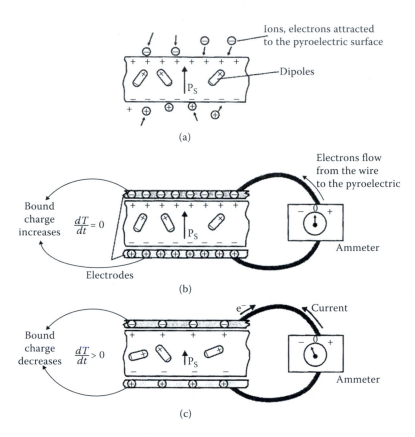

FIGURE 10.46 (a) Pyroelectric crystal with an intrinsic dipole moment in an open circuit state, (b) the same pyroelectric crystal in a short circuit state at a fixed temperature, (c) the same pyroelectric crystal in a short circuit state under heating—a change in the dipole moment generate electric current flowing the external circuit. (From Lang, S.B., *Phys. Today* (August), 31–35, 2005. With permission.)

Because the units of D are in C/m^2, the units of pyroelectric coefficient (p) will be C m^{-2}K^{-1}. If we apply an electric field (E) or a stress (X) to a material, the dipole moment will also change.

The dielectric displacement (D) for a material is given by

$$D = \varepsilon_0 E + P_{total} \tag{10.86}$$

The total polarization originates from the induced polarization ($P_{induced}$) and ferroelectric polarization (P_s). Thus, we can write

$$D = \varepsilon_0 E + P_{induced} + P_s \tag{10.87}$$

Now, $(\varepsilon_0 E + P_{induced}) = \varepsilon E$, where ε is the permittivity of the pyroelectric material; we can rewrite Equation 10.87 as follows:

$$D = P_s + \varepsilon E \tag{10.88}$$

Taking the derivative of both sides with respect to temperature (T), we get

$$\left(\frac{\partial D}{\partial T}\right) = \left(\frac{\partial P_s}{\partial T}\right) + E\left(\frac{\partial \varepsilon}{\partial T}\right) \tag{10.89}$$

The true pyroelectric coefficient (p) is defined as

$$p = \left(\frac{\partial P_s}{\partial T} \right)_{X,E}$$ (10.90)

In this definition, the constant stress (X) means that the material is not clamped and is free to expand or contract. Thus, this definition of the true pyroelectric coefficient includes both the so-called primary pyroelectric effect and the *secondary pyroelectric effect*. In the secondary pyroelectric effect, the changes in temperature or electric field cause a change in the dimension of the material (i.e., contraction or expansion due to thermal expansion or via the direct piezoelectric effect). Then the strain (x) is not constant. The change in dimensions, from either thermal expansion/contraction or the piezoelectric effect of the applied electric field, causes changes in both the dipole moment and the bound charge density, and this causes the development of a pyroelectric current or pyroelectric voltage.

Secondary pyroelectric effect should *not* be confused with a correction to the true pyroelectric coefficient (p) because of the temperature dependence of permittivity—that is, ($\partial \varepsilon / \partial T$) in Equation 10.91. We can write a *generalized pyroelectric coefficient* (p_g) as

$$p_g = p + E \left(\frac{\partial \varepsilon}{\partial T} \right)$$ (10.91)

The effect of the change in dielectric permittivity with temperature ($\partial \varepsilon / \partial T$) also results in the development of a change in the bound charge density, which can induce the development of a current. This is *not* considered the true pyroelectric effect because this change in the dielectric constant with temperature occurs in all dielectrics and not just in polar dielectrics. In ferroelectrics, the term ($\partial \varepsilon / \partial T$) can be quite high; this effect is comparable to the true pyroelectric effect.

To accurately measure the primary pyroelectric effect, we must consider the effect of a temperature change on spontaneous polarization (P_s) while maintaining a constant strain (x); that is, the material is not allowed to change its dimension and is rigidly clamped. The primary pyroelectric effect is defined as the change in saturation polarization (P_s) with temperature (T) when we keep the sample dimensions constant. This means we do not include any effects of the change in dielectric displacement (D) due to a change in the dielectric permittivity (ε) with temperature (T) or electric field (E).

Usually, it is very difficult to isolate and measure only the primary pyroelectric effect. The secondary pyroelectric effect can be calculated using the thermal expansion coefficient, Young's modulus, and the piezoelectric coefficient. The total pyroelectric effect (due to both the primary and secondary effects) typically is measured. The values of the pyroelectric coefficients for some materials are shown in Table 10.7; the total pyroelectric coefficients represent the algebraic sum of the primary and secondary coefficients.

For a pyroelectric crystal or poled material with a metalized electrode area A (perpendicular to the poling axis), the pyroelectric current is given by

$$I = A \left(\frac{dp_s}{dt} \right)$$ (10.92)

This can be rewritten as follows:

$$I = A \left(\frac{dp_s}{dt} \right) \left(\frac{dT}{dt} \right)$$ (10.93)

TABLE 10.7

Pyroelectric coefficients of Some Materials

Material	Primary Pyroelectric Coefficient ($\mu cm^{-2} \cdot K^{-1}$)	Secondary Pyroelectric Coefficient ($\mu cm^{-2} \cdot K^{-1}$)	Total Pyroelectric Coefficient ($\mu cm^{-2} \cdot K^{-1}$)
Ferroelectrics			
Poled $BaTiO_3$	−260	+60	−200
Poled $PbZr_{0.95}Ti_{0.05}O_3$	−305.7	+37.7	−268
Poled $PbTiO_3$			−180
Single crystal			
$LiNbO_3$	−95.8	+12.8	−83
$LiTaO_3$	−175	−1	−176
$PbGe_3O_{11}$	−110.5	+15.5	−95
$Ba_2NaNb_5O_{11}$	−141.7	+41.7	−100
$Sr_{0.5}Ba_{0.5}Nb_2O_6$	−502	−48	−550
$(CH_2CF_2)_n$	−14	−13	−27
Triglycine sulfate	+60	−330	−270
Not ferroelectrics			
Single crystal			
CdSe	−2.94	−0.56	−3.5
CdS	−3.0	−1.0	−4.0
ZnO	−6.9	−2.5	−9.4
Tourlamine	−0.48	−3.52	−4.0
$Li_2SO_4{:}2H_2O$	+60.2	+26.1	+86.3

Source: Lang, S.B., *Phys Today* (August), 31–35, 2005; Gupta, M.C. and Ballato, J., *The Handbook of Photonics*, CRC Press, Boca Raton, FL, 2007; Herbert, J.M., *Ferroelectric Transducers and Sensors*, Gordon and Breach Science Publishers, New York, 1982.

From this equation and Equation 10.90, we get

$$I = A \times p \times \left(\frac{dT}{dt} \right) \tag{10.94}$$

If the temperature change (ΔT) is small, then the pyroelectric coefficient (p) can be assumed to be constant over a small temperature range. Under these conditions, the pyroelectric current will be proportional to the rate of change of temperature (heating or cooling, as in Equation 10.94).

10.20.1 PYROELECTRIC DETECTORS FOR INFRARED DETECTION AND IMAGING

An important application of pyroelectric materials is to detect infrared (IR) radiation. There are two *windows* in the atmosphere in which IR radiation can travel through air without significant absorption (primarily by water vapor) and without being scattered by dust particles. One is in the range of 3–5 μm, and the other between 8 and 14 μm. At a temperature of ~300 K, objects and warm-blooded animals emit heat in the form of IR radiation of a wavelength (λ) ~10 μm. Pyroelectrics can be used to make IR detectors. We can use IR radiation detectors to get a thermal image of an intruder hiding behind a wall or tree, as long as there is a temperature difference between the body temperature of the intruder and his or her surroundings. This is one approach to making pyroelectric IR detection and imaging systems. Such pyroelectric material-based IR detectors are known as *thermal detectors.*

Another approach to making IR detectors is to use narrow-bandgap semiconductors (such as mercury cadmium telluride or HCT). These detectors, also known as *quantum detectors* or *photon detectors,* operate on the principle that an electron from the valence band is promoted to the conduction band by absorbing a photon (IR radiation, in this case). Such detectors are extremely sensitive in the far-IR range. However, detectors based on narrow-bandgap semiconductors require cooling to liquid nitrogen temperatures (~ 77 K). If we do not cool the narrow-bandgap semiconductor, there is enough thermal energy at room temperature to flood the conduction band simply by the thermal excitation of electrons from the valence band into the conduction band.

PROBLEMS

10.1 In Figure 10.17, the coercive voltage is ~3.0 V. What will the thickness of the film be if the corresponding coercive field is 70 kV/cm?

10.2 For ferroelectric materials, is the saturation polarization (P_s) a microstructure-sensitive property? Explain.

10.3 For ferroelectric materials, is the coercive field a microstructure-sensitive property? Explain.

10.4 What will happen if a piezoelectric ferroelectric material used in a device, such as an actuator, is exposed to temperatures that are close to the Curie temperature or Curie-temperature range?

10.5 Explain how the piezoelectric and electrostriction effects can be used to create mirrors, whose surface can be adjusted using an electric voltage (known as adaptive optics). What materials will you use for this application, and why?

10.6 PZN and PZN-8%-PT (PZN-8PT) single crystals of relaxor ferroelectric materials oriented in the (001) direction were investigated by Park et al. (1996) for their potential as electromechanical actuators. The percent strain developed in Direction 3 is shown as a function of the electric field, also in Direction 3 (Figure 10.47). From these data, show that the d_{33} coefficients are approximately 1700 and 1150 pC/N for PZN-8PT and PZN samples, respectively, as marked on Figure 10.47. What would the g_{33} values be if the dielectric constants for PZN-8PT and PZN materials are 4200 and 3600, respectively?

10.7 The electric field–strain response for the PZN and PZN-8PT samples is linear. The dielectric loss (tan δ) values for these samples are ~0.008 and 0.012 and are considered rather low.

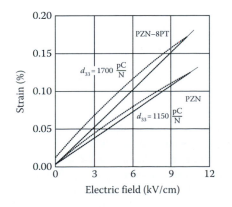

FIGURE 10.47 Strain–electric field data for PZN and PZN (001)-oriented single crystals. (From Park, S.-E., et al., *IEEE. Int. Symp. Appl. Ferroelectr.,* 1, 79–82, 1996. With permission.)

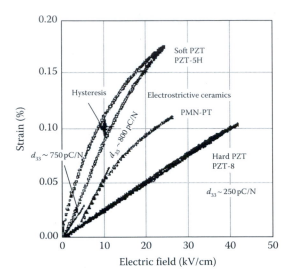

FIGURE 10.48 Strain–electric field relationship for lead magnesium niobate-lead titanate (PMN-PT) and soft and hard zirconium titanate (PZT). (From Park, S.-E. and Shrout T.R., *J. Appl. Phys.*, 82(4), 1804–1811, 1997. With permission.)

How is the linear relationship between the electric field–strain response and the dielectric loss to which it is connected? Explain.

10.8 Ultrahigh strain actuators based on relaxor ferroelectrics were investigated by Park and Shrout (1997). The data on PMN-PT and soft and hard PZT samples are shown in Figure 10.48.

    a. Show that for soft PZT samples, the d_{33} coefficient is ~700 pC/N. Use the initial linear portion of the data shown in Figure 10.48.

    b. What does the nonlinear nature of the electric field–strain curve tell us about the domain motion in these materials?

    c. As we can see from the data, there is a significant hysteresis in the strain–electric field behavior. What does this say about the suitability of this material for making an accurate positioning device?

    d. Would the dielectric losses in soft PZT-SH be relatively low or high? What is the problem if the dielectric losses are high?

10.9 From the data shown in Figure 10.48, show that the d_{33} coefficient for the hard PZT sample will be ~250 pC/N. What is the advantage of this material if it is used as an actuator or positioning device, compared to the soft PZT material?

10.10 From the data shown in Figure 10.48, show that the d_{33} coefficient for the electrostrictive PMN-PT ceramic sample is ~800 pC/N.

10.11 There is virtually no hysteresis in PMN-PT. Thus, the dielectric losses will be expected to be smaller than those for the soft PZT sample, and the micropositioning that can be achieved using these materials will be quite accurate. What are the limiting factors for using this material for micropositioning applications? What will happen if we attempt to induce larger strains using higher electric fields (e.g., to go up to 0.15% strain)?

10.12 A poled piezoelectric cylinder made using PZT-DOD type I material, of 5-mm diameter and 12-mm height, is subjected to a force of 700 N. What is the voltage generated across the height of the cylinder? What is the electrical energy (in millijoules) stored when the cylinder is under open-circuit conditions? What is the extra electrical energy stored when the cylinder undergoes additional compression under closed-circuit conditions? Use the properties of PZT-DOD type I included in Table 10.3.

10.13 A piezoelectric PVDF thin film is 4 cm in length, 2 cm in width, and 10 pm in thickness. Assume that the film is poled in the thickness direction. What will be the change in the dimension of the film along its length if a voltage of 150 V is applied in the thickness direction? What will be the change in the thickness of the film under the same conditions? Assume that $d_{31} = 25$ pC/N and $d_{33} = -35$ pC/N.

10.14 A piezoelectric plate is made from a poled polycrystalline DOD type II ceramic (Table 10.3). The original dimensions of the plate are $9 \times 4 \times 1$ mm, and those of the electrodes are 9×4 mm. If a voltage of 50 V is applied along the thickness direction, what are the new dimensions of this plate?

10.15 Piezoelectrics are utilized in dental and bone surgery. This involves using ultrasonic vibrations to cut tissue. Which piezoelectric effect, direct or converse, is used for this application? A frequency of 25 to 29 kHz is used to make micromovements of about 60–210 μm. These movements cut only mineralized tissue. In this frequency range, the neurovascular and other soft tissues are not cut (Labanca et al. 2008).

10.16 Vibrations in civil structures, such as buildings and bridges, can be controlled using piezoelectric materials formed into multilayer actuators or amplified piezoelectric actuators (Song et al. 2006). How do you think such actuators can be used for the control of vibrations in these structures?

10.17 BaTiO₃, one of the first ferroelectric ceramic materials developed, is widely used to make multilayer capacitors. However, this material has no widespread applications as a piezoelectric. Explain.

10.18 Piezoelectric fiber macrocomposites are used in some tennis rackets. Explain how such piezoelectrics work to reduce the transmission of vibrations from the racket to the player's elbow.

10.19 Can we use a lead-free PZT material that has been optimized for outstanding piezoelectric properties as a pyroelectric detector? Explain.

10.20 Find out how the focusing problems associated with the Hubble space telescopes were corrected using piezoelectric actuators. Explain the mechanism qualitatively.

GLOSSARY

Aging: The small level of randomization undergone with time by domains in a freshly poled piezoelectric after the poling electric field has been removed. This causes a decrease in the piezoelectric properties with time.

Butterfly loop: A diagram showing the development of strain in a ferroelectric material (Figure 10.26b).

Closed-circuit compliance (s^E): The compliance of a material when the electrodes providing the field for the piezoelectric effect are short-circuited (i.e., $E = 0$). This is the same as short-circuit compliance and is related to open-circuit compliance (s^D) by the following equation

$$s^D = s^E[1 - k^2]$$

Coercive field (e_c): The electric field necessary to cause domains in a ferroelectric with some remnant polarization to be randomized again in order to obtain a state of zero net polarization (Figure 10.15).

Connectivity: A particular geometrical arrangement of the different phases in a composite; this affects many properties (but not all) of a composite (Figure 10.43).

Converse piezoelectric effect: Generation of a mechanical strain by applying an electric field to a poled piezoelectric material; also known as the motor effect (Figure 10.25), it results in mechanical strain and is characterized by the d coefficient. The d coefficient, also known as the piezoelectric charge constant, is equal to the polarization induced per unit stress.

Curie-point shifters: Compounds added to a ferroelectric to lower its Curie-temperature.

Curie-point suppressors: Compounds added to a ferroelectric to broaden its Curie transition so that the dielectric constant is relatively stable with temperature.

Curie temperature: The temperature at which a ferroelectric material transforms into a centro-symmetric paraelectric form. Relaxor ferroelectrics have a Curie-temperature range.

Curie–Weiss law: The variation in the dielectric constant of the paraelectric phase at temperatures above T_0 (or T_c), which is written as

$$\varepsilon = \varepsilon_0 + \frac{C}{(T - T_0)} \left(\text{applicable for } T > T_0 \right)$$

or

$$\varepsilon = \frac{C}{(T - T_0)} \left(\text{applicable for } T > T_0 \right)$$

Also stated as "the inverse of dielectric susceptibility (χ_e) varies linearly with temperatures above T_c or T_0."

Depoling: The randomization of previously aligned domains by the application of higher temperatures or stress. Piezoelectric activity can cease to exist, and it limits the applications of piezoelectrics to relatively low temperatures.

Direct piezoelectric effect: Generation of a voltage or charge by applying a stress to a poled piezoelectric—also known as the generator effect and characterized by the g coefficients (Figure 10.25) as well as the piezoelectric voltage constant. It is equal to the strain induced per unit dielectric displacement (D) applied.

Electromechanical coupling coefficient (k): The ratio of piezoelectric energy density (W_{12}) stored in a material to the product of the electrical- (W_1) and mechanical-energy (W_2) densities stored; it is given by

$$k^2 = \frac{W_{12}}{W_1 W_2}$$

Electrostriction: The elastic strain (x) induced by the application of an electric field, which is proportional to the *square* of the electric field (E). This effect is seen in all materials and is embedded in the strain developed in ferroelectric piezoelectric materials.

Ferroelectric domains: A region of a ferroelectric material in which the polarization of all cells is in the same direction (Figure 10.6).

Ferroelectrics: Materials that show a spontaneous and reversible polarization.

Generalized pyroelectric coefficient (p_g): A pyroelectric coefficient that includes the primary and secondary coefficients, in addition to a correction needed because of the dependence of permittivity on temperature. It is given by the following equation:

$$p_g = p + E \left(\frac{\partial \varepsilon}{\partial T} \right) \tag{10.95}$$

Generator effect: See **Direct piezoelectric effect**.

Hard piezoelectrics: Compositions of lead zirconium titanate (PZT; often acceptor-doped) or other materials that offer lower piezoelectric coefficients, low dielectric losses, a higher coercive field, and little or no hysteresis in the strain–electric field relationship. These materials are well-suited for micropositioning and other applications that may require higher electric fields.

Hydrostatic piezoelectric coefficients: Piezoelectric coefficients under hydrostatic conditions that are designated with a subscript (e.g., d_h or g_h); the product $d_h \times g_h$ is the figure of merit that is important for applications including underwater sonar probes, hydrophones, and imaging applications.

Hysteresis loop: The trace of change in polarization or dielectric displacement for a ferroelectric material as a function of the electric field (Figure 10.15).

Morphotropic phase boundary (MPB): The boundary on a phase diagram across which there is a change in the crystal structure of a piezoelectric material; for compositions of ceramics, such as PZT, which are at or near the MPB, the dielectric properties are maximized. Recent research has raised questions about the existence of such a boundary in the PZT system.

Nonlinear dielectrics: Materials in which the polarization developed is not linearly related to the electric field. The dielectric constant of these materials is field-dependent; these include ferroelectrics and other materials, such as water, in which molecules have a permanent dipole moment.

Open-circuit compliance (s^D): The compliance of a material under open-circuit conditions, related to the short-circuit compliance by the following equation

$$s^D = s^E[1 - k^2].$$

Paraelectric phase: The high-temperature phase derived from an originally ferroelectric material's parent phase, which now has no dipole moment per unit cell. This phase behaves as a linear dielectric.

Photon detectors: Infrared (IR) detectors based on the creation of an electron–hole pair in narrow-band gap semiconductors by absorption of IR radiation. They are also known as quantum detectors and are fundamentally different from thermal detectors.

Piezoelectric: A material that develops an electrical voltage or charge when subjected to stress and a relatively large strain when subjected to an electric field.

Piezoelectric charge constant: The d coefficient that describes the converse piezoelectric effect. This is the strain generated per unit of electric field applied or the induced polarization per unit stress; the units are m/V or C/N.

Piezoelectric voltage constant: The g coefficient that describes the direct piezoelectric effect. This is the induced electric field per unit of stress or strain induced per unit of dielectric displacement (D) applied.

Poled ferroelectric: A ferroelectric material in which the domains are aligned in a particular direction; only a poled ferroelectric shows a measurable piezoelectric effect.

Poling: The process of applying an electric field to a ferroelectric or piezoelectric material in order to cause the alignment of domains. This process typically is carried out at higher temperatures, often using a heated oil bath.

Primary pyroelectric effect: The change in dielectric displacement (D) or saturation polarization (P_s) as a result of the temperature change while maintaining a constant strain (x), that is, with the material clamped.

Pseudocubic: The cubic crystal structure polymorph of materials such as $BaTiO_3$.

Pyroelectric coefficient (p): The change in dielectric displacement (D) caused by a unit change in temperature (θ).

Pyroelectric current: The current that flows when the temperature of a pyroelectric material is changing.

Quantum detectors: Pyroelectric infrared detectors. See **Thermal detectors**.

Relaxor ferroelectrics: Ceramic materials with high dielectric constants, high electrostriction coefficients, high-frequency dispersion, and a broad Curie transition, such as lead magnesium niobate (PMN). They are useful in actuator applications.

Remnant polarization (P_r): The polarization that remains after removing the electric field from a ferroelectric material that has been subjected to sufficiently high fields to cause saturation.

Saturation polarization (P_s): The maximum possible ferroelectric polarization that can be obtained for a given ferroelectric material.

Secondary pyroelectric effect: The change in the spontaneous polarization of a material with a change in temperature (which causes thermal expansion or contraction) or applied electric field (which causes a change in dimension because of the piezoelectric effect); here, the strain is not constant.

Short-circuit compliance: See **Closed-circuit compliance**.

Soft piezoelectrics: Materials with smaller coercive fields and higher d_{33} values that exhibit easy domain switching, causing dielectric losses and hysteresis in the electric field–strain relationships. They are typically donor-doped, such as PZT.

Strain-tuned ferroelectrics: Ferroelectric materials or devices whose properties are altered because of strain created during their processing.

Thermal detectors: Infrared (IR) detectors based on the pyroelectric effect. These are different from the quantum or photon detectors, which are based on the generation of an electron–hole pair in narrow-bandgap semiconductors.

Transducer: A device that can convert one form of energy into another.

Unpoled state: A ferroelectric or piezoelectric material that has not been subjected to the poling process.

Virgin state: A ferroelectric or piezoelectric material that has not been subjected to any electric field and therefore has a random configuration of domains and no net polarization.

Weiss domains: See **Ferroelectric domains**.

REFERENCES

Ahart, M., M. Somayazulu, R. E. Cohen, P. Ganesh, P. Dera, H. K. Mao, R. J. Hemley, et al. 2008. Origin of morphotropic phase boundaries in ferroelectrics. *Nature* 451: 545–9.

Askeland, D., and P. Fulay. 2006. *The Science and Engineering of Materials*. Washington, DC: Thomson.

Buchanan, R. C. 2004. *Ceramic Materials for Electronics*. New York: Marcel Dekker.

Chen, W., X. Yao, and X. Wei. 2007. Relaxor behavior of (Sr, Ba, Bi)TiO$_3$ ferroelectric ceramic. *Solid State Commun* 141:84–8.

Cheng, Z., and Q. Zhang. 2008. Field-activated electroactive polymers. *MRS Bull* 33:183–7.

Cross, E. 2004. Lead-free at last. *Nature* 432:24–5.

Dietze, M., and M. Es-Souni. 2008. Structural and functional properties of screen-printed PZT-PVDF-TrFE composites. *Sens Actuators* A143:329–34.

Gupta, M. C., and J. Ballato. 2007. *The Handbook of Photonics*. Boca Raton, FL: CRC Press.

Gururaja, T. R. 1996. Piezoelectric transducers for medical ultrasonic imaging, in *Proceedings of the Eighth IEEE International Symposium on Applications of Ferroelectrics,* pp. 259–265.

Harvey, C. J., J. M. Pilcher, R. J. Eckersley, M. J. K. Blomley, and D. O. Cosgrove. 2002. Advances in ultrasound. *Clin Radiol* 57:157–77.

Hench, L. L., and J. K. West. 1990. *Principles of Electronic Ceramics*. New York: Wiley.

Herbert, J. M. 1982. *Ferroelectric Transducers and Sensors*. Gordon and Breach Science. New York.

Ignatiev, A., Y. Q. Xu, N. J. Wu, D. Liu. 1998. Pyroelectric, ferroelectric, and dielectric properties of Mn- and Sb-doped PZT thin films for uncooled IR detectors. *Mater Sci Eng B* 56(2):191–4.

Kepler, R. G. 1995. Ferroelectric, Pyroelectric, and Piezoelectric Properties of Poly(vinylidene Fluoride). In *Ferroelectric Polymers: Chemistry, Physics, and Applications,* ed. H. S. Nalwa. New York: Marcel Dekker.

Labanca, M., F. Azzola, R. Vinci, and L. F. Rodella. 2008. Piezoelectric surgery: Twenty years of use. *Br J Oral Maxillofac Surg* 46(4):265–9.

Lang, S. B. 2005. Pyroelectricity: From ancient curiosity to modern imaging tool. *Phys Today* (August): 31–5.

Mehling, V., C. Tsakmakis, and D. Gross. 2007. Phenomenological model for the macroscopical material behavior of ferroelectric ceramics. *J Mech Phys Solids* 55:2106–41.

Moheimani, S. O. R., and A. J. Fleming. 2006. *Piezoelectric Transducers for Vibration Control and Damping.* Berlin: Springer.

Morgan Technical Ceramics, *Guide to Piezoelectric and Dielectric Ceramics.* Available at http://www.morganelectroceram-ics.com/pzbook.html.

Moulson, A. J., and J. M. Herbert. 2003. *Electroceramics: Materials, Properties, Applications.* New York: Wiley.

Nalwa, H. S. 1995. *Ferroelectric Polymers: Chemistry, Physics, and Applications.* New York: Marcel Dekker.

Neelkanta, P. 1995. *Handbook of Electromagnetic Materials.* Boca Raton, FL: CRC Press.

Osone, S., K. Brinkman, Y. Shimojo, and T. Iijima. 2008. Ferroelectric and piezoelectric properties of $Pb(Zr_xTi_{1-x})\,O_3$ thick films prepared by chemical solution deposition process. *Thin Solid Films* 516:4325–4329.

Osone, S., K. Brinkman, Y. Shimojo, and T. Iijima. 2010. Ferroelectric and piezoelectric properties of $Pb(Zr_xTi_{1-x})\,O_3$ thick films prepared by chemical solution deposition process. *Thin Solid Films* 516:4325–9

Park, S.-E., M. L. Mulvihill, P. D. Lopath, M. Zipparo, and T. R. Shrout. 1996. Crystal growth and ferroelectric related properties of $(1-x)Pb(A_{1/3}\,Nb_{2/3})O_3-xPbTiO_3$ (A = Zn^{2+}, Mg^{2+}). *IEEE Int Symp Appl Ferroelectr* 1:79–82.

Park, S.-E., and T. R. Shrout. 1997. Ultrahigh strain and piezoelectric behavior in relaxor based ferroelectric single crystals. *J Appl Phys* 82(4):1804–11.

Pilgrim, S. *Piezoelectric Materials: An Unheralded Component*, 11th ed. FabTech. Available at http://www.fabtech.org/white_papers/_a/piezoelectric_materials_an_unheralded_component.

Schlom, D. G., L.-Q. Chen, C.-B. Eom, K. M. Rabe, S. K. Streiffer, and J.-M. Triscone. 2007. Strain tuning of ferroelectric thin films. *Annu Rev Mater Res* 37:589–626.

Schneider, G. A. 2007. Influence of electric field and mechanical stresses on the fracture of ferroelectrics. *Annu Rev Mater Res* 37:491–538.

Shrout, T. S., and S. J. Zhang. 2007. Lead-free piezoelectric ceramics: Alternatives for PZT? *J Electroceram* 19:111–24.

Sodano, H. A., G. Park, and D. J. Inman. 2004. An investigation into the performance of macro fiber composites for sensing and structural vibration applications. *Mech Syst Signal Process* 18:683–97.

Song, G., V. Sethi, and H.-N. Li. 2006. Vibration control of civil structures using piezoceramic smart materials: A review. *Eng Struct* 28:1513–24.

Xu, Y. 1991. *Ferroelectric Materials and Their Applications.* Amsterdam: North Holland.

11 Magnetic Materials

11.1 INTRODUCTION

The word *magnet* has its origin in a *magnetic material* known as magnetite, which is a form of iron oxide or lodestone used as a magnetic compass. Lodestone was mined in the province of Magnesia. It is believed that among the minerals found in Magnesia (a part of Macedonia), magnesium carbonate was white, manganese dioxide was brown, and the third magnetite was black iron oxide. The magnetite was probably the first material known to be magnetic.

As we will see, in reality, *all* materials in this world are magnetic, that is, they respond to magnetic fields in some fashion. One objective of this chapter is to introduce the fundamental concepts related to magnetic materials. In this regard, we will explore the origin of magnetism in materials. We will define different types of magnetism in materials that include ferromagnetic and ferrimagnetic materials. The second objective is to explore different technologies based on the use of magnetic materials, including those used in information storage (e.g., magnetic hard disks). We will also briefly mention materials called *multiferroics*. These materials simultaneously exhibit two or more switchable properties, such as ferroelectric and ferromagnetic behaviors. Since the origins of the ferroelectricity and the ferromagnetism are different, there is a huge amount of interest in the basic science of multiferroics. Also, the magnetically tunable dielectric polarization and the electrically tunable magnetization have a great potential to offer a new paradigm of the device physics.

Most concepts regarding magnetic materials and technologies would be better followed if you have already learned the basics of linear dielectric materials and ferroelectric materials from Chapter 10. You would start to recognize that many of the equations we would deal with here for magnetic materials are very similar to those used for ferroelectrics. Thus, the so-called *phenomenology* underlying the group of dielectrics and ferroelectrics and that of ferromagnetic and ferrimagnetic materials is similar. However, it is important to keep in mind that although many of the phenomena and equations appear very similar, the physical origins of magnetic and dielectric behavior are quite different. The former is related to the spin motions of electrons and the latter is related to the nonuniform distribution of electric charge or the surface-bound electrons.

11.2 ORIGIN OF MAGNETISM

All materials are magnetic. This means that every material responds to an externally applied magnetic field in a specific manner. The origin of magnetism lies in a very basic principle, that is, a moving electric charge (namely electric current) produces a magnetic field. This connection between an electrical current and magnetism was probably first noted by Hans Oersted, who discovered that a compass needle placed next to a current-carrying wire was deflected in a perpendicular direction. Thus, it was found that a current-carrying coil behaved similar to a bar magnet and produced a magnetic moment (μ), which was found to be equal to the product of the current (I) and the area (A) of the loop (Figure 11.1). The magnetic moment is a building block of the magnet and conceptually similar to a dipole moment of dielectric materials in chapter 7. The unit of magnetic moment is ampere square meter ($A \cdot m^2$).

There are three origins of the electric charge motion responsible for magnetism: (i) electron orbiting nucleus (due to orbital angular momentum of electrons), (ii) electron spin (due to spin angular momentum of electrons, which is the most important mechanism for magnetic materials), (iii) nuclear spin (small contribution in general). Except for limited cases, the magnetic behavior of materials is mainly developed from two types of electron motions, namely spin and orbital (Figure 11.2). These motions can be compared with the motion of the earth that spins around its own axis while simultaneously having an orbit around the sun.

These motions of the electrons create a spin magnetic moment and an orbital magnetic moment. Niels Bohr first considered the magnetic behavior of an atom based on the planetary or orbital motion of electrons that resulted in the generation of a magnetic field and proposed the so-called Bohr's theory of magnetism. Later on, it was shown that the spin magnetic moment plays a critical role in determining the magnetic properties of materials.

Let's first take a look at the orbital magnetic moment originating from the orbital angular momentum of an electron, since this is easy to understand intuitively. A charged particle rotating in an orbit creates the angular magnetic moment (μ), which is given by

$$\mu = \frac{q \times L}{2 \times m} \tag{11.1}$$

In this equation, q is the charge, L is the angular momentum, and m is the mass of the charged particle. In Equation 11.1, L is equal to $r \cdot m \cdot v$ (r: the radius of rotation in an orbit, m: the mass of a charged particle, v: the velocity of a charged particle in an orbit). Bohr showed that the angular

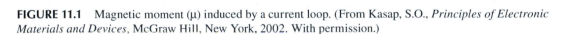

FIGURE 11.1   Magnetic moment (μ) induced by a current loop. (From Kasap, S.O., *Principles of Electronic Materials and Devices*, McGraw Hill, New York, 2002. With permission.)

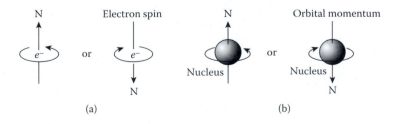

FIGURE 11.2   Spin motion (a) and orbital motion (b) of an electron. (From Askeland, D. and Fulay P., *The Science and Engineering of Materials*, Thomson, Washington, DC, 2006. With permission.)

momentum of an electron is quantized. In magnetism, a basic unit of magnetic moment is defined as shown in Equation 11.2. This is known as a *Bohr magneton* (μ_B):

$$\mu_B = \frac{h}{2\pi}\frac{q_e}{2m_e} = \hbar\frac{q_e}{2m_e} = 9.274 \times 10^{-24}\,\text{A}\cdot\text{m}^2 \tag{11.2}$$

In this equation, $\hbar = h/2\pi$, where h is the Planck's constant, q_e is the electronic charge, and m_e is the mass of the electron.

From Equation 11.1, another parameter known as the gyromagnetic ratio (γ) which relates the angular momentum to the magnetic moment is defined as:

$$\gamma = \frac{q}{2m} \tag{11.3}$$

The basic unit of charge that we often use is electronic charge ($q_e = 1.6 \times 10^{-19}$ C). Similarly, the spin magnetic moment from the spin motion of an electron can be expressed as $\mu = (-e/m)\,S$ in which S is the spin angular momentum.

Defining the Bohr magneton as a unit for the magnetic moment allows us to express the magnetic moment of an ion, atom, or a molecule. Now, let us see how we can calculate the total magnetic moment of a material. The total magnetic dipole moment per unit volume of a material is known as the *magnetization* (M). We can calculate the magnetization of a material by calculating the vector sum of different magnetizations of electrons and nuclei that make up the atom. The magnetization that develops in a material is similar to the development of polarization (P) in dielectrics, because the unit of the magnetic dipole moment is ampere square meter, and the unit of magnetization will be ampere square meter/cubic meter or ampere/meter.

The nucleus does *not* have a planetary motion. However, it does have a spin. There is a magnetic moment associated with the spin of the nucleus. However, the magnetic moment due to nuclear spin is very small ($\sim 10^{-3}\,\mu_B$) and can be ignored while calculating the total magnetic moment of an *atom* or an *ion*. Important applications where we make use of the nuclear magnetic moment are the techniques of nuclear magnetic resonance (NMR) and magnetic resonance imaging (MRI).

Thus, to calculate the total magnetic moment of an atom, we need to add up the magnetic moments of all the electrons. In doing so, we need to account for the magnetic moments due to both the angular motion and the spin of the electrons (Figure 11.2). The expressions for the magnetic moments involve concepts of quantum mechanics. Additional information regarding these concepts can be found in physics textbooks.

What is important is to recognize that most *free* atoms possess a net magnetic moment. The word free here means that atoms are not bonded to other atoms, that is, they are isolated and do not form part of any solid or liquid. Although most free atoms have a magnetic moment, most often but not always, when atoms bond to other atoms to form a solid or liquid, the electrons of different atoms interact, and the resultant material has no net magnetic moment. This is why, although all atoms contain electrons and each electron can be viewed as a magnet, most materials behave essentially as *nonmagnetic* materials.

Let us take a look at a few materials, such as transition elements (with incomplete 3d, 4d, or 5d orbitals), lanthanides (partially filled 4f orbital), or actinides (partially filled 5f orbital), wherein some of the electron shells are incomplete. In some of these materials, even after the atoms form a solid, the atoms in the resultant structure have a net effective magnetic moment. Similarly, ions of many elements may also have a net effective magnetic moment even when they are part of a material. In these solids with a partially-filled orbital, the magnetic moment resulting from the electron spin dominantly contributes to the magnetism and the role of the orbital magnetic moment is limited.

In manganese (Mn, atomic number $Z = 25$), for example, an electronic configuration is $1s^2 2s^2 2p^6$ $3s^2 3p^6 3d^5 4s^2$, and the orbitals from $1s^2$ to $3p^6$ are completely filled. The rules of quantum mechanics tell us that before filling up the 3d sublevel, the 4s level must be filled. Thus, after filling up the 3p orbital, we first add two electrons to the 4s level, and these are paired. This means that the only difference between the quantum numbers of the electrons in the 4s level is that one electron has an upward spin ($\uparrow$) whereas the other has a downward spin ($\downarrow$). In this case, five more electrons remain to fill the next energy level (3d). From quantum mechanics, the rule for filling the 3d subshell is that all electrons must remain unpaired until the subshell is half-filled; five electrons can thus have the same spin. However, they have different angular momentum which cancel each other. Therefore, in a manganese atom, there are five unpaired electrons with the same spin angular momentum. The spin of each unpaired electron produces a magnetic moment of one Bohr magneton. Therefore, a free Mn atom has a net magnetic moment of 5 μ_B (Table 11.1).

However, when Mn atoms form a Mn crystal, the magnetic moments of the spin motions cancel each other out. Thus, the atoms in a manganese *crystal* do *not* have a net magnetic moment. Similarly, in other compounds of Mn, bonding of Mn with other elements will have to be accounted for to find whether the resultant material will have any net magnetic moment. Most often, the spin and orbital magnetic moments are canceled out or *quenched* when elements form bonds with other atoms, resulting in zero net magnetic moment for most materials (paramagnetic in Figure 11.3). From a practical viewpoint, Mn is considered nonmagnetic, since the spin magnetic moments are randomly distributed. When we say that a material is not magnetic or that it is nonmagnetic, it means that the material has no net magnetic moment at zero magnetic field. Most materials that are considered nonmagnetic (e.g., copper, aluminum, silica, and polyethylene) exhibit a net magnetic moment only when they are exposed to an external magnetic field.

We now consider another example, namely iron (Fe). The electronic structure of an iron atom (atomic number Z = 26) is *expected* to be $1s^2 2s^2 2p^6 3s^2 3p^6 3d^8$. However, the actual electronic structure of iron is $1s^2 2s^2 2p^6 3s^2 3p^6 3d^6 4s^2$. This is because the 4s orbital (sublevel) is filled first. Thus, in this electronic configuration, the 3d orbital (capable of accepting 10 electrons) again remains incomplete. Out of these six electrons, the first five electrons in the 3d level can remain unpaired. However, one more electron is left over to fill the 3d level. Because the 3d orbital is now half-full, the sixth electron pairs up with one of the five electrons in the 3d level. Thus, in an iron (Fe) atom, four electrons in the 3d level remain unpaired (Table 11.1).

Each unpaired electron produces a magnetic dipole moment that is equal to a basic unit of magnetic moment, Bohr magneton (μ_B). Consequently, an iron atom has a net magnetic moment of 4 μ_B. In iron, when the atoms form a phase with a particular crystal structure, some, but not all, of these magnetic moments are quenched, that is, canceled. Therefore, a material such as iron has a net magnetic moment. In iron, all the magnetic moments of the atoms are in the same direction, due to an exchange interaction between neighboring spin magnetic moments. A material in which all the magnetic moments of atoms or ions are aligned in the same direction is known as a *ferromagnetic material*. It possesses strong net magnetic moments without an external magnetic field. (Figure 11.3).

TABLE 11.1
Pairing of 3d and 4s Electrons in Different Transition Elements

Element	Atomic Number (Z)	3d-Electron Spin Pairing					4s-Electron Spin Pairing
Sc	21	$\uparrow$					$\uparrow\downarrow$
Ti	22	$\uparrow$	$\uparrow$				$\uparrow\downarrow$
V	23	$\uparrow$	$\uparrow$	$\uparrow$			$\uparrow\downarrow$
Cr	24	$\uparrow$	$\uparrow$	$\uparrow$	$\uparrow$	$\uparrow$	$\uparrow$
Mn	25	$\uparrow$	$\uparrow$	$\uparrow$	$\uparrow$	$\uparrow$	$\uparrow\downarrow$
Fe	26	$\uparrow\downarrow$	$\uparrow$	$\uparrow$	$\uparrow$	$\uparrow$	$\uparrow\downarrow$
Co	27	$\uparrow\downarrow$	$\uparrow\downarrow$	$\uparrow$	$\uparrow$	$\uparrow$	$\uparrow\downarrow$
Ni	28	$\uparrow\downarrow$	$\uparrow\downarrow$	$\uparrow$	$\uparrow$	$\uparrow$	$\uparrow\downarrow$
Cu	29	$\uparrow\downarrow$	$\uparrow\downarrow$	$\uparrow\downarrow$	$\uparrow\downarrow$	$\uparrow\downarrow$	$\uparrow$

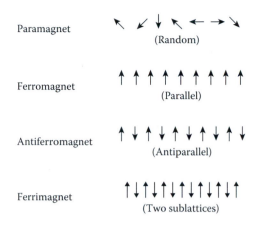

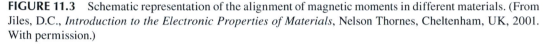

FIGURE 11.3 Schematic representation of the alignment of magnetic moments in different materials. (From Jiles, D.C., *Introduction to the Electronic Properties of Materials*, Nelson Thornes, Cheltenham, UK, 2001. With permission.)

In some materials (e.g., iron oxide [Fe_3O_4]), the ions at the different locations of the unit cell have magnetic moments that are aligned in opposite directions or are antiparallel. However, the magnetic moments of two sublattices are not completely canceled out, because two sublattices do not have the same magnitude of the magnetic moments (see Example 11.2). A material in which the magnetic moments of atoms or ions are antiparallel but are not canceled out is known as a *ferrimagnetic material*. The alignment of electron spins in ferrimagnetic materials is shown schematically in Figure 11.3.

Note that ferromagnetic or ferrimagnetic materials do *not* have to contain iron or other ferromagnetic elements (e.g., Ni, Co, Gd). For example, Cu_2MnAl, ZrZn, and InSb are ferromagnetic, even though the latter two are ferromagnetic only at very low temperatures (O'Handley 1999).

If the spin magnetic moments due to different ions or atoms are completely canceled out, the material is known as *antiferromagnetic* (e.g., Cr, α-Mn, and MnO). In Cr with a body-centered cubic (BCC) crystal structure, the magnetic moment of the center atom is canceled out by the magnetic moments of the corner atoms and the magnetic behavior is antiferromagnetic. Magnetic moments associated with some atoms and ions are shown in Table 11.2.

The following examples illustrate how to calculate the magnetic moments of ions.

Example 11.1: Magnetic Moments of Ferrous (Fe^{2+}) and Ferric (Fe^{3+}) Ions

The atomic number of iron (Fe) is 26. From Table 11.1, what are the magnetic moments of (a) ferrous (Fe^{2+}) and (b) ferric (Fe^{3+}) ions?

Solution

1. As seen earlier, the electronic structure of an iron atom is $1s^22s^22p^63s^23p^63d^64s^2$, with four unpaired electrons in the 3d level. When a ferrous ion (Fe^{2+}) is formed, two electrons are removed from an iron atom. This occurs by removing the two electrons that belong to the 4s energy level. Thus, the electronic configuration of the divalent ferrous (Fe^{2+}) ion is $1s^22s^22p^63s^23p^63d^6$. The first five electrons in the 3d level remain unpaired. The sixth electron is paired with one of the five electrons. This again leaves us the Fe^{2+} ion with four unpaired electrons. Consequently, the magnetic moment of a free ferrous (Fe^{2+}) ion is 4 μ_B.

2. For the ferric (Fe^{3+}) ion, we can think of starting with an Fe^{2+} ion and removing one more electron (or, starting with a neutral iron atom and taking out a total of three electrons). This third electron comes from one of the paired electrons in Fe^{2+} to give us a total of five unpaired electrons. Thus, the magnetic moment of a trivalent or ferric ion (Fe^{3+})

TABLE 11.2

Magnetic Moments (μ_B) of Some Atoms and Ions

Atom	
Iron	4
Cobalt	3
Nickel	2

Ion	
Fe^{2+} (ferrous)	4
Fe^{3+} (ferric)	5
Co^{2+}	3
Ni^{2+}	2
Mn^{2+}	5
Mg^{2+}, Zn^{2+}, and Li^+	0

Source: Goldman, A., *Handbook of Modern Ferromagnetic Materials*, Kluwer, Boston, 1999. With permission.

is 5 μ_B. In Example 11.2, we will see how the magnetic moments of the Fe^{2+} and Fe^{3+} ions can be used to calculate the total magnetization per unit volume in Fe_3O_4.

Example 11.2: Ferrimagnetism in Iron Oxide

From the magnetic moments of the ferrous and ferric ions, calculate the net magnetic moment per formula in ferrimagnetic iron oxide (Fe_3O_4).

Solution

Another way to write the formula for this material is $FeO:Fe_2O_3$. This representation distinguishes between the divalent and trivalent forms of iron. For one formula unit of this material, the number of unpaired electrons will be 4 (from Fe^{2+}) and 10 (five each from a total of two Fe^{3+}) ions. Thus, the total number of unpaired electrons is $4 + 10 = 14$ μ_B. However, this material is ferrimagnetic.

In Fe_3O_4, one-half of the ferric (Fe^{3+}) ions occupy the so-called tetrahedral sites (Chapter 2). The other half of the Fe^{3+} ions occupy the so-called octahedral sites. The spin magnetic moments of the ferric ions located in these different crystallographic sites are antiparallel, and they cancel each other out. The only magnetic moments remaining in Fe_3O_4 are due to the divalent ferrous ions that occupy the octahedral sites (same type of site that was occupied by one-half of the Fe^{3+} ions). Thus, in Fe_3O_4, the actual magnetic moment per unit formula will be only 4 μ_B since there is only one Fe^{2+} ion per formula unit (see Table 11.1).

11.3 MAGNETIZATION (*M*), FLUX DENSITY (*B*), MAGNETIC SUSCEPTIBILITY (χ_M), PERMEABILITY (μ), AND RELATIVE MAGNETIC PERMEABILITY (μ_r)

We now examine some of the relationships between the applied magnetic field (the cause) and the magnetization created (the effect). The equations will be similar to those we saw in Chapter 7 for dielectrics. However, there are fundamental and important differences between the origins of these phenomena. For example, ferromagnetic behavior is seen in amorphous materials, because the magnetic response is linked to the motion of electrons. Ferroelectricity, on the contrary, is seen only in crystalline materials or crystalline regions in materials such as polyvinylidene fluoride (PVDF). Ferromagnetism and ferrimagnetism are seen in materials that are either conductors or insulators,

TABLE 11.3

Equivalence between Properties and Phenomena of Dielectric and Magnetic Materials

Electric	Magnetic
Electric field (E)	Magnetizing field (H)
Polarization (P)	Magnetization (M)
Dielectric displacement (D)	Magnetic induction or flux density (B)
Dielectric susceptibility (χ_e)	Magnetic susceptibility (χ_m)
Dielectric constant (ε_r)	Relative magnetic permeability (μ_r)
Ferroelectricity	Ferromagnetism, ferrimagnetism
Linear dielectric behavior	Paramagnetism, diamagnetism
Relaxor ferroelectricity	Superparamagnetism
Electrostriction	Magnetostriction
Piezoelectricity	Piezomagnetism

whereas ferroelectric behavior is seen only in insulating materials. Some of these similarities and differences will become clearer as we discuss these materials in further detail.

When *magnetizing field* (H) is applied to a material, it creates magnetization (M). This is equivalent to the application of an electric field (E), which creates a polarization (P). When we use the term magnetizing field (H), we are referring to an *externally generated magnetic field*, which is created using either a current-carrying coil or a permanent magnet.

Magnetizing field (H) results in the creation of a *magnetic flux density* (B) or *magnetic induction* inside the material. When a magnetic flux (ϕ) is created by magnetizing field (H), there would be a certain number of magnetic flux lines per unit area. The number of flux lines per unit area is known as the magnetic flux density (B) or magnetic induction. Consider the flux density as the intensity of magnetic field at any given location in a magnetic material. The unit for flux density is Weber/square meter, written as Wb/m^2, which is also equivalent to Tesla.

The equivalent of the magnetic flux density generated due to the magnetizing field is the dielectric displacement (D) created by the application of an electric field. This equivalence between different electric and magnetic quantities is shown in Table 11.3. Before moving forward, note that both H and B are sometimes called magnetic field. This is because both H and B play roles that electric field (E) does in Maxwell's equations on the electric magnetic wave (Section 8.1). To prevent unnecessary confusion, we will call H and B as magnetizing field and magnetic flux density (or magnetic induction) in this book.

11.3.1 Magnetizing Field (H), Magnetization (M), and Flux Density (B)

Magnetizing field is often created by passing electrical current through a wire. Uniform magnetizing fields can be created using a toroidal solenoid (Figure 11.4a) or a solenoid made using an insulated conductive wire wrapped around a cylinder (Figure 11.4b). Consider a coil of length l with n turns, carrying a current I; then, the magnetizing field created is given by

$$H = \frac{nI}{l} \tag{11.4}$$

If $N = \dfrac{n}{l}$, then

$$H = N \times I \tag{11.5}$$

Equation 11.5 indicates that the unit of H is in amperes per meter (A/m).

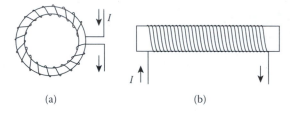

(a)   (b)

FIGURE 11.4 Illustration of (a) a toroidal solenoid and (b) a solenoid made using a current-carrying coil. (With kind permission from du Tremolet de Lacheisserie, E., et al., *Magnetism: Fundamentals*, Springer Science+Business Media, 2002.)

Recall the relationship between the dielectric displacement (*D*), polarization (*P*), and electric field (*E*), which is

$$D = \varepsilon_0 E + P \tag{11.6}$$

In magnetic materials, the magnetizing field (*H*) is the *cause*, and the *effect* is magnetization (*M*). The amount of magnetization (*M*) created within a material will depend on both the composition of the material, that is, how susceptible the material is to magnetizing field and the strength of magnetizing field.

We can show that the magnetic flux density created within the material (B_{int}) is related to the magnetization *M* and magnetizing field *H* by the following equation:

$$B_{int} = \mu_0 H + \mu_0 M \tag{11.7}$$

In this equation, the magnetic permeability of free space (μ_0) is given by

$$\mu_0 = 4\pi \times 10^{-7} \text{ Wb/A} \cdot \text{m or H/m}$$

Because the unit of magnetic induction (*B*) is Wb/m^2, and the unit of magnetizing field (*H*) is ampere/meter, the unit for magnetic permeability (μ) is Weber/ampere meter or Henry/meter (H/m). Also, note that in Section 11.2, the symbol μ has been used for the magnetic moment.

The first term on the right-hand side of Equation 11.7, that is, $\mu_0 H$, can be described as the magnetic induction B_0 that extends to the outside of the solenoid and is called the external magnetic induction.

Thus, the total magnetic induction created inside the material (B_{int}) is a sum of two factors, namely the externally applied induction and the magnetization created inside the material, that is, Equation 11.7.

In vacuum, there is no material, and hence, magnetization created inside a current-carrying coil is zero ($\mu_0 M = 0$). Then, the magnetic flux density (B_{int}) and the magnetizing field (*H*) are related by the following equation:

$$B_{int} = B_0 = \mu_0 H \text{ (for vacuum)} \tag{11.8}$$

We neglected the subscript B_{int} because there is no material inside the solenoid. This indicates that the magnetic flux inside the empty solenoid is the same as the external magnetic induction.

Now, consider placing different types of magnetic materials inside the solenoid and see how the magnetic flux density inside the solenoid (B_{int}) will change with a reference to B_0. For example, we can place a piece of copper or aluminum, silver, superconductor, or iron inside the solenoid and expose it to the magnetic field (*H*). Also, imagine that we have the magical ability to look into this material and observe the magnetic flux lines. We will observe that these different materials *react* to the external magnetic induction in different ways.

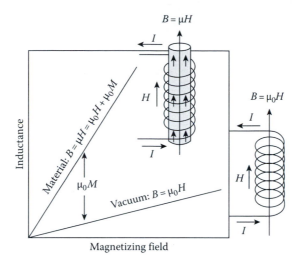

FIGURE 11.5 A current-carrying coil generates higher flux density (B) inside a magnetic material. (From Askeland, D. and Fulay P., *The Science and Engineering of Materials*, Thomson, Washington, DC, 2006. With permission.)

In the case of the superconductor, the flux density or induction inside the material is actually zero, that is, $B_{int} = 0$, because the induced magnetization ($\mu_0 M$) in the superconductor cancels B_0 (i.e., $\mu_0 M = -B_0$). This means that the entire magnetic flux is excluded from the inside of a superconductor in a superconducting state. For silver, the B_{int} or the flux density will be slightly less than B_0 ($B_0 + \mu_0 M < B_0$), that is, some of the flux is excluded from the inside of the material. Such materials in which the B_{int} is less than B_0 are known as *diamagnetic materials* which are the fifth class of magnetic materials in addition to the four classes summarized in Figure 11.3. Another way to state the diamagnetisim is to say that the flux lines will be diluted inside a diamagnetic material such as silver and completely excluded from the inside of a superconductor.

For materials such as copper and aluminum, we will see that the B_{int} will be slightly *greater* than B_0. These materials are known as *paramagnetic* materials. In these materials, the flux lines will be slightly concentrated inside the material.

For materials such as iron (or any other ferromagnetic or ferrimagnetic material), we will find that the flux lines are heavily concentrated, that is, the flux density inside the material (B_{int}) will be significantly higher than the external magnetic induction (B_0).

For the same value of the magnetizing field (H), the magnetic flux density (B_{int}) inside a ferromagnetic or ferrimagnetic material will be considerably higher than that for either vacuum or a diamagnetic or paramagnetic material (Figure 11.5 and Example 11.3). This is similar to the dielectric displacement (D) inside a parallel-plate capacitor being higher for a capacitor filled with a material having a high dielectric constant in comparison to the value for one filled with vacuum. In ferromagnetic and ferrimagnetic materials, the flux density caused by induced magnetization dominates (i.e., $\mu_0 M \gg \mu_0 H$).

We will discuss these different classes of materials in detail. For now, let us revert to see how the flux density created inside the material (B_{int}) relates to the external induction (B_0).

11.3.2 Magnetic Susceptibility (χ_M) and Magnetic Permeability (μ)

Similar to dielectrics, we define *magnetic susceptibility* (χ_m) as a property that relates magnetization M (the effect) to the magnetizing field H (the cause). Thus, magnetic susceptibility is defined as

$$\chi_m = \frac{M}{H} \tag{11.9}$$

Susceptibility values represent the tendency of a magnetic material to *fall for* the presence of magnetizing field.

For diamagnetic and paramagnetic materials, the internal flux density (B_{int}) created can be shown to be linearly proportional to the magnetizing field (H) according to the equation

$$B_{int} = \mu H \tag{11.10}$$

In this equation, the symbol μ is the magnetic permeability of a material. We note that similar to the dielectric constant (ε_r), magnetic permeability and susceptibility are tensors that relate two vectors. We will treat them here as scalar quantities. Equation 11.10 indicates that the magnetic induction created within a diamagnetic or paramagnetic material depends on the strength of the external magnetic field applied and the magnetic permeability of the material.

It is easier to express the magnetic permeability of a material relative to the magnetic permeability of free space. We define the *relative magnetic permeability* (μ_r) as the ratio μ/μ_0. We rewrite Equation 11.10 as follows:

$$B_{int} = \mu_0 \mu_r H \tag{11.11}$$

We can now establish a relationship between the magnetic susceptibility (χ_m) and the relative magnetic permeability (μ_r).

We start with Equation 11.7:

$$B_{int} = B_0 + \mu_0 M = \mu_0 H + \mu_0 M$$

For diamagnetic and paramagnetic materials, the magnetic field created *inside the material* (H_{int}) is approximately equal to the externally applied magnetic field (H).

From Equations 11.7 and 11.11, we get

$$B_{int} = \mu_0 \mu_r H = (\mu_0 H + \mu_0 M)$$

or, solving for M, we get

$$M = (\mu_r - 1)H \tag{11.12}$$

From Equations 11.9 and 11.12, we get a relationship between relative magnetic permeability (μ_r) and magnetic susceptibility (χ_m):

$$\chi_m = \mu_r - 1 \tag{11.13}$$

This is similar to dielectrics and can be written as follows:

$$\mu_r = 1 + \chi_m \tag{11.14}$$

11.3.3 DEMAGNETIZING FIELDS

Note that in deriving Equations 11.7 and 11.12, we assumed that for diamagnetic and paramagnetic materials, the *magnetizing field* created inside the material (H_{int}) filling the coil is approximately the same as the magnetizing field (H) that electric current generates inside an empty

---

TABLE 11.4

Demagnetizing Factors (N_d) for Different Shapes

Geometry	l/d	N_d
Toroid		0
Long cylinder		0
Cylinder	20	0.006
Cylinder	10	0.017
Cylinder	5	0.040
Cylinder	1	0.27
Sphere		0.333

Source: Buschow, K.H. and De Boer, F.R., *Physics of Magnetism and Magnetic Materials*, Kluwer, Boston, 2003. With permission.

---

coil (see Figure 11.5). This is *not* true for ferromagnetic and ferrimagnetic materials exhibiting large magnetization (*M*)—commonly known as magnetic materials. This is because inside these magnetic materials (when they are in their ferromagnetic or ferrimagnetic state), large magnetization produces internal demagnetizing field that partially nullify the magnetizing field (*H*) inside ferromagnetic or ferrimagnetics. This demagnetizing field causes the *magnetic flux lines* (showing *B*) *inside* the magnetic material to move in a direction *opposite* to that of the *magnetizing field lines* (showing *H*) inside the magnetic material. This can be shown using Maxwell's equations. Please pay careful attention to the arrows showing induction (*B*) in Figure 11.5.

Note that magnetizing field lines (depicting *H*) and flux lines (showing *B*) have a same direction outside the ferromagnetic or ferrimagnetic material.

However, the magnetizing field (*H*) and the magnetization (*M*) are antiparallel inside the ferromagnetic or ferrimagnetic material and magnetic induction (*B*) is actually reduced by a certain factor known as the *demagnetizing factor* (N_d). Instead of writing $B_{int} = \mu_0 H + \mu_0 M$ (Equation 11.7), as we did for diamagnetic and paramagnetic materials, we write the internal flux density for a ferromagnetic or ferrimagnetic material as

$$B_{int} = \mu_0 H + \mu_0 (1 - N_d) M \qquad (11.15)$$

$N_d M$ in Equation (11.15) is the demagnetizing field (H_d) which decreases the magnetic induction (*B*) by partially neutralizing the driving force (*H*) for the magnetization (*M*). The demagnetizing factor (N_d) is dependent on the geometry of magnetic materials (toroid, long cylinder, sphere, etc.) and ranges from 0 to 1 (Table 11.4).

As can be seen from these values of N_d, in many sample calculations, we often make use of toroid or long cylinders for which the N_d is zero or small so that the effect of the demagnetizing fields does not have to be considered. The demagnetizing factors indicate the ease with which magnetization within ferromagnetic or ferrimagnetic materials can be switched to different directions to decrease the magnetic flux density outside the materials.

11.3.4 FLUX DENSITY IN FERROMAGNETIC AND FERRIMAGNETIC MATERIALS

Thus, the magnetic flux density (*B*) induced in a magnetic material is related to the magnetizing field (*H*) by its magnetic permeability (μ) (Figure 11.5). Note that for convenience, in this diagram for ferromagnetic or ferrimagnetic materials, only the initial portion of the *B–H* relationship is shown. In reality, these materials are magnetically nonlinear (Figure 11.6). The more permeable

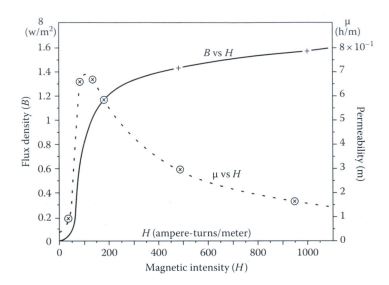

FIGURE 11.6 Dependence of magnetic permeability (μ) of iron on magnetizing field (H) (dashed curve). The development of initial magnetization is also shown (thick line) as the flux density. (From Jiles, D.C., *Introduction to Magnetism and Magnetic Materials*, Chapman and Hall, London, 1991. With permission.)

the material, the higher will be the flux density inside the material. This is what the term *magnetic permeability* means. In some ways, it is useful to think of magnetic permeability (μ) as being similar to electrical conductivity—the higher the magnetic permeability (μ), the higher the ability of a material to carry magnetic field.

The relative magnetic permeability of ferromagnetic and ferrimagnetic materials is large (~100–600,000) and field dependent. This is why, similar to ferroelectrics, ferromagnetic and ferrimagnetic materials are considered *nonlinear magnetic materials.*

The field dependence of the permeability (μ) of a magnetic field (Henry/meter) for a magnet of annealed iron (Fe) is shown in Figure 11.6.

The following example illustrates the effect of inserting a magnetic material on the flux density created within a solenoid.

Example 11.3: Flux Density for Air and Iron Core

The applied magnetic field (H) for a toroidal solenoid (Figure 11.4a) is given by $N \times I$, where N is the number of turns per meter and I is the current in amperes (Equation 11.5). (a) What is the flux density (B) created for a current of 0.5 A if the toroidal solenoid has an air core? (b) What is the flux density (B) if the core is filled with iron? Assume that the relative magnetic permeability of iron is 900 for the field generated using 0.5 A. The average circumference of the solenoid is 75 cm, and there are 1000 turns.

Solution

1. The average circumference of the solenoid winding is $I = 75$ cm, and there are 1000 turns. Therefore, the value of N (turns per meter) = 1000 turns/(0.75 m) = 1333.33.

    The magnetic field created is $H = N \times I = (1333.33) \times (0.5\ A) = 666.66$ A/m.

    The flux density generated from this field in air will be $(B_{air}) = (\mu_0 H) = (4\pi \times 10^{-7}\ H/m)$ (666.66 A/m) = 8.38×10^{-4} T. We assumed that the magnetic permeability of air (μ_{air}) is the same as that of vacuum (μ_0). A very small level of flux density (B) is generated inside the toroidal solenoid filled with air. For comparison, this is the type of flux density created by the earth's natural magnetic field.

2. Now the solenoid gap is filled with iron. The relative magnetic permeability (μ_r) of iron is 1000 for the level of magnetic field created inside the solenoid. In magnetic circuit design, such values of μ_r are usually obtained from data similar to those shown in Figure 11.6.

Therefore, for this scenario, the flux density created inside the solenoid will be (from Equation 11.11) $B = \mu_0 \mu_r H = (4\pi \times 10^{-7} \text{ H/m}) (1000) (666.66 \text{ A/m}) = 0.838 \text{ T}$. This is a relatively high flux density created inside the solenoid using iron, which is a ferromagnetic material.

11.4 CLASSIFICATION OF MAGNETIC MATERIALS

One way of classification described here is based on the relative values and signs of the magnetic susceptibility: diamagnetic, paramagnetic, antiferromagnetic, ferromagnetic, ferrimagnetic, superconductivity.

11.4.1 Diamagnetic Materials

In diamagnetic materials, when an external magnetic field is applied, the material tries to exclude the magnetic flux. Many materials (e.g., most metals, inert gases, and organic compounds) often classified as nonmagnetic actually exhibit a weak diamagnetic response. Diamagnetism is an inherent property of the orbital motion of individual electrons in a field (Goldman 1999). It is basically a Lenz's law–like effect, occurring at an atomic scale. The relative magnetic permeability (μ_r) of diamagnetic materials is <1. This means that the magnetic susceptibility (χ_m) is negative (Equation 11.13).

Note that atoms in the diamagnetic materials do not have net magnetic dipoles under zero magnetizing field. Only when a magnetizing field is applied, the magnetic moment is produced at an atomic level. In diamagnetic materials, each atom has an even number of valence electrons and there are no net magnetic moments in the atomic level. Then the applied magnetic field distorts electron orbital motion, leading to negative magnetic susceptibility (Table 11.5).

TABLE 11.5
Magnetic Susceptibilities of Some Elements

Element	Magnetic Susceptibility (χ_m)
Antimony (Sb)	-7.0×10^{-5}
Bismuth (Bi)	-1.7×10^{-1}
Copper	-0.94×10^{-5}
Lead	-1.7×10^{-5}
Silver	-2.6×10^{-5}
Aluminum	$+0.21 \times 10^{-4}$
Neodymium	3.0×10^{-3}
Palladium	8.2×10^{-1}
Platinum	2.9×10^{-5}
Oxygen (under NTP conditions)	17.9×10^{-7}

Source: Scott, W.T., *Physics of Electricity and Magnetism*, Wiley, New York, 1959; Goldman, A., *Handbook of Modern Ferromagnetic Materials*, Kluwer, Boston, MA, 1999. With permission.

Therefore, the diamagnetic materials reduce the effect of the magnetizing field. Since the magnetic moment of the diamagnetic materials comes from the precession motion of electrons, the susceptibility of the diamagnetic materials is independent of temperature. In contrast, other types of magnetic behaviors depend on temperature. Diamagnetic materials actually tend to expel the magnetic flux lines out from their interior. Certain materials known as superconductors are perfect diamagnetic materials. This means that in superconductors, the magnetic field can be completely expelled. This occurs when the temperature T of a superconductor is less than the so-called critical temperature (T_c), at which the superconducting state exists. This critical temperature (T_c) below which a material exhibits superconductivity is not related to the *Curie temperature* (T_c) of ferromagnetic and ferrimagnetic materials (see Section 11.6.1). The magnetic susceptibility of superconductors is $\chi_m = -1$. This important difference between a conductor and a superconductor is shown in Figure 11.7.

The expulsion of magnetic flux lines due to the diamagnetic behavior of a superconductor is known as the *Meissner effect*, which is the basis of magnetic levitation using superconductors. When a permanent magnet is brought near a superconductor (when it is in this superconducting state), it tries to expel the magnetic field. The result is that the magnet can float or levitate over the superconductor (Figure 11.8).

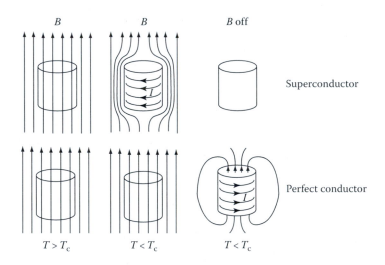

FIGURE 11.7   Expulsion of magnetic flux lines in a superconductor at $T < T_c$. In a conductor, the magnetic flux lines continue to penetrate. (From Kasap, S.O., *Principles of Electronic Materials and Devices*, McGraw Hill, New York, 2002. With permission.)

FIGURE 11.8   Illustration of Meissner effect in superconductors. (From http://www.wondermagnet.com/ images/super5.jpg. With permission.)

11.4.2 Paramagnetic Materials

When a paramagnetic material is placed inside the coil, the flux density inside (B_{int}) is slightly larger than B_0. This effect originates from the atoms or ions that have unpaired electrons, and it leads to a net magnetic moment for the free atom or an ion. The atoms or ions are known as *paramagnetic atoms* or *paramagnetic ions*. Examples include materials such as aluminum. The atoms in these materials have a permanent magnetic moment. When a magnetic field is applied, these magnetic moments get oriented and create a magnetization in the same direction as the applied field. Thus, in paramagnetic materials, the susceptibility is very small ($\chi_m \sim + 10^{-6} - 10^{-3}$) but positive. As temperature increases, the thermal motion becomes important and the susceptibility of the paramagnetic materials decreases.

This can be shown by a demonstration in which a stream of liquid oxygen is deflected by a strong permanent magnet held next to it (Featonby 2005). As a side note, the unpaired electrons that create a magnetic moment in oxygen are also related to the absorption of red light (~630-nm wavelength) and impart a blue color to liquid oxygen! Paramagnetic materials show a slight attraction in the presence of a permanent magnet. However, the paramagnetic response is so small that, for all practical purposes, these materials are also considered nonmagnetic. Another important point is that at the Curie temperature, ferromagnetic and ferrimagnetic materials transform into paramagnetic materials.

The relative value and sign of magnetic susceptibility (χ_m) or magnetic permeability (μ) are often used to classify magnetic materials. The relationships between magnetic field and magnetization for different materials are shown in Figure 11.9. Note that this diagram is not to scale. Also, it does *not* show the nonlinear nature of ferromagnetic and ferrimagnetic materials. In principle, if you apply a very high magnetic field, the inductance of the paramagnetic materials can be comparable to that of the ferromagnetic and ferromagnetic materials, since the inductance of the ferromagnetic and ferromagnetic materials is saturated at high magnetic field (Figure 11.13).

The magnetic susceptibility values for some diamagnetic and paramagnetic elements are presented in Table 11.5.

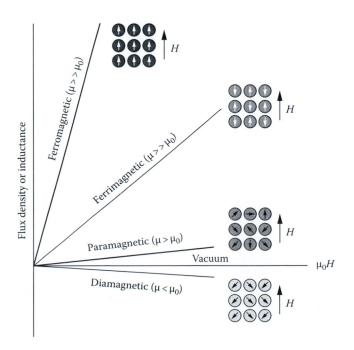

FIGURE 11.9 Magnetic permeability for different types of magnetic materials. Values are not to scale. (From Askeland, D. and Fulay P., *The Science and Engineering of Materials*, Thomson, Washington, DC, 2006. With permission.)

11.4.3 SUPERPARAMAGNETIC MATERIALS

Another technologically important class of materials is *superparamagnetic materials*. The superparamagnetism is seen in ferromagnetic and ferrimagnetic materials fabricated in the form of nanoparticles or nanoscale structures, that is, the grain size of a polycrystalline material becomes of the order of a few nanometers. For example, in the *bulk* form, BCC (Chapter 2) iron (Fe) is ferromagnetic. As we decrease the size of iron particles or grains from the bulk to a few micrometers and then down to a few nanometers, the thermal energy at room temperature $(\sim k_B T)$ becomes comparable with the magnetic energy. Such nanoparticles or nanoscale grains eventually become single domain. Thus, a nanoparticle or a nanosized grain ($\sim$5–10 nm) of iron or similar ferromagnetic or ferrimagnetic materials has a net dipole moment. However, if the energy associated with the magnetic moment of this single-domain nanoparticle or grain is comparable with the thermal energy, the magnetic dipole moment rotates rapidly at room temperature. Single-domain nanoparticles and nanograined materials of ferromagnetic and ferrimagnetic materials consequently behave as if they have no net magnetic moment. This behavior, in which a material that is originally either ferromagnetic or ferrimagnetic in its bulk form behaves similar to a paramagnetic material at a nanoscale level, is known as *superparamagnetism*. The materials in this nanoscale form that exhibit such behavior are known as superparamagnetic materials.

The existence of this effect imposes a limit on how small the grain size of magnetic nanostructures that are used for magnetic data storage can be.

The exact dimension or size at which the thermal energy can exceed the magnetic energy and render the material superparamagnetic depends on the composition, coercivity, and *magnetic anisotropy* of the material. However, in general terms, the superparamagnetic effects become dominant once the grain or particle size becomes of the order of less than 100 nm and more typically $\sim$10 nm.

Magnetic materials form the basis of different classes of what can be described as passively *smart materials*. A smart material is one which has properties that are controllable using an external stimulus (temperature, stress, electric field, magnetic field, etc.).

Materials known as *ferrofluids* are based on the use of superparamagnetic materials (Figure 11.10a). A ferrofluid is a stable dispersion of superparamagnetic particles (Figure 11.10b). Ferrofluids behave as *liquid magnets*, that is, when a permanent magnet is placed next to them, the entire material shows a body motion. The carrier fluid is typically an organic substance, but it could also be water. Because the particles are colloidal in nature and surface-active agents (surfactants) are used, the particles of magnetic materials used to make ferrofluids do not settle out easily or form agglomerates. Commercially available ferrofluids are made using iron, iron oxide, and other materials. Ferrofluids are used commercially as cooling media. The advantage in using them as such, as for example, for cooling of high-wattage speakers, is that this heat-transfer fluid can be held in place by the magnetic fields created by the permanent magnets used to make the audio speakers (Rosensweig et al. 2008).

The formation of *spikes* in a ferrofluid caused by its movement in response to a nonuniform magnetic field (created by a permanent magnet held below, as shown in Figure 11.10a) makes an interesting demonstration of the behavior of these materials.

Superparamagnetic nanoparticles of materials such as Fe_3O_4 have also been functionalized through surface modification so that biological species, molecules, cells, and so on can attach to them. These biological species can then be separated from the rest of the material using a permanent magnet.

Ferrofluids are distinctly *different* from the so-called *magnetorheological* (MR) *fluids* that are based on dispersions of larger (not superparamagnetic) and magnetically multidomain ferromagnetic or ferrimagnetic particles with very low coercivity. Applications of MR fluids include vibration control and have been commercialized (Figure 11.10c).

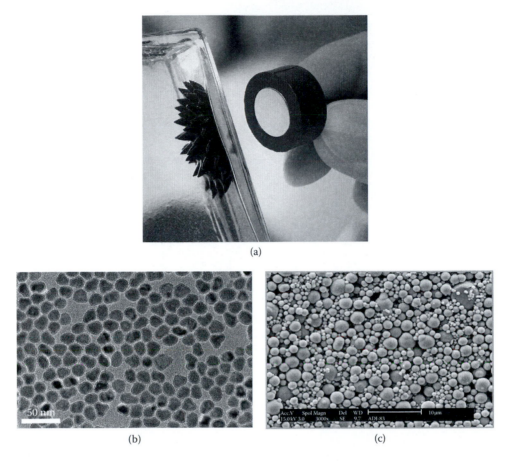

(a)

(b)   (c)

FIGURE 11.10   (a) Formation of spikes in a ferrofluid. (b) Superparamagnetic nanoparticles of iron oxide (Fe_3O_4). (Courtesy of Ferrotec Corporation and Dr. Kuldip Raj, Ferrotec Corporation, USA.) (c) Soft magnetic, multidomain iron (Fe) particles (~2 μm) used in magnetorheological fluids. (Courtesy of Pradeep Fulay, University of Pittsburgh, PA.)

11.4.4   Antiferromagnetic Materials

In antiferromagnetic materials, the individual atom or ion has a net spin magnetic moment, however, these magnetic moments are aligned in an antiparallel way. If then, the spin magnetic moments cancel out completely. This is because of the way by which the ions or atoms are arranged in a given crystal structure. Consequently, antiferromagnetic materials have no net magnetization in the absence of applied magnetic field (H). Chromium (Cr) and α-manganese (α-Mn) are examples of antiferromagnetic *elements*. An example of an antiferromagnetic *compound* is manganese oxide (MnO), in which the spin magnetic moments of Mn^{2+} ions located at different crystallographic sites cancel each other out (Figure 11.11). Note that in this compound, oxygen ions do not have a net magnetic moment. However, through a quantum mechanical interaction, they mediate the magnetic moments of Mn^{2+} ions. In the antiferromagnetic materials, susceptibility (χ_m) is slightly larger than 0 and dependent upon the applied field intensity.

If we heat an antiferromagnetic material to a high temperature, eventually all the antiferromagnetic coupling will be destroyed and the material will become paramagnetic. When we cool an antiferromagnetic material from a high temperature, known as the *Néel temperature* (T_N or θ_N), at which it is in a paramagnetic state, the antiferromagnetic coupling will set in on the sublattice. This will be discussed later in Section 11.5 (Figure 11.12c).

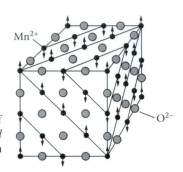

FIGURE 11.11 Antiferromagnetic coupling of magnetic moments of Mn^{2+} ions in MnO. (From Askeland, D. and Fulay P., *The Science and Engineering of Materials*, Thomson, Washington, DC, 2006. With permission.)

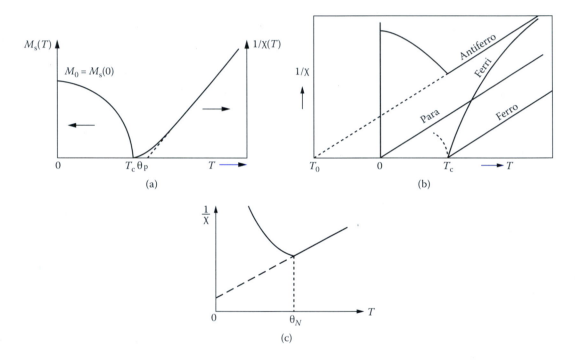

FIGURE 11.12 (a) Curie–Weiss law behavior for ferromagnetic materials. Note that the Curie–Weiss law is the variation of $1/\chi$ at $T > T_c$. (With kind permission from du Tremolet de Lacheisserie, E., et al., *Magnetism: Fundamentals*, Springer Science+Business Media, 2002.) (b) Comparison of $1/\chi$ versus T for ferromagnetic, ferrimagnetic, antiferromagnetic, and paramagnetic materials. (From Goldman, A., *Handbook of Modern Ferromagnetic Materials*, Kluwer, Boston, MA, 1999. With permission.) (c) Néel temperature (T_N or θ_N) for antiferromagnetic materials. (From Goldman, A., *Handbook of Modern Ferromagnetic Materials*, Kluwer, Boston, MA, 1999. With permission.)

11.4.5 FERROMAGNETIC AND FERRIMAGNETIC MATERIALS

These materials have the strongest response to an externally applied magnetic field. In other words, these materials are very susceptible to magnetic fields. In general, when we refer to a magnetic material, it is understood that this refers to a ferromagnetic or ferrimagnetic material (Figure 11.12a). The term nonmagnetic material is commonly used for diamagnetic or paramagnetic materials.

The most important feature of ferromagnetic and ferrimagnetic materials is that there is net magnetization even without magnetizing field, due to the spontaneous alignment of the magnetic moments in solids (Figure 11.3). Therefore, ferromagnetic and ferrimagnetic materials are characterized by a *magnetic hysteresis loop*, similar to the polarization–electric field loops exhibited by ferroelectric materials. The terminology used for a magnetic hysteresis loop is very similar to that used

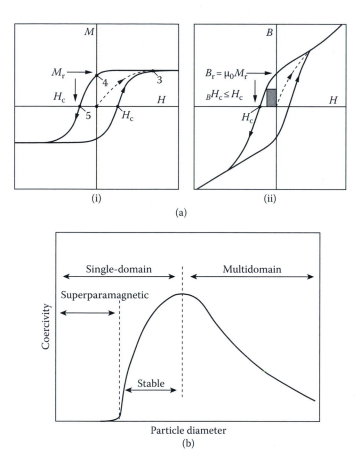

FIGURE 11.13 (a) Typical magnetic hysteresis loops of ferromagnetic and ferrimagnetic materials shown here as (i) M–H loop and (ii) B–H loop, respectively. (From Skomski, R. and J.Coey M.D, eds., *Permanent Magnetism*, Institute of Physics, Bristol, UK, 1999. With permission.) (b) Schematic illustration showing transition of multidomain particles or grains to single-domain and superparamagnetic particles or grains. (From Hadjipanayis, G.C., *J. Magn. Magn. Mater.*, 200. 373, 1991. With permission.)

for ferroelectrics and is shown in Figure 11.13a. Note that in Figure 11.9, we have shown only the initial portion of the hysteresis loop. In Figure 11.13a, the nonlinear nature of these materials is quite clear. Furthermore, in general, the magnetic susceptibility values of ferromagnetic and ferrimagnetic materials will be very high compared with those of diamagnetic and paramagnetic materials.

The hysteresis loop is the trace of magnetization (M) developed in these materials as a function of magnetizing field (H). Note that a ferromagnetic and a ferrimagnetic material are not distinguished simply by examining the shape of the hysteresis loop.

It is a common practice to show the values of $\mu_0 M$ (Tesla) instead of the value of magnetization M (ampere/meter). The quantity $\mu_0 M$ is known as the *magnetic polarization* (J). Often, the hysteresis loop is also shown as the magnetic induction (B) versus the magnetizing field (H). Again similar to linear dielectrics, for linear magnetic materials (i.e., diamagnetic and paramagnetic), by definition, the magnetic susceptibility (χ_m) is independent of the applied field (H). However, for ferromagnetic and ferrimagnetic materials, the magnetic susceptibility or permeability is indeed dependent on the applied field (Figure 11.13a).

Similar to ferroelectrics (Chapter 10), ferromagnetic and ferrimagnetic materials exhibit the presence of *magnetic domains*. A magnetic domain is a region of a material in which the magnetic moments have a coupling that results in net magnetization. A ferromagnetic or ferrimagnetic material typically consists of many magnetic domains. If the material is polycrystalline, typically, there will be many magnetic domains within each grain. A change in the magnetization of the hysteresis loop in Figure 11.13a is due to the reversal of the magnetic domains. As magnetizing field increases, domains whose magnetization direction is parallel to magnetizing field direction becomes larger and other domains shrink. As mentioned in Section 11.4.3, if a particle or grain becomes too small, it becomes a single-domain body and eventually becomes a superparamagnetic material (Figure 11.13b). In addition, notice the concomitant changes in coercivity of these materials.

The maximum possible magnetization we can get in a ferromagnetic or ferrimagnetic material is known as the *saturation magnetization* (M_s). This is also sometimes shown with the symbol $J_s = \mu_0 M_s$. It is defined as the maximum possible magnetic dipole moment per unit volume of a given material. Very often, the saturation magnetization is expressed as the flux density (B) that it creates in a material. Because $\mu_0 M \gg \mu_0 H$ for ferromagnetic and ferrimagnetic materials, the saturation magnetization is often expressed as $B_{sat} \sim \mu_0 M_s$ (Figure 11.13a).

Note that in literature, coercivity is sometimes written as H_{cB} or bH_c. The subscript or prefix B shows that this is the coercivity for the induction–magnetic field (B–H) loop. If the coercivity symbols appear as H_{cM} or mH_c, it indicates the coercivity value for the magnetization–magnetic field (M–H) loop. For materials with $\mu_0 H_{cM} > B_r$, the magnitude of H_{cM} is larger than H_{cB} (du Trémolet de Lacheisserie et al. 2002).

Sometimes, especially in the older literature on magnetic materials, you will encounter cgs and emu units for magnetic properties. For example, magnetization is expressed in Gauss. One Tesla is 10,000 Gauss. Conversions of different units for the various magnetic properties are listed in Table 11.6. The conversion factors provided are for conversion of a Gaussian quantity to the corresponding unit in the SI system.

The following examples illustrate how the magnetic moment of free atoms or ions can be used to calculate the saturation magnetization in ferromagnetic and ferrimagnetic materials, and how the results compare with the measured values.

Example 11.4: Saturation Magnetization (M_s) and Magnetic Polarization (J_s) of BCC Iron

The BCC form of iron (Fe) is ferromagnetic.

1. If the room-temperature unit-cell lattice constant of this structure is 2.866 Å, what will be the saturation magnetic polarization of BCC iron?
2. Compare this with the experimental value of 2.1 T, which is the room-temperature saturation magnetization of this form of iron.
3. What is the "effective" magnetic moment of iron atoms in a BCC iron crystal?

Solution

1. As shown in Figure 11.14, in BCC iron, we have eight iron atoms at the corners and one atom at the center of the unit cell. Each of the eight corner atoms counts only as one-eighth because it is shared among a total of eight unit cells (Chapter 2). The atom at the cube center counts as one. Thus, the actual number of atoms per unit cell of BCC iron is two.

   Each iron atom has a magnetic moment of 4 μ_B because there are four unpaired electrons per atom (Table 11.1).

   Thus, the total dipole moment per unit cell of BCC iron will be = (2 atoms/unit cell) × 4 μ_B = 8 × 11.274 × 10^{-24} A · m²/unit cell.

   The volume of each unit cell in BCC iron = (2.866 × 10^{-10} m)³ = 23.544 × 10^{-30} m³. Now, BCC iron is ferromagnetic. This means that all the magnetic dipole moments associated with all the atoms point in the same direction.

   Thus, the magnetization (M_s) will be equal to the total magnetization per unit volume.

TABLE 11.6
Units for the Various Magnetic Properties

Quantity	Symbol	Gaussian System	Conversion Factor	SI System
Magnetic flux	ϕ	Mx, Gem^2	10^{-8}	Wb · Vs
Magnetic flux density, magnetic induction	B	G	10^{-4}	T, Wb · m^{-2}
Magnetic potential difference, magnetomotoric force	U, F	Gb (gilbert)	$10/4\pi$	A
Magnetic field strength (also unit of coercivity H_c)	H	Oe	$1000/4\pi$	A · m^{-1}
Volume magnetization	$4\pi M$	G	$1000/4\pi$	A · m^{-1}
Volume magnetization	M	emu cm^{-3}, G	1000	A · m^{-1}
Magnetic polarization	J	emu cm^{-3}, G	$4\pi \times 10^{-1}$	T, Wb·m^{-2}
Mass magnetization	M, σ	emu g^{-1}, G cm^{3}g^{-1}	1	A · m^2 · kg^{-1}, J·T^{-1} · kg^{-1}
Magnetic moment	m	emu, erg G^{-1}	1/1000	A · m^2, J · T^{-1}
Magnetic dipole moment	J	emu, erg G^{-1}	$4\pi \times 10^{-10}$	Wb · m vs · m
Volume susceptibility	χ, κ	Dimensionless, emu cm^{-3}	4π	Dimensionless
Mass susceptibility	χ, κ	emu g^{-1}, cm^{-3} g^{-1}	$4\pi \times 1000$	m^3 · kg^{-1}
Molar susceptibility	χ_{mol}	emu mol^{-1}, cm^3 mol^{-1}	$4\pi \times 10^{-6}$	m^3 · mol^{-1}
Permeability, $\mu = \mu_0\mu_I$	μ	$\mu^* = \mu_r$	$4\pi \times 10^{-7}$	H · m^{-1}, Vs · A^{-1}·m^{-1}
Relative permeability, μ/μ_0	μ_r	Dimensionless	1	Dimensionless
Energy density, energy product	w	erg cm^{-3}	1/10	J · m^{-3}
Demagnetization factor	N_d	Dimensionless	$1/4\pi$	Dimensionless

Source: Bloor, D. et al., eds., *Encyclopedia of Advanced Materials*, vol. 4, Pergamon Press, Oxford, UK, 1994. With permission.

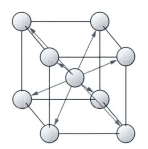

FIGURE 11.14   Crystal structure of body-centered cubic iron.

$$M_s = \frac{8 \times 9.274 \times 10^{-24} \text{ A} \cdot \text{m}^2}{23.544 \times 10^{-30} \text{ m}^3} = 3.15 \times 10^6 \text{ A/m}$$

Magnetization is often expressed as $\mu_0 M_s$, rather than as M_s, because then, it has units of Tesla, and it is easier to compare $\mu_0 M$ with $\mu_0 H$ (see Example 11.6).

2. The M_s value of 3.15×10^6 A/m is equivalent to $J_s = \mu_0 M_s = 3.95$ T. The measured value of saturation magnetization is 2.1 T. This difference is due to cancelation of some of the magnetic spin moments when the "free" iron atoms get together and form a BCC crystal.

3. Because the actual value of magnetization is 2.1 T, we can back-calculate the "effective" magnetic moment of an iron atom in BCC iron. Let the effective magnetic moment be x Bohr magnetons per iron atom in BCC iron. Then, the total saturation magnetization will be

$$M_s = \frac{x\dfrac{\text{magnetons}}{\text{atom}} \times \dfrac{\text{atoms}}{\text{unit cell}} \times 9.27 \times 10^{-24} \dfrac{\text{A/m}}{\text{Bohr magneton}}}{\left(23.544 \times 10^{-30}\right)\dfrac{\text{m}^3}{\text{unit cell}}}$$

Therefore, $M_s = 7.87 \times x \times 10^5$ A/m. This corresponds to the magnetic polarization $J_s = \mu_0 M_s = (7.87 \times x \times 10^5 \text{ A/m}) (4\pi \times 10^{-7} \text{ Wb/A} \cdot \text{m}) = (0.9895 \times x)$ T.

This is given as 2.1 T, which means $x = 2.122$.

Thus, although a free iron atom has a magnetic moment of 4 μ_B, an iron atom in a BCC iron crystal behaves as if its magnetic moment is only 2.122 μ_B.

Similar calculations can be done for other materials as well. The properties of some magnetic materials, including the effective magnetic moments of atoms, expressed as the number of Bohr magnetons per atom or per formula unit (n_B/formula unit), are shown in Table 11.7. Compare these values with those listed in Table 11.2.

Example 11.5: Saturation Magnetization in Ceramic Ferrites

Ceramic ferrites are useful magnetic materials. Iron oxide (Fe_3O_4) is one example of a ceramic ferrite (Figure 11.15a). This material shows ferrimagnetic behavior, characterized by the antiferromagnetic coupling of the magnetic moments of Fe^{3+} ions on different crystallographic sites (Example 11.2). These crystallographic arrangements are shown in a subcell of Fe_3O_4 (Figure 11.15b). Oxygen ions mediating these magnetic interactions are not shown for the sake of clarity. (a) Calculate the saturation magnetization ($\mu_0 M_s$) for Fe_3O_4 if the lattice constant (a_0) of the larger unit cell, which

TABLE 11.7

Magnetic Properties of Some Magnetic Materials

Material	Crystal Structure	$\mu_0 M_s$ (290 K) Tesla	M_s (290 K) emu/cm³	$\mu_0 M_s$ (0 K) Tesla	M_s (0 K) emu/cm³	n_B/Formula Unit ($T = 0$ K)	Curie Temperature (K)
Fe	BCC	2.1	1707	2.2	1707	2.22	1043
Co	HCP, FCC	1.8	1440	1.82	1446	1.72	1388
Ni	FCC	0.61	485	0.64	510	0.606	627
$Ni_{80}Fe_{20}$	FCC	1.0	800	1.17	930	1.0	–
Gd	HCP			2.6	2060	7.63	292
Dy	HCP			3.67	2920	10.2	88
CrO_2		0.65	515			2.03	386
$MnOFe_2O_3$	Spinel	0.51	410			5.0	573
$FeOFe_2O_3$	Spinel	0.6	480			4.1	858
$Y_3Fe_5O_{12}$	Garnet (YIG)	0.16	130	0.25	200	5.0	560
$Nd_2Fe_{14}B$	Tetragonal	1.6	1280				585
Amorphous $Fe_{80}B_{20}$	Amorphous	1.6	1260	1.9	1480	2.0	650

Source: O'Handley, R.C., *Modern Magnetic Materials, Principles, and Applications*, Wiley, New York, 1999. With permission.

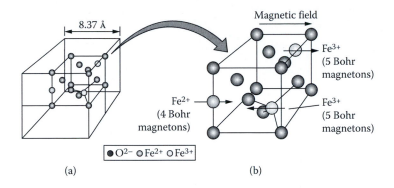

FIGURE 11.15 (a) Crystal structure of iron oxide and (b) arrangements of ferrous (Fe^{2+}) and ferric (Fe^{3+}) ions in a subcell of Fe_3O_4. (From Askeland, D. and Fulay P. *The Science and Engineering of Materials*, Thomson, Washington, DC, 2006. With permission.)

consists of eight smaller unit cells (Figure 11.15a), is 8.37 Å. (b) What will be the saturation flux density (B_s) for this material?

Solution

1. In Fe_3O_4, the magnetic moments of ferric ions (Fe^{3+}) (each Fe^{3+} has a magnetic moment of 5) canceled out because of the antiferromagnetic coupling of an equal number of Fe^{3+} ions located at both tetrahedral and octahedral sites. Thus, only the ferrous ions (Fe^{2+}) contribute to the net magnetic moment.

   In each subcell, the single Fe^{2+} contributes to the total magnetic moment. We have a total of eight subcells that make the larger unit cell (Figure 11.15a). The magnetic moment for each cell being 4 μ_B for eight unit cells, the total magnetic moment will be 32 μ_B. Under saturation conditions, all these magnetic moments will be aligned with the field due to the divalent ferrous ions and hence the total magnetization will be

$$M_s = (32 \times \mu_B) = \left(32 \text{ Bohr magnetons} \times 9.27 \times 10^{-24} \frac{A \cdot m^2}{\text{Bohr magneton}}\right)$$

The volume of the unit cell is $(8.37 \times 10^{-10} \text{ m})^3 = 5.86 \times 10^{-30} \text{ m}^3$, and the saturation magnetization per unit volume is

$$M_s = \frac{2.96 \times 10^{-22} \text{ A} \cdot m^2}{5.86 \times 10^{-30} \text{ m}^3} = 5.1 \times 10^5 \text{ A/m}$$

This can be expressed in Tesla as the magnetic polarization $J_s = \mu_0 M_s = (4\pi \times 10^{-7}$ Wb/A $\cdot$ m) $(5.1 \times 10^5$ A/m) = 0.64 T.

This is close to the experimentally measured value of ~0.6 T at 290 K (Table 11.7). For ferrites, the differences in the measured and calculated values of magnetization may be due to the incomplete quenching of the orbital magnetic moments, changes in the relative ratios of divalent and trivalent ions (i.e., changes in Fe^{2+}/Fe^{3+} ratio), and changes in the distribution of these ions among different tetrahedral and octahedral sites.

2. Now we can calculate the magnetic flux density under saturation (B_{sat}). The relationship between B and M is given by

$$B = \mu_0 H + \mu_0 M$$

For ferromagnetic and ferrimagnetic materials, the second term dominates, that is, $\mu_0 M \gg \mu_0 H$, especially under high fields. Thus, ignoring the first term, $B_{sat} \sim \mu_0 M_s = 0.64$ T.

TABLE 11.8

Magnetic Moments for Different Ferrites

Ferrite	Tetrahedral Sites A	Octahedral Sites E	Bohr Magnetons per Formula Unit
$NiFe_2O_4$	$Fe^{3+} \downarrow 5\,\mu_B$	$Ni^{2+} \uparrow 2\,\mu_B$	$2\mu_B$
		$Fe^{3+} \uparrow 5\,\mu_B$	
$MnFe_2O_4$	$Mn^{2+} \downarrow 0.8 \times 5\,\mu_B$	$Mn^{2+} \uparrow 0.2 \times 5\,\mu_B$	$5\mu_B$
	$Fe^{3+} \downarrow 0.2 \times 5\,\mu_B$	$Fe^{3+} \uparrow 0.8 \times 5\,\mu_B$	
		$Fe^{3+} \uparrow 5\,\mu B$	
$ZnFe_2O_4$	$Zn^{2+} \downarrow 0\,\mu_B$	$Fe^{3+} \uparrow 5\,\mu_B$	$0\mu_B$
		$Fe^{3+} \uparrow 5\,\mu_B$	
$Zn_xMn_{(1-x)}Fe_2O_4$	$Zn^{2+} \downarrow 0\,\mu_B$	$Fe^{3+} \uparrow 5\,\mu_B$	$(1 + x) \times 5\,\mu_B$
	$Mn^{2+} \downarrow (1-x) \times 5\,\mu_B$	$Fe^{3+} \uparrow 5\,\mu_B$	
$Zn_xNi_{(1-x)}Fe_2O_4$	$Zn^{2+} \downarrow 0\,\mu_B$	$Ni^{2+} \uparrow (1-x) \times 2\,\mu_B$	$(1 + 4x) \times 2\,\mu_B$
	$Fe^{3+} \downarrow (1-x) \times 5\,\mu_B$	$Fe^{3+} \uparrow 5\,\mu_B$	
		$Fe^{3+} \uparrow x \times 5\,\mu_B$	

Source: Reprinted from Fiorillo, F., *Measurement and Characterization of Magnetic Materials*, Copyright 2004, with permission from Elsevier.

In Table 11.8, the values of magnetic moments and the coupling of different moments for some ceramic ferrites are shown.

When the divalent ions go to only the tetrahedral site and the trivalent ions occupy only the octahedral site, the structure is known as a normal spinel structure. For example, zinc ferrite ($ZnFe_2O_4$) has a normal structure. However, as seen in the case of Fe_3O_4 (written as $FeFe_2O_4$), the trivalent ions (i.e., what are supposed to be the B-site cations) are evenly distributed across both tetrahedral and octahedral sites. This structure is known as the inverse spinel structure. Examples of inverse spinel ferrites include $FeFe_2O_4$, $NiFe_2O_4$, and $CoFe_2O_4$. Some spinels, such as $MnFe_2O_4$, may exhibit partly normal and inverse spinel structures.

Note that the addition of zinc ferrite (with zero net magnetic moment) to nickel ferrite or $MnFe_2O_4$ *increases* the net magnetic moment of these materials. This is counterintuitive and can be explained as follows. When zinc ions substitute for a part of the tetrahedral sites in either nickel or manganese ferrite, they *reduce* the antiparallel coupling between Fe^{3+} ions distributed on the tetrahedral sites in the inverse spinel structure. This, in turn, causes an *increase* in the net magnetization.

11.5   OTHER PROPERTIES OF MAGNETIC MATERIALS

11.5.1   Curie Temperature (T_c)

Ferromagnetic and ferrimagnetic materials undergo a transformation at high temperatures, by which they become paramagnetic. The temperature at which the spontaneous magnetization in a ferromagnetic or ferrimagnetic material vanishes is known as the Curie temperature. Similar to ferroelectrics (Chapter 10), the Curie–Weiss law describes the variation of magnetic susceptibility with temperature for levels above the Curie temperature, written as follows:

$$\frac{1}{\chi_m} = \frac{T - \theta_p}{C} \tag{11.16}$$

In this equation, χ_m is the magnetic susceptibility, T is the temperature, θ_p is the paramagnetic Curie temperature (often, slightly greater than T_c), and C is the Curie constant.

The variation in $1/\chi_m$ as a function of temperature *for $T > T_c$* is what the Curie–Weiss law represents. Note that this law does *not* apply to the variation of magnetization at temperatures below T_c. The variation of susceptibility above T_c (Curie–Weiss law) and the changes in magnetization *below T_c* are shown in Figure 11.12. Because the material behaves as a paramagnetic material above T_c, the Curie–Weiss law describes the variation in permeability and susceptibility of the paramagnetic phase. Curie temperature (Table 11.7) is very important from a technological perspective because, often, this parameter limits the highest temperature that can be used for a magnetic material.

For ferrimagnetic materials, the dependence of $1/\chi_m$ on temperature departs from linearity as we approach T_c. This is shown in Figure 11.12b.

In antiferromagnetic materials, there is no net magnetic moment because the magnetic moments cancel out. However, as we increase the temperature, the extent of *antiparallel coupling is reduced*. This causes the magnetic susceptibility to *increase* (i.e., $1/\chi_m$ decreases) with increase in temperature. This increase in the susceptibility continues until temperature increases to Néel temperature (T_N or θ_N), at which an antiferromagnetic material becomes paramagnetic (Figure 11.12c).

11.5.2 Magnetic Permeability (μ)

In addition to saturation magnetization, many other properties of ferromagnetic and ferrimagnetic materials, important for different applications, are often specified. These include the initial (low-field) and high-field permeabilities.

As can be seen from Figures 11.6 and 11.16, the magnetic permeability (μ) of ferromagnetic and ferrimagnetic materials depends strongly on the strength of the applied magnetic field (H).

At very low fields, the magnetic permeability is low because the applied field is not sufficient to cause any kind of domain growth. This is often referred to as the initial permeability (μ_i). As the field increases, the domains are aligned along the direction of the magnetic field. This is similar to the orientation of domains in ferroelectric materials. As the domains become oriented, a relatively sharp increase in magnetization occurs. This is why, in this region, the magnetic permeability increases significantly to μ_{max} (Figure 11.16). As the field approaches relatively high values, in which essentially all domains are aligned, the permeability begins to decrease again (Figures 11.6 and 11.16). This is the saturation region. What this means is that the material cannot accept any more flux, that is, it essentially becomes impermeable to the magnetic field.

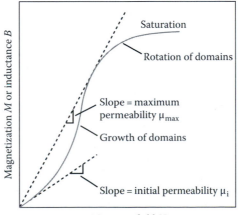

FIGURE 11.16 Field dependence of permeability for ferromagnetic and ferrimagnetic materials. (From Askeland, D. and Fulay P. *The Science and Engineering of Materials*, Thomson, Washington, DC, 2006. With permission.)

In the designing of magnetic circuits, often, care has to be taken not to create such *bottlenecks* that effectively will not allow any more magnetic flux to pass through. As shown in Figure 11.17, if the applied field is not enough to cause saturation, we get what is called a *minor loop*.

11.5.3 COERCIVE FIELD (H_c)

Similar to ferroelectrics, application of a certain level of *coercive field* ($\mu_0 H_c$ or H_c) is needed to remove any remnant magnetization in the material. This field is known as the *coercivity* (H_c) or coercive field of a magnetic material. The relative value of coercivity defines soft and hard magnetic materials. These materials are further discussed in Section 11.8.

The coercivity values define one way to classify ferromagnetic and ferrimagnetic materials as *hard* (see Section 11.9) or *soft magnetic materials* (see Section 11.11). A hard material is easily distinguished from a soft magnet by comparing the values of coercivity relative to the saturation magnetization (Figure 11.18).

Ferromagnetic and ferrimagnetic materials are used for flux enhancement or as materials that can store energy or information (data). The first application, that is, flux enhancement, requires high

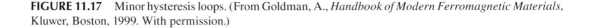

FIGURE 11.17　Minor hysteresis loops. (From Goldman, A., *Handbook of Modern Ferromagnetic Materials*, Kluwer, Boston, 1999. With permission.)

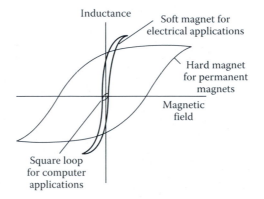

FIGURE 11.18　Typical shapes of hysteresis loops for soft and hard magnetic materials, diagram not to scale. (From Askeland, D. and Fulay P. *The Science and Engineering of Materials*, Thomson, Washington, DC, 2006. With permission.)

permeability and low coercivity (see Section 11.11). In general, materials that have a coercivity of less than ~5000 A/m are considered soft magnetic materials (see Section 11.11), whereas those with coercivity larger than approximately 10^4 A/m are known as hard magnetic materials (Figure 11.19). In the so-called cgs (emu) system of units, coercivity is expressed in Oersteds (Oe). One ampere/meter is equal to $4\pi \times 10^{-3}$ Oe.

Hard magnetic materials are also known as *permanent magnets* (see Section 11.9). Materials whose coercivity values range between a few hundred and ~10^4 A/m are known as *semihard magnetic materials*. These semihard materials are useful for magnetic storage of data (e.g., audio and video cassettes or magnetic hard disks used in computers and other electronic gadgets; see Section 11.12).

Similar to ferroelectrics, the coercivity (H_c) of magnetic materials is a microstructure-sensitive property. For essentially the same material composition, coercivity can be changed by orders of magnitude. Of course, we can change the composition and design an appropriate microstructure to develop a material with the desired magnetic properties. Currently, through such careful design, we have obtained extremely low coercivity magnetic materials, such as $Fe_{84}Zr_7B_9$ that has a coercivity of $\mu_0 H_c \sim 10^{-7}$ T. Note the negative sign in the exponent. On the contrary, we also have a material such as $Fe_{84}Nd_7B_9$ with very high coercivity of $\mu_0 H_c \sim 1$ T. We have a large range of coercivities in engineered magnetic materials. In real-world applications, the cost-to-performance ratio and many other factors, such as durability, weight, and mechanical properties, are also very important.

11.5.4 NUCLEATION AND PINNING CONTROL OF COERCIVITY

When magnetizing field is applied, the magnetic domains in a material are forced or coerced to orient in the direction of the field. The coercive field and the initial part of the B–H (M–H) loop depend on the mechanism by which the domains nucleate and grow. For example, consider a hypothetical hard magnetic material that has been subjected to a sufficiently high magnetic field to make it essentially a single domain. Now, if we start applying a field in the reverse direction, the field will try to align this single domain along the new direction. This will require nucleation of a new domain of an opposite

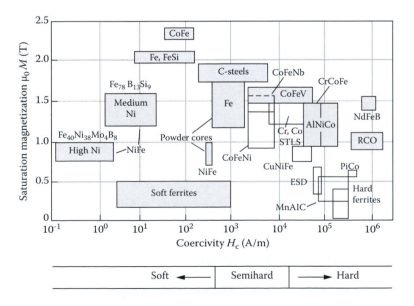

FIGURE 11.19 Relative coercivity values for classification of magnetic materials. (From Chin, G.Y., et al., In *Encyclopedia of Advanced Materials*, Pergamon Press, Oxford, 1994. With permission.)

magnetization and its domain walls. This situation is referred to as *nucleation-controlled coercivity.* A very good hard magnet, on application of magnetizing field in the reverse direction, will resist nucleation of domains as much as possible. If the coercivity is controlled by the nucleation of new domains, then the magnetizing process starting from the origin of the hysteresis loop (i.e., demagnetized state) to saturation (marked as the dotted curve in Figure 11.20a (ii)) is achieved using a much weaker magnetizing field, compared with the coercivity which is required to flip the magnetization direction of the single domain via the nucleation controlled process (marked as the solid line in Figure 11.20a (ii)). Most state-of-the art hard magnetic materials show nucleation-controlled coercivity (Figure 11.20b (i)).

Another way to enhance coercivity is to pin down the domain walls of the existing domains as much as possible. This is known as *pinning-controlled coercivity.* In this case, multiple domains can exist, but their domain wall motion is hindered by pinning. Typically, micro-structural defects, especially grain boundaries, can be pinning centers. If the magnetizing field required for proceeding from the origin of the hysteresis loop curve (demagnetized state) to saturation (marked as the dotted curve in Figure 11.20a (i)) is similar to the coercivity (marked as the solid line in Figure 11.20a (i)), the domain reversal is controlled by the domain-pinning process. An example of pinning-controlled coercivity is shown in Figure 11.20b (ii) for a Ce–Cu–Co–Fe magnetic material.

This is in contrast with the hysteresis loop for a Ce–Co–Cu–Fe magnet, in which defects that pin the domains control the coercivity. An understanding of the microstructural features that control the

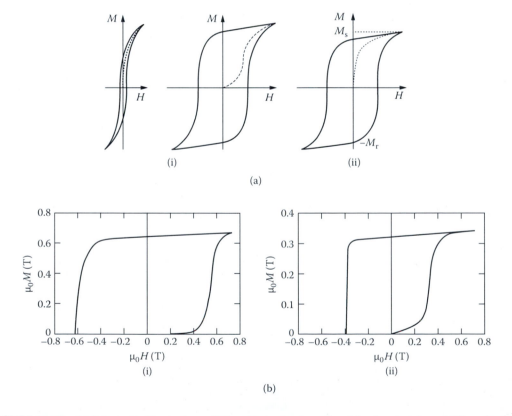

FIGURE 11.20 (a) Schematic illustrations of (i) wall pinning in soft and hard materials and (ii) nucleation-controlled coercivity in modern, hard magnetic materials. (b) (i) Nucleation-controlled coercivity in a ferrite and (ii) pinning-controlled coercivity in Ce–Co–Cu–Fe magnet; a difference in the magnetizing field between a dotted line and a solid line is smaller when the domain reversal occurs via the pinning-controlled process. (With kind permission from du Tremolet de Lacheisserie, E., et al., *Magnetism: Fundamentals*, Springer Science+Business Media, 2002.)

coercivity is important for the development of the hard magnetic materials, especially those used as permanent magnets (see Section 11.9) and magnetic recording materials (see Section 11.12).

11.5.5 Magnetic Anisotropy

The term magnetic anisotropy means that magnetic properties are dependent on the crystallographic direction. This property is critical for the development of permanent or hard magnetic materials. The coercivity of magnetic materials is related to the magnetic anisotropy in two ways. First, there is the *magnetocrystalline anisotropy*. Simply stated, this means that for a given single crystal material, magnetic properties, such as the coercive field, change depending on the crystallographic direction. The second type of magnetic anisotropy is known as *magnetoshape anisotropy* or *shape anisotropy*. This anisotropy means that if we have particles of two identical materials, with one particle being needle-like (acicular) and the other one nearly spherical, then the acicular particle will have a higher coercivity because of its shape. The cause of shape anisotropy can be traced to the difference in the demagnetization factors (due to differences in geometry; Table 11.4). For an elongated particle, the hard axis tends to line up along the longer axis of the particles. This effect is used in magnetic recording materials wherein elongated or needle-like particles of materials, such as iron or barium ferrites, are used (see Section 11.12). The elongated shape of these particles causes them to have higher coercivity, which helps with the retention of information stored.

Magnetocrystalline anisotropy is important in many applications of hard and soft materials and is discussed here in detail. For example, in magnetic materials used for data storage, we take advantage of the magnetocrystalline anisotropy by using oriented or textured thin films that possess higher coercivity and hence show better retention of data. On the contrary, we use grain-oriented steels for transformer cores so that the steel magnetizes easily along the easy directions.

Owing to the magnetocrystalline anisotropy present in an iron single crystal, the magnetization develops most easily along the [100] direction at smaller magnetizing field. The crystallographic direction in which the magnetization develops with a smaller coercivity is known as the easy-magnetization direction or the *easy axis*. This represents the direction(s) in which the average magnetic moment will be directed for a material with no application of an external magnetic field (Bloor et al. 1994). Alignment of magnetic moments along the easy-axis directions would minimize the free energy of the material. On the contrary, for a BCC iron single crystal, the [111] direction that represents a body diagonal will be the hard magnetization direction or *hard axis*. All the body diagonals, that is, the <111> family of directions, and not just the [111] direction, will represent the hard axes for BCC iron.

For cobalt (Co), the easy axis is the direction perpendicular to the hexagonal planes. The axes in the basal plane represent the hard directions (Chapter 2). The energy needed to rotate from the easy-magnetization direction is known as the magnetocrystalline anisotropy energy. This energy can be written as

$$E_a = K_1 \sin^2 \theta + K_2 \sin^4 \theta + K_3' \sin^6 \theta \cos 6\phi \quad \text{(for hexagonal crystals)} \tag{11.17}$$

For tetragonal crystals, the magnetocrystalline anisotropy energy is written as

$$E_a = K_1 \sin^2 \theta + K_2 \sin^4 \theta + K_3 \sin^4 \theta \cos 4\phi \quad \text{(for tetragonal crystals)} \tag{11.18}$$

In these equations, K_1, K_2, K_3, and K_3' are the coefficients of anisotropy. Usually, the coefficient K_1 dominates. The angle θ is that formed between the magnetization vector and the c axis. The angle ϕ is that between the projection of the magnetization vector on the basal plane and one of the a axes (Bloor et al. 1994). The coefficients of anisotropy are expressed in units of Joule/cubic meter or Gauss–Oersteds (G · Oe). The conversion factor is 1 mega G · Oe = 7.9577 kJ/m³.

The anisotropy field, K_1 values, and other properties of some hard magnetic materials are listed in Tables 11.9 and 10.

Note that the values of K_1 are listed as Megajoule/cubic meter, that is, the value of $SmCo_5$ is 17 MJ/m^3 or 17×10^6 J/m^3. The K_1 values of some materials are listed in Table 11.10.

In a ferromagnetic or ferrimagnetic material, the long-range spin ordering occurs because of what is called the *exchange interaction*, that is, the combination of the electrostatic coupling between electron orbitals and Pauli's exclusion principle from Chapter 2. In ferromagnetic materials, this interaction aligns the spins of neighboring atoms in the same direction (Figure 11.3). In ferrimagnetic

TABLE 11.9
Values of T_C, Anisotropy Field, K_1, and Saturation Magnetization for Some Hard Magnets

Compound	T_c (C)	$\mu_0 H_a$ (T)	K_1 (MJ m^{-3})	$\mu_0 M_s$ (T)
$SmCo_5$	720	40	17	1.05
Sm_2Co_{17}	823	6.5	3.3	1.30
$Nd_2Fe_{14}B$	312	6.7	5	1.60

Source: Jakubowicz, J., et al., *J. Magn. Magn. Mater.* 208(3), 163–168, 2000; Bloor, D., et al., eds., *Encyclopedia of Advanced Materials*, vol. 4, Pergamon Press, Oxford, UK, 1994.

Note: 1 MG · Oe = 7.9577 kJ/m^3.

TABLE 11.10
Approximate Values of Anisotropy Coefficient (K_1) for Some Magnetic Materials

Material	K_1 (kJ/m^3)
Iron	48
Nickel	−4.5
Cobalt	530
Fe_3O_4	−13
$CoFe_2O_4$	180–200
$NiFe_2O_4$	−6.9
$MgFe_2O_4$	−4
$MnFe_2O_4$	−4
$SmCo_5$	17,000
Sm_2Co_{17}	3300
$Nd_2Fe_{14}B$	4900
$BaFe_{12}O_{19}$	250
Cu_2MnAl	−0.47
$Y_3Fe_5O_{12}$	−2.5

Source: Reprinted from Fiorillo, F., *Measurement and Characterization of Magnetic Materials*, Copyright (2004), with permission from Elsevier; and With kind permission from du Tremolet de Lacheisserie, E., et al., *Magnetism: Fundamentals*, Springer Science+Business Media, 2002.

Note: 1 MG · Oe = 7.9577 kJ/m^3.

materials, this exchange interaction orients the spins of ions on different crystallographic sites in an antiparallel arrangement (Figure 11.15). In ceramic ferrites, this is known as the superexchange interaction because it is mediated by oxygen ions. The anisotropy, however, comes about because of the interactions between the electron orbitals and the potential associated with the atoms in the crystal (Bloor et al. 1994).

11.5.6 MAGNETIC DOMAIN WALLS

Normally, we would expect the magnetic moments to reverse their direction only when the applied magnetic field is equal to or greater than the so-called anisotropy field (H_a), which is given by

$$H_a = \frac{2K_1}{M_s} \tag{11.19}$$

This means that for an ideal magnetic material, domain switching must occur only at one value of the magnetizing field, that is, the hysteresis loop should be rectangular for the M (or J)–H loop and a parallelogram for the B–H loop (Figure 11.21a).

It has been experimentally observed that most permanent or hard magnetic materials (see Section 11.8) show magnetization reversal at fields that are only 10%–15% of the value of anisotropy field (H_a) (Buschow and De Boer 2003).

The reason for this phenomenon is that magnetic materials consist of magnetic domains, which are separated by domain walls known as *Bloch walls* (Figure 11.21b). The magnetization direction for adjacent domains is in opposite directions. This helps reduce the magnetostatic energy.

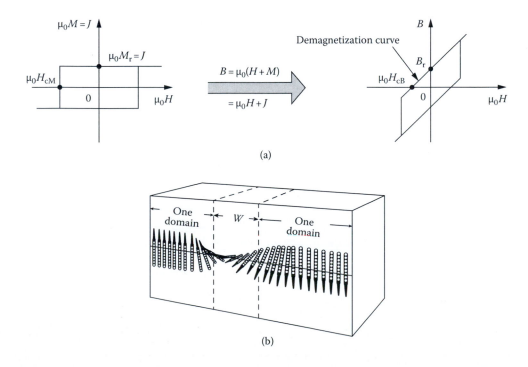

(a)

(b)

FIGURE 11.21 (a) M (or J)–H loop and B–H loop for an ideal ferromagnetic or ferrimagnetic material. (With kind permission from Springer Science+Business Media: *Magnetism: Fundamentals*, 2002, du Tremolet de Lacheisserie, E., D. Gignoux, and M. Schlenker.) (b) Illustration of change in magnetization across an 180° Bloch wall. (From Buschow, K.H. and De Boer F.R., *Physics of Magnetism and Magnetic Materials*, Kluwer, Boston, 2003. With permission.)

The magnetization direction changes gradually from one direction to an opposite direction through the thickness of the domain wall.

The thickness of the domain wall (W) in magnetic materials is often much greater than that of the ferroelectric domains in ferroelectric materials. The domain-wall thickness depends on the relative strengths of the anisotropy energy and the exchange energy. If the exchange energy is large, it tends to maintain the magnetic moment in the given direction and the magnetic dipoles rotate gradually in the domain wall, that is, the domain wall would be thicker.

If A is the average exchange energy and K_1 is the anisotropy energy constant, then it can be shown that the domain-wall thickness (W) is given by

$$W = \pi \sqrt{\frac{A}{K_1}} \tag{11.20}$$

and the domain-wall energy per unit area of a domain wall of width W is given by

$$E_{wall} = 2\pi \sqrt{K_1 A} \tag{11.21}$$

In deriving these equations, it is assumed that the effect of the demagnetization energy can be neglected.

If the anisotropy energy is large, then the domain walls would be thinner because the material prefers to undergo magnetization in the easy direction of magnetization. For example, in highly anisotropic materials such as tetragonal $Nd_2Fe_{14}B$ with high values of K_1, reaching ~4900×10^3 J/m³ (Tables 11.9 and 10), the domain-wall thickness is only ~5 nm. In materials such as BCC iron, the domain-wall thickness may be ~50 nm because of the relatively smaller anisotropy energy (smaller value of K_1, ~4.8×10^4 J/m³). Because the lattice constant (a_0) of iron is ~0.3 nm (Example 11.4), the domain-wall thickness will be greater than ~200 lattice spacings.

Thus, in many permanent magnetic materials (see Section 11.9), the observed value of coercivity is *significantly lower* than that expected from the value of the anisotropy field (H_a) because magnetization reversal occurs by the nucleation of Bloch walls and growth of reversed domains. This situation is somewhat similar to the yield stress (τ_{ys}) of metals and alloys. We know that based on the strengths of metallic and covalent bonds, most metallic materials should be extremely strong and brittle. However, in reality, their strength is much lower (almost 1000 times lower). Also, metals and alloys are relatively ductile, because mechanical deformation occurs by the motion of dislocations (known as slip) at much lower levels of stress and not by breaking of all the strong bonds (Chapter 2).

In some materials, such as thin films, magnetic domains can extend across the entire width of a sample. If there are Bloch walls between two domains that run the entire width of the sample, then in the wall region, the magnetization will be perpendicular to the plane of the film. This causes a considerable increase in the demagnetization energy. In this case, a *Néel wall* forms, in which the magnetic moment rotates within the plane of the film (Jiles 1991). This is shown in Figure 11.22. The formation of a Néel wall is seen in thin films because it occurs with lower energy.

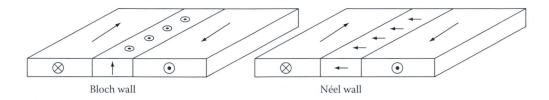

Bloch wall   Néel wall

FIGURE 11.22 Formation of Bloch and Néel walls in a magnetic thin film. (From Jiles, D.C., *Introduction to Magnetism and Magnetic Materials*, Chapman and Hall, London, 1991. With permission.)

Néel walls do not appear in bulk magnetic materials because they actually cause a higher demagnetization energy. The formation of Néel walls is favored in thin films when the thickness falls below a certain critical value.

11.5.7 180° AND NON-180° DOMAIN WALLS

Both Bloch and Néel walls are examples of what we call 180° domain walls. This means that the direction of magnetization in the domains separated by the walls is antiparallel. Accordingly, by definition, for these walls, the crystallographic directions will be equivalent (e.g., [100] and [$\bar{1}$00]). It is possible to have domain walls that are not 180°. For example, in cubic materials with the anisotropy energy constant greater than zero, it is possible to have 90° domain walls. An example of this is BCC iron, in which it is possible to get magnetization in adjacent domains in the [100] and [010] directions (Jiles 1991). In nickel (Ni), the anisotropy constant (K_1) is less than zero ($K_1 = -4.5 \times 10^3$ J/m³), and the result is domain walls of 71° and 109°. Note that for nickel, the easy-magnetization directions are all body diagonals (<111>; Figures 11.23 and 11.24).

Such 90° domains, shown in Figure 11.23, appear as closure domains in materials such as grain-oriented silicon-containing steel used in transformer cores. It is difficult to form 90° domain walls in materials such as cobalt (Co) because the directions in the basal plane are the hard magnetic directions, as shown in Figure 11.23.

One important significance of 90° walls is that they are stress-sensitive. For example, consider a domain wall between the [100] and [010] directions. If a tensile stress is applied to this material in the [100] direction, then the [100] domain is energetically favored. This causes the growth of [100] domains at the expense of the [010] domain. Consequently, 90° domain walls will move under a stress. This contributes to the strain produced by *magnetostriction* (see Section 11.7). However, if there are two 180° domains (say in the [100] and [$\bar{1}$00] directions), the energies of both will be reduced equally by the application of stress in the [100] direction. Thus, 180° domains will be insensitive to stress.

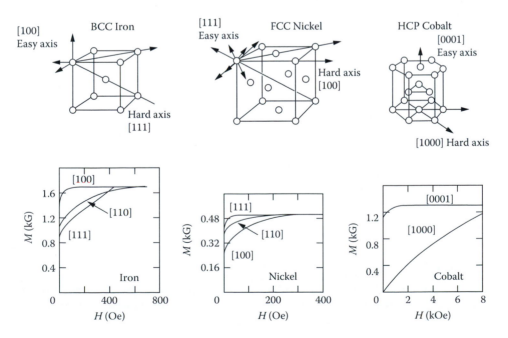

FIGURE 11.23 Crystal structures and magnetocrystalline anisotropy for Fe, Ni, and Co single crystals. (From O'Handley, R.C., *Modern Magnetic Materials, Principles, and Applications*, Wiley, New York, 1991. With permission.)

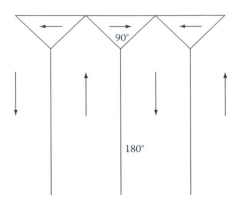

FIGURE 11.24 Illustration showing closure domains and 90° and 180° domain walls. (From Jiles, D.C., *Introduction to Magnetism and Magnetic Materials*, Chapman and Hall, Boca Raton, FL, 1991. With permission.)

11.5.8 Maximum Energy Products for Magnets

We saw the nature of the hysteresis loops for both ferromagnetic and ferrimagnetic materials in Figure 11.18. The first quadrant of the loop is important because it shows the dependence of permeability on the applied field. Once the material is saturated and the magnetic field is removed, the material is left in a state with a remnant magnetic induction (B_r; Figure 11.25), which makes the magnet permanent. The part of the loop in the second quadrant where the magnetic material is demagnetized from B_r to 0 is known as the *demagnetization curve* (Figure 11.25b). Magnetic circuit designs involving permanent magnets make use of this curve.

If we have a permanent magnet and we create a magnetic circuit with an air gap, then the magnetic induction decreases from B_r to some lower value, say to point P (Figure 11.26). The actual level of decrease depends on the geometry of the magnetic circuit and the demagnetization curve. The ratio of B to H at a point of operation along this curve is known as the permeance coefficient. The line drawn from the origin to the operating point (P) is known as the shearing line.

For permanent magnets, a frequently reported value is the so-called $(BH)_{max}$ *product*, also known as the *maximum energy product*. This is the maximum value of the product obtained by multiplying the corresponding B and H values on the demagnetization curve (Figure 11.26), which represents the maximum energy that can be stored within a given volume of a magnetic material.

The improvements in energy product and coercivity of permanent magnets are shown in Figure 11.27a. A useful conversion between different units for expressing the energy product is:

$$1 \text{ MG Oe} = \frac{10^6}{4\pi} \text{ erg/cc} = 7.96 \text{ kJ/m}^3 \tag{11.22}$$

11.5.9 Magnetic Losses

Magnetic losses refer to the energy dissipation that occurs when a magnetic material is subjected to a time-varying external magnetic field. They are similar to the dielectric losses occurring in dielectric materials (originating from different polarization mechanisms such as ferroelectric polarization). Magnetic losses are especially important in some applications, such as the use of different materials (e.g., iron–silicon steels) for transformers. The energy dissipation that occurs in a magnetic material can be seen from the area under the B–H loop, represented as $\int H \cdot db$. In materials

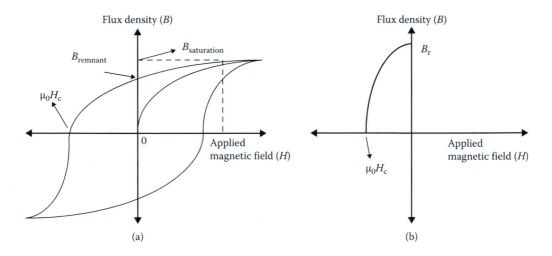

FIGURE 11.25   (a) Hysteresis loop at saturation and (b) demagnetization curve (second quadrant).

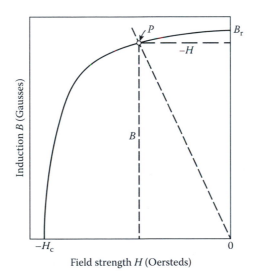

FIGURE 11.26   Demagnetization curve and a shearing line for $(BH)_{max}$. (From Bozroth, R. M., *Ferromagnetic Materials*, IEEE Press, Van Nostrand. Copyright (1955), by permission of IEEE.)

with higher electrical resistivity (e.g., ceramic ferrites), losses do not increase much with increase in the frequency of applied field. However, in magnetic materials that are good conductors (e.g., iron and steel), losses increase with increasing frequency. This is seen from the increase in the B–H loop area (Figure 11.27b).

The magnetic loss component can be treated mathematically in a way similar to that for dielectric losses using a complex dielectric constant. Magnetic permeability (μ) can be written as a complex number (μ^*) that has a real part (μ') and an imaginary part (μ''). The imaginary part is related to the magnetic losses that occur. The ratio of the imaginary and the real parts of magnetic permeability is defined as the loss tangent (tan δ). In some applications of ceramic ferrites used in high-frequency applications for microwave devices, characterization of such properties becomes very important. Magnetic losses due to eddy currents induced by the applied magnetic

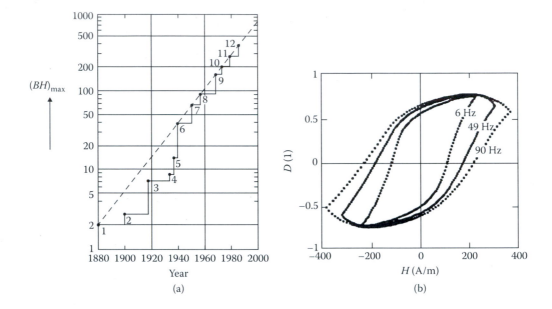

FIGURE 11.27 (a) Improvements in maximum energy product $(BH)_{max}$ in kJ/m³ over the years 1880 to 1980: 1, carbon steel; 2, tungsten steel; 3, cobalt steel; 4, Fe–Ni–Al alloy; 5, Ticonal II; 6, Ticonal G; 7, Ticonal GG; 8, Ticonal XX; 9, $SmCo_5$; 10, $(SmPr)Co_5$; 11, $Sm_2(Co_{0.85} F_{0.11} Mn_{0.04})_{17}$; and 12, $Nd_2Fe_{14}B$. (From Jiles, D.C., *Introduction to Magnetism and Magnetic Materials*, Chapman and Hall, Boca Raton, FL, 1991. With permission.) (b) Hysteresis loops for soft iron at different frequencies for induction up to 0.75 T. (With kind permission from du Tremolet de Lacheisserie, E., et al., *Magnetism: Fundamentals*, Springer Science+Business Media, 2002.)

field are known as *hysteresis losses*. Other components include excess losses due to domain motion and classical losses. For a detailed description of both origin and types of magnetic losses, the reader is directed to the work by Bloor et al. (1994). Losses due to eddy currents can be reduced by increasing the resistivity. This is the reason why silicon is added (up to ~3.5%) to iron to make the steel used in transformer cores. Addition of silicon also reduces the coercivity of the steel.

11.6 MAGNETOSTRICTION

We saw in Chapter 10 that every material responds to an electric field and shows electrostriction to some extent. The electrostriction strain is larger and more useful in materials that are more polarizable (e.g., relaxor ferroelectrics). Similarly, *magnetostriction* is seen in ferromagnetic and ferrimagnetic materials. Magnetostriction is defined as the development of strain in a material subjected to a magnetic field. A familiar example of the magnetostriction phenomenon is seen in magnetic steels used in transformers, which causes a *transformer hum*. This occurs as the magnetic materials in the transformer expand and contract to their original dimensions.

Magnetostriction in ferromagnetic and ferrimagnetic materials can come about spontaneously when there is a spontaneous alignment of magnetic moments at the Curie temperature. This is known as *spontaneous magnetostriction*.

Another form of magnetostriction is encountered when domains become reoriented in the presence of an external magnetic field. This is known as *Joule magnetostriction*. The extent of the magnetostriction effect is indicated by measuring the magnetostriction coefficient (λ). A material known as Terfenol-D ($Tb_{03}Dy_{07}Fe_{19}$) has a magnetostriction coefficient of ~1500 ppm, that is, 1500×10^{-6}, at room temperature. Other materials include $TbFe_2$, known as Terfenol, with a value of $\lambda_{111} = +2460 \times 10^{-6}$ (the highest reported so far), and $SmFe_2$ with $\lambda_{111} = -2100 \times 10^{-6}$ (Table 11.11).

The magnetostriction coefficient λ_{100} represents a change in length or saturation magnetostriction in the [100] direction when the magnetization is also along the [100] direction (after the material is cooled through the Curie temperature). The λ_{111} is defined similarly.

The acronym Terfenol comes from *Ter* for Terbium, Fe for iron, and NOL for the Naval Ordnance Laboratory that developed this material. Terfenol-D means that it has dysprosium (Dy) in it. Addition of Dy lowers the magnetocrystalline anisotropy, and this leads to better magnetostriction properties. These materials are being developed for a number of actuator applications, similar to the applications of piezoelectric materials. The cost of the elements such as Tb and Dy is high, making it difficult to create widespread applications of these materials. Current efforts are being directed to the development of relatively low-cost materials that can show large magnetostriction coefficients. Metglas™ ($Fe_{81}B_{13.5}Si_{3.5}C_2$) is one of the best isotropic magnetostrictive materials. Because magnetostriction basically involves causing strain using magnetic fields, many applications are similar to those of piezoelectrics (e.g., force sensors and sonar).

For polycrystalline materials with no texture or preferred grain orientation, the saturation magnetostriction coefficient (λ_s) is given by

$$\lambda_s = \frac{2}{5}\lambda_{100} + \frac{3}{5}\lambda_{111} \qquad (11.23)$$

In some materials such as Invar (an iron–nickel alloy), the thermal expansion effect is neutralized by the magnetostriction-induced contraction. This yields a material whose dimensions are essentially invariable with temperature, and hence the name Invar.

Equations and coefficients used are similar to those used for piezoelectrics (e.g., d_{33} and k_{33}). For example, the d_{33} coefficient of magnetostriction refers to the amount of strain produced by the application of unit magnetic field.

The following example illustrates the use of such coefficients.

Example 11.6: Terfenol-D Magnetostriction

The maximum d_{33} magnetostriction coefficient for Terfenol-D is 57×10^{-9} m/A (du Trémolet de Lacheisserie et al. 2002). A bar made from this material is 10 cm long and is exposed to a magnetic field (H) of 5 A/m. What will be the elongation produced in this bar?

TABLE 11.11

Room-Temperature Magnetostriction Coefficients of Some Materials

Material	$\lambda_{100}(10^{-6})$	$\lambda_{111}(10^{-6})$
Fe	24	−22
Ni	−51	−23
TbFe$_2$	—	2460
SmFe$_2$	—	−2100

Source: Buschow, K.H. and De Boer, F.R., *Physics of Magnetism and Magnetic Materials*, Kluwer, Boston, MA, 2003. With permission.

Solution

The d_{33} coefficient represents the strain produced per magnetic field. Thus, in this case, the strain produced $\frac{\Delta l}{l}$ will be given by

$$\lambda = 57 \times 10^{-9} = \frac{\text{Strain}}{\text{Magnetic field}} = \frac{\Delta l/l}{5 \text{ A/m}} = \frac{\Delta l}{5 \text{ A/m} \times 0.1 \text{m}}$$

Therefore, the increase in length will be = $(57 \times 10^{-9}) \times (5 \text{ A/m}) \times (0.1 \text{ m}) = 28.5$ nm.

11.7 SOFT AND HARD MAGNETIC MATERIALS

Ferromagnetic and ferrimagnetic materials are grouped into soft and hard materials, depending on the value of *coercivity* (Hc). The name shows the level of difficulty with which the magnetizing field changes the magnetization direction. As seen in Figure 11.19, materials with $H_c \sim 5000$ A/m are considered soft magnetic materials. Those with $H_c \sim >10^4$ A/m are considered hard or permanent magnets. A summary of some of these materials is shown in Figure 11.19.

The following example illustrates the coercivity magnitudes associated with soft and hard magnetic materials.

Example 11.7: Shapes of Hysteresis Loops: Soft and Hard Magnetic Materials

The coercivity (H_c) of a sample of a material known as grain-oriented steel is 300 A/m. The saturation magnetization ($\mu_0 M_s$) of this material is 2.1 T. A hysteresis loop for this material was measured to determine the properties.

1. If the coercivity of this material is also expressed in Tesla, how does the y-axis value of saturation magnetization compare with the x-axis value of coercivity?
2. What is the value of the coercivity of this material in Oersteds?
3. What is the value of the saturation magnetization of this grain-oriented steel in Gauss?
4. Repeat this calculation for a sample of a neodymium–iron–boron ($Nd_2Fe_{14}B$) permanent magnet, whose coercive field is 700,000 A/m, if the saturation magnetization is 1.2 T.

Solution

1. The coercivity is 300 A/m. To convert this value into Tesla, we multiply it by $\mu_0 = 4\pi \times 10^{-7}$ Wb/m · A.

   The coercivity value expressed in Tesla will be $\mu_0 H_c = (300$ A/m$)(4 \mu \times 10^{-7}$ Wb/m · A$) = 3.78 \times 10^{-4}$ Wb/m² or T. Compare this with $\mu_0 M_s = 2.1$ T. If we were to plot the hysteresis loop as $\mu_0 M$ versus $\mu_0 H$, the loop will look extremely slim.

2. The coercive field is expressed as ampere/meter or Oersteds (Oe). From Section 11.6.3 and Table 11.6, we know that 1 A/m = $4\pi \times 10^{-3}$ Oe. Thus, the coercivity of 300 A/m for this soft magnetic grain-oriented steel will be

$$= \left(300 \text{ A/m}\right) \times \left(4\pi \times 10^{-3} \frac{\text{Oe}}{\text{A/m}}\right) = 3.78 \text{ Oe}$$

3. The saturation magnetization of this grain-oriented steel is 2.1 T. One Tesla is 10,000 Gauss; thus, the saturation magnetization will be 21,000 Gauss.
4. For the $Nd_2Fe_{14}B$ magnet, the saturation magnetization is listed as 1.2 T. The coercivity of 700,000 A/m will be = $(700,000$ A/m$)(4\pi \times 10^{-7}$ Wb/m · A$) = 0.88$ Wb/m² or T.

Thus, for this permanent magnet, the hysteresis loop will appear square in shape. Also, because this material is stated to be a permanent magnet, we expect the remnant magnetization ($\mu_0 M_r$) and saturation magnetization ($\mu_0 M_s$) to be similar.

In the following sections, we discuss hard and soft magnetic materials in detail.

11.8 HARD MAGNETIC MATERIALS

Hard or permanent magnets are used in many applications to supply a permanent magnetic flux and to generate a static magnetic field. They are used in simple applications such as holding objects, electrical motors, and so on. Permanent magnets are also used for mechanical systems such as magnetic gears, bearings and shock absorbers.

The properties of some of the commercially available types of hard magnetic materials are summarized in Table 11.12. These include the H_c values for the J–H and B–H loops and the variations of B_r with temperature.

The anisotropy field, magnetocrystalline anisotropy coefficient (K_1), Curie temperature (T_c), and saturation magnetization for some hard magnetic materials were listed in Table 11.10.

The relative costs of some permanent magnet materials, expressed in US dollars/Joule versus the maximum energy per unit volume are shown in Figure 11.28.

From this diagram, it can be seen that the so-called *rare-earth magnets* (often based on NdFeB and SmCo) offer products with the highest energy. One problem with the use of rare-earth magnets

TABLE 11.12
Properties of Some Hard Magnetic Materials

Material	T_c (°C)	$(BH)_{max}$ (kJ m^{-3})	B_r (T)	dB_r/dT (%/deg.)	$_JH_c$ (kA m^{-1})	$_BH_c$ (kA · m^{-1})
Ferroxdure (SrFe$_{12}$O$_{19}$)	450	28	0.39	−0.2	275	265
Alnico$_4$	850	72	1.04	−0.015	–	124
SmCo$_5$	720	130–180	0.8–0.9	−0.01	1100–1500	600–670
Sm(Co, Fe, Cu, Zr)$_7$	800	200–240	0.95–1.15	−0.03	600–1300	600–900
Nd-Fe-B (sintered magnet)	310	200–280	1.0–1.2	−0.13	750–1500	600–850

Source: Buschow, K.H. and De Boer, F.R., *Physics of Magnetism and Magnetic Materials*, Kluwer, Boston, MA, 2003. With permission.

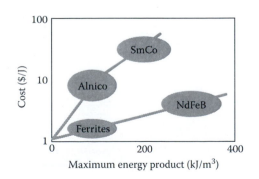

FIGURE 11.28 Relative cost of permanent magnets per unit energy versus maximum energy product. (With kind permission from du Tremolet de Lacheisserie, E., et al., *Magnetism: Fundamentals*, Springer Science+Business Media, 2002.)

is that their Curie temperature is low ($T_c \sim 300°C$). These materials also undergo surface oxidation, and this leads to a decrease in the coercivity. Magnets based on neodymium (e.g., $Nd_2Fe_{14}B$) are commonly known as *neomagnets*.

Alnico (pronounced al-knee-ko) magnets have lower coercivity. However, they have very good temperature stability, and because of their high Curie temperatures ($\sim 857°C$), they can be used up to $500°C$.

Ceramic ferrites offer the lowest cost per unit energy stored. Thus, ceramic ferrites are preferred for low-cost applications. The negative aspect of ceramic ferrites is that their saturation magnetization is low ($\sim 0.3–0.6$ T, Table 11.7), and hence the energy products are relatively low (Figure 11.28). Ceramic ferrite compositions can belong to both soft and hard categories.

The samarium–cobalt magnets ($SmCo_5$ and Sm_2Co_{17}) are expensive but high-performance magnets. Their Curie temperatures are high ($727°C$ for $SmCo_5$ and $827°C$ for Sm_2Co_{17}).

11.9 ISOTROPIC, TEXTURED (ORIENTED), AND BONDED MAGNETS

Permanent magnets are sometimes classified based on the orientation of magnetic domains. An *isotropic magnet* is a permanent magnet in which the spatial distribution of the easy directions of magnetizations is random. An isotropic magnet, in principle, is similar to an unpoled ferroelectric material. In *textured magnets,* also known as *oriented magnets,* most of the easy directions of magnetizations are oriented in a particular direction. Therefore, the distribution of orientations of the easy direction of magnetism inside the material is not random. This can be achieved by a process similar to the poling of ferroelectrics. In some cases, processing of a material (e.g., using a rolling process or extrusion) will cause orientation of different grains in a particular direction. Compared with isotropic magnets, the hysteresis loop of textured magnets is square (Figure 11.29) because after saturation, most domains prefer to remain aligned in specific directions. In Figure 11.29, the distribution of the easy axes is shown in the circles drawn for different stages of magnetization. For example, under saturation conditions, for both oriented and isotropic magnets, the easy axes are aligned along only one direction, which will be the direction of the applied magnetic field.

From the distribution of the directions of easy axes, we can see that for textured or oriented magnets, the remnant magnetization is much higher, typically $\sim 88\%–97\%$ of the saturation magnetization (du Trémolet de Lacheisserie et al. 2002). For isotropic magnets, the remnant magnetization is much lower ($\sim 50\%$ of the saturation magnetization).

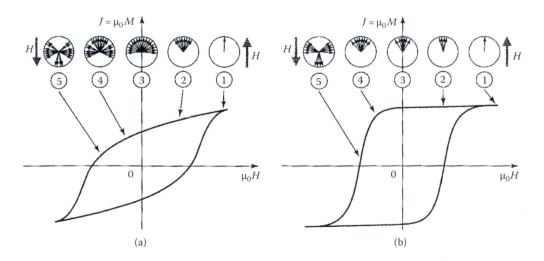

FIGURE 11.29 Hysteresis loops for (a) oriented and (b) isotropic magnets. (With kind permission from du Tremolet de Lacheisserie, E., et al., *Magnetism: Fundamentals*, Springer Science+Business Media, 2002.)

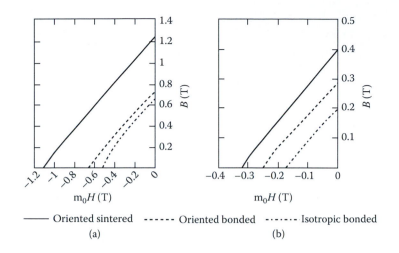

FIGURE 11.30 Demagnetization curves for (a) ceramic hard ferrites and (b) NdFeB magnets. (With kind permission from du Tremolet de Lacheisserie, E., et al., *Magnetism: Fundamentals*, Springer Science+Business Media, 2002.)

Bonded magnets are essentially composites made by dispersing magnetic powders in a polymer or metallic matrix. These are usually isotropic in nature because the easy axes within each particle (small crystal) are randomly arranged in the composite material. It is possible to orient the magnetic particles, for example, ferrite in a polymer, using a special processing technique. For example, oriented-bonded magnets are made with ceramic ferrites in a plastic using the calendaring process. Such materials, known as magnetic rubbers, have a saturation magnetization of ~0.25 T. In refrigerators, for example, we make use of these magnetic rubbers as door gaskets. Compared with the so-called *sintered magnets*, obtained by high-temperature sintering of metal or ceramic powders, the energy product and coercive fields of bonded magnets are lower. The sintered materials also could be oriented or isotropic.

The demagnetization curves for some ceramic hard ferrites and NdFeB magnets, which are (1) isotropic bonded, (2) oriented bonded, and (3) oriented and sintered, are shown in Figure 11.30.

11.10 SOFT MAGNETIC MATERIALS

Ceramic ferrites, which are alloys of iron containing nickel, cobalt, silicon, and some of the amorphous materials known as metallic glasses, constitute the most important classes of soft magnetic materials (Tables 11.13 and 11.14). The market share is dominated by electrical steels (especially nonoriented) as shown in Figure 11.31. The ranges of the properties of some technologically important magnetic materials are shown in Figure 11.32. These materials are used in transformers, electrical motors, generators, electromagnetic switches, and so on.

For many applications of soft magnetic materials, it is desirable to have materials that have high saturation magnetization (for lightweightness and compactness), very low coercivity and high resistivity (to minimize eddy current losses), and high permeability (to minimize the magnetic reluctance—equivalent of electrical resistance). No single material meets all these requirements. A choice is made depending on the technical specifications needed and the cost involved.

In grain-oriented steel, silicon is added (up to ~3.25%) to enhance resistivity and to reduce coercivity (Figure 11.32b). This reduces transformer core losses. Permeability of this steel depends on the crystallographic directions. As shown in Figure 11.23, for iron, which is the main ingredient of steels, magnetization is preferred along the [100] and [110] directions. As a result, the permeability of steel is higher in these directions. This is the reason why grain-oriented steels are preferred for making transformer cores; however, they also cost more.

TABLE 11.13

Properties of Some Soft Magnetic Materials

Names	Composition	μ_{max}	H_c (A/m)	J_s (T)
Fe	Fe 100	$3-50 \times 10^3$	1–100	2.16
NO Fe–Si	Fe(>96)–Si(<4)	$3-10 \times 10^3$	30–80	1.98–2.12
GO Fe–Si	Fe 97–Si 3	$20-80 \times 10^3$	4–15	2.03
Fe–Si 6.5%	Fe 93.5–Si 6.5	$5-30 \times 10^3$	10–40	1.80
Sintered powders	Fe 99.5–P 0.5	$0.2-2 \times 10^3$	100–500	1.65–1.95
Permalloy	Fe 16–Ni 79–Mo 5	5×10^5	0.4	0.80
Permendur	Fe 49–Co 49–V 2	2×10^3	100	2.4
Ferrites	$(Mn, Zn)O.Fe_2O_3$	3×10^3	20–80	0.2–0.5
Sendust	Fe 85–Si 95–Al 5.5	50×10^3	5	1.70
Amorphous (Fe-based)	$Fe_{78}B_{13}Si_{10}$	10^5	2	1.56
Amorphous (Co-based)	$Co_{17}Fe_4B_{15}Si_{10}$	5×10^5	0.5	0.86
Nanocrystalline	$Fe_{73.5}Cu_1Nb_3Si_{13.5}B_9$	10^5	0.5	1.2

Source: Reprinted from Fiorillo, F., *Measurement and Characterization of Magnetic Materials*, Copyright (2004), with permission from Elsevier; With kind permission from du Tremolet de Lacheisserie, E., et al., *Magnetism: Fundamentals*, Springer Science+Business Media, 2002.

Note: The composition is given in weight %; for amorphous and nanocrystalline alloys, it is expressed in atomic %. μ_{max}, maximum DC relative permeability; H_c, coercive field; J_s, saturation polarization at room temperature.

The properties of some commercially important soft magnetic materials are shown in Tables 11.13 (Fiorillo 2004) and Table 11.14. In Table 11.13, the maximum value of permeability (see Figure 11.16), coercive field (H_c in ampere/meter), and saturation polarization ($J_s = \mu_0 M_s$ in Tesla) are presented. All compositions are in weight %, except those for amorphous and nanocrystalline materials, which are in atomic %.

As discussed in Section 11.6.4, the coercivity of a magnetic material is controlled by the difficulty in nucleation or by the pinning of domains. In most hard magnetic materials, the nucleation of new domains controls the coercivity. In soft magnetic materials, defects such as grain boundaries affect the coercivity. Introduction of microstructural imperfections, which are of the order of domain-wall thickness, usually results in an increase in coercivity. The coercivity (H_c) of a magnetic material scales with the average grain size (d), as shown by the following equation:

$$H_c = H_{c,0} + \frac{\text{constant}}{d} \tag{11.24}$$

In this equation, $H_{c,0}$ is a constant that defines the baseline coercivity of a material and depends on factors such as intrinsic properties, stress, other defects such as inclusions, and so on (Bloor et al. 1994). In fact, a class of soft magnetic materials is made from amorphous materials. These materials are made in the form of metallic ribbons obtained by the rapid solidification of alloys. These materials are known as metallic glasses. Because there are no grain boundaries in these materials, the coercivity of these materials is very low (Table 11.15).

11.11   MAGNETIC DATA STORAGE MATERIALS

For storage of information using magnetic materials, the main idea has been to use the remnant magnetization as the basis for storing data. For this application, magnetically semihard materials

TABLE 11.14

Properties of Some Commonly Used Soft Magnetic Materials

Name	Composition	Permeability (μ_r)		Corrective (H_c) (A · m^{-1})	Retentivity (B_r) (T)	B_{max}	Resistivity ($\mu\Omega$ m)
		Initial	Maximum				
Ingot iron	99.8% Fe	150	5000	80	0.77	2.14	0.10
Low-carbon steel	99.5% Fe	200	4000	100		2.14	1.12
Silicon iron, unoriented	Fe–3% Si	270	8000	60		2.01	0.47
Silicon iron, grain-oriented	Fe–3% Si	1400	50,000	7	1.20	2.01	0.50
4750 alloy	Fe–48% Ni	11,000	80,000	2		1.55	0.48
4-79 permalloy	Fe–4% Mo–79% Ni	40,000	200,000	1		0.80	0.58
Superalloy	Fe–5% Mo–80% Ni	80,000	450,000	0.4		0.78	0.65
2V-permendur	Fe–2% V–49% Co	800	450,000	0.4		0.78	0.65
Supermendur	Fe–2% V–49% Co		100,000	16	2.00	2.30	0.40
Metglas[a]2650SC	$Fe_{81}B_{13.5}Si_{3.5}C_2$		300,000	3	1.46	1.61	1.35
Metglas[a]2650S-2	$Be_{78}B_{13}S_9$		600,000	2	1.35	1.56	1.37
MnZn ferrite	H5C2[b]	10,000		7	0.09	0.40	1.5×10^5
MnZn ferrite	H5E[b]	18,000		3	0.12	0.44	5×10^4
NiZn ferrite	K5[b]	290		80	0.25	0.33	2×10^{12}

Source: Chin, G.Y., et al., In *Encyclopedia of Advanced Materials*, vol. 1, Pergamon Press, Oxford, UK, 1994; Askeland, D. and Fulay, P., *The Science and Engineering of Materials*, Thomson, Washington, DC, 2006. With permission.

[a] Allied corporation trademark.

[b] TDK ferrite code.

Total production ≈ 12 million t/a (2006)
Estimated share

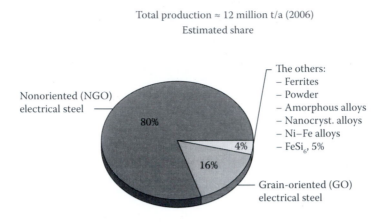

FIGURE 11.31   Market share of soft magnetic materials. (From Varga, L.K., and Davies H.A., *J. Magn. Magn. Mater.*, 320(20), 2411–2422, 2008. With permission.)

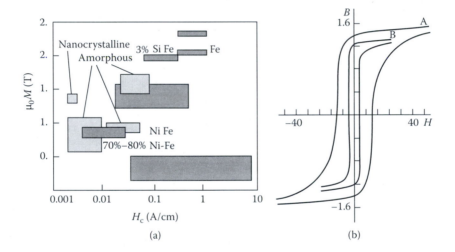

FIGURE 11.32   (a) Coercivity and saturation magnetization for some soft magnetic materials. (From Bloor, D., et al., eds. *Encyclopedia of Advanced Materials*, vol. 4, Pergamon Press, Oxford, UK, 1994. With permission.) (b) Hysteresis loops for pure iron and a grain-oriented steel with 97% Fe and 3% Si. (From Buschow, K.H., and De Boer F.R., *Physics of Magnetism and Magnetic Materials*, Kluwer, Boston, 2003. With permission.)

are needed because the information must be written, and we must be able to erase the information and rewrite it. There are two major types of magnetic media (Figure 11.33). One is in the form of thin films, used for making magnetic hard disks for personal computers and other electronic gadgets. The second technology involves magnetic tape media, such as audio and video tapes and computer disks. The magnetic tape technology, although available and useful, has encountered stiff competition with the advent of optically written media, such as DVDs and CDs, in addition to semiconductor-based flash memories widely used in digital cameras and other electronic hardware.

For magnetic tapes, cobalt-modified metallic iron, cobalt-modified gamma iron oxide (γ-Fe_2O_3), and barium ferrite ($BaFe_{12}O_{19}$) are some of the commonly used materials. Cobalt modification of gamma iron oxide leads to higher coercivity values that enhance data retention. Most particles used in this application are elongated to effectively take advantage of the magnetoshape anisotropy (Figure 11.34). The important properties of some magnetic particles used for magnetic tapes are included in Table 11.15.

TABLE 11.15
Properties of Some Materials Used in Magnetic Tape Media

Material Composition	Particle Length (µm)	Aspect Ratio	Magnetization (B_r)		Coercivity (H_c)		Surface Area (m^2/g)	Curie Temperature (T_c) (°C)
			Wb/m²	emu/cc	kA/m	Oe		
γ-Fe$_2$O$_3$	0.20	5:1	0.44	350	22–34	420	15–30	600
Co-γ-Fe$_2$O$_3$	0.20	6:1	0.48	380	30–75	940	20–35	700
CrO$_2$	0.20	10:1	0.50	400	30–75	950	18–55	125
Fe	0.15	10:1	1.40[a]	1100[a]	56–176	2200	20–60	770
Barium ferrite	0.05	0.02 um thick	0.40	320	56–240	3000	20–25	350

Source: Jorgensen, F., *Complete Handbook of Magnetic Recording*, 4th ed., McGraw Hill, New York, 1996. With permission.

[a] For overcoated, stable particles, use only 50%–80% of these values due to reduced volume of magnetic particles.

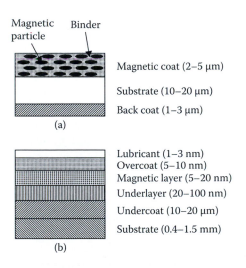

(a)

Magnetic particle Binder
Magnetic coat (2–5 µm)
Substrate (10–20 µm)
Back coat (1–3 µm)

(b)

Lubricant (1–3 nm)
Overcoat (5–10 nm)
Magnetic layer (5–20 nm)
Underlayer (20–100 nm)
Undercoat (10–20 µm)
Substrate (0.4–1.5 mm)

FIGURE 11.33 Schematic representation of (a) magnetic tape and (b) thin-film hard disk. (Reprinted from Li, Y. and Menon, A.K., *Encyclopedia of Materials: Science and Technology.* Copyright (2008), with permission from Elsevier.)

The technology to manufacture magnetic hard disk has advanced at a very rapid pace. Significant advances in the area of development of nanoscale materials, perpendicular recording, advanced read-and-write heads, devices based on giant magnetoresistance, spin electronics, and bit-patterned media have been made. These have led to remarkable advances related to the amount of information that can be stored on a one-inch square of a hard disk. The goal in the year 2008 was to be able to store an ultra-high density of 1 terabit in a one-inch square area (Tbit/in²). A terabit is one trillion bits. The hard disks today can store up to 200 Gbit/in². A gigabit is one billion bits.

There are also many other technologies such as magnetic random-access memories (MRAMs) that are being developed for information storage. Similarly, researchers have also been developing different classes of devices based on multiferroics for information storage. Multiferroics have strongly coupled electric, magnetic, and structural order parameters. In other words,

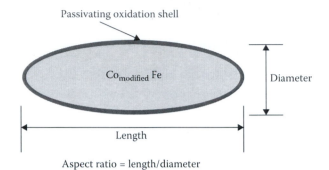

Aspect ratio = length/diameter

FIGURE 11.34 Schematic illustration of an advanced acicular particle of cobalt-modified iron. (From Bloor, D., et al., eds., *Encyclopedia of Advanced Materials*, vol. 4. Pergamon Press, Oxford, UK, 1994. With permission.)

the dielectric dipoles are produced by applying magnetic field or vice versa. The potential usefulness of multiferroic composites with superior ferroelectric or ferromagnetic properties is tremendous. Possible applications include the development of multiple-state memory elements, tunable ferromagnetic resonance devices controlled by electric field, and transducers with either magnetically modulated piezoelectricity or electrically modulated piezomagnetism. Examples of multiferroic materials are $RMnO_3$ and RMn_2O_5 single crystals (R: rare-earth elements) and thin films based on $BiFeO_3$. A physical mechanism of multiferroics is different from that of ferroelectricity and ferromagnetism. A general consensus on the mechanism of multiferroics is (i) that the magnetic spirals in frustrated ferromagnetic can induce the ferroelectricity by breaking the centrosymmetry and (ii) that the strain can manipulate the magnetic spirals by changing the spatial configuration of 3d orbitals. Including multiferroics, there are many rapidly evolving and very exciting developments occurring in the field of magnetic materials and devices which are challenging talented material scientists and engineers.

PROBLEMS

11.1 Comment on the following statement: "All materials in this world are magnetic."

11.2 What is the difference between a diamagnetic and paramagnetic material? Give an example of each type of material.

11.3 What is an antiferromagnetic material? Are there examples of antiferromagnetic coupling in ferromagnetic materials?

11.4 What is the difference between a ferromagnetic and a ferrimagnetic material?

11.5 Is it necessary for ferromagnetic or ferrimagnetic materials to contain ferromagnetic elements? Explain.

11.6 What is a superparamagnetic material?

11.7 What is a ferrofluid? Where are ferrofluids used commercially?

11.8 What is the basis of magnetic levitation using superconductors?

11.9 The coercive field for a neodymium iron boron–based magnet is reported to be 1.25 T; how much is this value in Oersted?

11.10 Bohr magneton (μ_B) is the basic unit for magnetic moment. It is based on the spin component of the magnetic moment of an electron that is aligned with an external magnetic field, and its value is given by $\mu_B = \hbar q_e/2m_e = 11.274 \times 10^{-24}$ A·m². Sometimes, it is necessary or useful to express the number n, which is the number of Bohr magnetons per mole of a material. Show that if MW is the molecular weight of a material and if x is the number of Bohr magnetons per atom, the value of Bohr magnetons per mole of a material is $(5.585 \times x)/(MW)$.

11.11 The magnetic moment of a nickel atom is listed as 0.606 Bohr magneton at 0 K (shown in Table 11.7). The value of saturation magnetization at 290 K is listed as 0.61 T. Nickel has a face-centered cubic (FCC) structure, and its lattice constant at 290 K is 3.5167 Å. What will be the effective magnetic moment of nickel atoms in FCC nickel at $T = 290$ K?

11.12 In Table 11.7, why are the values of saturation magnetization not listed for gadolinium (Gd) and dysprosium (Dy) at 290 K?

11.13 The exchange-energy (A) values for iron and nickel are 2.5×10^{-21} and 3.2×10^{-21} J, respectively. The anisotropy energies (K_1) for iron and nickel are 4.8×10^4 J/m^3 and -0.5×10^4 J/m^3, respectively (Jiles 1991). In which material would you expect the domain walls to be thicker? Why? How will the domain energies compare for iron and nickel?

11.14 What is a Bloch wall? How is it different from the Néel wall?

11.15 Why is the coercivity of permanent magnetic materials significantly lower than that expected from the value of the anisotropy field?

11.16 How is the coercivity of a polycrystalline soft magnetic material related to the grain size?

11.17 Why are 180° domain walls not stress-sensitive, whereas 90° domain walls are stress-sensitive?

11.18 What will be the flux density (B) created for a current of 1 A if a toroidal solenoid has an air core? Assume that the average circumference of the solenoid is 50 cm and that there are 1000 turns.

11.19 How much will the flux density (B) be if the core is filled with a ceramic ferrite of permeability 300? Assume that the average circumference of the solenoid is 50 cm and that there are 1000 turns. If the diameter of the solenoid core is 5 cm, what will be the flux in Wb?

11.20 The demagnetization curves for different commercially available permanent magnets are shown in Figure 11.35.
   a. Which magnets have the lowest and the highest coercivity values?
   b. Calculate the coercivity of the material Sm_2Co_{17} in Oersteds.
   c. Calculate the maximum energy product for $SmCo_5$ and $Nd_2Fe_{14}B$ (shown in Figure 11.35).

11.21 The properties of isotropic-bonded, oriented-bonded, and oriented-sintered magnets made from hard ceramic ferrites and NdFeB are shown in Figure 11.30.
   a. Why is the energy product for the oriented-sintered material (ferrite or NdFeB) better than that of either of the bonded magnets?
   b. Calculate the energy products for the oriented-sintered NdFeB and compare the value with those for oriented-bonded and isotropic-bonded NdFeB materials.

11.22 The lifting force (F) of a permanent magnet is given by the following equation:

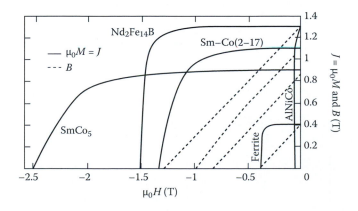

FIGURE 11.35 Demagnetization curves for some permanent magnets. (With kind permission from du Tremolet de Lacheisserie, E., et al., *Magnetism: Fundamentals*, Springer Science+Business Media, 2002.)

$$F = \frac{(\mu_0 M)^2 A}{2\mu_0} \qquad\qquad (11.25)$$

where A is the area of the magnet. Calculate the lifting force, in kN, for a permanent magnet with saturation magnetization of 1.5 T. Assume that the area is 0.25 m².

11.23 What types of materials are used for magnetic tape data storage? Why?

11.24 Permanent magnets have high coercivity; why can these not be used for magnetic data storage?

11.25 In Figure 11.32b, why is the maximum magnetization for the grain-oriented steel lower than that for pure iron?

11.26 What is the material used for the magnetic stripe on credit cards and automated teller machine cards used for banking?

11.27 What magnetic properties are important for a material to be used for transformer cores?

11.28 What is grain-oriented steel? What magnetic properties of steel are improved by producing grain-oriented steel? In what crystallographic directions are the grains oriented? Why?

11.29 Why is silicon added to grain-oriented steels?

11.30 The domain pattern in a rolled polycrystalline Ni–Fe alloy is shown in Figure 11.36.
   a. Are the grains oriented in this sample?
   b. Based on the domain geometry shown, is the material likely to have any net remnant magnetization?
   c. What will happen if this material is heated above its T_c and then "field-cooled," that is, annealed at temperatures below T_c by placing it in a magnetic field parallel to the rolling direction.

11.31 What are the different smart materials based on the use of magnetic materials?

11.32 What is a ferrofluid? How is it different from a magnetorheological (MR) fluid?

11.33 How does an MR fluid work in a magnetic shock absorber?

11.34 The magnetization curves for two MR fluids, one based on iron and another based on a ceramic ferrite, are shown in Figure 11.37.
   a. Calculate the initial permeabilities of both these MR fluids from the magnetization curves.
   b. Calculate the volume fraction of iron in the iron-based fluid. Assume that the saturation magnetization of iron is 2.1 T and that the MR fluid is made using a dispersion of iron particles in synthetic oil.
   c. Assuming that the ferrite-based fluid is made using iron oxide (Fe_3O_4), what is the volume fraction of this material in this MR fluid made using synthetic oil?

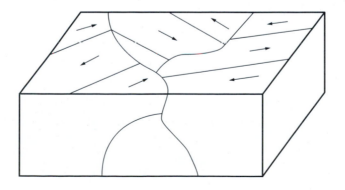

FIGURE 11.36 Domain pattern in rolled Ni–Fe alloy. (From Buschow, K.H. and De Boer, F.R., *Physics of Magnetism and Magnetic Materials*, Kluwer, Boston, 2003. With permission.)

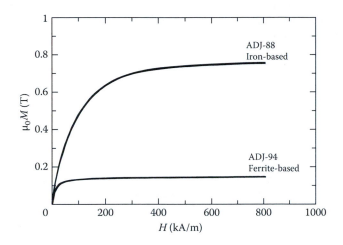

FIGURE 11.37 Magnetization curves for magnetorheological fluids. (From Fulay, P.P., et al., *Proc. MRS Symp. Mater. Smart Syst.*, 459, 99, 1997. With permission.)

11.35 If nickel ferrite has an inverse spinel structure, what will be the net magnetic moment of this structure per unit formula? Assume that the lattice constant of nickel ferrite is 8.37 Å.

11.36 In Table 11.8, zinc ferrite is shown to have zero magnetic moment. The magnetic moment per formula for $MnFe_2O_4$ is 5 μ_B and that of $NiFe_2O_4$ is 2 μ_B. Why does the initial addition of zinc ferrite to $MnFe_2O_4$ and $NiFe_2O_4$ *increase* the total magnetic moments of these materials? However, as increasing quantities of zinc ferrite are added, the magnetization eventually decreases. Explain.

11.37 What is magnetostriction? Explain how this phenomenon can be used to create ultrasonic waves and a nanopositioning device.

11.38 What phenomenon causes an electrical transformer to "hum"?

GLOSSARY

Antiferromagnetic material: A material that shows zero net magnetic moment because the magnetic moments of different ions or atoms cancel each other out completely.

$(BH)_{max}$ product: See **Maximum energy product**.

Bloch wall: A wall between two magnetic domains across which the magnetization direction changes by 180°. The thickness of a Bloch wall increases with increase in exchange energy (as the material tries to maintain the magnetization in the already-existing direction) and decreases with increased anisotropy (where the material prefers to magnetize in the easy direction of magnetization).

Bohr magneton: The unit in which the magnetic dipole moment of an electron, atom, or an ion is measured. A Bohr magneton is also equal to the component of the spin magnetic moment of an electron that is aligned with the applied magnetic field

$$\mu_B = \hbar \frac{q_e}{2m_e} = 9.274 \times 10^{-24} \, A \cdot m^2$$

Bonded magnets: Magnetic composites made by dispersing magnetic particles in a polymeric or metallic matrix, which are low-cost magnetic materials that could exhibit isotropic or oriented magentization.

Curie temperature (T_c): Temperature at which a ferromagnetic or ferrimagnetic material becomes paramagnetic. Note that this term is also used for ferroelectrics (Chapter 10).

Demagnetization curve: The second quadrant of the hysteresis loop in which a magnetic material moves from flux density B_r to zero. This curve is important for designing circuits involving permanent magnets.

Demagnetizing factor (N_d): A numerical factor that shows the extent of internal field created in a magnetic material. This field opposes the effect of the applied magnetic field.

Diamagnetism: Exclusion of the applied magnetic field from a material. Diamagnetic materials are defined by negative values of magnetic susceptibility, ranging from -10^{-6} to -1 (for superconductors). Most materials (e.g., most metals, inert gases, and organic compounds) that are classified as "nonmagnetic" are actually diamagnetic.

Domain: See **Magnetic domain**.

Domain wall: A surface across which the magnetization direction rotates. If the magnetization rotates $180°$ within a domain wall, the domain wall is known as a Bloch wall (Figure 11.21), and if it rotates such that it has a component normal to the wall, then it is called a Néel wall.

Easy axis: The crystallographic directions in which the magnetic moments in a material tend to line up naturally in the absence of any magnetic field, thus minimizing the free energy.

Exchange interaction: The long-range spin ordering occurring as a result of a quantum mechanical interaction, which is due to the combination of electrostatic coupling between electron orbitals and Pauli's exclusion principle. In ferromagnetic materials, this interaction causes the spin to be aligned in the same direction; and in ferrimagnetic materials, the magnetic moments due to the electron spin of ions at different crystallographic locations are antiparallel.

Ferrofluid: A stable dispersion of superparamagnetic particles, used as heat-transfer fluid; the carrier is typically an oil or organic fluid, but it could also be water.

Ferrimagnetic materials: Materials in which the magnetic moments of some of the atoms or ions are antiparallel but do not completely cancel out, thereby creating a net magnetic moment.

Ferromagnetic materials: Materials in which all the magnetic moments of atoms or ions remain in the same direction (e.g., Fe, Ni, Co, Gd, and alloy of Fe–Co).

Flux density (B): See **Magnetic flux density**.

Hard axis: Crystallographic directions along which it is difficult to reorient the magnetization direction.

Hard magnetic materials: Materials that have a high ($\sim > 10^4$ A/m) coercivity, also known as permanent magnets.

Hysteresis loop: See **Magnetic hysteresis loop**.

Hysteresis losses: Power loss in a magnetic material because of the occurrence of eddy currents induced by an applied magnetic field.

Induction (B): See **Magnetic flux density**.

Isotropic magnet: A magnet in which the spatial distribution of the magnetization directions is random.

Joule magnetostriction: The magnetostriction strain component caused by rotation of magnetic domains, which is different from the spontaneous magnetostriction that sets in when magnetic dipoles undergo a spontaneous alignment at temperatures near the T_c.

Liquid magnets: Materials showing an entire body motion when placed next to a permanent magnet. See also **Ferrofluids**.

Magnetic domain: Region of a ferromagnetic or ferrimagnetic material in which the magnetization (i.e., net magnetic moment due to different atoms or ions) is in the same direction.

Magnetic flux density (B): The magnetic flux created per unit area; unit is Weber/square meter or Tesla, which represents the strength of the magnetic field created inside a magnetic material; its level depends on the applied external field and the magnetic permeability of the magnetic material.

Magnetic induction: See **Magnetic flux density**.

Magnetic hysteresis loop: The trace of magnetization (M) developed in magnetic materials as a function of the applied magnetic field (H), plotted as $\mu_0 M$ versus H or flux density B versus H.

Magnetic levitation: See **Meissner effect**.

Magnetic losses: Energy dissipation occurring when a magnetic material is subjected to a time-varying external magnetic field.

Magnetization (M): The total magnetic moment per unit volume; unit is ampere/meter.

Magnetic permeability of free space (μ_0): A constant that relates the flux density B created by a magnetic field H in space. Its value is $\mu_0 = 4\pi \times 10^{-7}$ H/m or Wb/A·m.

Magnetic permeability (μ): A property that describes the ease with which magnetic flux lines are carried through a material. Ferromagnetic and ferrimagnetic materials have large magnetic permeability values, which are field dependent. The units are Henry/meter, which is the same as Weber/ampere·meter.

Magnetic polarization (J): The quantity $\mu_0 M$ that is shown using the symbol J.

Magnetocrystalline anisotropy: The dependence of magnetic properties, such as coercive field, on the crystallographic direction.

Magnetocrystalline anisotropy energy: The energy that is needed to rotate from the easy-magnetization direction to a given direction.

Magnetorheological (MR) fluids: Dispersions of noncolloidal and magnetically multidomain soft ferromagnetic or ferrimagnetic particles. Applications of MR fluids include vibration control.

Magnetoshape anisotropy: Anisotropy created by the shape of magnetic grains or particles; same as shape anisotropy; causes needle-like magnetic particles to exhibit higher coercivity, which is used in magnetic recording media.

Magnetostriction: A change in the dimension of a magnetic material after magnetization caused by either the development of spontaneous magnetization as the material goes through the Curie temperature or by application of a magnetic field, which causes changes in the domain configuration.

Maximum energy product: The maximum value of a product obtained by multiplying the corresponding B and H values on the demagnetization curve (Figure 11.26); also known as the $(BH)_{max}$ product; is a measure of the energy stored in a magnet of a given size.

Meissner effect: Expulsion of magnetic flux lines due to the diamagnetic behavior of a superconductor in a superconducting state, which is the basis of magnetic levitation using superconductors.

Minor loop: A magnetic hysteresis loop in which the applied fields do not reach saturation levels.

Multiferroics: Materials that possess two or more switchable properties, such as spontaneous polarization, spontaneous magnetization, or spontaneous strain.

Néel temperature (T_N or O_N): Temperature at which antiferromagnetic exchange–coupling interactions develop spontaneously in the sublattice (Figure 11.12c). In an antiferromagnetic material, these are overcome by *increasing* the temperature, due to which the material becomes paramagnetic. Note that below the T_N, because the antiferromagnetic coupling is reduced, the *susceptibility increases* (i.e., $1/\chi_m$ decreases).

Néel wall: A domain wall in which the magnetization rotates in the plane of the sample as we move from one direction of the magnetic moment to another (Figure 11.22); occurs in thin films and not in bulk magnetic materials.

Neomagnets: Permanent magnets based on neodymium (e.g., $Nd_2Fe_{14}B$).

Nonlinear magnetic materials: Ferromagnetic and ferrimagnetic materials below their Curie temperature.

Nucleation-controlled coercivity: When control of coercivity of a magnetic material is accomplished by making it difficult to nucleate new domain walls, the coercivity is said to be nucleation controlled. See also **Pinning-controlled coercivity**.

Oriented magnets: Magnetic materials in which the distribution of the easy axes of magnetization is not random. The hysteresis loops of these materials are squarer, and the remnant magnetization is higher; the same as textured magnets.

Paramagnetic atoms: Atoms that have a net magnetic moment (e.g., Al, Cu, Fe).

Paramagnetic ions: Ions that have a net magnetic moment (e.g., Fe^{2+}).

Paramagnetic materials: Materials with relatively small, but positive, susceptibility ($\sim +10^{-6}$–10^{-3}); for example, liquid oxygen, copper, and aluminum. Ferromagnetic and ferrimagnetic materials become paramagnetic above their Curie temperatures.

Permanent magnets: Materials that have a large coercivity ($\sim >10^4$ A/m); also known as hard magnetic materials.

Pinning-controlled coercivity: When coercivity values are controlled by controlling the concentration of defects or grain boundaries that can pin domains. In this case, domains already exist, and the focus is on pinning the domain walls. See also **Nucleation-controlled coercivity**.

Rare-earth magnets: Permanent magnets based on rare-earth elements (e.g., Sm, Co, and Nd).

Relative magnetic permeability (μ_r): Ratio of the magnetic permeability of a material to that of magnetic permeability of free space; has no units and is related to the susceptibility as shown by the following equation: $\mu_r = 1 + \chi_m$.

Saturation magnetization ($\mu_0 M_s$): The maximum possible magnetization in a ferromagnetic or ferrimagnetic material; the unit is Tesla.

Semihard magnetic materials: Materials whose coercivity values are between a few hundred and $\sim 10^4$ A/m (Figure 11.19).

Shape anisotropy: See **Magnetoshape anisotropy**.

Sintered magnets: Solid and relatively dense magnetic materials obtained by the compaction of metal or ceramic powders of a magnetic material, followed by high-temperature sintering. These materials could be either oriented or isotropic.

Smart material: A material whose properties are controllable using an external field or stimulus (e.g., piezoelectrics and magnetostrictive materials).

Soft magnetic materials: In general, materials with a coercivity less than ~ 5000 A/m are considered magnetically soft materials.

Spontaneous magnetostriction: Development of strain in ferromagnetic and ferrimagnetic materials, which occurs spontaneously when there is a spontaneous alignment of magnetic moments at the Curie temperature.

Superparamagnetism: A behavior in which a material that is originally either ferromagnetic or ferrimagnetic in its bulk form behaves as a paramagnetic material on a nanoscale.

Superparamagnetic materials: Nanoscale particles, grains, or structures made from materials that are ferromagnetic or ferrimagnetic in their bulk form but behave as paramagnetic materials because of randomization of magnetic interactions by thermal energy.

Textured magnets: See **Oriented magnets**.

REFERENCES

Askeland, D., and P. Fulay. 2006. *The Science and Engineering of Materials*. Washington, DC: Thomson.

Bloor, D., M. C. Flemings, R. J. Brook, S. Mahajan, and R. W. Cahn, eds. 1994. *Encyclopedia of Advanced Materials*, vol. 4. Oxford, UK: Pergamon Press.

Bozroth, R. M. 1955 (Reprinted 1993). *Ferromagnetic Materials*. Van Nostrand: IEEE Press.

Buschow, K. H., and F. R. De Boer. 2003. *Physics of Magnetism and Magnetic Materials*. Boston, MA: Kluwer.

Chin, G.Y., et al. 1994. Magnetic materials: An overview, basic concepts, magnetic measurements, and magnetostrictive materials. In *Encyclopedia of Advanced Materials*, vol. 1. eds. D. Bloor, M. C. Flemings, R. J. Brook, S. Mahajan, and R. W. Cahn, 1424. Oxford, UK: Pergamon Press.

Chu, Y.-H., L. W. Martin, M. B. Holcomb, and R. Ramesh. 2007. Controlling magnetism with multiferroics. *Mater Today* 10(10):16–23.

du Trémolet de Lacheisserie, E., D. Gignoux, and M. Schlenker. 2002. *Magnetism: Fundamentals.* Berlin: Springer.

Fiorillo, F. 2004. *Measurement and Characterization of Magnetic Materials.* Oxford: Elsevier.

Featonby, D. 2005. Experiments with neodymium magnets. *Phys Educ* 40(6): 505–8.

Fulay, P. P., A. D. Jatkar, and J. M. Ginder. 1997. Synthesis and properties of magnetorheological fluids for active vibration control. *Proc MRS Symp Mater Smart Syst* 459:99.

Goldman, A. 1999. *Handbook of Modern Ferromagnetic Materials.* Boston, MA: Kluwer.

Hadjipanayis, G. C. 1999. Nanophase hard magnets. *J Magn Magn Mater* 200:373.

Jakubowicz, J., M. Jurczyk, A. Handstein, D. Hinz, O. Gutfleisch, and K.-H. Müller. 2000. Temperature dependence of magnetic properties for nanocomposite $Nd_2(Fe,Co,M)_{14}B/\alpha$-Fe magnets. *J Magn Magn Mater* 208(3):163–8.

Jiles, D. C. 1991. *Introduction to Magnetism and Magnetic Materials.* London: Chapman and Hall.

Jiles, D. C. 2001. *Introduction to the Electronic Properties of Materials.* Cheltenham, UK: Nelson Thornes.

Jorgensen, F. 1996. *Complete Handbook of Magnetic Recording.* 4th ed. New York: McGraw Hill.

Kasap, S. O. 2002. *Principles of Electronic Materials and Devices.* New York: McGraw Hill.

Li, Y., and A. K. Menon. 2008. Magnetic recording technologies: Overview. In *Encyclopedia of Materials: Science and Technology*, eds. D. Bloor et al. Amsterdam: Elsevier.

O'Handley, R. C. 1999. *Modern Magnetic Materials, Principles, and Applications.* New York: Wiley.

Rosensweig, R. E., Y. Hirota, S. Tsuda, and K. Raj. 2008. Study of audio speakers containing ferrofluid. *J Phys Condens Matter* 20(204):147.

Scott, W. T. 1959. *Physics of Electricity and Magnetism.* New York: Wiley.

Skomski, R., and J. M. D. Coey, eds. 1999. *Permanent Magnetism.* Bristol, UK: Institute of Physics.

Varga, L. K., and H. A. Davies. 2008. Challenges in optimizing the magnetic properties of bulk soft magnetic materials. *J Magn Magn Mater* 320(20): 2411–22.

Index